This book given by
Mrs. Bernice Morecock in memory
of her husband

EARLE M. MORECOCK

1899-1982

Teacher & Administrator
RIT College of Engineering   1924-1964
First Dean, College of Engineering   1953
Dean Emeritus   1964

Wallace Memorial Library
Rochester Institute of Technology

AN INTRODUCTION TO
NUMERICAL ANALYSIS
FOR ELECTRICAL AND
COMPUTER ENGINEERS

AN INTRODUCTION TO NUMERICAL ANALYSIS FOR ELECTRICAL AND COMPUTER ENGINEERS

Christopher J. Zarowski
University of Alberta, Canada

WILEY-
INTERSCIENCE

A JOHN WILEY & SONS, INC. PUBLICATION

For general information on our other products and services, please contact our Customer Care
Department within the United States at 877-762-2974, outside the United States at 317-572-3993 or
fax 317-572-4002.

Wiley also publishes its books in a variety of electronic formats. Some content that appears in print,
however, may not be available in electronic format.

Library of Congress Cataloging-in-Publication Data:

Zarowski, Christopher J.
    An introduction to numerical analysis for electrical and computer engineers / Christopher
J. Zarowski.
       p. cm.
    Includes bibliographical references and index.
    ISBN 0-471-46737-5 (coth)
       1. Electric engineering—Mathematics. 2. Computer science—Mathematics. 3. Numerical
analysis I. Title.

TK153.Z37 2004
621.3'01'518—dc22

                                                                          2003063761

Printed in the United States of America.

10 9 8 7 6 5 4 3 2 1

In memory of my mother
Lilian
and of my father
Walter

CONTENTS

PREFACE

The subject of numerical analysis has a long history. In fact, it predates by centuries the existence of the modern computer. Of course, the advent of the modern computer in the middle of the twentieth century gave greatly added impetus to the subject, and so it now plays a central role in a large part of engineering analysis, simulation, and design. This is so true that no engineer can be deemed competent without some knowledge and understanding of the subject. Because of the background of the author, this book tends to emphasize issues of particular interest to electrical and computer engineers, but the subject (and the present book) is certainly relevant to engineers from all other branches of engineering.

Given the importance level of the subject, a great number of books have already been written about it, and are now being written. These books span a colossal range of approaches, levels of technical difficulty, degree of specialization, breadth versus depth, and so on. So, why should this book be added to the already huge, and growing list of available books?

To begin, the present book is intended to be a part of the students' first exposure to numerical analysis. As such, it is intended for use mainly in the second year of a typical 4-year undergraduate engineering program. However, the book may find use in later years of such a program. Generally, the present book arises out of the author's objections to educational practice regarding numerical analysis. To be more specific

1. Some books adopt a "grocery list" or "recipes" approach (i.e., "methods" at the expense of "analysis") wherein several methods are presented, but with little serious discussion of issues such as how they are obtained and their relative advantages and disadvantages. In this genre often little consideration is given to error analysis, convergence properties, or stability issues. When these issues are considered, it is sometimes in a manner that is too superficial for contemporary and future needs.

2. Some books fail to build on what the student is supposed to have learned prior to taking a numerical analysis course. For example, it is common for engineering students to take a first-year course in matrix/linear algebra. Yet, a number of books miss the opportunity to build on this material in a manner that would provide a good bridge from first year to more sophisticated uses of matrix/linear algebra in later years (e.g., such as would be found in digital signal processing or state variable control systems courses).

3. Some books miss the opportunity to introduce students to the now quite vital area of functional analysis ideas as applied to engineering problem solving. Modern numerical analysis relies heavily on concepts such as function spaces, orthogonality, norms, metrics, and inner products. Yet these concepts are often considered in a very ad hoc way, if indeed they are considered at all.

4. Some books tie the subject matter of numerical analysis far too closely to particular software tools and/or programming languages. But the highly transient nature of software tools and programming languages often blinds the user to the timeless nature of the underlying principles of analysis. Furthermore, it is an erroneous belief that one can successfully employ numerical methods solely through the use of "canned" software without any knowledge or understanding of the technical details of the contents of the can. While this does not imply the need to understand a software tool or program down to the last line of code, it does rule out the "black box" methodology.

5. Some books avoid detailed analysis and derivations in the misguided belief that this will make the subject more accessible to the student. But this denies the student the opportunity to learn an important mode of thinking that is a huge aid to practical problem solving. Furthermore, by cutting the student off from the language associated with analysis the student is prevented from learning those skills needed to read modern engineering literature, and to extract from this literature those things that are useful for solving the problem at hand.

The prospective user of the present book will likely notice that it contains material that, in the past, was associated mainly with more advanced courses. However, the history of numerical computing since the early 1980s or so has made its inclusion in an introductory course unavoidable. There is nothing remarkable about this. For example, the material of typical undergraduate signals and systems courses was, not so long ago, considered to be suitable only for graduate-level courses. Indeed, most (if not all) of the contents of any undergraduate program consists of material that was once considered far too advanced for undergraduates, provided one goes back far enough in time.

Therefore, with respect to the observations mentioned above, the following is a summary of some of the features of the present book:

1. An axiomatic approach to function spaces is adopted within the first chapter. So the book immediately exposes the student to function space ideas, especially with respect to metrics, norms, inner products, and the concept of orthogonality in a general setting. All of this is illustrated by several examples, and the basic ideas from the first chapter are reinforced by routine use throughout the remaining chapters.

2. The present book is not closely tied to any particular software tool or programming language, although a few MATLAB-oriented examples are presented. These may be understood without any understanding of MATLAB

(derived from the term *matrix laboratory*) on the part of the student, however. Additionally, a quick introduction to MATLAB is provided in Chapter 13. These examples are simply intended to illustrate that modern software tools implement many of the theories presented in the book, and that the numerical characteristics of algorithms implemented with such tools are not materially different from algorithm implementations using older software technologies (e.g., catastrophic convergence, and ill conditioning, continue to be major implementation issues). Algorithms are often presented in a Pascal-like pseudocode that is sufficiently transparent and general to allow the user to implement the algorithm in the language of their choice.

3. Detailed proofs and/or derivations are often provided for many key results. However, not all theorems or algorithms are proved or derived in detail on those occasions where to do so would consume too much space, or not provide much insight. Of course, the reader may dispute the present author's choices in this matter. But when a proof or derivation is omitted, a reference is often cited where the details may be found.

4. Some modern applications examples are provided to illustrate the consequences of various mathematical ideas. For example, chaotic cryptography, the CORDIC (*co*ordinate *r*otational *d*igital *c*omputing) method, and least squares for system identification (in a biomedical application) are considered.

5. The sense in which series and iterative processes converge is given fairly detailed treatment in this book as an understanding of these matters is now so crucial in making good choices about which algorithm to use in an application. Thus, for example, the difference between pointwise and uniform convergence is considered. Kernel functions are introduced because of their importance in error analysis for approximations based on orthogonal series. Convergence rate analysis is also presented in the context of root-finding algorithms.

6. Matrix analysis is considered in sufficient depth and breadth to provide an adequate introduction to those aspects of the subject particularly relevant to modern areas in which it is applied. This would include (but not be limited to) numerical methods for electromagnetics, stability of dynamic systems, state variable control systems, digital signal processing, and digital communications.

7. The most important general properties of orthogonal polynomials are presented. The special cases of Chebyshev, Legendre, and Hermite polynomials are considered in detail (i.e., detailed derivations of many basic properties are given).

8. In treating the subject of the numerical solution of ordinary differential equations, a few books fail to give adequate examples based on nonlinear dynamic systems. But many examples in the present book are based on nonlinear problems (e.g., the Duffing equation). Furthermore, matrix methods are introduced in the stability analysis of both explicit and implicit methods for nth-order systems. This is illustrated with second-order examples.

Analysis is often embedded in the main body of the text rather than being relegated to appendixes, or to formalized statements of proof immediately following a theorem statement. This is done to discourage attempts by the reader to "skip over the math." After all, skipping over the math defeats the purpose of the book.

Notwithstanding the remarks above, the present book lacks the rigor of a mathematically formal treatment of numerical analysis. For example, Lebesgue measure theory is entirely avoided (although it is mentioned in passing). With respect to functional analysis, previous authors (e.g., E. Kreyszig, *Introductory Functional Analysis with Applications*) have demonstrated that it is very possible to do this while maintaining adequate rigor for engineering purposes, and this approach is followed here.

It is largely left to the judgment of the course instructor about what particular portions of the book to cover in a course. Certainly there is more material here than can be covered in a single term (or semester). However, it is recommended that the first four chapters be covered largely in their entirety (perhaps excepting Sections 1.4, 3.6, 3.7, and the part of Section 4.6 regarding SVD). The material of these chapters is simply too fundamental to be omitted, and is often drawn on in later chapters.

Finally, some will say that topics such as function spaces, norms and inner products, and uniform versus pointwise convergence, are too abstract for engineers. Such individuals would do well to ask themselves in what way these ideas are more abstract than Boolean algebra, convolution integrals, and Fourier or Laplace transforms, all of which are standard fare in present-day electrical and computer engineering curricula.

Engineering past   Engineering present   Engineering future

Christopher Zarowski

1 Functional Analysis Ideas

1.1 INTRODUCTION

Many engineering analysis and design problems are far too complex to be solved without the aid of computers. However, the use of computers in problem solving has made it increasingly necessary for users to be highly skilled in (practical) mathematical analysis. There are a number of reasons for this. A few are as follows.

For one thing, computers represent data to finite precision. Irrational numbers such as π or $\sqrt{2}$ do not have an exact representation on a digital computer (with the possible exception of methods based on symbolic computing). Additionally, when arithmetic is performed, errors occur as a result of rounding (e.g., the truncation of the product of two n-bit numbers, which might be $2n$ bits long, back down to n bits). Numbers have a limited dynamic range; we might get overflow or underflow in a computation. These are examples of *finite-precision arithmetic effects*. Beyond this, computational methods frequently have sources of error independent of these. For example, an infinite series must be truncated if it is to be evaluated on a computer. The truncation error is something "additional" to errors from finite-precision arithmetic effects. In all cases, the sources (and sizes) of error in a computation must be known and understood in order to make sensible claims about the accuracy of a computer-generated solution to a problem.

Many methods are "iterative." Accuracy of the result depends on how many iterations are performed. It is possible that a given method might be very slow, requiring many iterations before achieving acceptable accuracy. This could involve much computer runtime. The obvious solution of using a faster computer is usually unacceptable. A better approach is to use mathematical analysis to understand why a method is slow, and so to devise methods of speeding it up. Thus, an important feature of analysis applied to computational methods is that of assessing how much in the way of computing resources is needed by a given method. A given computational method will make demands on computer memory, operations count (the number of arithmetic operations, function evaluations, data transfers, etc.), number of bits in a computer word, and so on.

A given problem almost always has many possible alternative solutions. Other than accuracy and computer resource issues, ease of implementation is also relevant. This is a human labor issue. Some methods may be easier to implement on a given set of computing resources than others. This would have an impact

An Introduction to Numerical Analysis for Electrical and Computer Engineers, by C.J. Zarowski
ISBN 0-471-46737-5 © 2004 John Wiley & Sons, Inc.

on software/hardware development time, and hence on system cost. Again, mathematical analysis is useful in deciding on the relative ease of implementation of competing solution methods.

The subject of numerical computing is truly vast. Methods are required to handle an immense range of problems, such as solution of differential equations (ordinary or partial), integration, solution of equations and systems of equations (linear or nonlinear), approximation of functions, and optimization. These problem types appear to be radically different from each other. In some sense the differences between them are true, but there are means to achieve some unity of approach in understanding them.

The branch of mathematics that (perhaps) gives the greatest amount of unity is sometimes called *functional analysis*. We shall employ ideas from this subject throughout. However, our usage of these ideas is not truly rigorous; for example, we completely avoid topology, and measure theory. Therefore, we tend to follow simplified treatments of the subject such as Kreyszig [1], and then only those ideas that are immediately relevant to us. The reader is assumed to be very comfortable with elementary linear algebra, and calculus. The reader must also be comfortable with complex number arithmetic (see Appendix 1.A *now* for a review if necessary). Some knowledge of electric circuit analysis is presumed since this will provide a source of applications examples later. (But application examples will also be drawn from other sources.) Some knowledge of ordinary differential equations is also assumed.

It is worth noting that an understanding of functional analysis is a tremendous aid to understanding other subjects such as quantum physics, probability theory and random processes, digital communications system analysis and design, digital control systems analysis and design, digital signal processing, fuzzy systems, neural networks, computer hardware design, and optimal design of systems. Many of the ideas presented in this book are also intended to support these subjects.

1.2   SOME SETS

Variables in an engineering problem often take on values from sets of numbers. In the present setting, the sets of greatest interest to us are (1) the *set of integers* $\mathbf{Z} = \{\ldots -3, -2, -1, 0, 1, 2, 3 \ldots\}$, (2) the set of *real numbers* $\mathbf{R}$, and (3) the *set of complex numbers* $\mathbf{C} = \{x + jy | j = \sqrt{-1}, x, y \in \mathbf{R}\}$. The set of nonnegative integers is $\mathbf{Z}^+ = \{0, 1, 2, 3, \ldots, \}$ (so $\mathbf{Z}^+ \subset \mathbf{Z}$). Similarly, the set of nonnegative real numbers is $\mathbf{R}^+ = \{x \in \mathbf{R} | x \geq 0\}$. Other kinds of sets of numbers will be introduced if and when they are needed.

If A and B are two sets, their *Cartesian product* is denoted by $A \times B = \{(a, b) | a \in A, b \in B\}$. The Cartesian product of n sets denoted $A_0, A_1, \ldots, A_{n-1}$ is $A_0 \times A_1 \times \cdots \times A_{n-1} = \{(a_0, a_1, \ldots, a_{n-1}) | a_k \in A_k\}$.

Ideas from matrix/linear algebra are of great importance. We are therefore also interested in sets of vectors. Thus, $\mathbf{R}^n$ shall denote the set of n-element vectors with real-valued components, and similarly, $\mathbf{C}^n$ shall denote the set of n-element vectors with complex-valued components. By default, we assume any vector x to be a column vector:

$$x = \begin{bmatrix} x_0 \\ x_1 \\ \vdots \\ x_{n-2} \\ x_{n-1} \end{bmatrix}. \tag{1.1}$$

Naturally, row vectors are obtained by transposition. We will generally avoid using bars over or under symbols to denote vectors. Whether a quantity is a vector will be clear from the context of the discussion. However, bars will be used to denote vectors when this cannot be easily avoided. The indexing of vector elements x_k will often begin with 0 as indicated in (1.1). Naturally, matrices are also important. Set $\mathbf{R}^{n \times m}$ denotes the set of matrices with n rows and m columns, and the elements are real-valued. The notation $\mathbf{C}^{n \times m}$ should now possess an obvious meaning. Matrices will be denoted by uppercase symbols, again without bars. If A is an $n \times m$ matrix, then

$$A = [a_{p,q}]_{p=0,\ldots,n-1,\ q=0,\ldots,m-1}. \tag{1.2}$$

Thus, the element in row p and column q of A is denoted $a_{p,q}$. Indexing of rows and columns again will typically begin at 0. The subscripts on the right bracket "]" in (1.2) will often be omitted in the future. We may also write a_{pq} instead of $a_{p,q}$ where no danger of confusion arises.

The elements of any vector may be regarded as the elements of a sequence of finite length. However, we are also very interested in sequences of infinite length. An *infinite sequence* may be denoted by $x = (x_k) = (x_0, x_1, x_2, \ldots)$, for which x_k could be either real-valued or complex-valued. It is possible for sequences to be *doubly infinite*, for instance, $x = (x_k) = (\ldots, x_{-2}, x_{-1}, x_0, x_1, x_2, \ldots)$.

Relationships between variables are expressed as mathematical functions, that is, *mappings* between sets. The notation $f | A \to B$ signifies that function f associates an element of set A with an element from set B. For example, $f | \mathbf{R} \to \mathbf{R}$ represents a function defined on the real-number line, and this function is also real-valued; that is, it maps "points" in $\mathbf{R}$ to "points" in $\mathbf{R}$. We are familiar with the idea of "plotting" such a function on the xy plane if $y = f(x)$ (i.e., $x, y \in \mathbf{R}$). It is important to note that we may regard sequences as functions that are defined on either the set $\mathbf{Z}$ (the case of doubly infinite sequences), or the set $\mathbf{Z}^+$ (the case of singly infinite sequences). To be more specific, if, for example, $k \in \mathbf{Z}^+$, then this number maps to some number x_k that is either real-valued or complex-valued. Since vectors are associated with sequences of finite length, they, too, may be regarded as functions, but defined on a finite subset of the integers. From (1.1) this subset might be denoted by $\mathbf{Z}_n = \{0, 1, 2, \ldots, n-2, n-1\}$.

Sets of functions are important. This is because in engineering we are often interested in mappings between sets of functions. For example, in electric circuits voltage and current waveforms (i.e., functions of time) are input to a circuit via voltage and current sources. Voltage drops across circuit elements, or currents through circuit elements are output functions of time. Thus, any circuit maps functions from an input set to functions from some output set. Digital signal processing systems do the same thing, except that here the functions are sequences. For example, a simple digital signal processing system might accept as input the sequence (x_n), and produce as output the sequence (y_n) according to

$$y_n = \frac{x_n + x_{n+1}}{2} \tag{1.3}$$

for which $n \in \mathbf{Z}^+$.

Some specific examples of sets of functions are as follows, and more will be seen later. The set of real-valued functions defined on the interval $[a, b] \subset \mathbf{R}$ that are n times *continuously* differentiable may be denoted by $C^n[a, b]$. This means that all derivatives up to and including order n exist and are continuous. If $n = 0$ we often just write $C[a, b]$, which is the set of continuous functions on the interval $[a, b]$. We remark that the notation $[a, b]$ implies inclusion of the endpoints of the interval. Thus, (a, b) implies that the endpoints a and b are not to be included [i.e., if $x \in (a, b)$, then $a < x < b$].

A polynomial in the *indeterminate* x of degree n is

$$p_n(x) = \sum_{k=0}^{n} p_{n,k} x^k. \tag{1.4}$$

Unless otherwise stated, we will always assume $p_{n,k} \in \mathbf{R}$. The indeterminate x is often considered to be either a real number or a complex number. But in some circumstances the indeterminate x is merely regarded as a "placeholder," which means that x is not supposed to take on a value. In a situation like this the polynomial coefficients may also be regarded as elements of a vector (e.g., $p_n = [p_{n,0} \ p_{n,1} \ \cdots \ p_{n,n}]^T$). This happens in digital signal processing when we wish to convolve[1] sequences of finite length, because the multiplication of polynomials is mathematically equivalent to the operation of sequence convolution. We will denote the set of all polynomials of degree n as $\mathbf{P}^n$. If x is to be from the interval $[a, b] \subset \mathbf{R}$, then the set of polynomials of degree n on $[a, b]$ is denoted by $\mathbf{P}^n[a, b]$. If $m < n$ we shall usually assume $\mathbf{P}^m[a, b] \subset \mathbf{P}^n[a, b]$.

1.3   SOME SPECIAL MAPPINGS: METRICS, NORMS, AND INNER PRODUCTS

Sets of objects (vectors, sequences, polynomials, functions, etc.) often have certain special mappings defined on them that turn these sets into what are commonly called *function spaces.* Loosely speaking, functional analysis is about the properties

---

[1] These days it seems that the operation of convolution is first given serious study in introductory signals and systems courses. The operation of convolution is fundamental to all forms of signal processing, either analog or digital.

of function spaces. Generally speaking, numerical computation problems are best handled by treating them in association with suitable mappings on well-chosen function spaces. For our purposes, the three most important special types of mappings are (1) metrics, (2) norms, and (3) inner products. You are likely to be already familiar with special cases of these really very general ideas.

The vector dot product is an example of an inner product on a vector space, while the Euclidean norm (i.e., the square root of the sum of the squares of the elements in a real-valued vector) is a norm on a vector space. The Euclidean distance between two vectors (given by the Euclidean norm of the difference between the two vectors) is a metric on a vector space. Again, loosely speaking, metrics give meaning to the concept of "distance" between points in a function space, norms give a meaning to the concept of the "size" of a vector, and inner products give meaning to the concept of "direction" in a vector space.[2]

In Section 1.1 we expressed interest in the sizes of errors, and so naturally the concept of a norm will be of interest. Later we shall see that inner products will prove to be useful in devising means of overcoming problems due to certain sources of error in a computation. In this section we shall consider various examples of function spaces, some of which we will work with later on in the analysis of certain computational problems. We shall see that there are many different kinds of metric, norm, and inner product. Each kind has its own particular advantages and disadvantages as will be discovered as we progress through the book.

Sometimes a quantity cannot be computed exactly. In this case we may try to estimate *bounds* on the size of the quantity. For example, finding the exact error in the truncation of a series may be impossible, but putting a bound on the error might be relatively easy. In this respect the concepts of supremum and infimum can be important. These are defined as follows.

Suppose we have $E \subset \mathbf{R}$. We say that E is *bounded above* if E has an *upper bound*, that is, if there exists a $B \in \mathbf{R}$ such that $x \leq B$ for all $x \in E$. If $E \neq \emptyset$ (*empty set*; set containing no elements) there is a *supremum* of E [also called a *least upper bound* (lub)], denoted

$$\sup E.$$

For example, suppose $E = [0, 1)$, then any $B \geq 1$ is an upper bound for E, but $\sup E = 1$. More generally, $\sup E \leq B$ for every upper bound B of E. Thus, the supremum is a "tight" upper bound. Similarly, E may be *bounded below*. If E has a *lower bound* there is a $b \in \mathbf{R}$ such that $x \geq b$ for all $x \in E$. If $E \neq \emptyset$, then there exists an *infimum* [also called a *greatest lower bound* (glb)], denoted by

$$\inf E.$$

For example, suppose now $E = (0, 1]$; then any $b \leq 0$ is a lower bound for E, but $\inf E = 0$. More generally, $\inf E \geq b$ for every lower bound b of E. Thus, the infimum is a "tight" lower bound.

---

[2]The idea of "direction" is (often) considered with respect to the concept of an orthogonal basis in a vector space. To define "orthogonality" requires the concept of an inner product. We shall consider this in various ways later on.

1.3.1 Metrics and Metric Spaces

In mathematics an axiomatic approach is often taken in the development of analysis methods. This means that we define a set of objects, a set of operations to be performed on the set of objects, and rules obeyed by the operations. This is typically how mathematical systems are constructed. The reader (hopefully) has already seen this approach in the application of Boolean algebra to the analysis and design of digital electronic systems (i.e., digital logic). We adopt the same approach here. We will begin with the following definition.

Definition 1.1: Metric Space, Metric   A *metric space* is a set X and a function $d | X \times X \rightarrow \mathbf{R}^+$, which is called a *metric* or *distance function* on X. If $x, y, z \in X$ then d satisfies the following axioms:

(M1) $d(x, y) = 0$ if and only if (iff) $x = y$.
(M2) $d(x, y) = d(y, x)$ (symmetry property).
(M3) $d(x, y) \leq d(x, z) + d(z, y)$ (triangle inequality).

We emphasize that X by itself cannot be a metric space until we define d. Thus, the metric space is often denoted by the pair (X, d). The phrase "if and only if" probably needs some explanation. In (M1), if you were told that $d(x, y) = 0$, then you must immediately conclude that $x = y$. Conversely, if you were told that $x = y$, then you must immediately conclude that $d(x, y) = 0$. Instead of the words "if and only if" it is also common to write

$$d(x, y) = 0 \Leftrightarrow x = y.$$

The phrase "if and only if" is associated with elementary logic. This subject is reviewed in Appendix 1.B. It is recommended that the reader study that appendix before continuing with later chapters.

Some examples of metric spaces now follow.

Example 1.1   Set $X = \mathbf{R}$, with

$$d(x, y) = |x - y| \tag{1.5}$$

forms a metric space. The metric (1.5) is what is commonly meant by the "distance between two points on the real number line." The metric (1.5) is quite useful in discussing the sizes of errors due to rounding in digital computation. This is because there is a norm on $\mathbf{R}$ that gives rise to the metric in (1.5) (see Section 1.3.2).

Example 1.2   The set of vectors $\mathbf{R}^n$ with

$$d(x, y) = \left[\sum_{k=0}^{n-1} [x_k - y_k]^2 \right]^{1/2} \tag{1.6a}$$

forms a (Euclidean) metric space. However, another valid metric on $\mathbf{R}^n$ is given by

$$d_1(x, y) = \sum_{k=0}^{n-1} |x_k - y_k|. \tag{1.6b}$$

In other words, we can have the metric space (X, d), or (X, d_1). These spaces are different because their metrics differ.

Euclidean metrics, and their related norms and inner products, are useful in posing and solving least-squares approximation problems. Least-squares approximation is a topic we shall consider in detail later.

Example 1.3   Consider the set of (singly) infinite, complex-valued, and bounded sequences

$$X = \{x = (x_0, x_1, x_2, \ldots) | x_k \in \mathbf{C}, |x_k| \le c(x)(\text{all } k)\}. \tag{1.7a}$$

Here $c(x) \ge 0$ is a bound that may depend on x, but not on k. This set forms a metric space that may be denoted by $l^\infty[0, \infty]$ if we employ the metric

$$d(x, y) = \sup_{k \in \mathbf{Z}^+} |x_k - y_k|. \tag{1.7b}$$

The notation $[0, \infty]$ emphasizes that the sequences we are talking about are only singly infinite. We would use $[-\infty, \infty]$ to specify that we are talking about doubly infinite sequences.

Example 1.4   Define $J = [a, b] \subset \mathbf{R}$. The set $C[a, b]$ will be a metric space if

$$d(x, y) = \sup_{t \in J} |x(t) - y(t)|. \tag{1.8}$$

In Example 1.1 the metric (1.5) gives the "distance" between points on the real-number line. In Example 1.4 the "points" are real-valued, continuous functions of $t \in [a, b]$. In functional analysis it is essential to get used to the idea that functions can be considered as points in a space.

Example 1.5   The set X in (1.7a), where we now allow $c(x) \to \infty$ (in other words, the sequence need not be bounded here), but with the metric

$$d(x, y) = \sum_{k=0}^{\infty} \frac{1}{2^{k+1}} \frac{|x_k - y_k|}{1 + |x_k - y_k|} \tag{1.9}$$

is a metric space. (Sometimes this space is denoted s.)

Example 1.6 Let p be a real-valued constant such that $p \geq 1$. Consider the set of complex-valued sequences

$$X = \left\{ x = (x_0, x_1, x_2, \ldots) | x_k \in \mathbf{C}, \sum_{k=0}^{\infty} |x_k|^p < \infty \right\}. \tag{1.10a}$$

This set together with the metric

$$d(x, y) = \left[\sum_{k=0}^{\infty} |x_k - y_k|^p \right]^{1/p} \tag{1.10b}$$

forms a metric space that we denote by $l^p[0, \infty]$.

Example 1.7 Consider the set of complex-valued functions on $[a, b] \subset \mathbf{R}$

$$X = \left\{ x(t) \left| \int_a^b |x(t)|^2 \, dt < \infty \right. \right\} \tag{1.11a}$$

for which

$$d(x, y) = \left[\int_a^b |x(t) - y(t)|^2 \, dt \right]^{1/2} \tag{1.11b}$$

is a metric. Pair (X, d) forms a metric space that is usually denoted by $L^2[a, b]$.

The metric space of Example 1.7 (along with certain variations) is very important in the theory of orthogonal polynomials, and in least-squares approximation problems. This is because it turns out to be an inner product space too (see Section 1.3.3). Orthogonal polynomials have a major role to play in the solution of least squares, and other types of approximation problem.

All of the metrics defined in the examples above may be shown to satisfy the axioms of Definition 1.1. Of course, at least in some cases, much effort might be required to do this. In this book we largely avoid making this kind of effort.

1.3.2 Norms and Normed Spaces

So far our examples of function spaces have been metric spaces (Section 1.3.1). Such spaces are not necessarily associated with the concept of a vector space. However, normed spaces (i.e., spaces with norms defined on them) are always associated with vector spaces. So, before we can define a norm, we need to recall the general definition of a vector space.

The following definition invokes the concept of a *field* of numbers. This concept arises in abstract algebra and number theory [e.g., 2, 3], a subject we wish to avoid considering here.[3] It is enough for the reader to know that $\mathbf{R}$ and $\mathbf{C}$ are fields under

---

[3]This avoidance is not to disparage abstract algebra. This subject is a necessary prerequisite to understanding concepts such as fast algorithms for digital signal processing (i.e., fast Fourier transforms, and fast convolution algorithms; e.g., see Ref. 4), cryptography and data security, and error control codes for digital communications.

the usual real and complex arithmetic operations. These are really the only fields that we shall work with. We remark, largely in passing, that rational numbers (set denoted **Q**) are also a field under the usual arithmetic operations.

Definition 1.2: Vector Space   A *vector space* (*linear space*) over a field K is a nonempty set X of elements $x, y, z, \ldots$ called *vectors* together with two algebraic operations. These operations are vector addition, and the multiplication of vectors by scalars that are elements of K. The following axioms must be satisfied:

(V1) If $x, y \in X$, then $x + y \in X$ (additive closure).

(V2) If $x, y, z \in X$, then $(x + y) + z = x + (y + z)$ (associativity).

(V3) There exists a vector in X denoted 0 (*zero vector*) such that for all $x \in X$, we have $x + 0 = 0 + x = x$.

(V4) For all $x \in X$, there is a vector $-x \in X$ such that $-x + x = x + (-x) = 0$. We call $-x$ the *negative of a vector*.

(V5) For all $x, y \in X$ we have $x + y = y + x$ (commutativity).

(V6) If $x \in X$ and $a \in K$, then the product of a and x is ax, and $ax \in X$.

(V7) If $x, y \in X$, and $a \in K$, then $a(x + y) = ax + ay$.

(V8) If $a, b \in K$, and $x \in X$, then $(a + b)x = ax + bx$.

(V9) If $a, b \in K$, and $x \in X$, then $ab(x) = a(bx)$.

(V10) If $x \in X$, and $1 \in K$, then $1x = x$ multiplication of a vector by a unit scalar; all fields contain a unit scalar (i.e., a number called "one").

In this definition, as already noted, we generally work only with $K = \mathbf{R}$, or $K = \mathbf{C}$. We represent the zero vector by 0 just as we also represent the scalar zero by 0. Rarely is there danger of confusion.

The reader is already familiar with the special instances of this that relate to the sets $\mathbf{R}^n$ and $\mathbf{C}^n$. These sets are vector spaces under Definition 1.2, where vector addition is defined to be

$$x + y = \begin{bmatrix} x_0 \\ x_1 \\ \vdots \\ x_{n-1} \end{bmatrix} + \begin{bmatrix} y_0 \\ y_1 \\ \vdots \\ y_{n-1} \end{bmatrix} = \begin{bmatrix} x_0 + y_0 \\ x_1 + y_1 \\ \vdots \\ x_{n-1} + y_{n-1} \end{bmatrix}, \tag{1.12a}$$

and multiplication by a field element is defined to be

$$ax = \begin{bmatrix} ax_0 \\ ax_1 \\ \vdots \\ ax_{n-1} \end{bmatrix}. \tag{1.12b}$$

The zero vector is $0 = [00 \cdots 00]^T$, and $-x = [-x_0 - x_1 \cdots - x_{n-1}]^T$. If $X = \mathbf{R}^n$ then the elements of x and y are real-valued, and $a \in \mathbf{R}$, but if $X = \mathbf{C}^n$ then the elements of x and y are complex-valued, and $a \in \mathbf{C}$. The metric spaces in Example 1.2 are therefore also vector spaces under the operations defined in (1.12a,b).

Some further examples of vector spaces now follow.

Example 1.8   Metric space $C[a, b]$ (Example 1.4) is a vector space under the operations

$$(x + y)(t) = x(t) + y(t), \quad (\alpha x)(t) = \alpha x(t), \tag{1.13}$$

where $\alpha \in \mathbf{R}$. The zero vector is the function that is identically zero on the interval $[a, b]$.

Example 1.9   Metric space $l^2[0, \infty]$ (Example 1.6) is a vector space under the operations

$$x + y = (x_0, x_1, \ldots) + (y_0, y_1, \ldots) = (x_0 + y_0, x_1 + y_1, \ldots),$$

$$\alpha x = (\alpha x_0, \alpha x_1, \ldots). \tag{1.14}$$

Here $\alpha \in \mathbf{C}$.

If $x, y \in l^2[0, \infty]$, then some effort is required to verify axiom (V1). This requires the *Minkowski inequality*, which is

$$\left[\sum_{k=0}^{\infty} |x_k + y_k|^p \right]^{1/p} \leq \left[\sum_{k=0}^{\infty} |x_k|^p \right]^{1/p} + \left[\sum_{k=0}^{\infty} |y_k|^p \right]^{1/p}. \tag{1.15}$$

Refer back to Example 1.6; here we employ $p = 2$, but (1.15) is valid for $p \geq 1$. Proof of (1.15) is somewhat involved, and so is omitted here. The interested reader can see Kreyszig [1, pp. 11–15].

We remark that the Minkowski inequality can be proved with the aid of the *Hölder inequality*

$$\sum_{k=0}^{\infty} |x_k y_k| \leq \left[\sum_{k=0}^{\infty} |x_k|^p \right]^{1/p} \left[\sum_{k=0}^{\infty} |y_k|^q \right]^{1/q} \tag{1.16}$$

for which here $p > 1$ and $\frac{1}{p} + \frac{1}{q} = 1$.

We are now ready to define a normed space.

Definition 1.3: Normed Space, Norm   A *normed space* X is a vector space with a norm defined on it. If $x \in X$ then the norm of x is denoted by

$$||x|| \quad \text{(read this as ``norm of } x\text{'').}$$

The norm must satisfy the following axioms:

(N1) $||x|| \geq 0$ (i.e., the norm is nonnegative).

(N2) $||x|| = 0 \Leftrightarrow x = 0$.

(N3) $||\alpha x|| = |\alpha| \, ||x||$. Here α is a scalar in the field of X (i.e., $\alpha \in K$; see Definition 3.2).

(N4) $||x + y|| \le ||x|| + ||y||$ (triangle inequality).

The normed space is vector space X together with a norm, and so may be properly denoted by the pair $(X, || \cdot ||)$. However, we may simply write X, and say "normed space X," so the norm that goes along with X is understood from the context of the discussion.

It is important to note that all normed spaces are also metric spaces, where the metric is given by

$$d(x, y) = ||x - y|| \quad (x, y \in X). \tag{1.17}$$

The metric in (1.17) is called the *metric induced by the norm*.

Various other properties of norms may be deduced. One of these is:

Example 1.10   Prove $| \, ||y|| - ||x|| \, | \le ||y - x||$.

Proof   From (N3) and (N4)

$$||y|| = ||y - x + x|| \le ||y - x|| + ||x||, \, ||x|| = ||x - y + y|| \le ||y - x|| + ||y||.$$

Combining these, we obtain

$$||y|| - ||x|| \le ||y - x||, \, ||y|| - ||x|| \ge -||y - x||.$$

The claim follows immediately.

We may regard the norm as a mapping from X to set $\mathbf{R}$: $|| \cdot |||X \to \mathbf{R}$. This mapping can be shown to be continuous. However, this requires generalizing the concept of continuity that you may know from elementary calculus. Here we define continuity as follows.

Definition 1.4: Continuous Mapping   Suppose $X = (X, d)$ and $Y = (Y, \overline{d})$ are two metric spaces. The mapping $T|X \to Y$ is said to be *continuous at a point* $x_0 \in X$ if for all $\epsilon > 0$ there is a $\delta > 0$ such that

$$\overline{d}(Tx, Tx_0) < \epsilon \quad \text{for all } x \text{ satisfying} \quad d(x, x_0) < \delta. \tag{1.18}$$

T is said to be *continuous* if it is continuous at every point of X.

Note that Tx is just another way of writing $T(x)$. $(\mathbf{R}, | \cdot |)$ is a normed space; that is, the set of real numbers with the usual arithmetic operations defined on it is a vector space, and the absolute value of an element of $\mathbf{R}$ is the norm of that element. If we identify Y in Definition 1.4 with metric space $(\mathbf{R}, |\cdot|)$, then (1.18) becomes

$$\overline{d}(Tx, Tx_0) = \overline{d}(||x||, ||x_0||) = |\ ||x|| - ||x_0||\ | < \epsilon, \quad d(x, x_0) = ||x - x_0|| < \delta.$$

To make these claims, we are using (1.17). In other words, X and Y are normed spaces, and we employ the metrics induced by their respective norms. In addition, we identify T with $||\cdot||$. Using Example 1.10, we obtain

$$|\ ||x|| - ||x_0||\ | \le ||x - x_0|| < \delta.$$

Thus, the requirements of Definition 1.4 are met, and so we conclude that norms are continuous mappings.

We now list some other normed spaces.

Example 1.11   The *Euclidean space* $\mathbf{R}^n$ and the *unitary space* $\mathbf{C}^n$ are both normed spaces, where the norm is defined to be

$$||x|| = \left[\sum_{k=0}^{n-1} |x_k|^2\right]^{1/2}. \tag{1.19}$$

For $\mathbf{R}^n$ the absolute value bars may be dropped.[4] It is easy to see that $d(x, y) = ||x - y||$ gives the same metric as in (1.6a) for space $\mathbf{R}^n$. We further remark that for $n = 1$ we have $||x|| = |x|$.

Example 1.12   The space $l^p[0, \infty]$ is a normed space if we define the norm to be

$$||x|| = \left[\sum_{k=0}^{\infty} |x_k|^p\right]^{1/p} \tag{1.20}$$

for which $d(x, y) = ||x - y||$ coincides with the metric in (1.10b).

Example 1.13   The sequence space $l^\infty[0, \infty]$ from Example 1.3 of Section 1.3.1 is a normed space, where the norm is defined to be

$$||x|| = \sup_{k \in \mathbf{Z}^+} |x_k|, \tag{1.21}$$

and this norm induces the metric of (1.7b).

---

[4]Suppose $z = x + jy$ ($j = \sqrt{-1}$, $x, y \in \mathbf{R}$) is some arbitrary complex number. Recall that $z^2 \ne |z|^2$ in general.

Example 1.14   The space $C[a, b]$ first seen in Example 1.4 is a normed space, where the norm is defined by

$$||x|| = \sup_{t \in J} |x(t)|. \tag{1.22}$$

Naturally, this norm induces the metric of (1.8).

Example 1.15   The space $L^2[a, b]$ of Example 1.7 is a normed space for the norm

$$||x|| = \left[\int_a^b |x(t)|^2 \, dt \right]^{1/2}. \tag{1.23}$$

This norm induces the metric in (1.11b).

The normed space of Example 1.15 is important in the following respect. Observe that

$$||x||^2 = \int_a^b |x(t)|^2 \, dt. \tag{1.24}$$

Suppose we now consider a resistor with resistance R. If the voltage drop across its terminals is $v(t)$ and the current through it is $i(t)$, we know that the instantaneous power dissipated in the device is $p(t) = v(t)i(t)$. If we assume that the resistor is a linear device, then $v(t) = Ri(t)$ via Ohm's law. Thus

$$p(t) = v(t)i(t) = Ri^2(t). \tag{1.25}$$

Consequently, the amount of energy delivered to the resistor over time interval $t \in [a, b]$ is given by

$$E = R \int_a^b i^2(t) \, dt. \tag{1.26}$$

If the voltage/current waveforms in our circuit containing R belong to the space $L^2[a, b]$, then clearly $E = R||i||^2$. We may therefore regard the square of the L^2 norm [given by (1.24)] of a signal to be the *energy of the signal*, provided the norm exists. This notion can be helpful in the optimal design of electric circuits (e.g., electric filters), and also of optimal electronic circuits. In analogous fashion, an element x of space $l^2[0, \infty]$ satisfies

$$||x||^2 = \sum_{k=0}^{\infty} |x_k|^2 < \infty \tag{1.27}$$

[see (1.10a) and Example 1.12]. We may consider $||x||^2$ to be the *energy of the single-sided sequence x*. This notion is useful in the optimal design of digital filters.

1.3.3   Inner Products and Inner Product Spaces

The concept of an inner product is necessary before one can talk about orthogonal bases for vector spaces. Recall from elementary linear algebra that orthogonal bases were important in representing vectors. From a computational standpoint, as mentioned earlier, orthogonal bases can have a simplifying effect on certain types of approximation problem (e.g., least-squares approximations), and represent a means of controlling numerical errors due to so-called ill-conditioned problems.

Following our axiomatic approach, consider the following definition.

Definition 1.5:   Inner Product Space, Inner Product   An *inner product space* is a vector space X with an *inner product* defined on it. The inner product is a mapping $\langle \cdot, \cdot \rangle | X \times X \to K$ that satisfies the following axioms:

(I1) $\langle x + y, z \rangle = \langle x, z \rangle + \langle y, z \rangle$.

(I2) $\langle \alpha x, y \rangle = \alpha \langle x, y \rangle$.

(I3) $\langle x, y \rangle = \langle y, x \rangle^*$.

(I4) $\langle x, x \rangle \geq 0$, and $\langle x, x \rangle = 0 \Leftrightarrow x = 0$.

Naturally, $x, y, z \in X$, and α is a scalar from the field K of vector space X. The asterisk superscript on $\langle y, x \rangle$ in (I3) denotes complex conjugation.[5]

If the field of X is not $\mathbf{C}$, then the operation of complex conjugation in (I3) is redundant.

All inner product spaces are also normed spaces, and hence are also metric spaces. This is because the inner product induces a norm on X

$$\|x\| = [\langle x, x \rangle]^{1/2} \tag{1.28}$$

for all $x \in X$. Following (1.17), the induced metric is

$$d(x, y) = \|x - y\| = [\langle x - y, x - y \rangle]^{1/2}. \tag{1.29}$$

Directly from the axioms of Definition 1.5, it is possible to deduce that (for $x, y, z \in X$ and $a, b \in K$)

$$\langle ax + by, z \rangle = a \langle x, z \rangle + b \langle y, z \rangle, \tag{1.30a}$$

$$\langle x, ay \rangle = a^* \langle x, y \rangle, \tag{1.30b}$$

and

$$\langle x, ay + bz \rangle = a^* \langle x, y \rangle + b^* \langle x, z \rangle. \tag{1.30c}$$

The reader should prove these as an exercise.

[5]If $z = x + yj$ is a complex number, then its conjugate is $z^* = x - yj$.

We caution the reader that not all normed spaces are inner product spaces. We may construct an example with the aid of the following example.

Example 1.16 Let x, y be from an inner product space. If $|| \cdot ||$ is the norm induced by the inner product, then $||x + y||^2 + ||x - y||^2 = 2(||x||^2 + ||y||^2)$. This is the *parallelogram equality*.

Proof Via (1.30a,c) we have

$$||x + y||^2 = \langle x + y, x + y \rangle = \langle x, x + y \rangle + \langle y, x + y \rangle$$
$$= \langle x, x \rangle + \langle x, y \rangle + \langle y, x \rangle + \langle y, y \rangle,$$

and
$$||x - y||^2 = \langle x - y, x - y \rangle = \langle x, x - y \rangle - \langle y, x - y \rangle$$
$$= \langle x, x \rangle - \langle x, y \rangle - \langle y, x \rangle + \langle y, y \rangle.$$

Adding these gives the stated result.

It turns out that the space $l^p[0, \infty]$ with $p \neq 2$ is not an inner product space. The parallelogram equality can be used to show this. Consider $x = (1, 1, 0, 0, \ldots)$, $y = (1, -1, 0, 0, \ldots)$, which are certainly elements of $l^p[0, \infty]$ [see (1.10a)]. We see that

$$||x|| = ||y|| = 2^{1/p}, ||x + y|| = ||x - y|| = 2.$$

The parallelogram equality is not satisfied, which implies that our norm does not come from an inner product. Thus, $l^p[0, \infty]$ with $p \neq 2$ cannot be an inner product space.

On the other hand, $l^2[0, \infty]$ **is** an inner product space, where the inner product is defined to be

$$\langle x, y \rangle = \sum_{k=0}^{\infty} x_k y_k^*. \tag{1.31}$$

Does this infinite series converge? Yes, it does. To see this, we need the *Cauchy–Schwarz inequality*.[6] Recall the Hölder inequality of (1.16). Let $p = 2$, so that $q = 2$. Then the Cauchy–Schwarz inequality is

$$\sum_{k=0}^{\infty} |x_k y_k| \leq \left[\sum_{k=0}^{\infty} |x_k|^2 \right]^{1/2} \left[\sum_{k=0}^{\infty} |y_k|^2 \right]^{1/2}. \tag{1.32}$$

---

[6]The inequality we consider here is related to the Schwarz inequality. We will consider the Schwarz inequality later on. This inequality is of immense practical value to electrical and computer engineers. It is used to derive the matched-filter receiver, which is employed in digital communications systems, to derive the uncertainty principle in quantum mechanics and in signal processing, and to derive the Cramér–Rao lower bound on the variance of parameter estimators, to name only three applications.

Now

$$|\langle x, y \rangle| = \left| \sum_{k=0}^{\infty} x_k y_k^* \right| \leq \sum_{k=0}^{\infty} |x_k y_k|. \tag{1.33}$$

The inequality in (1.33) follows from the triangle inequality for $| \cdot |$. (Recall that the absolute value operation is a norm on $\mathbf{R}$. It is also a norm on $\mathbf{C}$; if $z = x + jy \in \mathbf{C}$, then $|z| = \sqrt{x^2 + y^2}$.) The right-hand side of (1.32) is finite because x and y are in $l^2[0, \infty]$. Thus, from (1.33), $\langle x, y \rangle$ is finite. Thus, the series (1.31) converges.

It turns out that $C[a, b]$ is not an inner product space, either. But we will not demonstrate the truth of this claim here.

Some further examples of inner product spaces are as follows.

Example 1.17   The *Euclidean space* $\mathbf{R}^n$ is an inner product space, where the inner product is defined to be

$$\langle x, y \rangle = \sum_{k=0}^{n-1} x_k y_k. \tag{1.34}$$

The reader will recognize this as the *vector dot product* from elementary linear algebra; that is, $x \cdot y = \langle x, y \rangle$. It is well worth noting that

$$\langle x, y \rangle = x^T y. \tag{1.35}$$

Here the superscript T denotes *transposition*. So, x^T is a row vector. The inner product in (1.34) certainly induces the norm in (1.19).

Example 1.18   The *unitary space* $\mathbf{C}^n$ is an inner product space for the inner product

$$\langle x, y \rangle = \sum_{k=0}^{n-1} x_k y_k^*. \tag{1.36}$$

Again, the norm of (1.19) is induced by inner product (1.36). If H denotes the operation of complex conjugation and transposition (this is called *Hermitian transposition*), then

$$y^H = [y_0^* y_1^* \cdots y_{n-1}^*]$$

(row vector), and

$$\langle x, y \rangle = y^H x. \tag{1.37}$$

Example 1.19   The space $L^2[a, b]$ from Example 1.7 is an inner product space if the inner product is defined to be

$$\langle x, y \rangle = \int_a^b x(t) y^*(t) \, dt. \tag{1.38}$$

The norm induced by (1.38) is

$$\|x\| = \left[\int_a^b |x(t)|^2 \, dt \right]^{1/2}. \tag{1.39}$$

This in turn induces the metric in (1.11b).

Now we consider the concept of orthogonality in a completely general manner.

Definition 1.6: Orthogonality   Let x, y be vectors from some inner product space X. These vectors are *orthogonal* iff

$$\langle x, y \rangle = 0.$$

The orthogonality of x and y is symbolized by writing $x \perp y$. Similarly, for subsets $A, B \subset X$ we write $x \perp A$ if $x \perp a$ for all $a \in A$, and $A \perp B$ if $a \perp b$ for all $a \in A$, and $b \in B$.

If we consider the inner product space $\mathbf{R}^2$, then it is easy to see that $\langle [1 \, 0]^T, [0 \, 1]^T \rangle = 0$, so $[0 \, 1]^T$, and $[1 \, 0]^T$ are orthogonal vectors. In fact, these vectors form an orthogonal basis for $\mathbf{R}^2$, a concept we will consider more generally below. If we define the *unit vectors* $e_0 = [1 \, 0]^T$, and $e_1 = [0 \, 1]^T$, then we recall that any $x \in \mathbf{R}^2$ can be expressed as $x = x_0 e_0 + x_1 e_1$. (The extension of this reasoning to $\mathbf{R}^n$ for $n > 2$ should be clear.) Another example of a pair of orthogonal vectors would be $x = \frac{1}{\sqrt{2}}[1 \, 1]^T$, and $y = \frac{1}{\sqrt{2}}[1 \, -1]^T$. These too form an orthogonal basis for the space $\mathbf{R}^2$.

Define the functions

$$\phi(x) = \begin{cases} 0, & x < 0 \text{ and } x \geq 1 \\ 1, & 0 \leq x < 1 \end{cases} \tag{1.40}$$

and

$$\psi(x) = \begin{cases} 0, & x < 0 \text{ and } x \geq 1 \\ 1, & 0 \leq x < \frac{1}{2} \\ -1, & \frac{1}{2} \leq x < 1 \end{cases}. \tag{1.41}$$

Function $\phi(x)$ is called the *Haar scaling function*, and function $\psi(x)$ is called the *Haar wavelet* [5]. The function $\phi(x)$ is also called an *non-return-to-zero* (NRZ) *pulse*, and function $\psi(x)$ is also called a *Manchester pulse* [6]. It is easy to confirm that these pulses are elements of $L^2(\mathbf{R}) = L^2(-\infty, \infty)$, and that they are orthogonal, that is, $\langle \phi, \psi \rangle = 0$ under the inner product defined in (1.38). This is so because

$$\langle \phi, \psi \rangle = \int_{-\infty}^{\infty} \phi(x) \psi^*(x) \, dx = \int_0^1 \psi(x) \, dx = 0.$$

Thus, we consider ϕ and ψ to be elements in the inner product space $L^2(\mathbf{R})$, for which the inner product is

$$\langle x, y \rangle = \int_{-\infty}^{\infty} x(t) y^*(t) \, dt.$$

It turns out that the Haar wavelet is the simplest example of the more general class of *Daubechies wavelets*. The general theory of these wavelets first appeared in Daubechies [7]. Their development has revolutionized signal processing and many other areas.[7] The main reason for this is the fact that for any $f(t) \in L^2(\mathbf{R})$

$$f(t) = \sum_{n=-\infty}^{\infty} \sum_{k=-\infty}^{\infty} \langle f, \psi_{n,k} \rangle \psi_{n,k}(t), \qquad (1.42)$$

where $\psi_{n,k}(t) = 2^{n/2} \psi(2^n t - k)$. This doubly infinite series is called a *wavelet series expansion* for f. The coefficients $f_{n,k} = \langle f, \psi_{n,k} \rangle$ have finite energy. In effect, if we treat either k or n as a constant, then the resulting doubly infinite sequence is in the space $l^2[-\infty, \infty]$. In fact, it is also the case that

$$\sum_{n=-\infty}^{\infty} \sum_{k=-\infty}^{\infty} |f_{n,k}|^2 < \infty. \qquad (1.43)$$

It is to be emphasized that the ψ used in (1.42) could be (1.41), or it could be chosen from the more general class in Ref. 7. We shall not prove these things in this book, as the technical arguments are quite hard.

The wavelet series is presently not as familiar to the broader electrical and computer engineering community as is the *Fourier series*. A brief summary of the Fourier series is as follows. Again, rigorous proofs of many of the following claims will be avoided, though good introductory references to Fourier series are Tolstov [8] or Kreyszig [9]. If $f \in L^2(0, 2\pi)$, then

$$f(t) = \sum_{n=-\infty}^{\infty} f_n e^{jnt}, \quad j = \sqrt{-1}, \qquad (1.44)$$

where the *Fourier (series) coefficients* are given by

$$f_n = \frac{1}{2\pi} \int_0^{2\pi} f(t) e^{-jnt} \, dt. \qquad (1.45)$$

We may define

$$e_n(t) = \exp(jnt) \quad (t \in (0, 2\pi), \; n \in \mathbf{Z}) \qquad (1.46)$$

[7]For example, in digital communications the problem of designing good signaling pulses for data transmission is best treated with respect to wavelet theory.

so that we see

$$\langle f, e_n \rangle = \frac{1}{2\pi} \int_0^{2\pi} f(t) \left[e^{jnt} \right]^* dt = f_n. \tag{1.47}$$

The series (1.44) is the *complex Fourier series expansion* for f. Note that for $n, k \in \mathbf{Z}$

$$\exp[jn(t + 2\pi k)] = \exp[jnt] \exp[2\pi jnk] = \exp[jnt]. \tag{1.48}$$

Here we have used *Euler's identity*

$$e^{jx} = \cos x + j \sin x \tag{1.49}$$

and $\cos(2\pi k) = 1$, $\sin(2\pi k) = 0$. The function e^{jnt} is therefore 2π-periodic; that is, its period is 2π. It therefore follows that the series on the right-hand side of (3.40) is a 2π-periodic function, too. The result (1.48) implies that, although f in (1.44) is initially defined only on $(0, 2\pi)$, we are at liberty to "periodically extend" f over the entire real-number line; that is, we can treat f as one period of the periodic function

$$\tilde{f}(t) = \sum_{k \in \mathbf{Z}} f(t + 2\pi k) \tag{1.50}$$

for which $f(t) = \tilde{f}(t)$ for $t \in (0, 2\pi)$. Thus, series (1.44) is a way to represent periodic functions. Because $f \in L^2(0, 2\pi)$, it turns out that

$$\sum_{n=-\infty}^{\infty} |f_n|^2 < \infty \tag{1.51}$$

so that $(f_n) \in l^2[-\infty, \infty]$.

Observe that in (1.47) we have "redefined" the inner product on $L^2(0, 2\pi)$ to be

$$\langle x, y \rangle = \frac{1}{2\pi} \int_0^{2\pi} x(t) y^*(t) \, dt \tag{1.52}$$

which differs from (1.38) in that it has the factor $\frac{1}{2\pi}$ in front. This variation also happens to be a valid inner product on the vector space defined by the set in (1.11a). Actually, it is a simple example of a weighted inner product.

Now consider, for $n \neq m$

$$\langle e_n, e_m \rangle = \frac{1}{2\pi} \int_0^{2\pi} e^{jnt} e^{-jmt} \, dt = \frac{1}{2\pi j(n-m)} \left[e^{j(n-m)t} \right]_0^{2\pi}$$

$$= \frac{e^{2\pi j(n-m)} - 1}{2\pi j(n-m)} = \frac{1 - 1}{2\pi j(n-m)} = 0. \tag{1.53}$$

Similarly

$$\langle e_n, e_n \rangle = \frac{1}{2\pi} \int_0^{2\pi} e^{jnt} e^{-jnt} \, dt = \frac{1}{2\pi} \int_0^{2\pi} dt = 1. \tag{1.54}$$

So, e_n and e_m (if $n \neq m$) are orthogonal with respect to the inner product in (1.52).

From basic electric circuit analysis, periodic signals have finite power. Therefore, series (1.44) is a way to represent finite power signals.[8] We might therefore consider the space $L^2(0, 2\pi)$ to be the "space of finite power signals." From considerations involving the wavelet series representation of (1.42), we may consider $L^2(\mathbf{R})$ to be the "space of finite energy signals." Recall also the discussion at the end of Section 1.3.2 (last paragraph).

An example of a Fourier series expansion is the following.

Example 1.20   Suppose that

$$f(t) = \left\{ \begin{array}{ll} 1, & 0 < t < \pi \\ -1, & \pi \le t < 2\pi \end{array} \right. . \tag{1.55}$$

A sketch of this function is one period of a 2π-periodic square wave. The Fourier coefficients are given by (for $n \ne 0$)

$$
\begin{aligned}
f_n &= \frac{1}{2\pi} \int_0^{2\pi} f(t) e^{-jnt}\, dt = \frac{1}{2\pi} \left[\int_0^\pi e^{-jnt}\, dt - \int_\pi^{2\pi} e^{-jnt}\, dt \right] \\
&= \frac{1}{2\pi} \left[-\frac{1}{jn} \left[e^{-jnt} \right]_0^\pi + \frac{1}{jn} \left[e^{-jnt} \right]_\pi^{2\pi} \right] = \frac{1}{2\pi} \left[\frac{1 - e^{-jn\pi} - e^{-jn\pi} + 1}{jn} \right] \\
&= \frac{1}{\pi} \frac{1 - e^{-jn\pi}}{jn} = \frac{2}{\pi n} e^{-jn\pi/2} \frac{e^{jn\pi/2} - e^{-jn\pi/2}}{2j} = \frac{2}{\pi n} e^{-jn\pi/2} \sin\left(\frac{\pi n}{2} \right),
\end{aligned}
\tag{1.56}
$$

where we have made use of

$$\sin x = \frac{1}{2j}[e^{jx} - e^{-jx}]. \tag{1.57}$$

This is easily derived using the Euler identity in (1.49). For $n = 0$, it should be clear that $f_0 = 0$.

The coefficients f_n in (1.56) involve expressions containing j. Since $f(t)$ is real-valued, it therefore follows that we can rewrite the series expansion in such a manner as to avoid complex arithmetic. It is almost a standard practice to do this. We now demonstrate this process:

$$
\begin{aligned}
\sum_{n=-\infty}^{\infty} f_n e^{jnt} &= \frac{2}{\pi} \left[\sum_{n=1}^{\infty} \frac{1}{n} e^{-jn\pi/2} \sin\left(\frac{\pi}{2}n \right) e^{jnt} + \sum_{n=-\infty}^{-1} \frac{1}{n} e^{-jn\pi/2} \sin\left(\frac{\pi}{2}n \right) e^{jnt} \right] \\
&= \frac{2}{\pi} \left[\sum_{n=1}^{\infty} \frac{1}{n} e^{-jn\pi/2} \sin\left(\frac{\pi}{2}n \right) e^{jnt} + \sum_{n=1}^{\infty} \frac{1}{n} e^{jn\pi/2} \sin\left(\frac{\pi}{2}n \right) e^{-jnt} \right]
\end{aligned}
$$

[8]In fact, using phasor analysis and superposition, you can apply (1.44) to determine the steady-state output of a circuit for any periodic input (including, and especially, nonsinusoidal periodic functions). This makes the Fourier series very important in electrical/electronic circuit analysis.

$$= \frac{2}{\pi} \sum_{n=1}^{\infty} \frac{1}{n} \sin\left(\frac{\pi}{2}n\right) \left[e^{jnt} e^{-j\pi n/2} + e^{-jnt} e^{j\pi n/2}\right]$$

$$= \frac{4}{\pi} \sum_{n=1}^{\infty} \frac{1}{n} \cos\left[n\left(t - \frac{\pi}{2}\right)\right] \sin\left(\frac{\pi}{2}n\right)$$

Here we have used the fact that (see Appendix 1.A)

$$e^{jnt} e^{-j\pi n/2} + e^{-jnt} e^{j\pi n/2} = 2 \text{ Re } [e^{jnt} e^{-j\pi n/2}] = 2 \cos\left[n\left(t - \frac{\pi}{2}\right)\right].$$

This is so because if $z = x + jy$, then $z + z^* = 2x = 2 \text{ Re } [z]$. Since

$$\cos(\alpha + \beta) = \cos\alpha \cos\beta - \sin\alpha \sin\beta,$$

we have

$$\cos\left[n\left(t - \frac{\pi}{2}\right)\right] = \cos(nt) \cos\frac{\pi n}{2} + \sin(nt) \sin\frac{\pi n}{2}.$$

However, if n is an even number, then $\sin(\pi n/2) = 0$, and if n is an odd number, then $\cos(\pi n/2) = 0$. Therefore

$$\frac{4}{\pi} \sum_{n=1}^{\infty} \frac{1}{n} \cos\left[n\left(t - \frac{\pi}{2}\right)\right] \sin\left(\frac{\pi}{2}n\right)$$

$$= \frac{4}{\pi} \sum_{n=0}^{\infty} \frac{1}{2n+1} \sin\left[(2n+1)t\right] \sin^2\left[(2n+1)\frac{\pi}{2}\right],$$

but $\sin^2[(2n+1)\frac{\pi}{2}] = 1$, so finally we have

$$f(t) = \sum_{n=-\infty}^{\infty} f_n e^{jnt} = \frac{4}{\pi} \sum_{n=0}^{\infty} \frac{1}{2n+1} \sin[(2n+1)t].$$

It is important to note that the wavelet series and Fourier series expansions have something in common, in spite of the fact that they look quite different and indeed are associated with quite different function spaces. The common feature is that both representations involve the use of orthogonal basis functions. We are now ready to consider this in a general manner.

Begin by recalling from elementary linear algebra that a basis for a vector space such as $X = \mathbf{R}^n$ or $X = \mathbf{C}^n$ is a set of n vectors, say

$$B = \{e_0, e_1, \ldots, e_{n-1}\} \tag{1.58}$$

such that the elements e_k (basis vectors) are *linearly independent*. This means that no vector in the set can be expressed as a linear combination of any of the others.

In general, it is not necessary that $\langle e_k, e_n \rangle = 0$ for $n \neq k$. In other words, independence does not require orthogonality. However, if set B is a basis (orthogonal or otherwise) then for any $x \in X$ (vector space) there exists a set of coefficients from the field of the vector space, say, $b = \{b_0, b_1 \ldots, b_{n-1}\}$, such that

$$x = \sum_{k=0}^{n-1} b_k e_k. \tag{1.59}$$

We say that spaces $\mathbf{R}^n$ and $\mathbf{C}^n$ are of *dimension n*. This is a direct reference to the number of basis vectors in B. This notion generalizes.

Now let us consider a sequence space (e.g., $l^2[0, \infty]$). Suppose $x = (x_0, x_1, x_2, \ldots) \in l^2[0, \infty]$. Define the following unit vector sequences:

$$e_0 = (1, 0, 0, 0, \ldots), \quad e_1 = (0, 1, 0, 0, \ldots), \quad e_2 = (0, 0, 1, 0, \ldots), \text{ etc. } \tag{1.60}$$

Clearly

$$x = \sum_{k=0}^{\infty} x_k e_k. \tag{1.61}$$

It is equally clear that no vector e_k can be expressed as a linear combination of any of the others. Thus, the *countably infinite* set[9] $B = \{e_0, e_1, e_2, \ldots\}$ forms a basis for $l^2[0, \infty]$. The sequence space is therefore of *infinite dimension* because B has a countable infinity of members. It is apparent as well that, under the inner product defined in (1.31), we have $\langle e_n, e_m \rangle = \delta_{n-m}$. Sequence $\delta = (\delta_n)$ is called the *Krönecker delta sequence*. It is defined by

$$\delta_n = \begin{cases} 1, & n = 0 \\ 0, & n \neq 0 \end{cases}. \tag{1.62}$$

Therefore, the vectors in (1.60) are mutually orthogonal as well. So they happen to form an orthogonal basis for $l^2[0, \infty]$. Of course, this is not the only possible basis. In general, given a countably infinite set of vectors $\{e_k | k \in \mathbf{Z}^+\}$ [no longer necessarily those in (1.60)] that are linearly independent, and such that $e_k \in l^2[0, \infty]$, for any $x \in l^2[0, \infty]$ there will exist coefficients $a_k \in \mathbf{C}$ such that

$$x = \sum_{k=0}^{\infty} a_k e_k. \tag{1.63}$$

In view of the above, consider the following linearly independent set of vectors from some inner product space X:

$$B = \{e_k | e_k \in X, \quad k \in \mathbf{Z}\}. \tag{1.64}$$

---

[9]A set A is countably infinite if its members can be put into one-to-one (1–1) correspondence with the members of the set $\mathbf{Z}^+$. This is also equivalent to being able to place the elements of A into 1–1 correspondence with the elements of $\mathbf{Z}$.

Assume that this is a basis for X. In this case for any $x \in X$, there are coefficients a_k such that

$$x = \sum_{k \in \mathbf{Z}} a_k e_k. \tag{1.65}$$

We define the set B to be orthogonal iff for all $n, k \in \mathbf{Z}$

$$\langle e_n, e_k \rangle = \delta_{n-k}. \tag{1.66}$$

Assume that the elements of B in (1.64) satisfy (1.66). It is then easy to see that

$$\langle x, e_n \rangle = \left\langle \sum_k a_k e_k, e_n \right\rangle = \sum_k \langle a_k e_k, e_n \rangle \qquad \text{(using (I1))}$$

$$= \sum_k a_k \langle e_k, e_n \rangle \qquad \text{(using (I2))}$$

$$= \sum_k \delta_{k-n} a_k \qquad \text{(using (1.66))}$$

so finally we may say that

$$\langle x, e_n \rangle = a_n. \tag{1.67}$$

In other words, if the basis B is orthogonal, then

$$x = \sum_{k \in \mathbf{Z}} \langle x, e_k \rangle e_k. \tag{1.68}$$

Previous examples (e.g., Fourier series expansion) are merely special cases of this general idea. We see that one of the main features of an orthogonal basis is the ease with which we can obtain the coefficients a_k. Nonorthogonal bases are harder to work with in this respect. This is one of the reasons why orthogonal bases are so universally popular.

A few comments on terminology are in order here. Some would say that the condition (1.66) on B in (1.64) means that B is an *orthonormal set*, and we would say that condition

$$\langle e_n, e_k \rangle = \alpha_n \delta_{n-k}$$

is the condition for B to be an orthogonal set, where α_n is not necessarily unity (i.e., equal to one) for all n. However, in this book we often insist that orthogonal basis vectors be "normalized" so condition (1.66) holds.

We conclude the present section by considering the following theorem. It was mentioned in a footnote that the following *Schwarz inequality* (or variations of it) is of very great value in electrical and computer engineering.

Theorem 1.1: Schwarz Inequality   Let X be an inner product space, where $x, y \in X$. Then

$$|\langle x, y \rangle| \leq ||x|| \, ||y||. \tag{1.69}$$

Equality holds iff $\{x, y\}$ is a linearly dependent set.

Proof   If $y = 0$ then $\langle x, 0 \rangle = 0$, and (1.69) clearly holds in this special case. Let $y \neq 0$. For all scalars α in the field of X we must have [via inner product axioms and (1.30)]

$$0 \leq ||x - \alpha y||^2 = \langle x - \alpha y, x - \alpha y \rangle$$
$$= \langle x, x \rangle - \alpha^* \langle x, y \rangle - \alpha [\langle y, x \rangle - \alpha^* \langle y, y \rangle].$$

If we select $\alpha^* = \langle y, x \rangle / \langle y, y \rangle$, then the quantity in the brackets $[\cdot]$ vanishes. Thus

$$0 \leq \langle x, x \rangle - \frac{\langle y, x \rangle}{\langle y, y \rangle} \langle x, y \rangle = ||x||^2 - \frac{|\langle x, y \rangle|^2}{||y||^2}$$

[using $\langle x, y \rangle = \langle y, x \rangle^*$, i.e., axiom (I3)]. Rearranging, this yields

$$|\langle x, y \rangle|^2 \leq ||x||^2 ||y||^2,$$

and the result (1.69) follows (we must take positive square roots as $||x|| \geq 0$, and $|x| \geq 0$).

Equality holds iff $y = 0$, or else $||x - \alpha y||^2 = 0$, hence $x - \alpha y = 0$ [recall (N2)], so $x = \alpha y$, demonstrating linear dependence of x and y.

We may now see what Theorem 1.1 has to say when applied to the special case of a vector dot product.

Example 1.21   Suppose that X is the inner product space of Example 1.17. Since

$$|\langle x, y \rangle| = \left| \sum_{k=0}^{n-1} x_k y_k \right|$$

and $||x|| = \left[\sum_{k=0}^{n-1} x_k^2 \right]^{1/2}$, we have from Theorem 1.1 that

$$\left| \sum_{k=0}^{n-1} x_k y_k \right| \leq \left[\sum_{k=0}^{n-1} x_k^2 \right]^{1/2} \left[\sum_{k=0}^{n-1} y_k^2 \right]^{1/2}. \tag{1.70}$$

If $y_k = \alpha x_k$ $(\alpha \in \mathbf{R})$ for all $k \in \mathbf{Z}_n$, then

$$\left| \sum_{k=0}^{n-1} x_k y_k \right| = |\alpha| \sum_{k=0}^{n-1} x_k^2,$$

and $\left[\sum_{k=0}^{n-1} y_k^2\right]^{1/2} = |\alpha| \left[\sum_{k=0}^{n-1} x_k^2\right]^{1/2}$, hence

$$\left[\sum_{k=0}^{n-1} x_k^2\right]^{1/2} \left[\sum_{k=0}^{n-1} y_k^2\right]^{1/2} = |\alpha| \sum_{k=0}^{n-1} x_k^2.$$

Thus, (1.70) does indeed hold with equality when $y = \alpha x$.

1.4   THE DISCRETE FOURIER SERIES (DFS)

The subject of discrete Fourier series (DFS) and its relationship to the complex Fourier series expansion of Section 1.3.3 is often deferred to later courses (e.g., signals and systems), but will be briefly considered here as an additional example of an orthogonal series expansion.

The complex Fourier series expansion of Section 1.3.3 was for 2π-periodic functions defined on the real-number line. A similar series expansion exists for N-periodic sequences such as $\tilde{x} = (\tilde{x}_n)$; that is, for $N \in \{2, 3, 4, \ldots\} \subset \mathbf{Z}$, consider

$$\tilde{x}_n = \sum_{k \in \mathbf{Z}} x_{n+kN} \tag{1.71}$$

where $x = (x_n)$ is such that $x_n = 0$ for $n < 0$, and for $n \geq N$ as well. Thus, x is just one period of $\tilde{x}$. We observe that

$$\tilde{x}_{n+mN} = \sum_{k=-\infty}^{\infty} x_{n+mN+kN} = \sum_{k=-\infty}^{\infty} x_{n+(m+k)N} = \sum_{r=-\infty}^{\infty} x_{n+rN} = \tilde{x}_n$$

$(r = m + k)$. This confirms that $\tilde{x}$ is indeed N-periodic (i.e., periodic with period N). We normally assume in a context such as this that $x_n \in \mathbf{C}$. We also regard x as a vector: $x = [x_0 \ x_1 \ \cdots \ x_{N-1}]^T \in \mathbf{C}^N$. An inner product may be defined on the space of N-periodic sequences according to

$$\langle \tilde{x}, \tilde{y} \rangle = \langle x, y \rangle = y^H x \tag{1.72}$$

(recall Example 1.18), where $y \in \mathbf{C}^N$ is one period of $\tilde{y}$. We assume, of course, that $\tilde{x}$ and $\tilde{y}$ are bounded sequences so that (1.72) is well defined.

Now define $e_k = [e_{k,0} \ e_{k,1} \ \cdots \ e_{k,N-1}]^T \in \mathbf{C}^N$ according to

$$e_{k,n} = \exp\left[j\frac{2\pi}{N} kn\right], \tag{1.73}$$

where $n \in \mathbf{Z}_N$. The periodization of $e_k = (e_{k,n})$ is

$$\tilde{e}_{k,n} = \sum_{m \in \mathbf{Z}} e_{k,n+mN} \tag{1.74}$$

yielding $\tilde{e}_k = (\tilde{e}_{k,n})$. That (1.73) is periodic with period N with respect to index n is easily seen:

$$e_{k,n+mN} = \exp\left[j\frac{2\pi}{N}k(n+mN)\right] = \exp\left[j\frac{2\pi}{N}kn\right]\exp\left[j2\pi km\right] = e_{k,n}.$$

It can be shown (by exercise) that [using definition (1.72)]

$$\langle \tilde{e}_k, \tilde{e}_r \rangle = \langle e_k, e_r \rangle = \sum_{n=0}^{N-1} \exp\left[-j\frac{2\pi}{N}rn\right]\exp\left[j\frac{2\pi}{N}kn\right]$$

$$= \sum_{n=0}^{N-1} \exp\left[j\frac{2\pi}{N}(k-r)n\right] = \begin{cases} N, & k-r=0 \\ 0, & \text{otherwise} \end{cases}. \tag{1.75}$$

Thus, if we consider $(e_{k,n})$, and $(e_{r,n})$ with $k \neq r$ we find that these sequences are orthogonal, and so form an orthogonal basis for the vector space $\mathbf{C}^N$. From (1.75) we may write

$$\langle e_k, e_r \rangle = N\delta_{k-r}. \tag{1.76}$$

Thus, there must exist another vector $X = [X_0 \ X_1 \ \cdots \ X_{N-1}]^T \in \mathbf{C}^N$ such that

$$x_n = \frac{1}{N}\sum_{k=0}^{N-1} X_k \exp\left[j\frac{2\pi}{N}kn\right] \tag{1.77}$$

for $n \in \mathbf{Z}_N$. In fact

$$\langle x, e_r \rangle = \sum_{n=0}^{N-1} x_n e_{r,n}^*$$

$$= \frac{1}{N}\sum_{n=0}^{N-1}\left\{\sum_{k=0}^{N-1} X_k \exp\left[j\frac{2\pi}{N}kn\right]\right\}\exp\left[-j\frac{2\pi}{N}rn\right]$$

$$= \frac{1}{N}\sum_{k=0}^{N-1} X_k\left\{\sum_{n=0}^{N-1}\exp\left[j\frac{2\pi}{N}(k-r)n\right]\right\}$$

$$= \frac{1}{N}\sum_{k=0}^{N-1} X_k(N\delta_{k-r}) = X_r. \tag{1.78}$$

That is

$$X_k = \sum_{n=0}^{N-1} x_n \exp\left[-j\frac{2\pi}{N}kn\right] \tag{1.79}$$

for $k \in \mathbf{Z}_N$.

In (1.77) we see $x_{n+mN} = x_n$ for all $m \in \mathbf{Z}$. Thus, (x_n) in (1.77) is N-periodic, and so we have $\tilde{x}_n = \frac{1}{N} \sum_{k=0}^{N-1} X_k \exp\left[j\frac{2\pi}{N}kn\right]$ with X_k given by (1.79). Equation (1.77) is the *discrete Fourier series* (DFS) expansion for an N-periodic complex-valued sequence $\tilde{x}$ such as in (1.71). The *DFS coefficients* are given by (1.79). However, it is common practice to consider only $\tilde{x}_n$ for $n \in \mathbf{Z}_N$, which is equivalent to only considering the vector $x \in \mathbf{C}^N$. In this case the vector $X \in \mathbf{C}^N$ given by (1.79) is now called the *discrete Fourier transform* (DFT) of the vector x, and the expression in (1.77) is the *inverse DFT* (IDFT) of the vector X. We observe that the DFT, and the IDFT can be concisely expressed in matrix form, where we define the *DFT matrix*

$$F = \left[\exp\left(-j\frac{2\pi}{N}kn\right)\right]_{k,n\in\mathbf{Z}_N} \in \mathbf{C}^{N\times N}, \tag{1.80}$$

and we see from (1.77) that $F^{-1} = \frac{1}{N}F^*$ (*IDFT matrix*). Thus, $X = Fx$. We remark that the symmetry of F (i.e., $F = F^T$) means that either k or n in (1.80) may be interpreted as row or column indices.

The DFT has a long history, and its invention is now attributed to Gauss [10]. The DFT is of central importance to numerical computing generally, but has particularly great significance in digital signal processing as it represents a numerical approximation to the Fourier transform, and it can also be used to efficiently implement digital filtering operations via so-called *fast Fourier transform* (FFT) algorithms. The construction of FFT algorithms to efficiently compute $X = Fx$ (and $x = F^{-1}X$) is rather involved, and not within the scope of the present book. Simply note that the *direct* computation of the matrix-vector product $X = Fx$ needs N^2 complex multiplications and $N(N-1)$ complex additions. For $N = 2^p$ ($p \in \{1, 2, 3, \ldots\}$), which is called the *radix-2 case*, the algorithm of Cooley and Tukey [11] reduces the number of operations to something proportional to $N \log_2 N$, which is a substantial savings compared to N^2 operations with the direct approach when N is large enough. Essentially, the method in Ref. 11 implicitly factors F according to $F = F_p F_{p-1} \cdots F_1$, where the matrix factors $F_k \in \mathbf{C}^{N\times N}$ are *sparse* (i.e., contain many zero-valued entries). Note that multiplication by zero is not implemented in either hardware or software and so does not represent a computational cost in the practical implementation of the FFT algorithm. It is noteworthy that the algorithm of Ref. 11 also has a long history dating back to the work of Gauss, as noted by Heideman et al. [10]. It is also important to mention that fast algorithms exist for all possible $N \neq 2^p$ [4]. The following example suggests one of the important applications of the DFT/DFS.

Example 1.22 Suppose that $x_n = Ae^{j\theta n}$ with $\theta = \frac{2\pi}{N}m$ for $m = 1, 2, \ldots,$ $\frac{N}{2} - 1$ (N is assumed to be even here). From (1.79) using (1.75)

$$X_k = AN\delta_{m-k}. \tag{1.81}$$

Now suppose instead that $x_n = Ae^{-j\theta n}$, so similarly

$$X_k = A \sum_{n=0}^{N-1} \exp\left[-j\frac{2\pi}{N}n(m+k)\right]$$

$$= A \sum_{n=0}^{N-1} \exp\left[j\frac{2\pi}{N}n(N-m-k)\right] = AN\delta_{N-m-k}. \qquad (1.82)$$

Thus, if now $x_n = A\cos(\theta n) = \frac{1}{2}A[e^{j\theta n} + e^{-j\theta n}]$, then from (1.81) and (1.82), we must have

$$X_k = \frac{1}{2}AN[\delta_{m-k} + \delta_{N-m-k}]. \qquad (1.83)$$

We observe that $X_k = 0$ for all $k \neq m, N - m$, but that

$$X_m = \frac{1}{2}AN, \quad \text{and} \quad X_{N-m} = \frac{1}{2}AN.$$

Thus, X_k is nonzero only for indices $k = m$ and $k = N - m$ corresponding to the frequency of (x_n), which is $\theta = \frac{2\pi}{N}m$. The DFT/DFS is therefore quite useful in detecting "sinusoids" (also sometimes called "tone detection"). This makes the DFT/DFS useful in such applications as narrowband radar and sonar signal detection.

Can you explain the necessity (or, at least, the desirability) of the second equality in Eq. (1.82)?

APPENDIX 1.A   COMPLEX ARITHMETIC

Here we summarize the most important facts about arithmetic with complex numbers $z \in \mathbf{C}$ (set of complex numbers). You shall find this material very useful in electric circuits, as well as in the present book.

Complex numbers may be represented in two ways: (1) *Cartesian (rectangular) form* or (2) *polar form*. First we consider the Cartesian form.

In this case $z \in \mathbf{C}$ has the form $z = x + jy$, where $x, y \in \mathbf{R}$ (set of real numbers), and $j = \sqrt{-1}$. The complex conjugate of z is defined to be $z^* = x - jy$ (so $j^* = -j$).

Suppose that $z_1 = x_1 + jy_1$ and $z_2 = x_2 + jy_2$ are two complex numbers. Addition and subtraction are defined as

$$z_1 \pm z_2 = (x_1 \pm x_2) + j(y_1 \pm y_2)$$

[e.g., $(1 + 2j) + (3 - 5j) = 4 - 3j$, and $(1 + 2j) - (3 - 5j) = -2 + 7j$]. Using $j^2 = -1$, the product of z_1 and z_2 is

$$z_1 z_2 = (x_1 + jy_1)(x_2 + jy_2)$$

$$= x_1 x_2 + j^2 y_1 y_2 + jy_1 x_2 + jx_1 y_2$$

$$= (x_1 x_2 - y_1 y_2) + j(x_1 y_2 + x_2 y_1).$$

We note that

$$zz^* = (x + jy)(x - jy) = x^2 + y^2 = |z|^2,$$

so $|z| = \sqrt{x^2 + y^2}$ defines the *magnitude of z*. For example, $(1 + 2j)(3 - 5j) = 13 + j$. The quotient of z_1 and z_2 is defined to be

$$\frac{z_1}{z_2} = \frac{z_1 z_2^*}{z_2 z_2^*} = \frac{(x_1 + jy_1)(x_2 - jy_2)}{x_2^2 + y_2^2}$$

$$= \frac{(x_1 x_2 + y_1 y_2) + j(x_2 y_1 - x_1 y_2)}{x_2^2 + y_2^2}$$

$$= \frac{x_1 x_2 + y_1 y_2}{x_2^2 + y_2^2} + j \frac{x_2 y_1 - x_1 y_2}{x_2^2 + y_2^2},$$

where the last equality is z_1/z_2 in Cartesian form.

Now we may consider polar form representations. For $z = x + jy$, we may regard x and y as the x and y coordinates (respectively) of a point in the Cartesian plane (sometimes denoted $\mathbf{R}^2$).[10] We may therefore express these coordinates in polar form; thus, for any x and y we can write

$$x = r \cos \theta, \quad y = r \sin \theta,$$

where $r \geq 0$, and $\theta \in [0, 2\pi)$, or $\theta \in (-\pi, \pi]$. We observe that

$$x^2 + y^2 = r^2(\cos^2 \theta + \sin^2 \theta) = r^2,$$

so $|z| = r$.

Now recall the following Maclaurin series expansions (considered in greater depth in Chapter 3):

$$\sin x = \sum_{n=1}^{\infty} (-1)^{n-1} \frac{x^{2n-1}}{(2n-1)!}$$

$$\cos x = \sum_{n=1}^{\infty} (-1)^{n-1} \frac{x^{2n-2}}{(2n-2)!}$$

$$e^x = \sum_{n=1}^{\infty} \frac{x^{n-1}}{(n-1)!}$$

---

[10]This suggests that z may be equivalently represented by the column vector $[xy]^T$. The vector interpretation of complex numbers can be quite useful.

These series converge for $-\infty < x < \infty$. Observe the following:

$$e^{jx} = \sum_{n=1}^{\infty} \frac{(jx)^{n-1}}{(n-1)!} = \sum_{n=1}^{\infty} \left[\frac{(jx)^{(2n-1)-1}}{[(2n-1)-1]!} + \frac{(jx)^{(2n-1)}}{[2n-1]!} \right],$$

where we have split the summation into terms involving even n and odd n. Thus, continuing

$$e^{jx} = \sum_{n=1}^{\infty} \left[\frac{j^{2n-2} x^{2n-2}}{(2n-2)!} + \frac{j^{2n-1} x^{2n-1}}{(2n-1)!} \right]$$

$$= \sum_{n=1}^{\infty} j^{2n-2} \left[\frac{x^{2n-2}}{(2n-2)!} + j \frac{x^{2n-1}}{(2n-1)!} \right] \quad (j j^{2n-2} = j^{2n-1})$$

$$= \sum_{n=1}^{\infty} (-1)^{n-1} \frac{x^{2n-2}}{(2n-2)!} + j \sum_{n=1}^{\infty} (-1)^{n-1} \frac{x^{2n-1}}{(2n-1)!}$$

$$(j^{2n-2} = (j^2)^{n-1} = (-1)^{n-1})$$

$$= \cos x + j \sin x.$$

Thus, $e^{jx} = \cos x + j \sin x$. This is justification for *Euler's identity* in (1.49). Additionally, since $e^{-jx} = \cos x - j \sin x$, we have

$$e^{jx} + e^{-jx} = 2 \cos x, \quad e^{jx} - e^{-jx} = 2j \sin x.$$

These immediately imply that

$$\sin x = \frac{e^{jx} - e^{-jx}}{2j}, \quad \cos x = \frac{e^{jx} + e^{-jx}}{2}.$$

These identities allow for the conversion of expressions involving trig(onometric) functions into expressions involving exponentials, and vice versa. The necessity to do this is frequent. For this reason, they should be memorized, or else you should remember how to derive them "on the spot" when necessary.

Now observe that

$$r e^{j\theta} = r \cos \theta + j r \sin \theta,$$

so that if $z = x + jy$, then, because there exist r and θ such that $x = r \cos \theta$ and $y = r \sin \theta$, we may immediately write

$$z = r e^{j\theta}.$$

This is z in polar form. For example (assuming that θ is in radians)

$$1 + j = \sqrt{2}e^{j\pi/4}, \quad -1 + j = \sqrt{2}e^{3\pi j/4},$$
$$1 - j = \sqrt{2}e^{-j\pi/4}, \quad -1 - j = \sqrt{2}e^{-3\pi j/4}.$$

It can sometimes be useful to observe that

$$j = e^{j\pi/2}, \quad -j = e^{-j\pi/2}, \quad \text{and} \quad -1 = e^{\pm j\pi}.$$

If $z_1 = r_1 e^{j\theta_1}$, and $z_2 = r_2 e^{j\theta_2}$, then

$$z_1 z_2 = r_1 r_2 e^{j(\theta_1 + \theta_2)}, \quad \frac{z_1}{z_2} = \frac{r_1}{r_2} e^{j(\theta_1 - \theta_2)}.$$

In other words, multiplication and division of complex numbers is very easy when they are expressed in polar form.

Finally, some terminology. For $z = x + jy$, we call x the *real part* of z, and we call y the *imaginary part* of z. The notation is

$$x = \text{Re}\,[z], \quad y = \text{Im}\,[z].$$

That is, $z = \text{Re}\,[z] + j\,\text{Im}\,[z]$.

APPENDIX 1.B   ELEMENTARY LOGIC

Here we summarize the basic language and ideas associated with elementary logic as some of what is found here appears in later sections and chapters of this book. The concepts found here appear often in mathematics and engineering literature.

Consider two mathematical statements represented as P and Q. Each statement may be either true or false. Suppose that we know that if P is true, then Q is certainly true (allowing the possibility that Q is true even if P is false). Then we say that P implies Q, or Q is implied by P, or P is a *sufficient condition* for Q, or symbolically

$$P \Rightarrow Q \quad \text{or} \quad Q \Leftarrow P.$$

Suppose that if P is false, then Q is certainly false (allowing the possibility that Q may be false even if P is true). Then we say that P is implied by Q, or Q implies P, or P is a *necessary condition* for Q, or

$$P \Leftarrow Q \quad \text{or} \quad Q \Rightarrow P.$$

Now suppose that if P is true, then Q is certainly true, and if P is false, then Q is certainly false. In other words, P and Q are either both true or both false. Then we say that P implies and is implied by Q, or P is a *necessary and sufficient condition* for Q, or P and Q are *logically equivalent*, or P *if and only if* Q, or symbolically

$$P \Leftrightarrow Q.$$

A common abbreviation for "if and only if" is *iff*.

The *logical contrary* of the statement P is called "not P." It is often denoted by either $\overline{P}$ or $\sim P$. This is the statement that is true if P is false, or false if P is true. For example, if P is the statement "$x > 1$," then $\sim P$ is the statement "$x \le 1$." If P is the statement "$f(x) \ne 0$ for all $x \in \mathbf{R}$," then $\sim P$ is the statement "there is at least one $x \in \mathbf{R}$ for which $f(x) = 0$." We may write

$$x^4 - 5x^2 + 4 = 0 \Leftarrow x = 1 \quad \text{or} \quad x = 2,$$

but the converse is not true because $x^4 - 5x^2 + 4 = 0$ is a quartic equation possessing four possible solutions. We may write

$$x = 3 \Rightarrow x^2 = 3x,$$

but we cannot say $x^2 = 3x \Rightarrow x = 3$ because $x = 0$ is also possible.

Finally, we observe that

$$P \Rightarrow Q \text{ is equivalent to } \sim P \Leftarrow \sim Q,$$

$$P \Leftarrow Q \text{ is equivalent to } \sim P \Rightarrow \sim Q,$$

$$P \Leftrightarrow Q \text{ is equivalent to } \sim P \Leftrightarrow \sim Q;$$

that is, taking logical contraries reverses the directions of implication arrows.

REFERENCES

1. E. Kreyszig, *Introductory Functional Analysis with Applications*, Wiley, New York, 1978.
2. A. P. Hillman and G. L. Alexanderson, *A First Undergraduate Course in Abstract Algebra*, 3rd ed., Wadsworth, Belmont, CA, 1983.
3. R. B. J. T. Allenby, *Rings, Fields and Groups: An Introduction to Abstract Algebra*, Edward Arnold, London, UK, 1983.
4. R. E. Blahut, *Fast Algorithms for Digital Signal Processing*, Addison-Wesley, Reading, MA, 1985.
5. C. K. Chui, *Wavelets: A Mathematical Tool for Signal Analysis*. SIAM, Philadelphia, PA, 1997.
6. R. E. Ziemer and W. H. Tranter, *Principles of Communications: Systems, Modulation, and Noise*, 3rd ed., Houghton Mifflin, Boston, MA, 1990.
7. I. Daubechies, "Orthonormal Bases of Compactly Supported Wavelets," *Commun. Pure Appl. Math.* **41**, 909–996 (1988).

8. G. P. Tolstov, *Fourier Series* (transl. from Russian by R. A. Silverman), Dover Publications, New York, 1962.

9. E. Kreyszig, *Advanced Engineering Mathematics*, 4th ed., Wiley, New York, 1979.

10. M. T. Heideman, D. H. Johnson and C. S. Burrus, "Gauss and the History of the Fast Fourier Transform," *IEEE ASSP Mag.* **1**, 14–21 (Oct. 1984).

11. J. W. Cooley and J. W. Tukey, "An Algorithm for the Machine Calculation of Complex Fourier Series," *Math. Comput.*, **19**, 297–301 (April 1965).

PROBLEMS

1.1. (a) Find $a, b \in \mathbf{R}$ in

$$\frac{1 + 2j}{-3 - j} = a + bj.$$

(b) Find $r, \theta \in \mathbf{R}$ in

$$-3 + j = re^{j\theta}$$

(Of course, choose $r > 0$, and $\theta \in (-\pi, \pi]$.)

1.2. Solve for $x \in \mathbf{C}$ in the quadratic equation

$$x^2 - 2r \cos\theta x + r^2 = 0.$$

Here $r \geq 0$, and $\theta \in (-\pi, \pi]$. Express your solution in polar form.

1.3. Let θ, and ϕ be arbitrary angles (so $\theta, \phi \in \mathbf{R}$). Show that

$$(\cos\theta + j\sin\theta)(\cos\phi + j\sin\phi) = \cos(\theta + \phi) + j\sin(\theta + \phi).$$

1.4. Prove the following theorem. Suppose $z \in \mathbf{C}$ such that

$$z = r\cos\theta + jr\sin\theta$$

for which $r = |z| > 0$, and $\theta \in (-\pi, \pi]$. Let $n \in \{1, 2, 3, \ldots\}$ (i.e., n is a positive integer). The n different nth roots of z are given by

$$r^{1/n}\left[\cos\left(\frac{\theta + 2\pi k}{n}\right) + j\sin\left(\frac{\theta + 2\pi k}{n}\right)\right],$$

for $k = 0, 1, 2, \ldots, n - 1$.

1.5. State whether the following are true or false:

(a) $|x| < 2 \Rightarrow x < 2$

(b) $|x| < 3 \Leftarrow 0 < x < 3$

(c) $x - y > 0 \Rightarrow x > y > 0$

(d) $xy = 0 \Rightarrow x = 0$ and $y = 0$

(e) $x = 10 \Leftarrow x^2 = 10x$

Explain your answer in all cases.

1.6. Consider the function

$$f(x) = \begin{cases} -x^2 + 2x + 1, & 0 \le x < 1 \\ x^2 - 2x + \frac{3}{2}, & 1 < x \le 2 \end{cases}.$$

Find

$$\sup_{x \in [0,2]} f(x), \quad \inf_{x \in [0,2]} f(x).$$

1.7. Suppose that we have the following polynomials in the indeterminate x:

$$a(x) = \sum_{k=0}^{n} a_k x^k, \quad b(x) = \sum_{j=0}^{m} b_j x^j.$$

Prove that

$$c(x) = a(x)b(x) = \sum_{l=0}^{n+m} c_l x^l,$$

where

$$c_l = \sum_{k=0}^{n} a_k b_{l-k}.$$

[*Comment:* This is really asking us to prove that discrete convolution is mathematically equivalent to polynomial multiplication. It explains why the MATLAB routine for multiplying polynomials is called *conv*. Discrete convolution is a fundamental operation in digital signal processing, and is an instance of something called *finite impulse response* (FIR) filtering. You will find it useful to note that $a_k = 0$ for $k < 0$, and $k > n$, and that $b_j = 0$ for $j < 0$, and $j > m$. Knowing this allows you to manipulate the summation limits to achieve the desired result.]

1.8. Recall Example 1.5. Suppose that $x_k = 2^{k+1}$, and that $y_k = 1$ for $k \in \mathbf{Z}^+$. Find the sum of the series $d(x, y)$. (*Hint:* Recall the theory of geometric series. For example, $\sum_{k=0}^{N} \alpha^k = \frac{1-\alpha^{N+1}}{1-\alpha}$ if $\alpha \ne 1$.)

1.9. Prove that if $x \ne 1$, then

$$S_n = \sum_{k=1}^{n} k x^{k-1}$$

is given by

$$S_n = \frac{1 - (n+1)x^n + nx^{n+1}}{(1-x)^2}.$$

What is the formula for S_n when $x = 1$? (*Hint:* Begin by showing that $S_n - xS_n = 1 + x + x^2 + \cdots + x^{n-1} - nx^n$.)

1.10. Recall Example 1.1. Prove that $d(x, y)$ in (1.5) satisfies all the axioms for a metric.

1.11. Recall Example 1.18. Prove that $\langle x, y \rangle$ in (1.36) satisfies all the axioms for an inner product.

1.12. By direct calculation, show that if x, y, z are elements from an inner product space, then

$$||z - x||^2 + ||z - y||^2 = \tfrac{1}{2}||x - y||^2 + 2||z - \tfrac{1}{2}(x+y)||^2$$

(Appolonius' identity).

1.13. Suppose $x, y \in \mathbf{R}^3$ (three-dimensional Euclidean space) such that

$$x = [1 \quad 1 \quad 1]^T, \quad y = [1 \quad -1 \quad 1]^T.$$

Find all vectors $z \in \mathbf{R}^3$ such that $\langle x, z \rangle = \langle y, z \rangle = 0$.

1.14. The complex Fourier series expansion method as described is for $f \in L^2(0, 2\pi)$. Find the complex Fourier series expansion for $f \in L^2(0, T)$, where $0 < T < \infty$ (i.e., the interval on which f is defined is now of arbitrary length).

1.15. Consider again the complex Fourier series expansion for $f \in L^2(0, 2\pi)$. Specifically, consider Eq. (1.44). If $f(t) \in \mathbf{R}$ for all $t \in (0, 2\pi)$, then show that $f_n = f_{-n}^*$. [The sequence (f_n) is *conjugate symmetric*.] Use this to show that for suitable $a_n, b_n \in \mathbf{R}$ (all n) we have

$$\sum_{n=-\infty}^{\infty} f_n e^{jnt} = a_0 + \sum_{n=1}^{\infty} [a_n \cos(nt) + b_n \sin(nt)].$$

How are the coefficients a_n and b_n related to f_n ? (Be very specific. There is a simple formula.)

1.16. (a) Suppose that $f \in L^2(0, 2\pi)$, and that specifically

$$f(t) = \begin{cases} 1, & 0 < t < \pi \\ j, & \pi \le t < 2\pi \end{cases}.$$

Find f_n in Eq. (1.44) using (1.45); that is, find the complex Fourier series expansion for $f(t)$. Make sure that you appropriately simplify your series expansion.

(b) Show how to use the result in Example 1.20 to find the complex Fourier series expansion for $f(t)$ in (a).

1.17. This problem is about finding the Fourier series expansion for the waveform at the output of a full-wave rectifier circuit. This circuit is used in AC/DC (alternating/direct-current) converters. Knowledge of the Fourier series expansion gives information to aid in the design of such converters.

(a) Find the complex Fourier series expansion of

$$f(t) = \left| \sin\left(\frac{2\pi}{T_1}t\right) \right| \in L^2\left(0, \frac{T_1}{2}\right).$$

(b) Find the sequences (a_n), and (b_n) in

$$f(t) = a_0 + \sum_{n=1}^{\infty} \left[a_n \cos\left(\frac{2\pi n}{T}t\right) + b_n \sin\left(\frac{2\pi n}{T}t\right) \right]$$

for $f(t)$ in (a). You need to consider how T is related to T_1.

1.18. Recall the definitions of the Haar scaling function and Haar wavelet in Eqs. (1.40) and (1.41), respectively. Define $\phi_{k,n}(t) = 2^{k/2}\phi(2^k t - n)$, and $\psi_{k,n}(t) = 2^{k/2}\psi(2^k t - n)$. Recall that $\langle f(t), g(t) \rangle = \int_{-\infty}^{\infty} f(t)g^*(t)\,dt$ is the inner product for $L^2(\mathbf{R})$.

(a) Sketch $\phi_{k,n}(t)$, and $\psi_{k,n}(t)$.

(b) Evaluate the integrals

$$\int_{-\infty}^{\infty} \phi_{k,n}^2(t)\,dt, \quad \text{and} \quad \int_{-\infty}^{\infty} \psi_{k,n}^2(t)\,dt.$$

(c) Prove that

$$\langle \phi_{k,n}(t), \phi_{k,m}(t) \rangle = \delta_{n-m}.$$

1.19. Prove the following version of the Schwarz inequality. For all $x, y \in X$ (inner product space)

$$|\operatorname{Re}[\langle x, y \rangle]| \leq \|x\|\,\|y\|$$

with equality iff $y = \beta x$, and $\beta \in \mathbf{R}$ is a constant.

[*Hint:* The proof of this one is not quite like that of Theorem 1.1. Consider $\langle \alpha x + y, \alpha x + y \rangle \geq 0$ with $\alpha \in \mathbf{R}$. The inner product is to be viewed as a quadratic in α.]

1.20. The following result is associated with the proof of the uncertainty principle for analog signals.

Prove that for $f(t) \in L^2(\mathbf{R})$ such that $|t|f(t) \in L^2(\mathbf{R})$ and $f^{(1)}(t) = df(t)/dt \in L^2(\mathbf{R})$, we have the inequality

$$\left| \operatorname{Re}\left[\int_{-\infty}^{\infty} tf(t)[f^{(1)}(t)]^*\,dt \right] \right|^2 \leq \left[\int_{-\infty}^{\infty} |tf(t)|^2\,dt \right] \left[\int_{-\infty}^{\infty} |f^{(1)}(t)|^2\,dt \right].$$

1.21. Suppose $e_k = [e_{k,0} \ e_{k,1} \ \cdots \ e_{k,N-2} \ e_{k,N-1}]^T \in \mathbf{C}^N$, where

$$e_{k,n} = \exp\left[j\frac{2\pi}{N}kn\right]$$

and $k \in \mathbf{Z}_N$. If $x, y \in \mathbf{C}^N$ recall that $\langle x, y \rangle = \sum_{k=0}^{N-1} x_k y_k^*$. Prove that $\langle e_k, e_r \rangle = N\delta_{k-r}$. Thus, $B = \{e_k | k \in \mathbf{Z}_N\}$ is an orthogonal basis for $\mathbf{C}^N$. Set B is important in digital signal processing because it is used to define the *discrete Fourier transform*.

2 Number Representations

2.1 INTRODUCTION

In this chapter we consider how numbers are represented on a computer largely with respect to the errors that occur when basic arithmetical operations are performed on them. We are most interested here in so-called *rounding errors* (also called *roundoff errors*). Floating-point computation is emphasized. This is due to the fact that most numerical computation is performed with floating-point numbers, especially when numerical methods are implemented in high-level programming languages such as C, Pascal, FORTRAN, and C++. However, an understanding of floating-point requires some understanding of fixed-point schemes first, and so this case will be considered initially. In addition, fixed-point schemes are used to represent integer data (i.e., subsets of **Z**), and so the fixed-point representation is important in its own right. For example, the exponent in a floating-point number is an integer.

The reader is assumed to be familiar with how integers are represented, and how they are manipulated with digital hardware from a typical introductory digital electronics book or course. However, if this is not so, then some review of this topic appears in Appendix 2.A. The reader should study this material *now* if necessary.

Our main (historical) reference text for the material of this chapter is Wilkinson [1]. However, Golub and Van Loan [4, Section 2.4] is also a good reference. Golub and Van Loan [4] base their conventions and results in turn on Forsythe et al. [5].

2.2 FIXED-POINT REPRESENTATIONS

We now consider fixed-point fractions. We must do so because the mantissa in a floating-point number is a fixed-point fraction.

We assume that fractions are $t + 1$ digits long. If the number is in binary, then we usually say "$t + 1$ bits" long instead. Suppose, then, that x is a $(t + 1)$-bit fraction. We shall write it in the form

$$(x)_2 = x_0.x_1x_2 \cdots x_{t-1}x_t \quad (x_k \in \{0, 1\}). \tag{2.1}$$

An Introduction to Numerical Analysis for Electrical and Computer Engineers, by C.J. Zarowski
ISBN 0-471-46737-5 © 2004 John Wiley & Sons, Inc.

The notation $(x)_2$ means that x is in base-2 (binary) form. More generally, $(x)_r$ means that x is expressed as a base-r number (e.g., if $r = 10$ this would be the decimal representation). We use this notation to emphasize which base we are working with when necessary (e.g., to avoid ambiguity). We shall assume that (2.1) is a two's complement fraction. Thus, bit x_0 is the *sign bit*. If this bit is 1, we interpret the fraction to be negative; otherwise, it is nonnegative. For example, $(1.1011)_2 = (-0.3125)_{10}$. [To take the two's complement of $(1.1011)_2$, first complement every bit, and then add $(0.0001)_2$. This gives $(0.0101)_2 = (0.3125)_{10}$.] In general, for the case of a $(t + 1)$-bit two's complement fraction, we obtain

$$-1 \le x \le 1 - 2^{-t}. \tag{2.2}$$

In fact

$$(-1)_{10} = (1.\underbrace{00\ldots00}_{t \; bits})_2, \quad (1 - 2^{-t})_{10} = (0.\underbrace{11\ldots11}_{t \; bits})_2. \tag{2.3}$$

We may regard (2.2) as specifying the *dynamic range* of the $(t + 1)$-bit two's complement fraction representation scheme. Numbers beyond this range are not represented. Justification of (2.2) [and (2.3)] would follow the argument for the conversion of two's complement integers into decimal integers that is considered in Appendix 2.A.

Consider the set $\{x \in \mathbf{R} | -1 \le x \le 1 - 2^{-t}\}$. In other words, x is a real number within the limits imposed by (2.2), but it is not necessarily equal to a $(t + 1)$-bit fraction. For example, $x = \sqrt{2} - 1$ is in the range (2.2), but it is an irrational number, and so does not possess an exact $(t + 1)$-bit representation. We may choose to approximate such a number with $t + 1$ bits. Denote the $(t + 1)$-bit approximation of x as $Q[x]$. For example, $Q[x]$ might be the approximation to x obtained by selecting an element from set

$$B = \{b_n = -1 + 2^{-t}n | n = 0, 1, \ldots, 2^{t+1} - 1\} \subset \mathbf{R} \tag{2.4}$$

that is the closest to x, where distance is measured by the metric in Example 1.1. Note that each number in B is representable as a $(t + 1)$-bit fraction. In fact, B is the entire set of $(t + 1)$-bit two's complement fractions. Formally, our approximation is given by

$$Q[x] = \underset{n \in \{0, 1, \ldots, 2^{t+1} - 1\}}{\mathrm{argmin}} |x - b_n|. \tag{2.5}$$

The notation "argmin" means "let $Q[x]$ be the b_n for the n in the set $\{0, 1, \ldots, 2^{t+1} - 1\}$ that minimizes $|x - b_n|$." In other words, we choose the *arg*ument b_n that *min*imizes the distance to x. Some reflection (and perhaps considering some simple examples for small t) will lead the reader to conclude that the error in this approximation satisfies

$$|x - Q[x]| \le 2^{-(t+1)}. \tag{2.6}$$

The error $\epsilon = x - Q[x]$ is called *quantization error*. Equation (2.6) is an upper bound on the size (norm) of this error. In fact, in the notation of Chapter 1, if $||x|| = |x|$, then $||\epsilon|| = ||x - Q[x]|| \leq 2^{-(t+1)}$. We remark that our quantization method is not unique. There are many other methods, and these will generally lead to different bounds.

When we represent the numbers in a computational problem on a computer, we see that errors due to quantization can arise even before we perform any operations on the numbers at all. However, errors will also arise in the course of performing basic arithmetic operations on the numbers. We consider the sources of these now.

If x, y are coded as in (2.1), then their sum might not be in the range specified by (2.2). This can happen only if x and y are either both positive or both negative. Such a condition is *fixed-point overflow*. (A test for overflow in two's complement integer addition appears in Appendix 2.A, and it is easy to modify it for the problem of overflow testing in the addition of fractions.) Similarly, overflow can occur when a negative number is subtracted from a positive number, or if a positive number is subtracted from a negative number. A test for this case is possible, too, but we omit the details. Other than the problem of overflow, no errors can occur in the addition or subtraction of fractions.

With respect to fractions, rounding error arises only when we perform multiplication and division. We now consider errors in these operations.

We will deal with multiplication first. Suppose that x and y are represented according to (2.1). Suppose also that $x_0 = y_0 = 0$. It is easy to see that the product of x and y is given by

$$
p = xy = \left(\sum_{k=0}^{t} x_k 2^{-k} \right) \left(\sum_{n=0}^{t} y_n 2^{-n} \right)
$$
$$
= (x_0 + x_1 2^{-1} + \cdots + x_t 2^{-t})(y_0 + y_1 2^{-1} + \cdots + y_t 2^{-t})
$$
$$
= x_0 y_0 + (x_0 y_1 + x_1 y_0)2^{-1} + \cdots + x_t y_t 2^{-2t}. \tag{2.7}
$$

This implies that the product is a $(2t + 1)$-bit number. If we allow x and y to be either positive or negative, then the product will also be $2t + 1$ bits long. Of course, one of these bits is the sign bit. If we had to multiply several numbers together, we see that the product wordsize would grow in some proportion to the number of factors in the product. The growth is clearly very rapid, and no practical computer could sustain this for very long. We are therefore forced in general to *round off* the product p back down to a number that is only $t + 1$ bits long. Obviously, this will introduce an error.

How should the rounding be done? There is more than one possibility (just as there is more than one way to quantize). Wilkinson [1, p. 4] suggests the following. Since the product p has the form

$$
(p)_2 = p_0.p_1 p_2 \cdots p_{t-1} p_t p_{t+1} \cdots p_{2t} \quad (p_k \in \{0, 1\}) \tag{2.8}
$$

we may add $2^{-(t+1)}$ to this product, and then simply discard the last t bits of the resulting sum (i.e., the bits indexed $t+1$ to $2t$). For example, suppose $t = 4$, and consider

$$0.00111111 = p$$

$$\underline{+0.00001000 = 2^{-5}}$$

$$0.01000111$$

Thus, the rounded product is $(0.0100)_2$. The error involved in rounding in this manner is not higher in magnitude than $\frac{1}{2}2^{-t} = 2^{-(t+1)}$. Define the result of the rounding operation to be $fx[p] = fx[xy]$, so then

$$|p - fx[p]| \leq \tfrac{1}{2}2^{-t}. \tag{2.9}$$

[For the previous example, $p = (0.00111111)_2$, and so $fx[p] = (0.0100)_2$.] It is natural to measure the sizes of errors in the same way as we measured the size of quantization errors earlier. Thus, (2.9) is an upper bound on the size of the error due to rounding a product. As with quantization, other rounding methods would generally give other bounds. We remark that Wilkinson's suggestion amounts to "ordinary rounding."

Finally, we consider fixed-point division. Again, suppose that x and y are represented as in (2.1), and consider the quotient $q = x/y$. Obviously, we must avoid $y = 0$. Also, the quotient will not be in the permitted range given by (2.2) unless $|y| \geq |x|$. This implies that when fixed-point division is implemented either the *dividend* x or the *divisor* y need to be scaled to meet this restriction. Scaling is multiplication by a power of 2, and so should be implemented to reduce rounding error. We do not consider the specifics of how to achieve this. Another problem is that x/y may require an infinite number of bits to represent it. For example, suppose

$$q = \frac{(0.0010)_2}{(0.0110)_2} = \frac{(0.125)_{10}}{(0.375)_{10}} = \left(\frac{1}{3}\right)_{10} = (0.\overline{01})_2.$$

The bar over 01 denotes the fact that this pattern repeats indefinitely. Fortunately, the same recipe for the rounding of products considered above may also be used to round quotients. If $fx[q]$ again denotes the result of applying this procedure to q, then

$$|q - fx[q]| \leq \tfrac{1}{2}2^{-t}. \tag{2.10}$$

We see that the difficulties associated with division in fixed-point representations means that fixed-point arithmetic should, if possible, not be used to implement algorithms that require division. This forces us to either (1) employ floating-point representations or (2) develop algorithms that solve the problem without the need for division operations.

Both strategies are employed in practice. Usually choice 1 is easier.

2.3 FLOATING-POINT REPRESENTATIONS

In the previous section we have seen that fixed-point numbers are of very limited dynamic range. This poses a major problem in employing them in engineering computations since obviously we desire to work with numbers far beyond the range in (2.2). Floating-point representations provide the definitive solution to this problem. We remark (in passing) that the basic organization of a floating-point arithmetic unit [i.e., digital hardware for floating-point addition and subtraction appears in Ref. 2 (see pp. 295–306)]. There is a standard IEEE format for floating-point numbers. We do not consider this standard here, but it is summarized in Ref. 2 (see pp. 304–306). Some of the technical subtleties associated with the IEEE standard are considered by Higham [6].

Following Golub and Van Loan [4, p. 61], the set **F** (subset of **R**) of floating-point numbers consists of numbers of the form

$$x = x_0.x_1x_2\cdots x_{t-1}x_t \times r^e, \tag{2.11}$$

where x_0 is a sign bit (which means that we can replace x_0 by $\pm$; this is done in Ref. 4), and r is the base of the representation [typically $r = 2$ (binary), or $r = 10$ (decimal); we will emphasize $r = 2$]. Therefore, $x_k \in \{0, 1, \ldots, r - 2, r - 1\}$ for $1 \le k \le t$. These are the digits (bits if $r = 2$) of the *mantissa*. We therefore see that the mantissa is a fraction. [1] It is important to note that $x_1 \ne 0$, and this has implications with regard to how operations are performed and the resulting rounding errors. We call e the *exponent*. This is an integer quantity such that $L \le e \le U$. For example, we might represent e as an n-bit two's complement integer. We will assume this unless otherwise specified in what follows. This would imply that $(e)_2 = e_{n-1}e_{n-2}\cdots e_1e_0$, and so

$$-2^{n-1} \le e \le 2^{n-1} - 1 \tag{2.12}$$

(see Appendix A for justification). For nonzero $x \in \mathbf{F}$, then

$$m \le |x| \le M, \tag{2.13a}$$

where

$$m = r^{L-1}, \quad M = r^U(1 - r^{-t}). \tag{2.13b}$$

Equation (2.13) gives the *dynamic range* for the floating-point representation. With $r = 2$ we see that the total wordsize for the floating-point number is $t + n + 1$ bits.

In the absence of rounding errors in a computation, our numbers may initially be from the set

$$G = \{x \in \mathbf{R}|m \le |x| \le M\} \cup \{0\}. \tag{2.14}$$

---

[1]Including the sign bit the mantissa is (for $r = 2$) $t + 1$ bits long. Frequently in what follows we shall refer to it as being only t bits long. This is because we are ignoring the sign bit, which is always understood to be present.

This set is analogous to the set $\{x \in \mathbf{R}| -1 \le x \le 1 - 2^{-t}\}$ that we saw in the previous section in our study of fixed-point quantization effects. Again following Golub and Van Loan [4], we may define a mapping (operator) $fl|G \to \mathbf{F}$. Here $c = fl[x]$ $(x \in G)$ is obtained by choosing the closest $c \in \mathbf{F}$ to x. As you might expect, distance is measured using $|| \cdot || = | \cdot |$, as we did in the previous section. Golub and Van Loan call this *rounded arithmetic* [4], and it coincides with the rounding procedure described by Wilkinson [1, pp. 7–11].

Suppose that x and y are two floating-point numbers (i.e., elements of $\mathbf{F}$) and that "*op*" denotes any of the four basic arithmetic operations (addition, subtraction, multiplication, or division). Suppose $|x \ op \ y| \notin G$. This implies that either $|x \ op \ y| > M$ (floating-point *overflow*), or $0 < |x \ op \ y| < m$ (floating-point *underflow*) has occurred. Under normal circumstances an *arithmetic fault* such as overflow will not happen unless an unstable procedure is being performed. The issue of "numerical stability" will be considered later. Overflows typically cause runtime error messages to appear. The underflow arithmetic fault occurs when a number arises that is not zero, but is too small to represent in the set $\mathbf{F}$. This usually poses less of a problem than overflow. [2] However, as noted before, we are concerned mainly with rounding errors here. If $|x \ op \ y| \in G$, then we assume that the computer implementation of $x \ op \ y$ will be given by $fl[x \ op \ y]$. In other words, the operator fl models rounding effects in floating-point arithmetic operations. We remark that where floating-point arithmetic is concerned, rounding error arises in all four arithmetic operations. This contrasts with fixed-point arithmetic wherein rounding errors arise only in multiplication and division.

It turns out that for the floating-point rounding procedure suggested above

$$fl[x \ op \ y] = (x \ op \ y)(1 + \epsilon), \tag{2.15}$$

where

$$|\epsilon| \le \tfrac{1}{2} r^{1-t} (= 2^{-t} \quad \text{if} \quad r = 2). \tag{2.16}$$

We shall justify this only for the case $r = 2$. Our arguments will follow those of Wilkinson [1, pp. 7–11].

Let us now consider the addition of the base-2 floating-point numbers

$$x = \underbrace{x_0.x_1 \cdots x_t}_{=m_x} \times 2^{e_x} \tag{2.17a}$$

and

$$y = \underbrace{y_0.y_1 \cdots y_t}_{=m_y} \times 2^{e_y}, \tag{2.17b}$$

and we assume that $|x| > |y|$. (If instead $|y| > |x|$, then reverse the roles of x and y.) If $e_x - e_y > t$, then

$$fl[x + y] = x. \tag{2.18}$$

---

[2]Underflows are simply set to zero on some machines.

For example, if $t = 4$, and $x = 0.1001 \times 2^4$, and $y = 0.1110 \times 2^{-1}$, then to add these numbers, we must shift the bits in the mantissa of one of them so that both have the same exponent. If we choose y (usually shifting is performed on the smallest number), then $y = 0.00000111 \times 2^4$. Therefore, $x + y = 0.10010111 \times 2^4$, but then $fl[x + y] = 0.1001 \times 2^4 = x$.

Now if instead we have $e_x - e_y \leq t$, we divide y by $2^{e_x - e_y}$ by shifting its mantissa $e_x - e_y$ positions to the right. The sum $x + 2^{e_y - e_x} y$ is then calculated exactly, and requires $\leq 2t$ bits for its representation. The sum is multiplied by a power of 2, using left or right shifts to ensure that the mantissa is properly normalized [recall that for x in (2.11) we must have $x_1 \neq 0$]. Of course, the exponent must be modified to account for the shift of the bits in the mantissa. The $2t$-bit mantissa is then rounded off to t bits using fl. Because we have $|m_x| + 2^{e_y - e_x}|m_y| \leq 1 + 1 = 2$, the largest possible right shift is by one bit position. However, a left shift of up to t bit positions might be needed because of the cancellation of bits in the summation process. Let us consider a few examples. We will assume that $t = 4$.

Example 2.1   Let $x = 0.1001 \times 2^4$, and $y = 0.1010 \times 2^1$. Thus

$$0.10010000 \times 2^4$$

$$\underline{+0.00010100 \times 2^4}$$

$$0.10100100 \times 2^4$$

and the sum is rounded to 0.1010×2^4 (computed sum).

Example 2.2   Let $x = 0.1111 \times 2^4$, and $y = 0.1010 \times 2^2$. Thus

$$0.11110000 \times 2^4$$

$$\underline{+0.00101000 \times 2^4}$$

$$1.00011000 \times 2^4$$

but $1.00011000 \times 2^4 = 0.100011000 \times 2^5$, and this exact sum is rounded to 0.1001×2^5 (computed sum).

Example 2.3   Let $x = 0.1111 \times 2^{-4}$, and $y = -.1110 \times 2^{-4}$. Thus

$$0.11110000 \times 2^{-4}$$

$$\underline{-0.11100000 \times 2^{-4}}$$

$$0.00010000 \times 2^{-4}$$

but $0.00010000 \times 2^{-4} = 0.1000 \times 2^{-7}$, and this exact sum is rounded to 0.1000×2^{-7} (computed sum). Here there is much *cancellation* of the bits leading in turn to a large shift of the mantissa of the exact sum to the left. Yet, the computed sum is exact.

We observe that the *computed sum* is obtained by computing the *exact sum*, normalizing it so that the mantissa $s_0.s_1 \cdots s_{t-1}s_t s_{t+1} \cdots s_{2t}$ satisfies $s_1 = 1$ (i.e., $s_1 \neq 0$), and *then* we round it to t places (i.e., we apply fl). If the *normalized exact sum* is $s = m_s \times 2^{e_s} (= x + y)$, then the rounding error ϵ' is such that $|\epsilon'| \leq \frac{1}{2}2^{-t}2^{e_s}$. Essentially, the error ϵ' is due to rounding the mantissa (a fixed-point number) according to the method used in Section 2.2. Because of the form of m_s, $\frac{1}{2}2^{e_s} \leq |s| < 2^{e_s}$, and so

$$fl[x + y] = (x + y)(1 + \epsilon) \tag{2.19}$$

which is just a special case of (2.15). This expression requires further explanation, however. Observe that

$$\frac{|s - fl[s]|}{|s|} = \frac{|s - (s + \epsilon')|}{|s|} = \frac{|\epsilon'|}{|s|} \leq \frac{\frac{1}{2}2^{-t}2^{e_s}}{|s|}$$

which is the *relative error*[3] due to rounding. Because we have $\frac{1}{2}2^{e_s} \leq |s| < 2^{e_s}$, this error is biggest when $|s| = \frac{1}{2}2^{e_s}$, so therefore we conclude that

$$\frac{|s - fl[s]|}{|s|} \leq 2^{-t}. \tag{2.20}$$

From (2.19) $fl[s] = s + s\epsilon$, so that $|s - fl[s]| = |s||\epsilon|$, or $|\epsilon| = |s - fl[s]|/|s|$. Thus, $|\epsilon| \leq 2^{-t}$, which is (2.16). In other words, $|\epsilon'|$ is the absolute error, and $|\epsilon|$ is the relative error.

Finally, if $x = 0$ or $y = 0$ then no rounding error occurs: $\epsilon = 0$. Subtraction results do not differ from addition.

Now consider computing the product of x and y in (2.17). Since $x = m_x \times 2^{e_x}$, and $y = m_y \times 2^{e_y}$ with $x_1 \neq 0$, and $y_1 \neq 0$ we must have

$$\frac{1}{2}\frac{1}{2} \leq |m_x m_y| < 1. \tag{2.21}$$

This implies that it may be necessary to normalize the mantissa of the product with a shift to the left, and an appropriate adjustment of the exponent as well. The $2t$-bit mantissa of the product is rounded to give a t-bit mantissa. If $x = 0$, or $y = 0$ (or both x and y are zero), then the product is zero.

---

[3]In general, if a is the exact value of some quantity and $\hat{a}$ is some approximation to a, the *absolute error* is $||a - \hat{a}||$, while the *relative error* is

$$\frac{||a - \hat{a}||}{||a||}(a \neq 0).$$

The relative error is usually more meaningful in practice. This is because an error is really "big" or "small" only in relation to the size of the quantity being approximated.

We may consider a few examples. We will suppose $t = 4$. Begin with $x = 0.1010 \times 2^2$, and $y = 0.1111 \times 2^1$, so then

$$xy = 0.10010110 \times 2^3,$$

and so $fl[xy] = 0.1001 \times 2^3$ (computed product). If now $x = 0.1000 \times 2^4$, $y = 0.1000 \times 2^{-1}$, then, before normalizing the mantissa, we have

$$xy = 0.01000000 \times 2^3,$$

and after normalization we have

$$xy = 0.10000000 \times 2^2$$

so that $fl[xy] = 0.1000 \times 2^2$ (computed product). Finally, suppose that $x = 0.1010 \times 2^0$, and $y = 0.1010 \times 2^0$, so then the unnormalized product is

$$xy = 0.01100100 \times 2^0$$

for which the normalized product is

$$xy = 0.11001000 \times 2^{-1},$$

so finally $fl[xy] = 0.1101 \times 2^{-1}$ (computed product).

The application of fl to the normalized product will have exactly the same effect as it did in the case of addition (or of subtraction). This may be understood by recognizing that a $2t$-bit mantissa will "look the same" to operator fl regardless of how that mantissa was obtained. It therefore immediately follows that

$$fl[xy] = (xy)(1 + \epsilon), \tag{2.22}$$

which is another special case of (2.15), and $|\epsilon| \leq 2^{-t}$, which is (2.16) again.

Now consider the quotient x/y, for x and $y \neq 0$ in (2.17),

$$q = \frac{x}{y} = \frac{m_x \times 2^{e_x}}{m_y \times 2^{e_y}} = \frac{m_x}{m_y} \times 2^{e_x - e_y} = m_q \times 2^{e_q} \tag{2.23}$$

(so $m_q = m_x/m_y$, and $e_q = e_x - e_y$). The arithmetic unit in the machine has an accumulator that we assume contains m_x and which is "double length" in that it is $2t$ bits long. Specifically, this accumulator initially stores $x_0.x_1 \cdots x_t \underbrace{0 \cdots 0}_{t \text{ bits}}$. If $|m_x| > |m_y|$ the number in the accumulator is shifted one place to the right, and so e_q is increased by one (i.e., incremented). The number in the accumulator is then divided by m_y in such a manner as to give a correctly rounded t-bit result. This implies that the computed mantissa of the quotient, say, $m_q = q_0.q_1 \cdots q_t$, satisfies the normalization condition $q_1 = 1$, so that $\frac{1}{2} \leq |m_q| < 1$. Once again we must have

$$fl\left[\frac{x}{y}\right] = \frac{x}{y}(1 + \epsilon) \tag{2.24}$$

such that $|\epsilon| \leq 2^{-t}$. Therefore, (2.15) and (2.16) are now justified for all instances of *op*.

We complete this section with a few examples. Suppose $x = 0.1010 \times 2^2$, and $y = 0.1100 \times 2^{-2}$, then

$$q = \frac{x}{y} = \frac{0.1010 \times 2^2}{0.1100 \times 2^{-2}} = \frac{0.10100000 \times 2^2}{0.1100 \times 2^{-2}}$$

$$= \frac{0.10100000}{0.1100} \times 2^4 = 0.11010101 \times 2^4$$

so that $fl[q] = 0.1101 \times 2^4$ (computed quotient). Now suppose that $x = 0.1110 \times 2^3$, and $y = 0.1001 \times 2^{-2}$, and so

$$q = \frac{x}{y} = \frac{0.1110 \times 2^3}{0.1001 \times 2^{-2}} = \frac{0.01110000 \times 2^4}{0.1001 \times 2^{-2}}$$

$$= \frac{0.01110000}{0.1001} \times 2^6 = 0.11000111 \times 2^6$$

so that $fl[q] = 0.1100 \times 2^6$ (computed quotient).

Thus far we have emphasized *ordinary rounding*, but an alternative implementation of fl is to use *chopping*. If $x = \pm\left(\sum_{k=1}^{\infty} x_k 2^{-k}\right) \times 2^e$, then, for chopping operator fl, we have $fl[x] = \pm\left(\sum_{k=1}^{t} x_k 2^{-k}\right) \times 2^e$ (chopping x to $t + 1$ bits including the sign bit). Thus, the absolute error is

$$|\epsilon'| = |x - fl[x]| = \left(\sum_{k=t+1}^{\infty} x_k 2^{-k}\right) 2^e \leq 2^e \sum_{k=t+1}^{\infty} 2^{-k}$$

(as $x_k = 1$ for all $k > t$), but since $\sum_{k=t+1}^{\infty} 2^{-k} = 2^{-t}$, we must have

$$|\epsilon'| = |x - fl[x]| \leq 2^{-t} 2^e,$$

and so the relative error for chopping is

$$|\epsilon| = \frac{|x - fl[x]|}{|x|} \leq \frac{2^{-t} e^e}{\frac{1}{2} 2^e} = 2^{-t+1}$$

(because we recall that $|x| \geq \frac{1}{2} 2^e$). We see that the error in chopping is somewhat bigger than the error in rounding, but chopping is somewhat easier to implement.

2.4   ROUNDING EFFECTS IN DOT PRODUCT COMPUTATION

Suppose $x, y \in \mathbf{R}^n$. We recall from Chapter 1 (and from elementary linear algebra) that the vector dot product is given by

$$\langle x, y \rangle = x^T y = y^T x = \sum_{k=0}^{n-1} x_k y_k. \tag{2.25}$$

This operation occurs in matrix–vector product computation (e.g., $y = Ax$, where $A \in \mathbf{R}^{n \times n}$), digital filter implementation (i.e., computing discrete-time convolution), numerical integration, and other applications. In other words, it is so common that it is important to understand how rounding errors can affect the accuracy of a computed dot product.

We may regard dot product computation as a recursive process. Thus

$$s_{n-1} = \sum_{k=0}^{n-1} x_k y_k = \sum_{k=0}^{n-2} x_k y_k + x_{n-1} y_{n-1} = s_{n-2} + x_{n-1} y_{n-1}.$$

So

$$s_k = s_{k-1} + x_k y_k \tag{2.26}$$

for $k = 0, 1, \ldots, n - 1$, and $s_{-1} = 0$. Each arithmetic operation in (2.26) is a separate floating-point operation and so introduces its own error into the overall calculation. We would like to obtain a general expression for this error. To begin, we may model the computation process according to

$$\hat{s}_0 = fl[x_0 y_0]$$
$$\hat{s}_1 = fl[\hat{s}_0 + fl[x_1 y_1]]$$
$$\hat{s}_2 = fl[\hat{s}_1 + fl[x_2 y_2]]$$
$$\vdots$$
$$\hat{s}_{n-2} = fl[\hat{s}_{n-3} + fl[x_{n-2} y_{n-2}]]$$
$$\hat{s}_{n-1} = fl[\hat{s}_{n-2} + fl[x_{n-1} y_{n-1}]]. \tag{2.27}$$

From (2.15) we may write

$$\hat{s}_0 = (x_0 y_0)(1 + \delta_0)$$
$$\hat{s}_1 = [\hat{s}_0 + (x_1 y_1)(1 + \delta_1)](1 + \epsilon_1)$$
$$\hat{s}_2 = [\hat{s}_1 + (x_2 y_2)(1 + \delta_2)](1 + \epsilon_2)$$
$$\vdots$$

$$\hat{s}_{n-2} = [\hat{s}_{n-3} + (x_{n-2}y_{n-2})(1 + \delta_{n-2})](1 + \epsilon_{n-2})$$

$$\hat{s}_{n-1} = [\hat{s}_{n-2} + (x_{n-1}y_{n-1})(1 + \delta_{n-1})](1 + \epsilon_{n-1}), \qquad (2.28)$$

where $|\delta_k| \le 2^{-t}$ (for $k = 0, 1, \ldots, n - 1$), and $|\epsilon_k| \le 2^{-t}$ (for $k = 1, 2, \ldots,$ $n - 1$), via (2.16). It is possible to write[4]

$$\hat{s}_{n-1} = \sum_{k=0}^{n-1} x_k y_k (1 + \gamma_k) = s_{n-1} + \sum_{k=0}^{n-1} x_k y_k \gamma_k, \qquad (2.29)$$

where

$$1 + \gamma_k = (1 + \delta_k) \prod_{j=k}^{n-1} (1 + \epsilon_j)(\epsilon_0 = 0). \qquad (2.30)$$

Note that the Π notation means, for example

$$\prod_{k=0}^{n} x_k = x_0 x_1 x_2 \cdots x_{n-1} x_n, \qquad (2.31)$$

where Π is the symbol to compute the product of all x_k for $k = 0, 1, \ldots, n$. The similarity to how we interpret Σ notation should therefore be clear.

The absolute value operator is a norm on **R**, so from the axioms for a norm (recall Definition 1.3), we must have

$$|s_{n-1} - \hat{s}_{n-1}| = |x^T y - fl[x^T y]| \le \sum_{k=0}^{n-1} |x_k y_k||\gamma_k|. \qquad (2.32)$$

In particular, obtaining this involves the repeated use of the triangle inequality. Equation (2.32) thus represents an upper bound on the absolute error involved in computing a vector dot product. Of course, the notation $fl[x^T y]$ symbolizes the floating-point approximation to the exact quantity $x^T y$. However, the bound in (2.32) is incomplete because we need to appropriately bound the numbers γ_k.

To obtain the bound we wish involves using the following lemma.

Lemma 2.1: We have

$$1 + x \le e^x, \quad x \ge 0 \qquad (2.33a)$$

$$e^x \le 1 + 1.01x, \quad 0 \le x \le .01. \qquad (2.33b)$$

---

[4]Equation (2.29) is most easily arrived at by considering examples for small n, for instance

$$\hat{s}_3 = x_0 y_0 (1 + \delta_0)(1 + \epsilon_0)(1 + \epsilon_1)(1 + \epsilon_2)(1 + \epsilon_3) + x_1 y_1 (1 + \delta_1)(1 + \epsilon_1)(1 + \epsilon_2)(1 + \epsilon_3)$$

$$+ x_2 y_2 (1 + \delta_2)(1 + \epsilon_2)(1 + \epsilon_3) + x_3 y_3 (1 + \delta_3)(1 + \epsilon_3),$$

and using such examples to "spot the pattern."

Proof   Begin with consideration of (2.33a). Recall that for $-\infty < x < \infty$

$$e^x = \sum_{n=0}^{\infty} \frac{x^n}{n!}. \tag{2.34}$$

Therefore

$$e^x = 1 + x + \sum_{n=2}^{\infty} \frac{x^n}{n!}$$

so that

$$1 + x = e^x - \sum_{n=2}^{\infty} \frac{x^n}{n!},$$

but the terms in the summation are all nonnegative, so (2.33a) follows immediately.

Now consider (2.33b), which is certainly valid for $x = 0$. The result will follow if we prove

$$\frac{e^x - 1}{x} \leq 1.01 \quad (x \neq 0).$$

From (2.34)

$$\frac{e^x - 1}{x} = \sum_{m=0}^{\infty} \frac{x^m}{(m+1)!} = 1 + \sum_{m=1}^{\infty} \frac{x^m}{(m+1)!}$$

so we may also equivalently prove instead that

$$\sum_{m=1}^{\infty} \frac{x^m}{(m+1)!} \leq 0.01$$

for $0 < x \leq 0.01$. Observe that

$$\sum_{m=1}^{\infty} \frac{x^m}{(m+1)!} = \frac{1}{2}x + \frac{1}{6}x^2 + \frac{1}{24}x^3 + \cdots \leq \frac{1}{2}x + x^2 + x^3 + x^4 + \cdots$$

$$= \frac{1}{2}x + \sum_{k=2}^{\infty} x^k = \frac{1}{2}x + \sum_{k=0}^{\infty} x^k - 1 - x$$

$$= \frac{1}{1-x} - \frac{1}{2}x - 1 = \frac{1}{2}x\frac{1+x}{1-x}.$$

It is not hard to verify that

$$\frac{1}{2}x\frac{1+x}{1-x} \leq 0.01$$

for $0 < x \leq 0.01$. Thus, (2.33b) follows.

If $n = 1, 2, 3, \ldots$, and if $0 \leq nu \leq 0.01$, then

$$(1 + u)^n \leq (e^u)^n \quad \text{[via (2.33a)]}$$
$$\leq 1 + 1.01nu \quad \text{[via (2.33b)]}. \tag{2.35}$$

Now if $|\delta_i| \leq u$ for $i = 0, 1, \ldots, n - 1$ then

$$\prod_{i=0}^{n-1}(1 + \delta_i) \leq \prod_{i=0}^{n-1}(1 + |\delta_i|) \leq (1 + u)^n$$

so via (2.35)

$$\prod_{i=0}^{n-1}(1 + \delta_i) \leq 1 + 1.01nu, \tag{2.36}$$

where we must emphasize that $0 \leq nu \leq 0.01$. Certainly there is a δ such that

$$1 + \delta = \prod_{i=0}^{n-1}(1 + \delta_i), \tag{2.37}$$

and so from (2.36), $|\delta| \leq 1.01nu$. If we identify γ_k with δ in (2.33) for all k, then

$$|\gamma_k| \leq 1.01nu \tag{2.38}$$

for which we consider $u = 2^{-t}$ [because in (2.30) both $|\epsilon_i|$ and $|\delta_i| \leq 2^{-t}$]. Using (2.38) in (2.32), we obtain

$$|x^T y - fl[x^T y]| \leq 1.01nu \sum_{k=0}^{n-1} |x_k y_k|, \tag{2.39}$$

but $\sum_{k=0}^{n-1} |x_k y_k| = \sum_{k=0}^{n-1} |x_k||y_k|$, and this may be symbolized as $|x|^T |y|$ (so that $|x| = [|x_0||x_1| \cdots |x_{n-1}|]^T$). Thus, we may rewrite (2.39) as

$$|x^T y - fl[x^T y]| \leq 1.01nu|x|^T |y|. \tag{2.40}$$

Observe that the relative error satisfies

$$\frac{|x^T y - fl[x^T y]|}{|x^T y|} \leq 1.01nu \frac{|x|^T |y|}{|x^T y|}. \tag{2.41}$$

The bound in (2.41) may be quite large if $|x|^T |y| \gg |x^T y|$. This suggests the possibility of a large relative error. We remark that since $u = 2^{-t}$, $nu \leq 0.01$ will hold in all practical cases unless n is very large (a typical value for t is $t = 56$).

The potentially large relative errors indicated by the analysis we have just made are a consequence of the details of how the dot product was calculated. As noted on p. 65 of Ref. 4, the use of a double-precision accumulator to compute the dot product can reduce the error dramatically. Essentially, if x and y are floating-point vectors with t-bit mantissas, the "running sum" s_k [of (2.26)] is built up in an accumulator with a $2t$-bit mantissa. Multiplication of two t-bit numbers can be stored exactly in a double-precision variable. The large dynamic floating-point range limits the likelihood of overflow/underflow. Only when final sum s_{n-1} is written to a single-precision memory location will there be a rounding error. It therefore follows that when this alternative procedure is employed, we get

$$fl[x^T y] = x^T y(1 + \delta) \tag{2.42}$$

for which $|\delta| \approx 2^{-t} \ (= u)$. Clearly, this is a big improvement.

The material of this section shows

1. The analysis required to obtain insightful bounds on errors can be quite arduous.
2. Proper numerical technique can have a dramatic effect in reducing errors.
3. Proper technique can be revealed by analysis.

The following example illustrates how the bound on rounding error in dot product computation may be employed.

Example 2.4 Assume the existence of a square root function such that $fl[\sqrt{x}] = \sqrt{x}(1 + \epsilon)$ and $|\epsilon| \le u$. We use the algorithm that corresponds to the bound of Eq. (2.40) to compute $x^T x$ ($x \in \mathbf{R}^n$), and then use this to give an algorithm for $||x|| = \sqrt{x^T x}$. This can be expressed in the form of pseudocode:

```
s_{-1} := 0;
for k := 0 to n - 1 do begin
    s_k := s_{k-1} + x_k^2;
    end;
||x|| := \sqrt{s_{n-1}};
```

We will now obtain a bound on the relative error due to rounding in the computation of $||x||$. We will use the fact that $\sqrt{1 + x} \le 1 + x$ (for $x \ge 0$).

Now

$$\epsilon_1 = \frac{fl[x^T x] - x^T x}{x^T x} \implies fl[x^T x] = x^T x(1 + \epsilon_1),$$

and via (2.41)

$$|\epsilon_1| \le 1.01 n u \frac{|x|^T |x|}{|x^T x|}$$

$$= 1.01 n u \frac{||x||^2}{||x||^2} = 1.01 n u$$

($|x|^T |x| = \sum_{k=0}^{n-1} |x_k|^2 = \sum_{k=0}^{n-1} x_k^2 = ||x||^2$, and $|||x||^2| = ||x||^2$). So in "shorthand" notation, $fl[\sqrt{fl[x^T x]}] \equiv fl[||x||]$, and

$$fl[||x||] = \sqrt{x^T x}\sqrt{1 + \epsilon_1}(1 + \epsilon) = ||x||\sqrt{1 + \epsilon_1}(1 + \epsilon),$$

and $\sqrt{1 + \epsilon_1} \leq 1 + \epsilon_1$, so

$$fl[||x||] \leq ||x||(1 + \epsilon_1)(1 + \epsilon).$$

Now $(1 + \epsilon_1)(1 + \epsilon) = 1 + \epsilon_1 + \epsilon + \epsilon_1\epsilon$, implying that

$$||x||(1 + \epsilon_1)(1 + \epsilon) = ||x|| + ||x||(\epsilon_1 + \epsilon + \epsilon_1\epsilon)$$

so therefore

$$fl[||x||] \leq ||x|| + ||x||(\epsilon_1 + \epsilon + \epsilon_1\epsilon),$$

and thus

$$\left| \frac{fl[||x||] - ||x||}{||x||} \right| \leq |\epsilon_1 + \epsilon + \epsilon_1\epsilon| \leq u + 1.01nu + 1.01nu^2$$

$$= u[1 + 1.01n + 1.01nu].$$

Of course, we have used the fact that $|\epsilon| \leq u$.

2.5   MACHINE EPSILON

In Section 2.3 upper bounds on the error involved in applying the operator fl were derived. Specifically, we found that the relative error satisfies

$$|\epsilon'| = \frac{|x - fl[x]|}{|x|} \leq \begin{cases} 2^{-t} & \text{(rounding)} \\ 2^{-t+1} & \text{(chopping)} \end{cases}. \tag{2.43}$$

As suggested in Section 2.4, these bounds are often denoted by u; that is, $u = 2^{-t}$ for rounding, and $u = 2^{-t+1}$ for chopping. The bound u is often called the *unit roundoff* [4, Section 2.4.2].

The details of how floating-point arithmetic is implemented on any given computing machine may not be known or readily determined by the user. Thus, u may not be known. However, an "experimental" approach is possible. One may run a simple program to "estimate" u, and the estimate is the *machine epsilon*, denoted ϵ_M. The machine epsilon is defined to be the difference between 1.0 and the next biggest floating-point number [6, Section 2.1]. Consequently, $\epsilon_M = 2^{-t+1}$. A pseudocode to compute ϵ_M is as follows:

```
stop := 1;
eps := 1.0;
while stop == 1 do begin
      eps := eps/2.0;
      x := 1.0 + eps;
      if x ≤ 1.0
         begin
```

```
      stop := 0;
      end;
    end;
eps := 2.0 * eps;
```

This code may be readily implemented as a MATLAB routine. MATLAB stores eps ($= \epsilon_M$) as a built-in constant, and the reader may wish to test the code above to see if the result agrees with MATLAB eps (as a programming exercise).

In this book we shall (unless otherwise stated) regard machine epsilon and unit roundoff as practically interchangeable.

APPENDIX 2.A   REVIEW OF BINARY NUMBER CODES

This appendix summarizes typical methods used to represent integers in binary. Extension of the results in this appendix to fractions is certainly possible. This material is normally to be found in introductory digital electronics books. The reader is here assumed to know Boolean algebra. This implies that the reader knows that + can represent either algebraic addition, or the logical OR operation. Similarly, xy might mean the logical AND of the Boolean variables x and y, or it might mean the arithmetic product of the real variables x and y. The context must be considered to ascertain which meaning applies.

Below we speak of "complements." These are used to represent negative integers, and also to facilitate arithmetic with integers. We remark that the results of this appendix are presented in a fairly general manner. Thus, the reader may wish, for instance, to see numerical examples of arithmetic using two's complement (2's comp.) codings. The reader can consult pp. 276–280 of Ref. 2 for such examples. Almost any other books on digital logic will also provide a source of numerical examples [3].

We may typically interpret a bit pattern in one of four ways, assuming that the bit pattern is to represent a number (negative or nonnegative integer). An example of this is as follows, and it provides a summary of common representations (e.g., for $n = 3$ bits):

Bit Pattern	Unsigned Integer	2's Comp.	1's Comp.	Sign Magnitude
0 0 0	0	0	0	0
0 0 1	1	1	1	1
0 1 0	2	2	2	2
0 1 1	3	3	3	3
1 0 0	4	-4	-3	-0
1 0 1	5	-3	-2	-1
1 1 0	6	-2	-1	-2
1 1 1	7	-1	-0	-3

In the four coding schemes summarized in this table, the interpretation of the bit pattern is always the same when the most significant bit (MSB) is zero. A similar table for $n = 4$ appears in Hamacher et al. [2, see p. 271].

Note that, philosophically speaking, the table above implies that a bit pattern can have more than one meaning. It is up to the engineer to decide what meaning it should have. Of course, this will be a function of purpose. Presently, our purpose is that bit patterns should have meaning with respect to the problems of numerical computing; that is, bit patterns must represent numerical information.

The relative merits of the three signed number coding schemes illustrated in the table above may be summarized as follows:

Coding Scheme	Advantages	Disadvantages
2's complement	Simple adder/subtracter circuit Only one code for 0	Circuit for finding the 2's comp. more complex than circuit for finding the 1's comp.
1's complement	Easy to obtain the 1's comp. of a number	Circuit for addition and subtraction more complex than for the 2's comp. adder/subtracter Two codes for 0
Sign magnitude	Intuitively obvious code	Has the most complex adder/subtracter circuit Two codes for 0

The following is a summary of some formulas associated with arithmetic (i.e., addition and subtraction) with r's and $(r - 1)$'s complements. In binary arithmetic $r = 2$, while in decimal arithmetic $r = 10$. We emphasize the case $r = 2$.

Let A be an n-digit base-r number (integer)

$$A = A_{n-1}A_{n-2}\cdots A_1 A_0$$

where $A_k \in \{0, 1, \ldots, r - 2, r - 1\}$. Digit A_{n-1} is the most significant digit (MSD), while digit A_0 is the least significant digit (LSD). Provided that A is not negative (i.e., is unsigned), we recognize that to convert A to a base-10 representation (i.e., ordinary decimal number) requires us to compute

$$\sum_{k=0}^{n-1} A_k r^k.$$

If A is allowed to be a negative integer, the usage of this summation needs modification. This is considered below.

The r's complement of A is defined to be

$$r\text{'s complement of } A = A^* = \begin{cases} r^n - A, & A \neq 0 \\ 0, & A = 0 \end{cases} \qquad (2.A.1)$$

The $(r-1)$'s complement of A is defined to be

$$(r-1)\text{'s complement of } A = \overline{A} = (r^n - 1) - A \qquad (2.A.2)$$

It is important not to confuse the bar over the A in (2.A.2) with the Boolean NOT operation, although for the special case of $r = 2$ the bar will denote complementation of each bit of A; that is, for $r = 2$

$$\overline{A} = \overline{A}_{n-1}\overline{A}_{n-2}\cdots\overline{A}_1\overline{A}_0$$

where the bar now denotes the logical NOT operation. More generally, if A is a base-r number

$$\overline{A} = (r-1) - A_{n-1} \quad (r-1) - A_{n-2} \quad \cdots (r-1) - A_1 \quad (r-1) - A_0$$

Thus, to obtain $\overline{A}$, each digit of A is subtracted from $r - 1$. As a consequence, comparing (2.A.1) and (2.A.2), we see that

$$A^* = \overline{A} + 1 \qquad (2.A.3)$$

where the plus denotes algebraic addition (which takes place in base r).

Now we consider the three (previously noted) different methods for coding integers when $r = 2$:

1. Sign-magnitude coding
2. One's complement coding
3. Two's complement coding

In all three of these coding schemes the most significant bit (MSB) is the *sign bit*. Specifically , if $A_{n-1} = 0$, the number is nonnegative, and if $A_{n-1} = 1$, the number is negative. It can be shown that when the complement (either one's or two's) of a binary number is taken, this is equivalent to placing a minus sign in front of the number. As a consequence, when given a binary number $A = A_{n-1}A_{n-2}\cdots A_1 A_0$ coded according to one of these three schemes, we may convert that number to a base-10 integer according to the following formulas:

1. *Sign-Magnitude Coding.* The sign-magnitude binary number $A = A_{n-1}A_{n-2}$ $\cdots A_1 A_0$ ($A_k \in \{0, 1\}$) has the base-10 equivalent

$$A = \begin{cases} \displaystyle\sum_{i=0}^{n-2} A_i 2^i, & A_{n-1} = 0 \\[2em] \displaystyle-\sum_{i=0}^{n-2} A_i 2^i, & A_{n-1} = 1 \end{cases} \qquad (2.A.4)$$

With this coding scheme there are two codings for zero:

$$(0)_{10} = (\underbrace{000\cdots00}_{n})_2 = (\underbrace{100\cdots00}_{n})_2$$

2. *One's Complement Coding.* In this coding we represent $-A$ as $\overline{A}$. The one's complement binary number $A = A_{n-1}A_{n-2}\cdots A_1A_0$ $(A_k \in \{0, 1\})$ has the base-10 equivalent

$$A = \begin{cases} \displaystyle\sum_{i=0}^{n-2} A_i 2^i & , A_{n-1} = 0 \\ \displaystyle-\sum_{i=0}^{n-2} \overline{A}_i 2^i & , A_{n-1} = 1 \end{cases} \tag{2.A.5}$$

With this coding scheme there are also two codes for zero:

$$(0)_{10} = (\underbrace{000\cdots00}_{n})_2 = (\underbrace{111\cdots11}_{n})_2$$

3. *Two's Complement Coding.* In this coding we represent $-A$ as $A^* \, (= \overline{A} + 1)$. The two's complement binary number $A = A_{n-1}A_{n-2}\cdots A_1A_0$ $(A_k \in \{0, 1\})$ has the base-10 equivalent

$$A = -2^{n-1}A_{n-1} + \sum_{i=0}^{n-2} A_i 2^i \tag{2.A.6}$$

The proof is as follows. If $A_{n-1} = 0$, then $A \geq 0$ and immediately the base-10 equivalent is $A = \sum_{i=0}^{n-2} A_i 2^i$ (via the procedure for converting a number in base-2 to one in base-10), which is (2.A.6) for $A_{n-1} = 0$. Now, if $A_{n-1} = 1$, then $A < 0$, and so if we take the two's complement of A we must get $|A|$:

$$|A| = \overline{A} + 1$$
$$= (1 - A_{n-1})(1 - A_{n-2})\cdots(1 - A_1)(1 - A_0) + \underbrace{00\cdots01}_{n}$$
$$= \sum_{i=0}^{n-1}(1 - A_i)2^i + 1$$
$$= 2^{n-1}(1 - A_{n-1}) + \sum_{i=0}^{n-2}(1 - A_i)2^i + 1$$
$$= \sum_{i=0}^{n-2} 2^i + 1 - \sum_{i=0}^{n-2} A_i 2^i \quad (A_{n-1} = 1)$$

$$= \frac{1 - 2^{n-1}}{1 - 2} + 1 - \sum_{i=0}^{n-2} A_i 2^i \left(\text{via} \sum_{i=0}^{n} a^i = \frac{1 - a^{n+1}}{1 - a} \right)$$

$$= 2^{n-1} - \sum_{i=0}^{n-2} A_i 2^i$$

and so $A = -2^{n-1} + \sum_{i=0}^{n-2} A_i 2^i$, which is (2.A.6) for $A_{n-1} = 1$. In this coding scheme there is only one code for zero:

$$(0)_{10} = (\underbrace{000 \cdots 00}_{n})_2$$

When n-bit integers are added together, there is the possibility that the sum may not fit in n bits. This is *overflow*. The condition is easy to detect by monitoring the signs of the operands and the sum. Suppose that x and y are n-bit two's complement coded integers, so that the sign bits of these operands are x_{n-1} and y_{n-1}. Suppose that the sum is denoted by s, implying that the sign bit is s_{n-1}. The Boolean function that tests for overflow of $s = x + y$ (algebraic sum of x and y) is

$$T = x_{n-1} y_{n-1} \overline{s}_{n-1} + \overline{x}_{n-1} \overline{y}_{n-1} s_{n-1}.$$

The first term will be logical 1 if the operands are negative while the sum is positive. The second term will be logical 1 if the operands are positive but the sum is negative. Either condition yields $T = 1$, thus indicating an overflow. A similar test may be obtained for subtraction, but we omit this here.

The following is both the procedure and the justification of the procedure for adding two's complement coded integers.

Theorem 2.A.1: Two's Complement Addition   If A and B are n-bit two's complement coded numbers, then compute $A + B$ (the sum of A and B) as though they were unsigned numbers, discarding any carryout.

Proof   Suppose that $A > 0$, $B > 0$; then $A + B$ will generate no carryout from the bit position $n - 1$ since $A_{n-1} = B_{n-1} = 0$ (i.e., the sign bits are zero-valued), and the result will be correct if $A + B < 2^{n-1}$. (If this inequality is not satisfied, then the sign bit will be one, indicating a negative answer, which is wrong. This amounts to an overflow.)

Suppose that $A \geq B > 0$; then

$$A + (-B) = A + B^* = A + 2^n - B = 2^n + A - B,$$

and if we discard the carryout, this is equivalent to subtracting 2^n (because the carryout has a weight of 2^n). Doing this yields $A + (-B) = A - B$.

Similarly

$$(-A) + B = A^* + B = 2^n - A + B = 2^n + B - A,$$

and discarding the carry out yields $(-A) + B = B - A$.

Again, suppose that $A \geq B > 0$, then

$$(-A) + (-B) = A^* + B^* = 2^n - A + 2^n - B = 2^n + [2^n - (A + B)]$$
$$= 2^n + (A + B)^*$$

so discarding the carryout gives $(-A) + (-B) = (A + B)^*$, which is the desired two's complement representation of $-(A + B)$, provided $A + B \leq 2^{n-1}$. (If this latter inequality is not satisfied, then we have an overflow.)

The procedure for subtraction (and its justification) follows similarly. We omit these details.

REFERENCES

1. J. H. Wilkinson, *Rounding Errors in Algebraic Processes*, Prentice-Hall, Englewood Cliffs, NJ, 1963.
2. V. C. Hamacher, Z Vranesic, and S. G. Zaky, *Computer Organization*, 3rd ed., McGraw-Hill, New York, 1990.
3. J. F. Wakerly, *Digital Design Principles and Practices*, Prentice-Hall, Englewood Cliffs, NJ, 1990.
4. G. H. Golub and C. F. Van Loan, *Matrix Computations*, 2nd ed., Johns Hopkins Univ. Press, Baltimore, MD, 1989.
5. G. E. Forsythe, M. A. Malcolm, and C. B. Moler, *Computer Methods for Mathematical Computations*, Prentice-Hall, Englewood Cliffs, NJ, 1977.
6. N. J. Higham, *Accuracy and Stability of Numerical Algorithms*. SIAM, Philadelphia, PA, 1996.

PROBLEMS

2.1. Let $fx[x]$ denote the operation of reducing x (a fixed-point binary fraction) to $t + 1$ bits (including the sign bit) according to the Wilkinson rounding (ordinary rounding) procedure in Section 2.2. Suppose that $a = (0.1000)_2$, $b = (0.1001)_2$, and $c = (0.0101)_2$, so $t = 4$ here. In arithmetic of unlimited precision, we always have $a(b + c) = ab + ac$. Suppose that a practical computing machine applies the operator $fx[\cdot]$ after every arithmetic operation.

(a) Find $x = fx[fx[ab] + fx[ac]]$.

(b) Find $y = fx[afx[b + c]]$.

Do you obtain $x = y$?

This problem shows that the order of operations in an algorithm implemented on a practical computer can affect the answer obtained.

2.2. Recall from Section 2.2 that

$$q = \left(\tfrac{1}{3}\right)_{10} = (0.\overline{01})_2.$$

Find the absolute error in representing q as a $(t+1)$-bit binary number. Find the relative error. Assume both ordinary rounding and chopping (defined at the end of Section 2.3 with respect to floating-point arithmetic).

2.3. Recall that we define a floating-point number in base r to have the form

$$x = \underbrace{x_0.x_1x_2\cdots x_{t-1}x_t}_{=f} \times r^e,$$

where $x_0 \in \{+, -\}$ (sign digit), $x_k \in \{0, 1, \ldots, r-1\}$ for $k = 1, 2, \ldots, t$, e is the exponent (a signed integer), and $x_1 \neq 0$ (so $r^{-1} \leq |f| < 1$) if $x \neq 0$. Show that for $x \neq 0$

$$m \leq |x| \leq M,$$

where for $L \leq e \leq U$, we have

$$m = r^{L-1}, \quad M = r^U(1 - r^{-t}).$$

2.4. Suppose $r = 10$. We may consider the *result* of a decimal arithmetic operation in the floating-point representation to be

$$x = \pm \left(\sum_{k=1}^{\infty} x_k 10^{-k} \right) \times 10^e.$$

(a) If $fl[x]$ is the operator for *chopping*, then

$$fl[x] = (\pm.x_1x_2 \cdots x_{t-1}x_t) \times 10^e,$$

thus, all digits x_k for $k > t$ are forced to zero.

(b) If $fl[x]$ is the operator for *rounding* then it is defined as follows. Add $\underbrace{0.00\cdots 01}_{t+1 \text{ digits}}$ to the mantissa if $x_{t+1} \geq 5$, but if $x_{t+1} < 5$, the mantissa is unchanged. Then all digits x_k for $k > t$ are forced to zero.

Show that the *absolute error for chopping* satisfies the upper bound

$$|x - fl[x]| \leq 10^{-t}10^e,$$

and that the *absolute error for rounding* satisfies the upper bound

$$|x - fl[x]| \leq \tfrac{1}{2}10^{-t}10^e.$$

Show that the *relative errors* satisfy

$$|\epsilon| = \frac{|x - fl[x]|}{|x|} \leq \begin{cases} 10^{1-t} & \text{(chopping)} \\ \frac{1}{2}10^{1-t} & \text{(rounding)} \end{cases}.$$

2.5. Suppose that $t = 4$ and $r = 2$ (i.e., we are working with floating-point binary numbers). Suppose that we have the operands

$$x = 0.1011 \times 10^{-3}, \quad y = -0.1101 \times 10^2.$$

Find $x + y$, $x - y$, and xy. Clearly show the steps involved.

2.6. Suppose that $A \in \mathbf{R}^{n \times n}$, $x \in \mathbf{R}^n$, and that $fl[Ax]$ represents the result of computing the product Ax on a floating-point computer. Define $|A| = [|a_{i,j}|]_{i,j=0,1,...,n-1}$, and $|x| = [|x_0||x_1| \cdots |x_{n-1}|]^T$. We have

$$fl[Ax] = Ax + e,$$

where $e \in \mathbf{R}^n$ is the error vector. Of course, e models the rounding errors involved in the actual computation of product Ax on the computer. Justify the bound

$$|e| \leq 1.01 nu|A||x|.$$

2.7. Explain why a conditional test such as

```
if x ≠ y then begin
  f := f/(x − y);
  end;
```

is unreliable.

(*Hint:* Think about dynamic range limitations in floating-point arithmetic.)

2.8. Suppose that $x = [x_0 x_1 \cdots x_{n-1}]^T$ is a real-valued vector, $||x||_\infty = \max_{0 \leq k \leq n-1} |x_k|$, and that we wish to compute $||x||_2 = \left[\sum_{k=0}^{n-1} x_k^2 \right]^{1/2}$. Explain the advantages, and disadvantages of the following algorithm with respect to computational efficiency (number of arithmetic operations, and comparisons), and dynamic range limitations in floating-point arithmetic:

```
m := ||x||∞;
s := 0;
for k := 0 to n − 1 do begin
   s := s + (xk/m)²;
   end;
||x||2 := m√s;
```

Comments regarding computational efficiency may be made with respect to the pseudocode algorithm in Example 2.4.

2.9. Recall that for $x^2 + bx + c = 0$, the roots are

$$x_1 = \frac{-b + \sqrt{b^2 - 4c}}{2}, \quad x_2 = \frac{-b - \sqrt{b^2 - 4c}}{2}.$$

If $b = -0.3001$, $c = 0.00006$, then the "exact" roots for this set of parameters are

$$x_1 = 0.29989993, \quad x_2 = 2.0006673 \times 10^{-4}.$$

Let us compute the roots using four-digit (i.e., $t = 4$) decimal (i.e., $r = 10$) floating-point arithmetic, where, as a result of rounding quantization b, and c are replaced with their approximations

$$\bar{b} = -0.3001 = b, \quad \bar{c} = 0.0001 \neq c.$$

Compute $\bar{x}_2$, which is the approximation to x_2 obtained using $\bar{b}$ and $\bar{c}$ in place of b and c. Show that the relative error is

$$\left| \frac{x_2 - \bar{x}_2}{x_2} \right| \approx 0.75$$

(i.e., the relative error is about 75%). (*Comment:* This is an example of *catastrophic cancellation.*)

2.10. Suppose $a, b \in \mathbf{R}$, and $x = a - b$. Floating-point approximations to a and b are $\hat{a} = fl[a] = a(1 + \epsilon_a)$ and $\hat{b} = fl[b] = b(1 + \epsilon_b)$, respectively. Hence the floating-point approximation to x is $\hat{x} = \hat{a} - \hat{b}$. Show that the relative error is of the form

$$|\epsilon| = \left| \frac{x - \hat{x}}{x} \right| \leq \alpha \frac{|a| + |b|}{|a - b|}.$$

What is α? When is $|\epsilon|$ large?

2.11. For $a \neq 0$, the quadratic equation $ax^2 + bx + c = 0$ has roots given by

$$x_1 = \frac{-b + \sqrt{b^2 - 4ac}}{2a}, \quad x_2 = \frac{-b - \sqrt{b^2 - 4ac}}{2a}.$$

For $c \neq 0$, quadratic equation $cx^2 + bx + a = 0$ has roots given by

$$x_1' = \frac{-b + \sqrt{b^2 - 4ac}}{2c}, \quad x_2' = \frac{-b - \sqrt{b^2 - 4ac}}{2c}.$$

(a) Show that $x_1 x_2' = 1$ and $x_2 x_1' = 1$.
(b) Using the result from Problem 2.10, explain accuracy problems that can arise in computing either x_1 or x_2 when $b^2 \gg |4ac|$. Can you use the result in part (a) to alleviate the problem? Explain.

3 Sequences and Series

3.1 INTRODUCTION

Sequences and series have a major role to play in computational methods. In this chapter we consider various types of sequences and series, especially with respect to their convergence behavior. A series might converge "mathematically," and yet it might not converge "numerically" (i.e., when implemented on a computer). Some of the causes of difficulties such as this will be considered here, along with possible remedies.

3.2 CAUCHY SEQUENCES AND COMPLETE SPACES

It was noted in the introduction to Chapter 1 that many computational processes are "iterative" (the Newton–Raphson method for finding the roots of an equation, iterative methods for linear system solution, etc.). The practical effect of this is to produce sequences of elements from function spaces. The sequence produced by the iterative computation is only useful if it *converges*. We must therefore investigate what this means.

In Chapter 1 it was possible for sequences to be either singly or doubly infinite. Here we shall assume sequences are singly infinite unless specifically stated to the contrary.

We begin with the following (standard) definition taken from Kreyszig [1, pp. 25–26]. Examples of applications of the definitions to follow will be considered later.

Definition 3.1: Convergence of a Sequence, Limit   A sequence (x_n) in a metric space $X = (X, d)$ is said to *converge*, or to *be convergent* iff there is an $x \in X$ such that

$$\lim_{n \to \infty} d(x_n, x) = 0. \tag{3.1}$$

The element x is called the *limit* of (x_n) (i.e., *limit of the sequence*), and we may state that

$$\lim_{n \to \infty} x_n = x. \tag{3.2}$$

An Introduction to Numerical Analysis for Electrical and Computer Engineers, by C.J. Zarowski
ISBN 0-471-46737-5   © 2004 John Wiley & Sons, Inc.

We say that (x_n) converges to x or has a limit x. If (x_n) is not convergent, then we say that it is a *divergent sequence*, or is simply *divergent*.

A shorthand expression for (3.2) is to write $x_n \to x$. We observe that sequence (x_n) is defined to converge (or not) with respect to a particular metric here denoted d (recall the axioms for a metric space from Chapter 1). We remark that it is possible that, for some (x_n) in some set X, the sequence might converge with respect to one metric on the set, but might not converge with respect to another choice of metric. It must be emphasized that the limit x must be an element of X in order for the sequence to be convergent.

Suppose, for example, that $X = (0, 1] \subset \mathbf{R}$, and consider the sequence $x_n = \frac{1}{n+1} (n \in \mathbf{Z}^+)$. Suppose also that $d(x, y) = |x - y|$. The sequence (x_n) does not converge in X because the sequence "wants to go to 0." But 0 is not in X. So the sequence does not converge. (Of course, the sequence converges in $X = \mathbf{R}$ with respect to our present choice of metric.)

It can be difficult in practice to ascertain whether a particular sequence converges according to Definition 3.1. This is because the limit x may not be known in advance. In fact, this is almost always the case in computing applications of sequences. Sometimes it is therefore easier to work with the following:

Definition 3.2: Cauchy Sequence, Complete Space   A sequence (x_n) in a metric space $X = (X, d)$ is called a *Cauchy sequence* iff for all $\epsilon > 0$ there is an $N(\epsilon) \in \mathbf{Z}^+$ such that

$$d(x_m, x_n) < \epsilon \tag{3.3}$$

for all $m, n > N(\epsilon)$. The space X is a *complete space* iff every Cauchy sequence in X converges.

We often write N instead of $N(\epsilon)$, because N may depend on our choice of ϵ. It is possible to prove that any convergent sequence is also Cauchy.

We remark that, if in fact the limit is known (or at least strongly suspected), then applying Definition 3.1 may actually be easier than applying Definition 3.2.

We see that under Definition 3.2 the elements of a Cauchy sequence get closer to each other as n and m increase. Establishing the "Cauchiness" of a sequence does not require knowing the limit of the sequence. This, at least in principle, simplifies matters. However, a big problem with this definition is that there are metric spaces X in which not all Cauchy sequences converge. In other words, there are *incomplete* metric spaces. For example, the space $X = (0, 1]$ with $d(x, y) = |x - y|$ is not complete. Recall that we considered $x_n = 1/(n + 1)$. This sequence is Cauchy,[1] but the limit is 0, which is not in X. Thus, this sequence is a nonconvergent Cauchy sequence. Thus, the space $(X, |\cdot|)$ is not complete.

---

[1] We see that

$$d(x_m, x_n) = \left| \frac{1}{m+1} - \frac{1}{n+1} \right| \leq \left| \frac{n-m}{nm} \right| = \left| \frac{1}{m} - \frac{1}{n} \right|.$$

A more subtle example of an incomplete metric space is the following. Recall space $C[a, b]$ from Example 1.4. Assume that $a = 0$ and $b = 1$, and now choose the metric to be

$$d(x, y) = \int_0^1 |x(t) - y(t)| \, dt \qquad (3.4)$$

instead of Eq. (1.8). Space $C[0, 1]$ with the metric (3.4) is not complete. This may be shown by considering the sequence of continuous functions illustrated in Fig. 3.1. The functions $x_m(t)$ in Fig. 3.1a form a Cauchy sequence. (Here we assume $m \geq 1$, and is an integer.) This is because $d(x_m, x_n)$ is the area of the triangle in Fig. 3.1b, and for any $\epsilon > 0$, we have

$$d(x_m, x_n) < \epsilon$$

whenever $m, n > 1/(2\epsilon)$. (Suppose $n \geq m$ and consider that $d(x_m, x_n) = \frac{1}{2}(\frac{1}{m} - \frac{1}{n}) \leq \frac{1}{2m} < \epsilon$.) We may see that this Cauchy sequence does not converge. Observe that we have

$$x_m(t) = 0 \quad \text{for} \quad t \in [0, \tfrac{1}{2}], \qquad x_m(t) = 1 \quad \text{for} \quad t \in [a_m, 1],$$

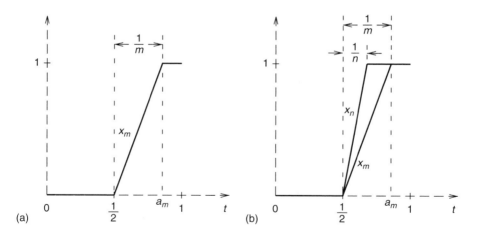

Figure 3.1 A Cauchy sequence of functions in $C[0, 1]$.

For any $\epsilon > 0$ we may find $N(\epsilon) > 0$ such that for $n \geq m > N(\epsilon)$

$$\frac{1}{m} - \frac{1}{n} < \epsilon.$$

If $n < m$, the roles of n and m may be reversed. The conditions of Definition 3.2 are met and so the sequence is Cauchy.

where $a_m = \frac{1}{2} + \frac{1}{m}$. Therefore, for all $x \in C[0, 1]$,

$$
\begin{aligned}
d(x_m, x) &= \int_0^1 |x_m(t) - x(t)| \, dt \\
&= \int_0^{1/2} |x(t)| \, dt + \int_{1/2}^{a_m} |x_m(t) - x(t)| \, dt + \int_{a_m}^1 |1 - x(t)| \, dt.
\end{aligned}
$$

The integrands are all nonnegative, and so each of the integrals on the right-hand side are nonnegative, too. Thus, to say that $d(x_m, x) \to 0$ implies that each integral approaches zero. Since $x(t)$ is continuous, it must be the case that

$$
x(t) = 0 \quad \text{for} \quad t \in [0, \tfrac{1}{2}), \quad x(t) = 1 \quad \text{for} \quad t \in (\tfrac{1}{2}, 1].
$$

However, this is not possible for a continuous function. In other words, we have a contradiction. Hence, (x_n) does not converge (i.e., has no limit in $X = C[0, 1]$). Again, we have a Cauchy sequence that does not converge, and so $C[0, 1]$ with the metric (3.4) is not complete.

This example also shows that a sequence of continuous functions may very well possess a discontinuous limit. Actually, we have seen this phenomenon before. Recall the example of the Fourier series in Chapter 1 (see Example 1.20). In this case the series representation of the square wave was made up of terms that are all continuous functions. Yet the series converges to a discontinuous limit. We shall return to this issue again later.

So now, some metric spaces are not complete. This means that even though a sequence is Cauchy, there is no guarantee of convergence. We are therefore faced with the following questions:

1. What metric spaces are complete?
2. Can they be "completed" if they are not?

The answer to the second question is "Yes." Given an incomplete metric space, it is always possible to complete it. We have seen that a Cauchy sequence does not converge when the sequence tends toward a limit that does not belong to the space; thus, in a sense, the space "has holes in it." Completion is the process of filling in the holes. This amounts to adding the appropriate elements to the set that made up the incomplete space. However, in general, this is a technically difficult process to implement in many cases, and so we will never do this. This is a job normally left to mathematicians.

We will therefore content ourselves with answering the first question. This will be done simply by listing complete metric spaces that are useful to engineers:

1. Sets $\mathbf{R}$ and $\mathbf{C}$ with the metric $d(x, y) = |x - y|$ are complete metric spaces.

2. Recall Example 1.3. The space $l^\infty[0, \infty]$ with the metric

$$d(x, y) = \sup_{k \in \mathbf{Z}^+} |x_k - y_k| \qquad (3.5)$$

is a complete metric space. (A proof of this claim appears in Ref. 1, p. 34.)

3. The Euclidean space $\mathbf{R}^n$ and the unitary space $\mathbf{C}^n$ both with metric

$$d(x, y) = \left[\sum_{k=0}^{n-1} |x_k - y_k|^2 \right]^{1/2} \qquad (3.6)$$

are complete metric spaces. (Proof is on p. 33 of Ref. 1.)

4. Recall Example 1.6. Fixing p, the space $l^p[0, \infty]$ such that $1 \le p < \infty$ is a complete metric space. Here we recall that the metric is

$$d(x, y) = \left[\sum_{k=0}^{\infty} |x_k - y_k|^p \right]^{1/p}. \qquad (3.7)$$

5. Recall Example 1.4. The set $C[a, b]$ with the metric

$$d(x, y) = \sup_{t \in [a,b]} |x(t) - y(t)| \qquad (3.8)$$

is a complete metric space. (Proof is on pp. 36–37 of Ref. 1.)

The last example is interesting because the special case $C[0, 1]$ with metric (3.4) was previously shown to be incomplete. Keeping the same set but changing the metric from that in (3.4) to that in (3.8) changes the situation dramatically.

In Chapter 1 we remarked on the importance of the metric space $L^2[a, b]$ (recall Example 1.7). The space is important as the "space of finite energy signals on the interval $[a, b]$." (A "finite power" interpretation was also possible.) An important special case of this was $L^2(\mathbf{R}) = L^2(-\infty, \infty)$. Are these metric spaces complete? Our notation implicitly assumes that the set (1.11a) (Chapter 1) contains the so-called *Lebesgue integrable functions* on $[a, b]$. In this case the space $L^2[a, b]$ is indeed complete with respect to the metric

$$d(x, y) = \left[\int_a^b |x(t) - y(t)|^2 \, dt \right]^{1/2}. \qquad (3.9)$$

Lebesgue integrable functions[2] have a complicated mathematical structure, and we have promised to avoid any measure theory in this book. It is enough for the reader

---

[2]One of the "simplest" introductions to these is Rudin [2]. However, these functions appear in the last chapter [2, Chapter 11]. Knowledge of much of the previous chapters is prerequisite to studying Chapter 11. Thus, the effort required to learn measure theory is substantial.

to assume that the functions in $L^2[a, b]$ are the familiar ones from elementary calculus.[3]

The complete metric spaces considered in the two previous paragraphs also happen to be normed spaces; recall Section 1.3.2. This is because the metrics are all induced by suitable norms on the spaces. It therefore follows that these spaces are complete normed spaces. Complete normed spaces are called *Banach spaces*.

Some of the complete normed spaces are also inner product spaces. Again, this follows because in those cases an inner product is defined that induced the norm. Complete inner product spaces are called *Hilbert spaces*. To be more specific, the following spaces are Hilbert spaces:

1. The Euclidean space $\mathbf{R}^n$ and the unitary space $\mathbf{C}^n$ along with the inner product

$$\langle x, y \rangle = \sum_{k=0}^{n-1} x_k y_k^* \tag{3.10}$$

are both Hilbert spaces.
2. The space $L^2[a, b]$ with the inner product

$$\langle x, y \rangle = \int_a^b x(t) y^*(t)\, dt \tag{3.11}$$

is a Hilbert space. [This includes the special case $L^2(\mathbf{R})$.]
3. The space $l^2[0, \infty]$ with the inner product

$$\langle x, y \rangle = \sum_{k=0}^{\infty} x_k y_k^* \tag{3.12}$$

is a Hilbert space.

We emphasize that (3.10) induces the metric (3.6), (3.11) induces the metric (3.9), and (3.12) induces the metric (3.7) (but only for case $p = 2$; recall from Chapter 1 that $l^p[0, \infty]$ is not an inner product space when $p \neq 2$). The three Hilbert spaces listed above are particularly important because of the fact, in part, that elements in these spaces have (as we have already noted) either finite energy or finite power interpretations. Additionally, least-squares problems are best posed and solved within these spaces. This will be considered later.

Define the set (of natural numbers) $\mathbf{N} = \{1, 2, 3, \ldots\}$. We have seen that sequences of continuous functions may have a discontinuous limit. An extreme example of this phenomenon is from p. 145 of Rudin [2].

---

[3]These "familiar" functions are called *Riemann integrable* functions. These functions form a proper subset of the Lebesgue integrable functions.

Example 3.1   For $n \in \mathbf{N}$ define

$$x_n(t) = \lim_{m \to \infty} [\cos(n!\pi t)]^{2m}.$$

When $n!t$ is an integer, then $x_n(t) = 1$ (simply because $\cos(\pi k) = \pm 1$ for $k \in \mathbf{Z}$). For all other values of t, we must have $x_n(t) = 0$ (simply because $|\cos t| < 1$ when t is not an integral multiple of π). Define

$$x(t) = \lim_{n \to \infty} x_n(t).$$

If t is *irrational*, then $x_n(t) = 0$ for all n. Suppose that t is *rational*; that is, suppose $t = p/q$ for which $p, q \in \mathbf{Z}$. In this case $n!t$ is an integer when $n \geq q$ in which case $x(t) = 1$. Consequently, we may conclude that

$$x(t) = \lim_{n \to \infty} \lim_{m \to \infty} [\cos(n!\pi t)]^{2m} = \begin{cases} 0, & t \text{ is irrational} \\ 1, & t \text{ is rational} \end{cases}. \qquad (3.13)$$

We have mentioned (in footnote 3, above) that Riemann integrable functions are a proper subset of the Lebesgue integrable functions. It turns out that $x(t)$ in (3.13) is Lebesgue integrable, but not Riemann integrable. In other words, you cannot use elementary calculus to find the integral of $x(t)$ in (3.13). Of course, $x(t)$ is a very strange function. This is typical; that is, functions that are not Riemann integrable are usually rather strange, and so are not commonly encountered (by the engineer). It therefore follows that we do not need to worry much about the more general class of Lebesgue integrable functions.

Limiting processes are potentially dangerous. This is illustrated by a very simple example.

Example 3.2   Suppose $n, m \in \mathbf{N}$. Define

$$x_{m,n} = \frac{m}{m + n}.$$

(This is a *double sequence*. In Chapter 1 we saw that these arise routinely in wavelet theory.) Treating n as a fixed constant, we obtain

$$\lim_{m \to \infty} x_{m,n} = 1$$

so

$$\lim_{n \to \infty} \lim_{m \to \infty} x_{m,n} = 1.$$

Now instead treat m as a fixed constant so that

$$\lim_{n \to \infty} x_{m,n} = 0$$

which in turn implies that

$$\lim_{m \to \infty} \lim_{n \to \infty} x_{m,n} = 0.$$

Interchanging the order of the limits has given two completely different answers.

Interchanging the order of limits clearly must be done with great care.
  The following example is simply another illustration of how to apply Definition 3.2.

Example 3.3 Define

$$x_n = 1 + \frac{(-1)^n}{n+1} \quad (n \in \mathbf{Z}^+).$$

This is a sequence in the metric space $(\mathbf{R}, |\cdot|)$. This space is complete, so we need not know the limit of the sequence to determine whether it converges (although we might guess that the limit is $x = 1$). We see that

$$d(x_m, x_n) = \left| \frac{(-1)^m}{m+1} - \frac{(-1)^n}{n+1} \right| \le \frac{1}{m+1} + \frac{1}{n+1} \le \frac{1}{m} + \frac{1}{n},$$

where the triangle inequality has been used. If we assume [without loss of generality (commonly abbreviated w.l.o.g.)] that $n \ge m > N(\epsilon)$ then

$$\frac{1}{m} + \frac{1}{n} \le \frac{2}{m} < \epsilon.$$

So, for a given $\epsilon > 0$, we select $n \ge m > 2/\epsilon$. The sequence is Cauchy, and so it must converge.

  We close this section with mention of Appendix 3.A. Think of the material in it as being a very big applications example. This appendix presents an introduction to *coordinate rotation digital computing* (CORDIC). This is an application of a particular class of Cauchy sequence (called a *discrete basis*) to the problem of performing certain elementary operations (e.g., vector rotation, computing sines and cosines). The method is used in application-specific integrated circuits (ASICs), gate arrays, and has been used in pocket calculators. Note that Appendix 3.A also illustrates a useful series expansion which is expressed in terms of the discrete basis.

3.3 POINTWISE CONVERGENCE AND UNIFORM CONVERGENCE

The previous section informed us that sequences can converge in different ways, assuming that they converge in any sense at all. We explore this issue further here.

Definition 3.3: Pointwise Convergence   Suppose that $(x_n(t))$ $(n \in \mathbf{Z}^+)$ is a sequence of functions for which $t \in S \subset \mathbf{R}$. We say that the sequence converges *pointwise* iff there is an $x(t)$ $(t \in S)$ so that for all $\epsilon > 0$ there is an $N = N(\epsilon, t)$ such that

$$|x_n(t) - x(t)| \leq \epsilon \tag{3.14}$$

for $n \geq N$. We call x the limit of (x_n) and write

$$x(t) = \lim_{n \to \infty} x_n(t) \quad (t \in S). \tag{3.15}$$

We emphasize that under this definition N may depend on both ϵ and t. We may contrast Definition 3.3 with the following definition.

Definition 3.4: Uniform Convergence   Suppose that $(x_n(t))$ $(n \in \mathbf{Z}^+)$ is a sequence of functions for which $t \in S \subset \mathbf{R}$. We say that the sequence converges *uniformly* iff there is an $x(t)$ $(t \in S)$ so that for all $\epsilon > 0$ there is an $N = N(\epsilon)$ such that

$$|x_n(t) - x(t)| \leq \epsilon \tag{3.16}$$

for $n \geq N$. We call x the limit of (x_n) and write

$$x(t) = \lim_{n \to \infty} x_n(t) \quad (t \in S). \tag{3.17}$$

We emphasize that under this definition N never depends on t, although it may depend on ϵ. It is apparent that a uniformly convergent sequence is also pointwise convergent. However, the converse is not true; that is, a pointwise convergent sequence is not necessarily uniformly convergent. This distinction is important in understanding the convergence behavior of series as well as of sequences. In particular, it helps in understanding convergence phenomena in Fourier (and wavelet) series expansions.

In contrast with the definitions of Section 3.2, under Definitions 3.3 and 3.4 the elements of (x_n) and the limit x need not reside in the same function space. In fact, we do not ask what function spaces they belong to at all. In other words, the definitions of this section represent a different approach to convergence analysis.

As with the Definition 3.1, direct application of Definitions 3.3 and 3.4 can be quite difficult since the limit, assuming it exists, is not often known in advance (i.e., a priori) in practice. Therefore, we would hope for a convergence criterion similar to the idea of Cauchy convergence in Section 3.2 (Definition 3.2). In fact, we have the following theorem (from Rudin [2, pp. 147–148]).

Theorem 3.1: The sequence of functions (x_n) defined on $S \subset \mathbf{R}$ converges uniformly on S iff for all $\epsilon > 0$ there is an N such that

$$|x_m(t) - x_n(t)| \leq \epsilon \tag{3.18}$$

for all $n, m \geq N$.

This is certainly analogous to the Cauchy criterion seen earlier. (We omit the proof.)

Example 3.4 Suppose that (x_n) is defined according to

$$x_n(t) = \frac{1}{nt + 1}, \quad t \in (0, 1) \text{ and } n \in \mathbf{N}.$$

A sketch of $x_n(t)$ for various n appears in Fig. 3.2. We see that ("by inspection") $x_n \to 0$. But consider for all $\epsilon > 0$

$$|x_n(t) - 0| = \frac{1}{nt + 1} \le \epsilon$$

which implies that we must have

$$n \ge \frac{1}{t}\left(\frac{1}{\epsilon} - 1\right) = N$$

so that N is a function of both t and ϵ. Convergence is therefore pointwise, and is not uniform.

Other criteria for uniform convergence may be established. For example, there is the following theorem (again from Rudin [2, p. 148]).

Theorem 3.2: Suppose that

$$\lim_{n \to \infty} x_n(t) = x(t) \quad (t \in S).$$

Define

$$M_n = \sup_{t \in S} |x_n(t) - x(t)|.$$

Then $x_n \to x$ uniformly on S iff $M_n \to 0$ as $n \to \infty$.

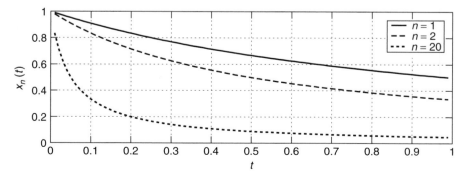

Figure 3.2 A plot of typical sequence elements for Example 3.4; here, $t \in [0.01, 0.99]$.

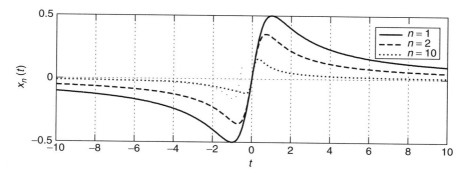

Figure 3.3   A plot of typical sequence elements for Example 3.5.

The proof is really an immediate consequence of Definition 3.4, and so is omitted here.

Example 3.5   Suppose that

$$x_n(t) = \frac{t}{1 + nt^2}, \quad t \in \mathbf{R} \text{ and } n \in \mathbf{N}.$$

A sketch of $x_n(t)$ for various n appears in Fig. 3.3. We note that

$$\frac{dx_n(t)}{dt} = \frac{(1 + nt^2) \cdot 1 - t \cdot (2nt)}{[1 + nt^2]^2} = \frac{1 - nt^2}{[1 + nt^2]^2} = 0$$

for $t = \pm \frac{1}{\sqrt{n}}$. We see that

$$x_n\left(\pm\frac{1}{\sqrt{n}}\right) = \pm\frac{1}{2\sqrt{n}}.$$

We also see that $x_n \to 0$. So then

$$M_n = \sup_{t \in \mathbf{R}} |x_n(t)| = \frac{1}{2\sqrt{n}}.$$

Clearly, $M_n \to 0$ as $n \to \infty$. Therefore, via Theorem 3.2, we immediately conclude that $x_n \to x$ uniformly on the real number line.

3.4   FOURIER SERIES

The Fourier series expansion was introduced briefly in Chapter 1, where the behavior of this series with respect to its convergence properties was not mentioned. In this section we shall demonstrate the pointwise convergence of the Fourier series by the analysis of a particular example. Much of what follows is from Walter [17]. However, of necessity, the present treatment is not so rigorous.

Suppose that

$$g(t) = \frac{1}{\pi}(\pi - t), \quad 0 < t < 2\pi. \tag{3.19}$$

The reader is strongly invited to show that this has Fourier series expansion

$$g(t) = \frac{2}{\pi} \sum_{k=1}^{\infty} \frac{1}{k} \sin(kt). \tag{3.20}$$

The procedure for doing this closely follows Example 1.20. In our analysis to follow, it will be easier to work with

$$f(t) = \frac{\pi}{2} g(t) = \sum_{k=1}^{\infty} \frac{1}{k} \sin(kt). \tag{3.21}$$

Define the *sequence of partial sums*

$$S_n(t) = \sum_{k=1}^{n} \frac{1}{k} \sin(kt) \quad (n \in \mathbf{N}). \tag{3.22}$$

So we infer that

$$\lim_{n \to \infty} S_n(t) = f(t), \tag{3.23}$$

but we do not know *in what sense* the partial sums tend to $f(t)$. Is convergence pointwise, or uniform?

We shall need the special function

$$D_n(t) = \frac{1}{\pi}\left[\frac{1}{2} + \sum_{k=1}^{n} \cos(kt)\right] = \frac{1}{2\pi} \frac{\sin(n + \frac{1}{2})t}{\sin(\frac{1}{2}t)}. \tag{3.24}$$

This function is called the *Dirichlet kernel*. The second equality in (3.24) is not obvious. We will prove it. Consider that

$$\sin\left(\frac{1}{2}t\right)(\pi D_n(t)) = \frac{1}{2}\sin\left(\frac{1}{2}t\right) + \sum_{k=1}^{n} \sin\left(\frac{1}{2}t\right)\cos(kt)$$

$$= \frac{1}{2}\sin\left(\frac{1}{2}t\right) + \frac{1}{2}\sum_{k=1}^{n}\sin\left(k + \frac{1}{2}\right)t + \frac{1}{2}\sum_{k=1}^{n}\sin\left(\frac{1}{2} - k\right)t$$

$$= \frac{1}{2}\sin\left(\frac{1}{2}t\right) + \frac{1}{2}\sum_{k=1}^{n}\sin\left(k + \frac{1}{2}\right)t - \frac{1}{2}\sum_{k=1}^{n}\sin\left(k - \frac{1}{2}\right)t,$$

$$\tag{3.25}$$

where we have used the identity $\sin a \cos b = \frac{1}{2}\sin(a+b) + \frac{1}{2}\sin(a-b)$. By expanding the sums and looking for cancellations

$$\sum_{k=1}^{n} \sin\left(k+\frac{1}{2}\right)t - \sum_{k=1}^{n} \sin\left(k-\frac{1}{2}\right)t = \sin\left(n+\frac{1}{2}\right)t - \sin\left(\frac{1}{2}t\right). \quad (3.26)$$

Applying (3.26) in (3.25), we obtain

$$\sin\left(\frac{1}{2}t\right)(\pi D_n(t)) = \frac{1}{2}\sin\left(n+\frac{1}{2}\right)t$$

so immediately

$$D_n(t) = \frac{1}{2\pi}\frac{\sin(n+\frac{1}{2})t}{\sin(\frac{1}{2}t)},$$

and this establishes (3.24). Using the identity $\sin(a+b) = \sin a \cos b + \cos a \sin b$, we may also write

$$D_n(t) = \frac{1}{2\pi}\left[\frac{\sin(nt)\cos(\frac{1}{2}t)}{\sin(\frac{1}{2}t)} + \cos(nt)\right]. \quad (3.27)$$

For $t > 0$, using the form of the Dirichlet kernel in (3.27), we have

$$\pi \int_0^t D_n(x)\,dx = \int_0^t \left[\frac{\sin(nx)\cos(\frac{1}{2}x)}{2\sin(\frac{1}{2}x)} + \frac{1}{2}\cos(nx)\right]dx$$

$$= \int_0^t \frac{\sin(nx)}{x}\,dx + \int_0^t \sin(nx)\left[\frac{1}{2}\frac{\cos(\frac{1}{2}x)}{\sin(\frac{1}{2}x)} - \frac{1}{x}\right]dx$$

$$+ \frac{1}{2}\int_0^t \cos(nx)\,dx. \quad (3.28)$$

We are interested in what happens when t is a small positive value, but n is large. To begin with, it is not difficult to see that

$$\lim_{n\to\infty} \frac{1}{2}\int_0^t \cos(nx)\,dx = \lim_{n\to\infty}\frac{1}{2n}\sin(nt) = 0. \quad (3.29)$$

Less clearly

$$\lim_{n\to\infty} \int_0^t \sin(nx)\left[\frac{1}{2}\frac{\cos(\frac{1}{2}x)}{\sin(\frac{1}{2}x)} - \frac{1}{x}\right]dx = 0 \quad (3.30)$$

(take this for granted). Through a simple change of variable

$$I(nt) = \int_0^t \frac{\sin(nx)}{x} \, dx = \int_0^{nt} \frac{\sin x}{x} \, dx. \tag{3.31}$$

In fact

$$\int_0^\infty \frac{\sin x}{x} \, dx = \frac{\pi}{2}. \tag{3.32}$$

This is not obvious, either. The result may be found in integral tables [18, p. 483]. In other words, even for very small t, $I(nt)$ does not go to zero as n increases. Consequently, using (3.29), (3.30), and (3.31) in (3.28), we have (for big n)

$$\pi \int_0^t D_n(x) \, dx \approx I(nt). \tag{3.33}$$

The results in the previous paragraph help in the following manner. Begin by noting that

$$S_n(t) = \sum_{k=1}^n \frac{1}{k} \sin(kt) = \sum_{k=1}^n \int_0^t \cos(kx) \, dx = \int_0^t \left[\sum_{k=1}^n \cos(kx) \right] dx$$

$$= \int_0^t \left[\frac{1}{2} + \sum_{k=1}^n \cos(kx) \right] dx - \frac{1}{2} t = \pi \int_0^t D_n(x) \, dx - \frac{1}{2} t \quad \text{(via (3.24))}. \tag{3.34}$$

So from (3.33)

$$S_n(t) \approx I(nt) - \tfrac{1}{2} t. \tag{3.35}$$

Define the sequence $t_n = \frac{1}{n} \pi$. Consequently

$$S_n(t_n) \approx I(\pi) - \frac{1}{2n} \pi. \tag{3.36}$$

As $n \to \infty$, $t_n \to 0$, and $S_n(t_n) \to I(\pi)$. We can say that for big n

$$S_n(0+) \approx I(\pi). \tag{3.37}$$

Now, $f(0+) = \frac{\pi}{2}$, so for big n

$$\frac{S_n(0+)}{f(0+)} \approx \frac{2}{\pi} \int_0^\pi \frac{\sin x}{x} \, dx \approx 1.18. \tag{3.38}$$

Numerical integration is needed to establish this. This topic is the subject of a later chapter, however.

We see that the sequence of partial sums $S_n(0+)$ converges to a value bigger than $f(0+)$ as $n \to \infty$. The approximation $S_n(t)$ therefore tends to "overshoot" the true value of $f(t)$ for small t. This is called the *Gibbs phenomenon*, or *Gibbs overshoot*. We observe that $t = 0$ is the place where $f(t)$ has a discontinuity. This tendency of the Fourier series to overshoot near discontinuities is entirely typical.

We note that $f(\pi) = S_n(\pi) = 0$ for all $n \geq 1$. Thus, for any $\epsilon > 0$

$$|f(\pi) - S_n(\pi)| \leq \epsilon$$

for all $n \geq 1$. The previous analysis for $t = 0+$, and this one for $t = \pi$ show that N (in the definitions of convergence) depends on t. Convergence of the Fourier series is therefore pointwise and not uniform. Generally, the Gibbs phenomenon is a symptom of pointwise convergence.

We remark that the Gibbs phenomenon has an impact in the signal processing applications of series expansions. Techniques for signal compression and signal enhancement are often based on series expansions. The Gibbs phenomenon can degrade the quality of decompressed or reconstructed signals. The phenomenon is responsible for "ringing artifacts." This is one reason why the convergence properties of series expansions are important to engineers.

Figure 3.4 shows a plot of $f(t)$, $S_n(t)$ and the *error*

$$E_n(t) = S_n(t) - f(t). \tag{3.39}$$

The reader may confirm (3.38) directly from the plot in Fig. 3.4a.

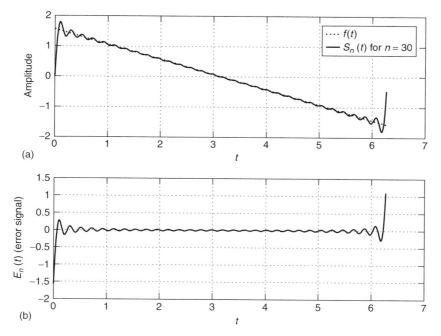

(a)

(b)

Figure 3.4   Plots of the Fourier series expansion for $f(t)$ in (3.21), $S_n(t)$ [of (3.22)] for $n = 30$, and the error $E_n(t) = S_n(t) - f(t)$.

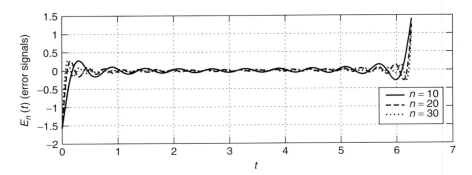

Figure 3.5 $E_n(t)$ of (3.39) for different values of n.

We conclude this section by remarking that

$$\lim_{n \to \infty} \int_0^{2\pi} |E_n(t)|^2 \, dt = 0, \tag{3.40}$$

that is, the energy of the error goes to zero in the limit as n goes to infinity. However, the amplitude of the error in the vicinity of a discontinuity remains unchanged in the limit. This is more clearly seen in Fig. 3.5, where the error is displayed for different values of n. Of course, this fact agrees with our analysis. Equation (3.40) is really a consequence of the fact that (recalling Chapter 1) $S_n(t)$ and $f(t)$ are both in the space $L^2(0, 2\pi)$. Rigorous proof of this is quite tough, and so we omit the proof entirely.

3.5 TAYLOR SERIES

Assume that $f(x)$ is real-valued and that $x \in \mathbf{R}$. One way to define the derivative of $f(x)$ at $x = x_0$ is according to

$$f^{(1)}(x) = \frac{df(x)}{dx}\Big|_{x=x_0} = \lim_{x \to x_0} \frac{f(x) - f(x_0)}{x - x_0}. \tag{3.41}$$

The notation is

$$f^{(n)}(x) = \frac{d^n f(x)}{dx^n} \tag{3.42}$$

(so $f^{(0)}(x) = f(x)$). From (3.41), we obtain

$$f(x) \approx f(x_0) + f^{(1)}(x_0)(x - x_0). \tag{3.43}$$

But how good is this approximation? Can we obtain a more accurate approximation to $f(x)$ if we know $f^{(n)}(x_0)$ for $n > 1$? Again, what is the accuracy of the resulting approximation? We consider these issues in this section.

Begin by recalling the following theorem.

Theorem 3.3: Mean-Value Theorem   If $f(x)$ is continuous for $x \in [a, b]$ with a continuous derivative for $x \in (a, b)$, then there is a number $\xi \in (a, b)$ such that

$$\frac{f(b) - f(a)}{b - a} = f^{(1)}(\xi). \tag{3.44}$$

Therefore, if $a = x_0$ and $b = x$, we must have

$$f(x) = f(x_0) + f^{(1)}(\xi)(x - x_0). \tag{3.45}$$

This expression is "exact," and so is in contrast with (3.43). Proof of Theorem 3.3 may be found in, for example, Bers [19, p. 636]. Theorem 3.3 generalizes to the following theorem.

Theorem 3.4: Generalized Mean-Value Theorem   Suppose that $f(x)$ and $g(x)$ are continuous functions on $x \in [a, b]$. Assume $f^{(1)}(x)$, and $g^{(1)}(x)$ exist and are continuous, and $g^{(1)}(x) \neq 0$ for $x \in (a, b)$. There is a number $\xi \in (a, b)$ such that

$$\frac{f(b) - f(a)}{g(b) - g(a)} = \frac{f^{(1)}(\xi)}{g^{(1)}(\xi)}. \tag{3.46}$$

Once again the proof is omitted, but may be found in Bers [19, p. 637].

The tangent to $f(x)$ at $x = x_0$ is given by $t(x) = f(x_0) + f^{(1)}(x_0)(x - x_0)$ ($t(x_0) = f(x_0)$ and $t^{(1)}(x_0) = f^{(1)}(x_0)$). We wish to consider $t(x)$ to be an approximation to $f(x)$, so the error is

$$f(x) - t(x) = e(x)$$

or

$$f(x) = f(x_0) + f^{(1)}(x_0)(x - x_0) + e(x). \tag{3.47}$$

Thus

$$\frac{e(x)}{x - x_0} = \frac{f(x) - f(x_0)}{x - x_0} - f^{(1)}(x_0).$$

But

$$\lim_{x \to x_0} \frac{f(x) - f(x_0)}{x - x_0} = f^{(1)}(x_0)$$

so immediately

$$\lim_{x \to x_0} \frac{e(x)}{x - x_0} = 0. \tag{3.48}$$

From (3.47) $e(x_0) = 0$, and also from (3.47), we obtain

$$f^{(1)}(x) = f^{(1)}(x_0) + e^{(1)}(x), \tag{3.49}$$

so $e^{(1)}(x_0) = 0$. From (3.49), $f^{(2)}(x) = e^{(2)}(x)$ (so we now assume that $f(x)$ has a second derivative, and we will also assume that it is continuous). Theorem 3.4 has the following corollary.

Corollary 3.1   Suppose that $f(a) = g(a) = 0$, so for all $b \neq a$ there is a $\xi \in (a, b)$ such that

$$\frac{f(b)}{g(b)} = \frac{f^{(1)}(\xi)}{g^{(1)}(\xi)}. \tag{3.50}$$

Now apply this corollary to $f(x) = e(x)$, and $g(x) = (x - x_0)^2$, with $a = x_0$, $b = x$. Thus, from (3.50)

$$\frac{e(x)}{(x - x_0)^2} = \frac{e^{(1)}(\tau)}{2(\tau - x_0)} \tag{3.51}$$

($\tau \in (x_0, x)$). Apply the corollary once more to $f(\tau) = e^{(1)}(\tau)$, and $g(\tau) = 2(\tau - x_0)$:

$$\frac{e^{(1)}(\tau)}{2(\tau - x_0)} = \frac{1}{2}e^{(2)}(\xi) = \frac{1}{2}f^{(2)}(\xi) \tag{3.52}$$

($\xi \in (x_0, \tau)$). Apply (3.51) in (3.52)

$$\frac{e(x)}{(x - x_0)^2} = \frac{1}{2}f^{(2)}(\xi)$$

or (for some $\xi \in (x_0, x)$)

$$e(x) = \tfrac{1}{2}f^{(2)}(\xi)(x - x_0)^2. \tag{3.53}$$

If $|f^{(2)}(t)| \leq M_2$ for $t \in (x_0, x)$, then

$$|e(x)| \leq \frac{1}{2 \cdot 1}M_2|x - x_0|^2 \tag{3.54}$$

and

$$f(x) = f(x_0) + f^{(1)}(x_0)(x - x_0) + e(x), \tag{3.55}$$

for which (3.54) is an upper bound on the size of the error involved in approximating $f(x)$ using (3.43).

Example 3.6   Suppose that $f(x) = \sqrt{x}$; then

$$f^{(1)}(x) = \frac{1}{2\sqrt{x}}, \quad f^{(2)}(x) = -\frac{1}{4}\frac{1}{x\sqrt{x}}.$$

Suppose that $x_0 = 1$, and $x = x_0 + \delta x = 1 + \delta x$. Thus, via (3.55)

$$\sqrt{x} = \sqrt{1 + \delta x} = 1 + f^{(1)}(1)\delta x + e(x) = 1 + \tfrac{1}{2}\delta x + e(x),$$

so if, for example, $|\delta x| < \frac{3}{4}$, then $|f^{(2)}(x)| \le 2$ (for $x \in (\frac{1}{4}, \frac{7}{4})$) so $M_2 = 2$, and so

$$|e(x)| \le (\delta x)^2$$

via (3.54). This bound may be compared to the following table of values:

δx	$\sqrt{1 + \delta x}$	$1 + \frac{1}{2}\delta x$	$e(x)$	$(\delta x)^2$
$-\frac{3}{4}$	0.5000	0.6250	-0.1250	0.5625
$-\frac{1}{2}$	0.7071	0.7500	-0.0429	0.2500
0	1.0000	1.0000	0.0000	0.0000
$\frac{1}{2}$	1.2247	1.2500	-0.0253	0.2500
$\frac{3}{4}$	1.3229	1.3750	-0.0521	0.5625

It is easy to see that indeed $|e(x)| \le (\delta x)^2$.

We mention that Corollary 3.1 leads to *l'Hôpital's rule*. It therefore allows us to determine

$$\lim_{x \to a} \frac{f(x)}{g(x)}$$

when $f(a) = g(a) = 0$. We now digress briefly to consider this subject. The rule applies if $f(x)$ and $g(x)$ are continuous at $x = a$, if $f(x)$ and $g(x)$ have continuous derivatives at $x = a$, and if $g^{(1)}(x) \ne 0$ near $x = a$, except perhaps at $x = a$. l'Hôpital's rule is as follows. If $\lim_{x \to a} f(x) = \lim_{x \to a} g(x) = 0$ and $\lim_{x \to a} \frac{f^{(1)}(x)}{g^{(1)}(x)}$ exists, then

$$\lim_{x \to a} \frac{f(x)}{g(x)} = \lim_{x \to a} \frac{f^{(1)}(x)}{g^{(1)}(x)}. \tag{3.56}$$

The rationale is that from Corollary 3.1 for all $x \ne a$ there is a $\xi \in (a, b)$ such that $\frac{f(x)}{g(x)} = \frac{f^{(1)}(\xi)}{g^{(1)}(\xi)}$. So, if x is close to a then ξ must also be close to a, and $\frac{f^{(1)}(\xi)}{g^{(1)}(\xi)}$ is close to its limit. l'Hôpital's rule is also referred to as "the rule for evaluating the indeterminate form $\frac{0}{0}$." If it happens that $f^{(1)}(a) = g^{(1)}(a) = 0$, then one may attempt l'Hôpital's rule yet again; that is, if $\lim_{x \to a} \frac{f^{(2)}(x)}{g^{(2)}(x)}$ exists, then

$$\lim_{x \to a} \frac{f(x)}{g(x)} = \lim_{x \to a} \frac{f^{(2)}(x)}{g^{(2)}(x)}.$$

Example 3.7   Consider

$$\lim_{x \to 0} \frac{\sin x - e^x + 1}{x^2} = \lim_{x \to 0} \frac{\dfrac{d}{dx}[\sin x - e^x + 1]}{\dfrac{d}{dx}[x^2]}$$

$$= \lim_{x \to 0} \frac{\cos x - e^x}{2x} = \lim_{x \to 0} \frac{-\sin x - e^x}{2} = -\frac{1}{2}$$

for which the rule has been applied twice. Now consider instead

$$\lim_{x \to 0} \frac{1 - 2x}{2 + 4x} = \lim_{x \to 0} \frac{\frac{d}{dx}[1 - 2x]}{\frac{d}{dx}[2 + 4x]}$$

$$= \lim_{x \to 0} \frac{-2}{4} = -\frac{1}{2}.$$

This is *wrong* ! l'Hôpital's rule does not apply here because $f(0) = 1$, and $g(0) = 2$ (i.e., we do not have $f(0) = g(0) = 0$ as needed by the theory).

The rule can be extended to cover other indeterminate forms (e.g., $\frac{\infty}{\infty}$). For example, consider

$$\lim_{x \to 0} x \log_e x = \lim_{x \to 0} \frac{\log_e x}{\frac{1}{x}} = \lim_{x \to 0} \frac{\frac{1}{x}}{-\frac{1}{x^2}}$$

$$= \lim_{x \to 0} (-x) = 0.$$

An interesting case is that of finding

$$\lim_{x \to \infty} \left(1 + \frac{1}{x}\right)^x.$$

This is an indeterminate of the form 1^∞. Consider

$$\lim_{x \to \infty} \log_e \left(1 + \frac{1}{x}\right)^x = \lim_{x \to \infty} \frac{\log_e \left(1 + \frac{1}{x}\right)}{\frac{1}{x}}$$

$$= \lim_{x \to \infty} \frac{\frac{1}{1 + \frac{1}{x}} \left(-\frac{1}{x^2}\right)}{-\frac{1}{x^2}} = \lim_{x \to \infty} \frac{1}{1 + \frac{1}{x}} = 1.$$

The logarithm and exponential functions are continuous functions, so it happens to be the case that

$$1 = \lim_{x \to \infty} \log_e \left(1 + \frac{1}{x}\right)^x = \log_e \left[\lim_{x \to \infty} \left(1 + \frac{1}{x}\right)^x\right],$$

that is, the limit and the logarithm can be interchanged. Thus

$$e^1 = e^{\log_e\left[\lim_{x\to\infty}\left(1+\frac{1}{x}\right)^x\right]}$$

so finally we have

$$\lim_{x\to\infty}\left(1+\frac{1}{x}\right)^x = e. \tag{3.57}$$

More generally, it can be shown that

$$\lim_{n\to\infty}\left(1+\frac{x}{n}\right)^n = e^x. \tag{3.58}$$

This result has various applications, including some in probability theory relating to Poisson and exponential random variables [20]. An alternative derivation of (3.57) appears on pp. 64–65 of Rudin [2], but involves the use of the Maclaurin series expansion for e. We revisit the Maclaurin series for e^x later.

We have demonstrated that for suitable $\xi \in (x_0, x)$

$$f(x) = f(x_0) + f^{(1)}(x_0)(x - x_0) + \tfrac{1}{2}f^{(2)}(\xi)(x - x_0)^2$$

(recall (3.55)). Define

$$p(x) = f(x_0) + f^{(1)}(x_0)(x - x_0) + \tfrac{1}{2}f^{(2)}(x_0)(x - x_0)^2 \tag{3.59}$$

so this is some approximation to $f(x)$ near $x = x_0$. Equation (3.43) is a linear approximation to $f(x)$, and (3.59) is a quadratic approximation to $f(x)$. Once again, we wish to consider the error

$$f(x) - p(x) = e(x). \tag{3.60}$$

We note that

$$p(x_0) = f(x_0), \quad p^{(1)}(x_0) = f^{(1)}(x_0), \quad p^{(2)}(x_0) = f^{(2)}(x_0). \tag{3.61}$$

In other words, the approximation to $f(x)$ in (3.59) matches the function and its first two derivatives at $x = x_0$. Because of (3.61), via (3.60)

$$e(x_0) = e^{(1)}(x_0) = e^{(2)}(x_0) = 0, \tag{3.62}$$

and so via (3.59) and (3.60)

$$e^{(3)}(x) = f^{(3)}(x) \tag{3.63}$$

(because $p^{(3)}(x) = 0$ since $p(x)$ is a quadratic in x). As in the derivation of (3.53), we may repeatedly apply Corollary 3.1:

$$\frac{e(x)}{(x - x_0)^3} = \frac{e^{(1)}(t_1)}{3(t_1 - x_0)^2} \qquad \text{for } t_1 \in (x_0, x)$$

$$\frac{e^{(1)}(t_1)}{3(t_1 - x_0)^2} = \frac{e^{(2)}(t_2)}{3 \cdot 2(t_2 - x_0)} \qquad \text{for } t_2 \in (x_0, t_1)$$

$$\frac{e^{(2)}(t_2)}{3 \cdot 2(t_2 - x_0)} = \frac{e^{(3)}(\xi)}{3 \cdot 2} \qquad \text{for } \xi \in (x_0, t_2),$$

which together yield

$$\frac{e(x)}{(x - x_0)^3} = \frac{f^{(3)}(\xi)}{3 \cdot 2} \quad \text{for } \xi \in (x_0, x)$$

or

$$e(x) = \frac{1}{3 \cdot 2 \cdot 1} f^{(3)}(\xi)(x - x_0)^3 \tag{3.64}$$

for some $\xi \in (x_0, x)$. Thus

$$f(x) = f(x_0) + f^{(1)}(x_0)(x - x_0) + \frac{1}{2 \cdot 1} f^{(2)}(x_0)(x - x_0)^2 + e(x). \tag{3.65}$$

Analogously to (3.54), if $|f^{(3)}(t)| \leq M_3$ for $t \in (x_0, x)$, then we have the error bound

$$|e(x)| \leq \frac{1}{3 \cdot 2 \cdot 1} M_3 |x - x_0|^3. \tag{3.66}$$

We have gone from a linear approximation to $f(x)$ to a quadratic approximation to $f(x)$. All of this suggests that we may generalize to a degree n polynomial approximation to $f(x)$. Therefore, we define

$$p_n(x) = \sum_{k=0}^{n} p_{n,k}(x - x_0)^k, \tag{3.67}$$

where

$$p_{n,k} = \frac{1}{k!} f^{(k)}(x_0). \tag{3.68}$$

Then

$$f(x) = p_n(x) + e_{n+1}(x), \tag{3.69}$$

where the error term is

$$e_{n+1}(x) = \frac{1}{(n + 1)!} f^{(n+1)}(\xi)(x - x_0)^{n+1} \tag{3.70}$$

for suitable $\xi \in (x_0, x)$. We call $p_n(x)$ the *Taylor polynomial of degree n*. This polynomial is the approximation to $f(x)$, and the error $e_{n+1}(x)$ in (3.70) can be formally obtained by the repeated application of Corollary 3.1. These details are omitted. Expanding (3.69), we obtain

$$f(x) = f(x_0) + f^{(1)}(x_0)(x - x_0) + \frac{1}{2!} f^{(2)}(x_0)(x - x_0)^2$$

$$+ \cdots + \frac{1}{n!} f^{(n)}(x_0)(x - x_0)^n + \frac{1}{(n+1)!} f^{(n+1)}(\xi)(x - x_0)^{n+1} \quad (3.71)$$

which is the familiar *Taylor formula* for $f(x)$. We remark that

$$f^{(k)}(x_0) = p_n^{(k)}(x_0) \quad (3.72)$$

for $k = 0, 1, \ldots, n - 1, n$. So we emphasize that the approximation $p_n(x)$ to $f(x)$ is based on forcing $p_n(x)$ to match the first n derivatives of $f(x)$, as well as enforcing $p_n(x_0) = f(x_0)$. If $|f^{(n+1)}(t)| \leq M_{n+1}$ for all $t \in I$ [interval I contains (x_0, x)], then

$$|e_{n+1}(x)| \leq \frac{1}{(n+1)!} M_{n+1} |x - x_0|^{n+1}. \quad (3.73)$$

If all derivatives of $f(x)$ exist and are continuous, then we have the *Taylor series expansion* of $f(x)$, namely, the infinite series

$$f(x) = \sum_{k=0}^{\infty} \frac{1}{k!} f^{(k)}(x_0)(x - x_0)^k. \quad (3.74)$$

The *Maclaurin series expansion* is a special case of (3.74) for $x_0 = 0$:

$$f(x) = \sum_{k=0}^{\infty} \frac{1}{k!} f^{(k)}(0) x^k. \quad (3.75)$$

If we retain only terms $k = 0$ to $k = n$ in the infinite series (3.74) and (3.75), we know that $e_{n+1}(x)$ gives the error in the resulting approximation. This error may be called the *truncation error* (since it arises from truncation of the infinite series to a finite number of terms). Now we consider some examples.

First recall the *binomial theorem*

$$(a + x)^n = \sum_{k=0}^{n} \binom{n}{k} x^k a^{n-k}, \quad (3.76)$$

where

$$\binom{n}{k} = \frac{n!}{k!(n-k)!}. \quad (3.77)$$

In (3.76) we emphasize that $n \in \mathbf{Z}^+$. But we can use Taylor's formula to obtain an expression for $(a + x)^\alpha$ when $\alpha \neq 0$, and α is not necessarily an element of $\mathbf{Z}^+$. Let us consider the special case

$$f(x) = (1 + x)^\alpha$$

for which (if $k \geq 1$)

$$f^{(k)}(x) = \alpha(\alpha - 1)(\alpha - 2) \cdots (\alpha - k + 1)(1 + x)^{\alpha - k}. \tag{3.78}$$

These derivatives are guaranteed to exist, provided $x > -1$. We will assume this restriction always applies. So, in particular

$$f^{(k)}(0) = \alpha(\alpha - 1)(\alpha - 2) \cdots (\alpha - k + 1) \tag{3.79}$$

giving the Maclaurin expansion

$$(1 + x)^\alpha = 1 + \sum_{k=1}^{n} \frac{1}{k!}[\alpha(\alpha - 1) \cdots (\alpha - k + 1)]x^k$$

$$+ \frac{1}{(n + 1)!}[\alpha(\alpha - 1) \cdots (\alpha - n)](1 + \xi)^{\alpha - n - 1}x^{n+1} \tag{3.80}$$

for some $\xi \in (x_0, x)$. We may extend the definition (3.77), that is, define

$$\binom{\alpha}{0} = 1, \quad \binom{\alpha}{k} = \frac{1}{k!}\alpha(\alpha - 1) \cdots (\alpha - k + 1)(k \geq 1) \tag{3.81}$$

so that (3.80) becomes

$$(1 + x)^\alpha = \underbrace{\sum_{k=0}^{n} \binom{\alpha}{k} x^k}_{=p_n(x)} + \underbrace{\binom{\alpha}{n + 1}(1 + \xi)^{\alpha - n - 1}x^{n+1}}_{=e_{n+1}(x)}, \tag{3.82}$$

for $x > -1$.

Example 3.8   We wish to compute $[1.03]^{1/3}$ with $n = 2$ in (3.82), and to estimate the error involved in doing so. We have $x = 0.03$, $\alpha = \frac{1}{3}$, and $\xi \in (0, .03)$. Therefore from (3.82) $[1 + x]^{1/3}$ is approximated by the Taylor polynomial

$$p_2(x) = 1 + \binom{\frac{1}{3}}{1} x + \binom{\frac{1}{3}}{2} x^2 = 1 + \frac{1}{3}x - \frac{1}{9}x^2$$

so

$$[1.03]^{1/3} \approx p_2(0.03) = 1.009900000$$

but $[1.03]^{1/3} = 1.009901634$, so $e_3(x) = 1.634 \times 10^{-6}$. From (3.82)

$$e_3(x) = \begin{pmatrix} \frac{1}{3} \\ 3 \end{pmatrix} (1+\xi)^{\frac{1}{3}-3}x^3 = \frac{5}{3^4}(1+\xi)^{-8/3}x^3,$$

and so $e_3(0.03) = \frac{5}{3^4}\frac{3^3}{10^6}(1+\xi)^{-8/3} = \frac{5}{3}\times 10^{-6}(1+\xi)^{-8/3}$. Since $0 < \xi < 0.03$, we have

$$1.5403 \times 10^{-6} < e_3(.03) < 1.6667 \times 10^{-6}.$$

The actual error is certainly within this range.

If $f(x) = \frac{1}{1+x}$, and if $x_0 = 0$, then

$$\frac{1}{1+x} = \underbrace{\sum_{k=0}^{n}(-1)^k x^k}_{=p_n(x)} + \underbrace{\frac{(-1)^{n+1}x^{n+1}}{1+x}}_{=r(x)}. \tag{3.83}$$

This may be seen by recalling that

$$\sum_{k=0}^{n}\alpha^k = \frac{1-\alpha^{n+1}}{1-\alpha} \quad (\alpha \neq 1). \tag{3.84}$$

So $\sum_{k=0}^{n}(-1)^k x^k = \frac{1-(-1)^{n+1}x^{n+1}}{1+x}$, and thus

$$\sum_{k=0}^{n}(-1)^k x^k + \frac{(-1)^{n+1}x^{n+1}}{1+x} = \frac{1-(-1)^{n+1}x^{n+1}}{1+x} + \frac{(-1)^{n+1}x^{n+1}}{1+x} = \frac{1}{1+x}.$$

This confirms (3.83). We observe that the remainder term $r(x)$ in (3.83) is not given by $e_{n+1}(x)$ in (3.82). We have obtained an exact expression for the remainder using elementary methods.

Now, from (3.83), we have

$$\frac{1}{1+t} = 1 - t + t^2 - t^3 + \cdots + (-1)^{n-1}t^{n-1} + \frac{(-1)^n t^n}{1+t},$$

and we see immediately that

$$\log_e(1+x) = \int_0^x \frac{dt}{1+t} = x - \frac{1}{2}x^2 + \frac{1}{3}x^3$$

$$+ \cdots + \frac{(-1)^{n-1}x^n}{n} + \underbrace{(-1)^n \int_0^x \frac{t^n}{1+t}dt}_{=r(x)}. \tag{3.85}$$

For $x > 0$, and $0 < t < x$ we have $\frac{1}{1+t} < 1$, implying that

$$0 < \int_0^x \frac{t^n}{1+t}dt \le \int_0^x t^n\, dt = \frac{x^{n+1}}{n+1} \quad (x > 0).$$

For $-1 < x < 0$ with $x < t < 0$, we have

$$\frac{1}{1+t} < \frac{1}{1+x} = \frac{1}{1-|x|}$$

so

$$\left| \int_0^x \frac{t^n}{1+t}dt \right| \le \frac{1}{1-|x|}\left| \int_0^x t^n\, dt \right| = \frac{1}{1-|x|}\left| \frac{x^{n+1}}{n+1} \right| = \frac{|x|^{n+1}}{(1-|x|)(n+1)}.$$

Consequently, we may conclude that

$$|r(x)| \le \begin{cases} \dfrac{1}{n+1}x^{n+1}, & x \ge 0 \\[2mm] \dfrac{|x|^{n+1}}{(1-|x|)(n+1)}, & -1 < x \le 0 \end{cases}. \tag{3.86}$$

Equation (3.85) gives us a means to compute logarithms, and (3.86) gives us a bound on the error.

Now consider (3.83) with x replaced by x^2:

$$\frac{1}{1+x^2} = \sum_{k=0}^n (-1)^k x^{2k} + \frac{(-1)^{n+1}x^{2n+2}}{1+x^2}. \tag{3.87}$$

Replacing n with $n-1$, replacing x with t, and expanding, this becomes

$$\frac{1}{1+t^2} = 1 - t^2 + t^4 - \cdots + (-1)^{n-1}t^{2n-2} + \frac{(-1)^n t^{2n}}{1+t^2},$$

where, on integrating, we obtain

$$\tan^{-1} x = \int_0^x \frac{dt}{1+t^2} = x - \frac{1}{3}x^3 + \frac{1}{5}x^5 - \frac{1}{7}x^7$$

$$+ \cdots + \frac{(-1)^{n-1}x^{2n-1}}{2n-1} + \underbrace{(-1)^n \int_0^x \frac{t^{2n}}{1+t^2}dt}_{=r(x)}. \tag{3.88}$$

Because $\frac{1}{1+t^2} \le 1$ for all $t \in \mathbf{R}$, it follows that

$$|r(x)| \le \frac{|x|^{2n+1}}{2n+1}. \tag{3.89}$$

We now have a method of computing π. Since $\frac{\pi}{4} = \tan^{-1}(1)$, we have

$$\frac{\pi}{4} = 1 - \frac{1}{3} + \frac{1}{5} - \frac{1}{7} + \cdots + \frac{(-1)^{n-1}}{2n-1} + r(1) \qquad (3.90)$$

and

$$r(1) \le \frac{1}{2n+1}. \qquad (3.91)$$

Using (3.90) to compute π is not efficient with respect to the number of arithmetic operations needed (i.e., it is not *computationally efficient*). This is because to achieve an accuracy of about $1/n$ requires about $n/2$ terms in the series [which follows from (3.91)]. However, if x is small (i.e., close to zero), then series (3.88) converges relatively quickly. Observe that

$$\tan^{-1}\frac{x+y}{1-xy} = \tan^{-1}x + \tan^{-1}y. \qquad (3.92)$$

Suppose that $x = \frac{1}{2}$, and $y = \frac{1}{3}$, then $\frac{x+y}{1-xy} = 1$, so

$$\frac{\pi}{4} = \tan^{-1}\left(\frac{1}{2}\right) + \tan^{-1}\left(\frac{1}{3}\right). \qquad (3.93)$$

It is actually faster to compute $\tan^{-1}(\frac{1}{2})$, and $\tan^{-1}(\frac{1}{3})$ using (3.88), and for these obtain π using (3.93) than to compute $\tan^{-1}(1)$ directly. In fact, this approach (a type of "divide and conquer" method) can be taken further by noting that

$$\tan^{-1}\left(\frac{1}{2}\right) = \tan^{-1}\left(\frac{1}{3}\right) + \tan^{-1}\left(\frac{1}{7}\right), \tan^{-1}\left(\frac{1}{3}\right) = \tan^{-1}\left(\frac{1}{5}\right) + \tan^{-1}\left(\frac{1}{8}\right)$$

implying that

$$\frac{\pi}{4} = 2\tan^{-1}\left(\frac{1}{5}\right) + \tan^{-1}\left(\frac{1}{7}\right) + 2\tan^{-1}\left(\frac{1}{8}\right). \qquad (3.94)$$

Now consider $f(x) = e^x$. Since $f^{(k)}(x) = e^x$ for all $k \in \mathbf{Z}^+$ we have for $x_0 = 0$ the Maclaurin series expansion

$$e^x = \sum_{k=0}^{\infty} \frac{x^k}{k!}. \qquad (3.95)$$

This is theoretically valid for $-\infty < x < \infty$. We have employed this series before in various ways. We now consider it as a computational tool for calculating e^x. Appendix 3.C is based on a famous example in Forsythe et al. [21, pp. 14–16]. This example shows that series expansions must be implemented on computers with rather great care. Specifically, Appendix 3.C shows what can happen when we compute e^{-20} by the direct implementation of the series (3.95). Using MATLAB as stated $e^{-20} \approx 4.1736 \times 10^{-9}$, which is based on keeping terms $k = 0$ to 88 (inclusive) of (3.95). Using additional terms will have no effect on the final answer as they are too small. However, the correct value is actually $e^{-20} = 2.0612 \times 10^{-9}$, as may be verified using the MATLAB exponential function, or using a typical pocket calculator. Our series approximation has resulted in an answer possessing no significant digits at all. What went wrong? Many of the terms in the series are orders of magnitude bigger than the final result and typically possess rounding errors about as big as the final answer. The phenomenon is called *catastrophic cancellation* (or *catastrophic convergence*). As Forsythe et al. [21] stated, "It is important to realize that this great cancellation is not the cause of error in the answer; it merely magnifies the error already present in the terms." Catastrophic cancellation can in principle be eliminated by carrying more significant digits in the computation. However, this is costly with respect to computing resources. In the present problem a cheap and very simple solution is to compute e^{20} using (3.95), and then take the reciprocal, i.e., use $e^{-20} = 1/e^{20}$.

An important special function is the *gamma function*:

$$\Gamma(z) = \int_0^\infty x^{z-1} e^{-x} \, dx. \tag{3.96}$$

Here, we assume $z \in \mathbf{R}$. This is an *improper integral* so we are left to wonder if

$$\lim_{M \to \infty} \int_0^M x^{z-1} e^{-x} \, dx$$

exists. It turns out that the integral (3.96) converges for $z > 0$, but diverges for $z \leq 0$. The proof is slightly tedious, and so we will omit it [22, pp. 273–274]. If $z = n \in \mathbf{N}$, then consider

$$\Gamma(n) = \int_0^\infty x^{n-1} e^{-x} \, dx. \tag{3.97}$$

Now

$$\Gamma(n+1) = \int_0^\infty x^n e^{-x} \, dx = \lim_{M \to \infty} \int_0^M x^n e^{-x} \, dx$$

$$= \lim_{M \to \infty} \left[-x^n e^{-x} \big|_0^M + n \int_0^M x^{n-1} e^{-x} \, dx \right]$$

(via $\int u \, dv = uv - \int v \, du$, i.e., *integration by parts*). Therefore

$$\Gamma(n+1) = n \int_0^\infty x^{n-1} e^{-x} \, dx = n\Gamma(n). \tag{3.98}$$

We see that $\Gamma(1) = \int_0^\infty e^{-x}\,dx = [-e^{-x}]_0^\infty = 1$. Thus, $\Gamma(n+1) = n!$. Using the gamma function in combination with (3.95), we may obtain *Stirling's formula*

$$n! \approx \sqrt{2\pi}\, n^{n+1/2} e^{-n} \tag{3.99}$$

which is a good approximation to $n!$ if n is big. The details of a rigorous derivation of this are tedious, so we give only an outline presentation. Begin by noting that

$$n! = \int_0^\infty x^n e^{-x}\,dx = \int_0^\infty e^{n\ln x - x}\,dx.$$

Let $x = n + y$, so

$$n! = e^{-n}\int_{-n}^\infty e^{n\ln(n+y)-y}\,dy.$$

Now, since $n\ln(n+y) = n\ln\left[n\left(1+\frac{y}{n}\right)\right] = n\ln n + n\ln\left(1+\frac{y}{n}\right)$, we have

$$n! = e^{-n}\int_{-n}^\infty e^{n\ln n + n\ln(1+\frac{y}{n})-y}\,dy$$

$$= e^{-n}n^n\int_{-n}^\infty e^{n\ln(1+\frac{y}{n})-y}\,dy.$$

Using (3.85), that is

$$\ln\left(1+\frac{y}{n}\right) = \frac{y}{n} - \frac{y^2}{2n^2} + \frac{y^3}{3n^3} - \cdots,$$

we have

$$n\ln\left(1+\frac{y}{n}\right) - y = -\frac{y^2}{2n} + \frac{y^3}{3n^2} - \cdots,$$

so

$$n! = n^n e^{-n}\int_{-n}^\infty e^{-\frac{y^2}{2n}+\frac{y^3}{3n^2}-\cdots}\,dy.$$

If now $y = \sqrt{n}v$, then $dy = \sqrt{n}\,dv$, and so

$$n! = n^{n+\frac{1}{2}}e^{-n}\int_{-\sqrt{n}}^\infty e^{-\frac{v^2}{2}+\frac{v^3}{3\sqrt{n}}-\cdots}\,dv.$$

So if n is big, then

$$n! \approx n^{n+\frac{1}{2}}e^{-n}\int_{-\infty}^\infty e^{-\frac{v^2}{2}}\,dv.$$

If we accept that

$$\int_{-\infty}^{\infty} e^{-x^2/2} \, dx = \sqrt{2\pi}, \tag{3.100}$$

then immediately we have

$$n! \approx \sqrt{2\pi} \, n^{n+\frac{1}{2}} e^{-n},$$

and the formula is now established. Stirling's formula is very useful in statistical mechanics (e.g., deriving the Fermi–Dirac distribution of fermion particle energies, and this in turn is important in understanding the operation of solid-state electronic devices at a physical level).

Another important special function is

$$g(x) = \frac{1}{\sqrt{2\pi\sigma^2}} \exp\left[-\frac{(x-m)^2}{2\sigma^2}\right], \quad -\infty < x < \infty \tag{3.101}$$

which is the *Gaussian function* (or *Gaussian pulse*). This function is of immense importance in probability theory [20], and is also involved in the uncertainty principle in signal processing and quantum mechanics [23]. A sketch of $g(x)$ for $m = 0$, with $\sigma^2 = 1$, and $\sigma^2 = 0.1$ appears in Fig. 3.6. For $m = 0$ and $\sigma^2 = 1$, the standard form pulse

$$f(x) = \frac{1}{\sqrt{2\pi}} e^{-x^2/2} \tag{3.102}$$

is sometimes defined [20]. In this case we observe that $g(x) = \frac{1}{\sigma} f\left(\frac{x-m}{\sigma}\right)$. We will show that

$$\int_{0}^{\infty} e^{-x^2} \, dx = \frac{1}{2}\sqrt{\pi}, \tag{3.103}$$

which can be used to obtain (3.100) by a simple change of variable. From [22] (p. 262) we have

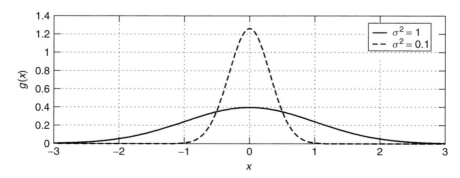

Figure 3.6 Plots of two Gaussian pulses, where $g(x)$ in (3.101) for $m = 0$, with $\sigma^2 = 1, 0.1$.

Theorem 3.5: Let $\lim_{x \to \infty} x^p f(x) = A$. Then

1. $\int_a^\infty f(x)\, dx$ converges if $p > 1$ and $-\infty < A < \infty$.
2. $\int_a^\infty f(x)\, dx$ diverges if $p \leq 1$ and $A \neq 0$ (A may be infinite).

We see that $\lim_{x \to \infty} x^2 e^{-x^2} = 0$ (perhaps via l'Hôpital's rule). So in Theorem 3.5 $f(x) = e^{-x^2}$, and $p = 2$, with $A = 0$, and so $\int_0^\infty e^{-x^2}\, dx$ converges. Define

$$I_M = \int_0^M e^{-x^2}\, dx = \int_0^M e^{-y^2}\, dy$$

and let $\lim_{M \to \infty} I_M = I$. Then

$$I_M^2 = \left(\int_0^M e^{-x^2}\, dx \right) \left(\int_0^M e^{-y^2}\, dy \right) = \int_0^M \int_0^M e^{-(x^2+y^2)}\, dx\, dy$$

$$= \int_{R_M} \int e^{-(x^2+y^2)}\, dx\, dy$$

for which R_M is the square $OABC$ in Fig. 3.7. This square has sides of length M. Since $e^{-(x^2+y^2)} > 0$, we obtain

$$\int_{R_L} \int e^{-(x^2+y^2)}\, dx\, dy \leq I_M^2 \leq \int_{R_U} \int e^{-(x^2+y^2)}\, dx\, dy, \tag{3.104}$$

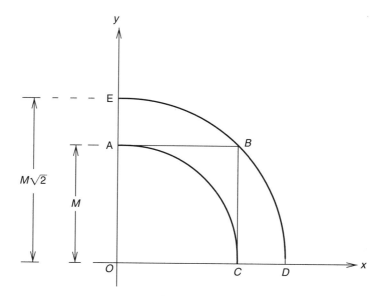

Figure 3.7   Regions used to establish (3.103).

where R_L is the region in the first quadrant bounded by a circle of radius M. Similarly, R_U is the region in the first quadrant bounded by a circle of radius $\sqrt{2}M$. Using polar coordinates, $r^2 = x^2 + y^2$ and $dx\,dy = r\,dr\,d\phi$, so (3.104) becomes

$$\int_{\phi=0}^{\pi/2} \int_{r=0}^{M} e^{-r^2} r\,dr\,d\phi \leq I_M^2 \leq \int_{\phi=0}^{\pi/2} \int_{r=0}^{\sqrt{2}M} e^{-r^2} r\,dr\,d\phi. \tag{3.105}$$

Since $-\frac{1}{2}\frac{d}{dx}e^{-x^2} = xe^{-x^2}$ we have $\int_0^M re^{-r^2}\,dr = -\frac{1}{2}[e^{-x^2}]_0^M = \frac{1}{2}[1 - e^{-M^2}]$. Thus, (3.105) reduces to

$$\frac{\pi}{4}[1 - e^{-M^2}] \leq I_M^2 \leq \frac{\pi}{4}[1 - e^{-2M^2}]. \tag{3.106}$$

If we now allow $M \to \infty$ in (3.106), then $I_M^2 \to \frac{\pi}{4}$, implying that $I^2 = \frac{\pi}{4}$, or $I = \frac{1}{2}\sqrt{\pi}$. This confirms (3.103).

In probability theory it is quite important to be able to compute functions such as the *error function*

$$\text{erf}(x) = \frac{2}{\sqrt{\pi}} \int_0^x e^{-t^2}\,dt. \tag{3.107}$$

This has wide application in digital communications system analysis, for example. No *closed-form*[4] expression for (3.107) exists. We may therefore try to compute (3.107) using series expansions. In particular, we may try working with the Maclaurin series expansion for e^x:

$$\text{erf}(x) = \frac{2}{\sqrt{\pi}} \int_0^x \left[\sum_{k=0}^{\infty} \frac{x^k}{k!}\right]_{x=-t^2} dt$$

$$= \frac{2}{\sqrt{\pi}} \sum_{k=0}^{\infty} \frac{(-1)^k}{k!} \int_0^x t^{2k}\,dt$$

$$= \frac{2}{\sqrt{\pi}} \sum_{k=0}^{\infty} \frac{(-1)^k x^{2k+1}}{k!(2k+1)}. \tag{3.108}$$

However, to arrive at this expression, we had to integrate an infinite series term by term. It is not obvious that we can do this. When is this justified?

A *power series* is any series of the form

$$f(x) = \sum_{k=0}^{\infty} a_k(x - x_0)^k. \tag{3.109}$$

Clearly, Taylor and Maclaurin series are all examples of power series. We have the following theorem.

---

[4]A closed-form expression is simply a "nice" formula typically involving more familiar functions such as sines, cosines, tangents, polynomials, and exponential functions.

Theorem 3.6: Given the power series (3.109), there is an $R \geq 0$ (which may be $R = +\infty$) such that the series is absolutely convergent for $|x - x_0| < R$, and is divergent for $|x - x_0| > R$. At $x - x_0 = R$ and at $x - x_0 = -R$, the series might converge or diverge.

Series (3.109) is *absolutely convergent* if the series

$$h(x) = \sum_{k=0}^{\infty} |a_k (x - x_0)^k| \qquad (3.110)$$

converges. We remark that absolutely convergent series are convergent. This means that if (3.110) converges, then (3.109) also converges. (However, the converse is not necessarily true.) We also have the following theorem.

Theorem 3.7: If

$$f(x) = \sum_{k=0}^{\infty} a_k (x - x_0)^k \quad \text{for} \quad |x - x_0| < R,$$

where $R > 0$ is the *radius of convergence* of the power series, then $f(x)$ is continuous and differentiable in the interval of convergence $x \in (x_0 - R, x_0 + R)$, and

$$f^{(1)}(x) = \sum_{k=1}^{\infty} k a_k (x - x_0)^{k-1}, \qquad (3.111a)$$

$$\int_{x_0}^{x} f(t)\, dt = \sum_{k=0}^{\infty} \frac{a_k}{k+1} (x - x_0)^{k+1}. \qquad (3.111b)$$

This series [Eq. (3.111a,b)] also has a radius of convergence R.

As a consequence of Theorem 3.7, Eq. (3.108) is valid for $-\infty < x < \infty$ (i.e., the radius of convergence is $R = +\infty$). This is because the Maclaurin expansion for e^x had $R = +\infty$.

Example 3.9   Here we will find an expression for the error involved in truncating the series for erf(x) in (3.108).

From (3.71) for some $\xi \in [0, x]$ (interval endpoints may be included because of continuity of the function being approximated)

$$e^x = \underbrace{\sum_{k=0}^{n} \frac{x^k}{k!}}_{= p_n(x)} + e_n(x),$$

where

$$e_n(x) = \frac{1}{(n+1)!} e^\xi x^{n+1}.$$

Thus, where $x = -t^2$, so for some ξ such that $-t^2 \leq \xi \leq 0$

$$e^{-t^2} = p_n(-t^2) + e_n(-t^2),$$

and hence

$$\mathrm{erf}(x) = \underbrace{\frac{2}{\sqrt{\pi}} \int_0^x p_n(-t^2)\, dt}_{=q_n(x)} + \underbrace{\frac{2}{\sqrt{\pi}} \int_0^x e_n(-t^2)\, dt}_{=\epsilon_n(x)},$$

where the degree n polynomial

$$q_n(x) = \frac{2}{\sqrt{\pi}} \int_0^x \left[\sum_{k=0}^n \frac{(-1)^k t^{2k}}{k!} \right] dt = \frac{2}{\sqrt{\pi}} \sum_{k=0}^n \frac{(-1)^k x^{2k+1}}{k!(2k+1)}$$

is the approximation, and we are interested in the error

$$\epsilon_n(x) = \mathrm{erf}(x) - q_n(x) = \frac{2}{\sqrt{\pi}} \int_0^x e_n(-t^2)\, dt.$$

Clearly

$$\epsilon_n(x) = \frac{2}{\sqrt{\pi}} \frac{(-1)^{n+1}}{(n+1)!} \int_0^x t^{2n+2} e^\xi\, dt,$$

where we recall that ξ depends on t in that $-t^2 \leq \xi \leq 0$. There is an *integral mean-value theorem*, which states that for $f(t), g(t) \in C[a, b]$ (and $g(t)$ does not change sign on the interval $[a, b]$) there is a $\zeta \in [a, b]$ such that

$$\int_a^b g(t) f(t)\, dt = f(\zeta) \int_a^b g(t)\, dt.$$

Thus, there is a $\zeta \in [-x^2, 0]$, giving

$$\epsilon_n(x) = \frac{2}{\sqrt{\pi}} e^\zeta \frac{(-1)^{n+1}}{(n+1)!} \frac{x^{2n+3}}{2n+3}.$$

Naturally the error expression in Example 3.9 can be used to estimate how many terms one must keep in the series expansion (3.108) in order to compute $\mathrm{erf}(x)$ to a desired accuracy.

3.6   ASYMPTOTIC SERIES

The Taylor series expansions of Section 3.5 might have a large radius of convergence, but practically speaking, if x is sufficiently far from x_0, then many many terms may be needed in a computer implementation to converge to the correct solution with adequate accuracy. This is highly inefficient. Also, if many terms are to be retained, then rounding errors might accumulate and destroy the result. In other words, Taylor series approximations are really effective only for x sufficiently close to x_0 (i.e., "small x"). We therefore seek expansion methods that give good approximations for large values of the argument x. These are called the *asymptotic expansions*, or *asymptotic series*. This section is just a quick introduction based mainly on Section 19.15 in Kreyszig [24]. Another source of information on asymptotic expansions, although applied mainly to problems involving differential equations, appears in Lakin and Sanchez [25].

Asymptotic expansions may take on different forms. That is, there are different "varieties" of such expansions. (This is apparent in Ref. 25.) However, we will focus on the following definition.

Definition 3.5:   A series of the form

$$\sum_{k=0}^{\infty} \frac{c_k}{x^k} \tag{3.112}$$

for which $c_k \in \mathbf{R}$ (real-valued constants), and $x \in \mathbf{R}$ is called an *asymptotic expansion*, or *asymptotic series*, of a function $f(x)$, which is defined for all sufficiently large x if, for every $n \in \mathbf{Z}^+$

$$\left[f(x) - \left(\sum_{k=0}^{n} \frac{c_k}{x^k} \right) \right] x^n \to 0 \quad \text{as} \quad x \to \infty, \tag{3.113}$$

and we shall then write

$$f(x) \sim \sum_{k=0}^{\infty} \frac{c_k}{x^k}.$$

It is to be emphasized that the series (3.112) need not converge for any x. The condition (3.113) suggests a possible method of finding sequence (c_k). Specifically

$$f(x) - c_0 \to 0 \quad \text{or} \quad c_0 = \lim_{x \to \infty} f(x),$$

$$\left[f(x) - c_0 - \frac{c_1}{x} \right] x \to 0 \quad \text{or} \quad c_1 = \lim_{x \to \infty} [f(x) - c_0] x,$$

$$\left[f(x) - c_0 - \frac{c_1}{x} - \frac{c_2}{x^2} \right] x^2 \to 0 \quad \text{or} \quad c_2 = \lim_{x \to \infty} \left[f(x) - c_0 - \frac{c_1}{x} \right] x^2,$$

or in general

$$c_n = \lim_{x \to \infty} \left[f(x) - \sum_{k=0}^{n-1} \frac{c_k}{x^k} \right] x^n \tag{3.114}$$

for $n \geq 1$. However, this recursive procedure is seldom practical for generating more than the first few series coefficients. Of course, in some cases this might be all that is needed. We remark that Definition 3.5 can be usefully extended according to

$$f(x) \sim g(x) + h(x) \left[\sum_{k=0}^{\infty} \frac{c_k}{x^k} \right] \tag{3.115}$$

for which

$$\frac{f(x) - g(x)}{h(x)} \sim \sum_{k=0}^{\infty} \frac{c_k}{x^k}. \tag{3.116}$$

The single most generally useful method for getting (c_k) is probably to use "integration by parts." This is illustrated with examples.

Example 3.10   Recall erf(x) from (3.107). We would like to evaluate this function for large x [whereas the series in (3.108) is better suited for small x; see the error expression in Example 3.9]. In this regard it is preferable to work with the *complementary error function*

$$\text{erfc}(x) = 1 - \text{erf}(x) = \frac{2}{\sqrt{\pi}} \int_x^{\infty} e^{-t^2} \, dt. \tag{3.117}$$

We observe that erf$(\infty) = 1$ [via (3.103)]. Now let $\tau = t^2$, so that $dt = \frac{1}{2}\tau^{-1/2}d\tau$. With this change of variable

$$\text{erfc}(x) = \frac{1}{\sqrt{\pi}} \int_{x^2}^{\infty} \tau^{-\frac{1}{2}} e^{-\tau} \, d\tau. \tag{3.118}$$

Now observe that via integration by parts, we have

$$\int_{x^2}^{\infty} \tau^{-\frac{1}{2}} e^{-\tau} \, d\tau = -\tau^{-1/2} e^{-\tau} \Big|_{x^2}^{\infty} - \frac{1}{2} \int_{x^2}^{\infty} \tau^{-\frac{3}{2}} e^{-\tau} \, d\tau$$

$$= \frac{1}{x} e^{-x^2} - \frac{1}{2} \int_{x^2}^{\infty} \tau^{-\frac{3}{2}} e^{-\tau} \, d\tau,$$

$$\int_{x^2}^{\infty} \tau^{-\frac{3}{2}} e^{-\tau} \, d\tau = -\tau^{-3/2} e^{-\tau} \Big|_{x^2}^{\infty} - \frac{3}{2} \int_{x^2}^{\infty} \tau^{-\frac{5}{2}} e^{-\tau} \, d\tau$$

$$= \frac{1}{x^3} e^{-x^2} - \frac{3}{2} \int_{x^2}^{\infty} \tau^{-\frac{5}{2}} e^{-\tau} \, d\tau,$$

and so on. We observe that this process of successive integration by parts has generated integrals of the form

$$F_n(x) = \int_{x^2}^{\infty} \tau^{-(2n+1)/2} e^{-\tau} \, d\tau \tag{3.119}$$

for $n \in \mathbf{Z}^+$, and we see that $\mathrm{erfc}(x) = \frac{1}{\sqrt{\pi}} F_0(x)$. So, if we apply integration by parts to $F_n(x)$, then

$$F_n(x) = \int_{x^2}^{\infty} \tau^{-(2n+1)/2} e^{-\tau} \, d\tau$$

$$= -\tau^{-(2n+1)/2} e^{-\tau} \big|_{x^2}^{\infty} - \frac{2n+1}{2} \int_{x^2}^{\infty} \tau^{-(2n+3)/2} e^{-\tau} \, d\tau$$

$$= x^{-(2n+1)} e^{-x^2} - \frac{2n+1}{2} \int_{x^2}^{\infty} \tau^{-(2n+3)/2} e^{-\tau} \, d\tau$$

so that we have the recursive expression

$$F_n(x) = \frac{1}{x^{2n+1}} e^{-x^2} - \frac{2n+1}{2} F_{n+1}(x) \tag{3.120}$$

which holds for $n \in \mathbf{Z}^+$. This may be rewritten as

$$e^{x^2} F_n(x) = \frac{1}{x^{2n+1}} - \frac{2n+1}{2} e^{x^2} F_{n+1}(x). \tag{3.121}$$

Repeated application of (3.121) yields

$$e^{x^2} F_0(x) = \frac{1}{x} - \frac{1}{2} e^{x^2} F_1(x),$$

$$e^{x^2} F_0(x) = \frac{1}{x} - \frac{1}{2x^3} + \frac{1 \cdot 3}{2^2} e^{x^2} F_2(x),$$

$$e^{x^2} F_0(x) = \frac{1}{x} - \frac{1}{2x^3} + \frac{1 \cdot 3}{2^2 x^5} - \frac{1 \cdot 3 \cdot 5}{2^8} e^{x^2} F_3(x),$$

and so finally

$$e^{x^2} F_0(x) = \underbrace{\left[\frac{1}{x} - \frac{1}{2x^3} + \frac{1 \cdot 3}{2^2 x^5} - \cdots + (-1)^{n-1} \frac{1 \cdot 3 \cdots (2n-3)}{2^{n-1} x^{2n-1}} \right]}_{=S_{2n-1}(x)}$$

$$+ (-1)^n \frac{1 \cdot 3 \cdots (2n-1)}{2^n} e^{x^2} F_n(x). \tag{3.122}$$

From this it appears that our asymptotic expansion is

$$e^{x^2} F_0(x) \sim \frac{1}{x} - \frac{1}{2x^3} + \frac{1 \cdot 3}{2^2 x^5} - \cdots + (-1)^{n-1} \frac{1 \cdot 3 \cdots (2n-3)}{2^{n-1} x^{2n-1}} + \cdots \quad (3.123)$$

However, this requires confirmation. Define $K_n = (-2)^{-n}[1 \cdot 3 \cdots (2n-1)]$. From (3.122), we have

$$[e^{x^2} F_0(x) - S_{2n-1}(x)]x^{2n-1} = K_n e^{x^2} x^{2n-1} F_n(x). \quad (3.124)$$

We wish to show that for any fixed $n = 1, 2, 3, \ldots$ the expression in (3.124) on the right of the equality goes to zero as $x \to \infty$. In (3.119) we have

$$\frac{1}{\tau^{(2n+1)/2}} \leq \frac{1}{x^{2n+1}}$$

for all $\tau \geq x^2$, which gives the bound

$$F_n(x) = \int_{x^2}^{\infty} \frac{e^{-\tau}}{\tau^{(2n+1)/2}} d\tau \leq \frac{1}{x^{2n+1}} \int_{x^2}^{\infty} e^{-\tau} d\tau = \frac{e^{-x^2}}{x^{2n+1}}. \quad (3.125)$$

But this implies that

$$|K_n|e^{x^2} x^{2n-1} F_n(x) \leq |K_n|e^{x^2} x^{2n-1} \frac{e^{-x^2}}{x^{2n+1}} = \frac{|K_n|}{x^2}$$

and $\frac{|K_n|}{x^2} \to 0$ for $x \to \infty$. Thus, immediately, (3.123) is indeed the asymptotic expansion for $e^{x^2} F_0(x)$. Hence

$$\text{erfc}(x) \sim \frac{1}{\sqrt{\pi}} e^{-x^2} \left[\frac{1}{x} - \frac{1}{2x^3} + \frac{1 \cdot 3}{2^2 x^5} - \cdots + (-1)^{n-1} \frac{1 \cdot 3 \cdots (2n-3)}{2^{n-1} x^{2n-1}} + \cdots \right].$$
$$(3.126)$$

We recall from Section 3.4 that the integral $\int_0^x \frac{\sin t}{t} dt$ was important in analyzing the Gibbs phenomenon in Fourier series expansions. We now consider asymptotic approximations to this integral.

Example 3.11   The *sine integral* is

$$\text{Si}(x) = \int_0^x \frac{\sin t}{t} dt, \quad (3.127)$$

and the *complementary sine integral* is

$$\text{si}(x) = \int_x^{\infty} \frac{\sin t}{t} dt. \quad (3.128)$$

It turns out that si$(0) = \frac{\pi}{2}$, which is shown on p. 277 of Spiegel [22]. We wish to find an asymptotic series for Si(x). Since

$$\frac{\pi}{2} = \int_0^\infty \frac{\sin t}{t}\,dt = \int_0^x \frac{\sin t}{t}\,dt + \int_x^\infty \frac{\sin t}{t}\,dt = \text{Si}(x) + \text{si}(x), \qquad (3.129)$$

we will consider the expansion of si(x). If we integrate by parts in succession, then

$$\int_x^\infty t^{-1} \sin t\,dt = \frac{1}{x}\cos x - 1 \cdot \int_x^\infty \frac{1}{t^2}\cos t\,dt,$$

$$\int_x^\infty t^{-2} \cos t\,dt = -\frac{1}{x^2}\sin x + 2 \cdot \int_x^\infty \frac{1}{t^3}\sin t\,dt,$$

$$\int_x^\infty t^{-3} \sin t\,dt = \frac{1}{x^3}\cos x - 3 \cdot \int_x^\infty \frac{1}{t^4}\cos t\,dt,$$

$$\int_x^\infty t^{-4} \cos t\,dt = -\frac{1}{x^4}\sin x + 4 \cdot \int_x^\infty \frac{1}{t^5}\sin t\,dt,$$

and so on. If $n \in \mathbf{N}$, then we may define

$$s_n(x) = \int_x^\infty t^{-n} \sin t\,dt, \quad c_n(x) = \int_x^\infty t^{-n} \cos t\,dt. \qquad (3.130)$$

Therefore, for odd n

$$s_n(x) = \frac{1}{x^n}\cos x - n\, c_{n+1}(x), \qquad (3.131\text{a})$$

and for even n

$$c_n(x) = -\frac{1}{x^n}\sin x + n\, s_{n+1}(x). \qquad (3.131\text{b})$$

We observe that $s_1(x) = \text{si}(x)$. Repeated application of the recursions (3.131a,b) results in

$$s_1(x) = \frac{1 \cdot 1}{x}\cos x + \frac{1 \cdot 1}{x^2}\sin x - 1 \cdot 2 s_3(x)$$

$$= \frac{1 \cdot 1}{x}\cos x + \frac{1 \cdot 1}{x^2}\sin x - \frac{1 \cdot 2}{x^3}\cos x - \frac{1 \cdot 2 \cdot 3}{x^4}\sin x + 1 \cdot 2 \cdot 3 \cdot 4 s_5(x),$$

or in general

$$s_1(x) = \cos x \left[\frac{1 \cdot 1}{x} - \frac{1 \cdot 2}{x^3} + \cdots + (-1)^{n+1}\frac{(2n-2)!}{x^{2n-1}} \right]$$

$$+ \sin x \left[\frac{1 \cdot 1}{x^2} - \frac{1 \cdot 2 \cdot 3}{x^4} + \cdots + (-1)^{n+1}\frac{(2n-1)!}{x^{2n}} \right]$$

$$+ (-1)^n (2n)! s_{2n+1}(x) \qquad (3.132)$$

for $n \in \mathbf{N}$ (with $0! = 1$). From this, we obtain

$$[s_1(x) - S_{2n}(x)]x^{2n} = (-1)^n (2n)! x^{2n} s_{2n+1}(x) \tag{3.133}$$

for which $S_{2n}(x)$ is appropriately defined as those terms in (3.132) involving $\sin x$ and $\cos x$. It is unclear whether

$$\lim_{x \to \infty} x^{2n} s_{2n+1}(x) = 0 \tag{3.134}$$

for any $n \in \mathbf{N}$. But if we accept (3.134), then

$$\mathrm{si}(x) \sim S_{2n}(x) = \cos x \left[\frac{1 \cdot 1}{x} - \frac{1 \cdot 2}{x^3} + \cdots + (-1)^{n+1} \frac{(2n-2)!}{x^{2n-1}} \right]$$
$$+ \sin x \left[\frac{1 \cdot 1}{x^2} - \frac{1 \cdot 2 \cdot 3}{x^4} + \cdots + (-1)^{n+1} \frac{(2n-1)!}{x^{2n}} \right]. \tag{3.135}$$

Figure 3.8 shows plots of $\mathrm{Si}(x)$ and the asymptotic approximation $\frac{\pi}{2} - S_{2n}(x)$ for $n = 1, 2, 3$. We observe that the approximations are good for "large x," but poor for "small x," as we would expect. Moreover, for small x, the approximations are

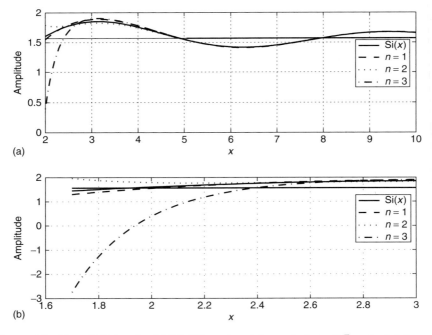

Figure 3.8 Plot of $\mathrm{Si}(x)$ using MATLAB routine sinint, and plots of $\frac{\pi}{2} - S_{2n}(x)$ for $n = 1, 2, 3$. The horizontal solid line is at height $\pi/2$.

better for smaller n. This is also reasonable. Our plots therefore constitute informal verification of the correctness of (3.135), in spite of potential doubts about (3.134).

Another approach to finding (c_k) is based on the fact that many special functions are solutions to particular differential equations. If the originating differential equation is known, it may be used to generate sequence (c_k). However, this section was intended to be relatively brief, and so this method is omitted. The interested reader may see Kreyszig [24].

3.7   MORE ON THE DIRICHLET KERNEL

In Section 3.4 the Dirichlet kernel was introduced in order to analyze the manner in which Fourier series converge in the vicinity of a discontinuity of a 2π-periodic function. However, this was only done with respect to the special case of $g(t)$ in (3.19). In this section we consider the Dirichlet kernel $D_n(t)$ in a more general manner.

We begin by recalling the complex Fourier series expansion from Chapter 1 (Section 1.3.3). If $f(t) \in L^2(0, 2\pi)$, then

$$f(t) = \sum_{n=-\infty}^{\infty} f_n e^{jnt}, \tag{3.136}$$

where $f_n = \langle f, e_n \rangle$, with $e_n(t) = e^{jnt}$, and

$$\langle x, y \rangle = \frac{1}{2\pi} \int_0^{2\pi} x(t) y^*(t) \, dt. \tag{3.137}$$

An *approximation* to $f(t)$ is the truncated Fourier series expansion

$$f_L(t) = \sum_{n=-L}^{L} f_n e^{jnt} \in L^2(0, 2\pi). \tag{3.138}$$

The approximation error is

$$\epsilon_L(t) = f(t) - f_L(t) \in L^2(0, 2\pi). \tag{3.139}$$

We seek a general expression for $\epsilon_L(t)$ that is hopefully more informative than (3.139). The overall goal is to generalize the error analysis approach seen in Section 3.4. Therefore, consider

$$\epsilon_L(t) = f(t) - \sum_{n=-L}^{L} \langle f, e_n \rangle e_n(t)$$

[via $f_n = \langle f, e_n \rangle$, and (3.138) into (3.139)]. Thus

$$
\epsilon_L(t) = f(t) - \sum_{n=-L}^{L} \left\{ \frac{1}{2\pi} \int_0^{2\pi} f(x) e^{-jnx} \, dx \right\} e_n(t)
$$

$$
= f(t) - \frac{1}{2\pi} \int_0^{2\pi} f(x) \left\{ \sum_{n=-L}^{L} e^{jn(t-x)} \right\} dx \tag{3.140}
$$

Since

$$
\sum_{n=-L}^{L} e^{jn(t-x)} = 1 + 2 \sum_{n=1}^{L} \cos[n(t-x)] \tag{3.141}
$$

(show this as an exercise), via (3.24), we obtain

$$
1 + 2 \sum_{n=1}^{L} \cos[n(t-x)] = \frac{\sin[(L + \frac{1}{2})(t-x)]}{\sin[\frac{1}{2}(t-x)]} = 2\pi D_L(t-x). \tag{3.142}
$$

Immediately, we see that

$$
\epsilon_L(t) = f(t) - \int_0^{2\pi} f(x) D_L(t-x) \, dx, \tag{3.143}
$$

where also (recall (3.139))

$$
f_L(t) = \int_0^{2\pi} f(x) D_L(t-x) \, dx. \tag{3.144}
$$

Equation (3.144) is an alternative *integral form* of the approximation to $f(t)$ originally specified in (3.138). The integral in (3.144) is really an example of something called a *convolution integral*. The following example will demonstrate how we might apply (3.143).

Example 3.12 Suppose that

$$
f(t) = \begin{cases} \sin t, & 0 < t < \pi \\ 0, & \pi \le t < 2\pi \end{cases}.
$$

Note that $f(t)$ is continuous for all t, but that $f^{(1)}(t) = df(t)/dt$ is not continuous everywhere. For example, $f^{(1)}(t)$ is not continuous at $t = \pi$. Plots of $f_L(t)$ for various L (see Fig. 3.10) suggest that $f_L(t)$ converges most slowly to $f(t)$ near $t = \pi$. Can we say something about the rate of convergence?

Therefore, consider

$$f_L(\pi) = \frac{1}{2\pi} \int_0^\pi \sin x \frac{\sin[(L + \frac{1}{2})(\pi - x)]}{\sin[\frac{1}{2}(\pi - x)]} \, dx. \tag{3.145}$$

Now

$$\sin\left[\frac{1}{2}(\pi - x)\right] = \sin\left(\frac{\pi}{2}\right) \cos\left(\frac{1}{2}x\right) - \cos\left(\frac{\pi}{2}\right) \sin\left(\frac{1}{2}x\right) = \cos\left(\frac{1}{2}x\right),$$

and since $\sin x = 2\sin(\frac{1}{2}x)\cos(\frac{1}{2}x)$ so (3.145) reduces to

$$\begin{aligned}
f_L(\pi) &= \frac{1}{\pi} \int_0^\pi \sin\left(\frac{1}{2}x\right) \sin\left[\left(L + \frac{1}{2}\right)(\pi - x)\right] dx \\
&= \frac{1}{2\pi} \int_0^\pi \left\{\cos\left[(L+1)x - \left(L + \frac{1}{2}\right)\pi\right] - \cos\left[Lx - \left(L + \frac{1}{2}\right)\pi\right]\right\} dx \\
&= \frac{1}{2\pi(L+1)} \left[\sin\left[(L+1)x - \left(L + \frac{1}{2}\right)\pi\right]\right]_0^\pi \\
&\quad - \frac{1}{2\pi L} \left[\sin\left[Lx - \left(L + \frac{1}{2}\right)\pi\right]\right]_0^\pi \\
&= \frac{1}{2\pi(L+1)}\left[1 + \sin\left(L + \frac{1}{2}\right)\pi\right] - \frac{1}{2\pi L}\left[-1 + \sin\left(L + \frac{1}{2}\right)\pi\right] \\
&= \frac{1}{2\pi} \frac{2L + 1 - (-1)^L}{L(L+1)}. \tag{3.146}
\end{aligned}$$

We see that, as expected

$$\lim_{L \to \infty} f_L(\pi) = 0.$$

Also, (3.146) gives $\epsilon_L(\pi) = -f_L(\pi)$, and this is the exact value for the approximation error at $t = \pi$ for all L. Furthermore

$$|\epsilon_L(\pi)| \propto \frac{1}{L}$$

for large L, and so we have a measure of the *rate of convergence* of the error, at least at the point $t = \pi$. (Symbol "$\propto$" means "proportional to.")

We remark that $f(t)$ in Example 3.12 may be regarded as the voltage drop across the resistor R in Fig. 3.9. The circuit in Fig. 3.9 is a simple half-wave

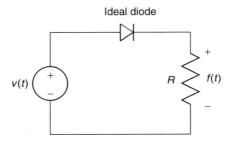

Figure 3.9 An electronic circuit interpretation for Example 3.12; here, $v(t) = \sin(t)$ for all $t \in \mathbf{R}$.

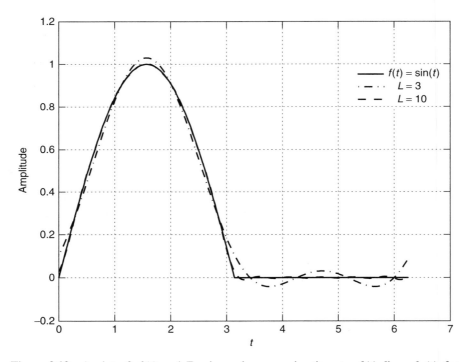

Figure 3.10 A plot of $f(t)$ and Fourier series approximations to $f(t)$ [i.e., $f_L(t)$ for $L = 3, 10$].

rectifier circuit. The reader ought to verify as an exercise that

$$f_L(t) = \frac{1}{\pi} + \frac{1}{2}\sin t + \sum_{n=2}^{L} \frac{1 + (-1)^n}{\pi(1 - n^2)}\cos nt. \tag{3.147}$$

Plots of $f_L(t)$ for $L = 3, 10$ versus the plot of $f(t)$ appear in Fig. 3.10.

3.8 FINAL REMARKS

We have seen that sequences and series might converge "mathematically" yet not "numerically." Essentially, we have seen three categories of difficulty:

1. Pointwise convergence of series leading to irreducible errors in certain regions of the approximation, such as the Gibbs phenomenon in Fourier expansions, which arises in the vicinity of discontinuities in the function being approximated

2. The destructive effect of rounding errors as illustrated by the catastrophic convergence of series

3. Slow convergence such as illustrated by the problem of computing π with the Maclaurin expansion for $\tan^{-1} x$

We have seen that some ingenuity may be needed to overcome obstacles such as these. For example, a divide-and-conquer approach helped in the problem of computing π. In the case of catastrophic convergence, the problem was solved by changing the computational algorithm. The problem of overcoming Gibbs overshoot is not considered here. However, it involves seeking uniformly convergent series approximations.

APPENDIX 3.A *CO*ORDINATE *R*OTATATION *DI*GITAL *C*OMPUTING (CORDIC)

3.A.1 Introduction

This appendix presents the basics of a method for computing "elementary functions," which includes the problem of rotating vectors in the plane, computing $\tan^{-1} x$, $\sin \theta$, and $\cos \theta$. The method to be used is called "*co*ordinate *r*otation *di*gital *c*omputing" (CORDIC), and was invented by Jack Volder [3] in the late 1950s. However, in spite of the age of the method, it is still important. The method is one of those great ideas that is able to survive despite technological changes. It is a good example of how a clever mathematical idea, if anything, becomes less obsolete with the passage of time.

The CORDIC method was, and is, desirable because it reduces the problem of computing apparently complicated functions, such as trig functions, to a succession of simple operations. Specifically, these simple operations are shifting and adding. In the 1950s it was a major achievement just to build systems that could add two numbers together because all that was available for use was vacuum tubes, and to a lesser degree, discrete transistors. However, even with the enormous improvements in computing technology that have occurred since then, it is still important to reduce complicated operations to simple ones. Thus, the CORDIC method has survived very well. For example, in 1980 [5] a special-purpose CORDIC VLSI (very large scale integration) chip was presented. More recent references to the method will be given later. Nowadays, the CORDIC method is more likely to be implemented with gate-array technology.

Since the CORDIC method involves the operations of shifting and adding only, once the mathematics of the method is understood, it is easy to build CORDIC computing hardware using the logic design methods considered in typical elementary digital logic courses or books.[5] Consideration of CORDIC computing also makes a connection between elementary mathematics courses (calculus and linear algebra) and computer hardware systems design, as well as the subject of numerical analysis.

3.A.2   The Concept of a Discrete Basis

The original paper of Volder [3] does not give a rigorous treatment of the mathematics of the CORDIC method. However, in this section (based closely on Schelin [6]), we will begin the process of deriving the CORDIC method in a mathematically rigorous manner. The central idea is to represent operands (e.g, the θ in $\sin\theta$) in terms of a *discrete basis*. When the discrete basis representation is combined with appropriate mathematical identities the CORDIC algorithm results.

Let $\mathbf{R}$ denote the set of real numbers. Everything we do here revolves around the following theorem (From Schelin [6]).

Theorem 3.A.1: Suppose that $\theta_k \in \mathbf{R}$ for $k \in \{0, 1, 2, 3, \ldots\}$ satisfy $\theta_0 \geq \theta_1 \geq \theta_2 \geq \cdots \geq \theta_n > 0$, and that

$$\theta_k \leq \sum_{j=k+1}^{n} \theta_j + \theta_n \quad \text{for} \quad 0 \leq k \leq n, \tag{3.A.1a}$$

and suppose $\theta \in \mathbf{R}$ satisfies

$$|\theta| \leq \sum_{j=0}^{n} \theta_j. \tag{3.A.1b}$$

If $\theta^{(0)} = 0$, and $\theta^{(k+1)} = \theta^{(k)} + \delta_k \theta_k$ for $0 \leq k \leq n$, where

$$\delta_k = \begin{cases} 1, & \text{if } \theta \geq \theta^{(k)} \\ -1, & \text{if } \theta < \theta^{(k)} \end{cases} \tag{3.A.1c}$$

then

$$|\theta - \theta^{(k)}| \leq \sum_{j=k}^{n} \theta_j + \theta_n \quad \text{for} \quad 0 \leq k \leq n, \tag{3.A.1d}$$

and so in particular $|\theta - \theta^{(n+1)}| \leq \theta_n$.

---

[5]The method is also easy to implement in an assembly language (or other low-level) programming environment.

Proof We may use proof by mathematical induction[6] on the index k. For $k = 0$

$$|\theta - \theta^{(0)}| = |\theta| \leq \sum_{j=0}^{n} \theta_j < \sum_{j=0}^{n} \theta_j + \theta_n$$

via (3.A.1b).

Assume that $|\theta - \theta^{(k)}| \leq \sum_{j=k}^{n} \theta_j + \theta_n$ is true, and consider $|\theta - \theta^{(k+1)}|$. Via (3.A.1c), δ_k and $\theta - \theta^{(k)}$ have the same sign, and so

$$|\theta - \theta^{(k+1)}| = |\theta - \theta^{(k)} - \delta_k \theta_k| = ||\theta - \theta^{(k)}| - \theta_k|.$$

Now, via the inductive hypothesis (i.e., $|\theta - \theta^{(k)}| \leq \sum_{j=k}^{n} \theta_j + \theta_n$), we have

$$|\theta - \theta^{(k)}| - \theta_k \leq \sum_{j=k}^{n} \theta_j + \theta_n - \theta_k = \sum_{j=k+1}^{n} \theta_j + \theta_n. \qquad (3.A.2a)$$

Via (3.A.1a), we obtain

$$-\left[\sum_{j=k+1}^{n} \theta_j + \theta_n\right] \leq -\theta_k$$

so that

$$-\left[\sum_{j=k+1}^{n} \theta_j + \theta_n\right] \leq |\theta - \theta^{(k)}| - \theta_k. \qquad (3.A.2b)$$

Combining (3.A.2a) with (3.A.2b) gives

$$|\theta - \theta^{(k+1)}| = ||\theta - \theta^{(k)}| - \theta_k| \leq \sum_{j=k+1}^{n} \theta_j + \theta_n$$

so that (3.A.1d) holds for k replaced by $k + 1$, and so (3.A.1d) holds via induction.

We will call this result *Schelin's theorem*. The set $\{\theta_k\}$ is called a *discrete basis* if it satisfies the restrictions given in the theorem.

In what follows we will interpret θ as an angle. However, note that θ in this theorem could be more general than this. Now suppose that we define

$$\theta_k = \tan^{-1} 2^{-k} \qquad (3.A.3)$$

for $k = 0, 1, \ldots, n$. Table 3.A.1 shows typical values for θ_k as given by (3.A.3). We see that good approximations to θ can be obtained from relatively small n (because from Schelin's theorem $|\theta - \theta^{(n+1)}| \leq \theta_n$).

[6]A brief introduction to proof by mathematical induction appears in Appendix 3.B.

**TABLE 3.A.1   Values
for Some Elements of
the Discrete Basis given
by Eq. (3.A.3)**

k	$\theta_k = \tan^{-1} 2^{-k}$
0	$45°$
1	$26.565°$
2	$14.036°$
3	$7.1250°$
4	$3.5763°$

Clearly, these satisfy $\theta_0 \geq \theta_1 \geq \cdots \geq \theta_n > 0$, which is one of the restrictions in Schelin's theorem.

The mean-value theorem of elementary calculus says that there exists a $\xi \in [x_0, x_0 + \Delta]$ such that

$$\left. \frac{df(x)}{dx} \right|_{x=\xi} = \frac{f(x_0 + \Delta) - f(x_0)}{\Delta}$$

so if $f(x) = \tan^{-1} x$, then $df(x)/dx = 1/(1 + x^2)$, and thus

$$\left. \frac{1}{1+x^2} \right|_{x \in [2^{-(k+1)}, 2^{-k}]} = \frac{\theta_k - \theta_{k+1}}{2^{-k} - 2^{-(k+1)}}$$

$$\Rightarrow \frac{\theta_k - \theta_{k+1}}{2^{-k} - 2^{-(k+1)}} \leq \left. \frac{1}{1+x^2} \right|_{x=2^{-(k+1)}}$$

because the slope of $\tan^{-1} x$ is largest at $x = 2^{-(k+1)}$, and this in turn implies

$$\theta_k - \theta_{k+1} \leq \frac{2^{-k} - 2^{-(k+1)}}{1 + 2^{-2(k+1)}} = \frac{2^{k+2} - 2^{k+1}}{1 + 2^{2(k+1)}} = \frac{2^{k+1}}{1 + 2^{2(k+1)}}$$

and

$$\theta_k \geq \frac{2^k}{1 + 2^{2k}}$$

because $\tan^{-1} x \geq x \frac{d}{dx} \tan^{-1} x = \frac{x}{1+x^2} (x \geq 0)$, for $k = 0, 1, \ldots, n$ (let $x = 2^{-k}$).[7]

[7] For $x \geq 0$

$$\frac{1}{1+x^2} \geq \frac{d}{dx} \frac{x}{1+x^2} = \frac{1-x^2}{(1+x^2)^2}$$

(certainly $1 \geq \frac{1-x^2}{1+x^2}$ for $x \geq 0$)

$$\Rightarrow \int_0^x \frac{1}{1+t^2} dt \geq \frac{x}{1+x^2} \Rightarrow \tan^{-1} x \geq \frac{x}{1+x^2}$$

Now, as a consequence of these results

$$\theta_k - \theta_n = (\theta_k - \theta_{k+1}) + (\theta_{k+1} - \theta_{k+2}) + \cdots + (\theta_{n-2} - \theta_{n-1}) + (\theta_{n-1} - \theta_n)$$

$$= \sum_{j=k}^{n-1} (\theta_j - \theta_{j+1})$$

$$\leq \sum_{j=k}^{n-1} \frac{2^{j+1}}{1 + 2^{2(j+1)}} \left(\theta_j - \theta_{j+1} \leq \frac{2^{j+1}}{1 + 2^{2(j+1)}} \right)$$

$$= \sum_{j=k+1}^{n} \frac{2^j}{1 + 2^{2j}}$$

$$\leq \sum_{j=k+1}^{n} \theta_j \left(\theta_j \geq \frac{2^j}{1 + 2^{2j}} \right)$$

implying that

$$\theta_k \leq \sum_{j=k+1}^{n} \theta_j + \theta_n$$

for $k = 0, 1, \ldots, n$, and thus (3.A.1a) holds for $\{\theta_k\}$ in (3.A.3), and so (3.A.3) is a concrete example of a discrete basis. If you have a pocket calculator handy then it is easy to verify that

$$\sum_{j=0}^{3} \theta_j = 92.73° > 90°,$$

so we will work only with angles θ that satisfy $|\theta| \leq 90°$. Thus, for such θ, there exists a sequence $\{\delta_k\}$ such that $\delta_k \in \{-1, +1\}$, where

$$\theta = \sum_{k=0}^{n} \delta_k \theta_k + \epsilon_{n+1} = \theta^{(n+1)} + \epsilon_{n+1}, \qquad (3.A.4)$$

where $|\epsilon_{n+1}| \leq \theta_n = \tan^{-1} 2^{-n}$. Equation (3.A.1c) gives us a way to find $\{\delta_k\}$. We can also write that any angle θ satisfying $|\theta| \leq \frac{\pi}{2}$ (radians) can be represented exactly as

$$\theta = \sum_{k=0}^{\infty} \delta_k \theta_k$$

for appropriately chosen *coordinates* $\{\delta_k\}$.

In the next section we will begin to see how useful it is to be able to represent angles in terms of the discrete basis given by (3.A.3).

3.A.3 Rotating Vectors in the Plane

No one would disagree that a basic computational problem in electrical and computer engineering is to find

$$z' = e^{j\theta}z, \quad \theta \in \mathbf{R}, \tag{3.A.5}$$

where $j = \sqrt{-1}$, $z = x + jy$, and $z' = x' + jy'$ $(x, y, x', y' \in \mathbf{R})$. Certainly, this problem arises in the phasor analysis of circuits, in computer graphics (to rotate objects, for example), and it also arises in digital signal processing (DSP) a lot.[8]

Thus, z and z' are complex variables. Expressing them in terms of their real and imaginary parts, we may rewrite (3.A.5) as

$$x' + jy' = (\cos\theta + j\sin\theta)(x + jy) = [x\cos\theta - y\sin\theta] + j[x\sin\theta + y\cos\theta],$$

and this can be further rewritten as the matrix–vector product

$$\begin{bmatrix} x' \\ y' \end{bmatrix} = \begin{bmatrix} \cos\theta & -\sin\theta \\ \sin\theta & \cos\theta \end{bmatrix} \begin{bmatrix} x \\ y \end{bmatrix} \tag{3.A.6}$$

which we recognize as the formula for rotating the vector $[xy]^T$ in the plane to $[x'y']^T$.

Recall the trigonometric identities

$$\sin\theta = \frac{\tan\theta}{\sqrt{1 + \tan^2\theta}}, \qquad \cos\theta = \frac{1}{\sqrt{1 + \tan^2\theta}}, \tag{3.A.7}$$

and so for $\theta = \theta_k$ in (3.A.3)

$$\sin\theta_k = \frac{2^{-k}}{\sqrt{1 + 2^{-2k}}}, \qquad \cos\theta_k = \frac{1}{\sqrt{1 + 2^{-2k}}}. \tag{3.A.8}$$

Thus, to rotate $x_k + jy_k$ by angle $\delta_k\theta_k$ to $x_{k+1} + jy_{k+1}$ is accomplished via

$$\begin{bmatrix} x_{k+1} \\ y_{k+1} \end{bmatrix} = \frac{1}{\sqrt{1 + 2^{-2k}}} \begin{bmatrix} 1 & -\delta_k 2^{-k} \\ \delta_k 2^{-k} & 1 \end{bmatrix} \begin{bmatrix} x_k \\ y_k \end{bmatrix}, \tag{3.A.9}$$

---

[8]In DSP it is often necessary to compute the discrete Fourier transform (DFT), which is an approximation to the Fourier transform (FT) and is defined by

$$X_n = \sum_{k=0}^{N-1} \exp\left(-j\frac{2\pi kn}{N}x_k\right),$$

where $\{x_k\}$ is the samples of some analog signal, i.e., $x_k = x(kT)$, where k is an integer and T is a positive constant; $x(t)$ for $t \in \mathbf{R}$ is the analog signal. Note that additional information about the DFT appeared in Section 1.4.

where we've used (3.A.6). From Schelin's theorem, $\theta^{(n+1)} \approx \theta$, so if we wish to rotate $x_0 + jy_0 = x + jy$ by $\theta^{(n+1)}$ to $x_{n+1} + jy_{n+1}$ ($\approx x' + jy'$) then via (3.A.9)

$$\begin{bmatrix} x_{n+1} \\ y_{n+1} \end{bmatrix} = \left(\prod_{k=0}^{n} \frac{1}{\sqrt{1 + 2^{-2k}}} \right) \begin{bmatrix} 1 & -\delta_n 2^{-n} \\ \delta_n 2^{-n} & 1 \end{bmatrix}$$

$$\cdots \begin{bmatrix} 1 & -\delta_1 2^{-1} \\ \delta_1 2^{-1} & 1 \end{bmatrix} \begin{bmatrix} 1 & -\delta_0 \\ \delta_0 & 1 \end{bmatrix} \begin{bmatrix} x_0 \\ y_0 \end{bmatrix}. \tag{3.A.10}$$

Define

$$K_n = \prod_{k=0}^{n} \frac{1}{\sqrt{1 + 2^{-2k}}}, \tag{3.A.11}$$

where $\prod_{k=0}^{n} \alpha_k = \alpha_n \alpha_{n-1} \alpha_{n-2} \cdots \alpha_1 \alpha_0$.

Consider

$$\begin{bmatrix} \hat{x}_{n+1} \\ \hat{y}_{n+1} \end{bmatrix} = \begin{bmatrix} 1 & -\delta_n 2^{-n} \\ \delta_n 2^{-n} & 1 \end{bmatrix} \cdots \begin{bmatrix} 1 & -\delta_1 2^{-1} \\ \delta_1 2^{-1} & 1 \end{bmatrix} \begin{bmatrix} 1 & -\delta_0 \\ \delta_0 & 1 \end{bmatrix} \begin{bmatrix} x_0 \\ y_0 \end{bmatrix},$$

$$\tag{3.A.12}$$

which is the same expression as (3.A.10) except that we have dropped the multiplication by K_n. We observe that to implement (3.A.12) requires only the simple (to implement in digital hardware, or assembly language) operations of shifting and adding. Implementing (3.A.12) rotates $x + jy$ by approximately the desired amount, but gives a solution vector that is a factor of $1/K_n$ longer than it should be. Of course, this can be corrected by multiplying $[\hat{x}_{n+1} \quad \hat{y}_{n+1}]^T$ by K_n if desired. Note that in some applications, this would not be necessary.

Note that

$$\begin{bmatrix} \cos\theta & -\sin\theta \\ \sin\theta & \cos\theta \end{bmatrix} \approx K_n \begin{bmatrix} 1 & -\delta_n 2^{-n} \\ \delta_n 2^{-n} & 1 \end{bmatrix}$$

$$\cdots \begin{bmatrix} 1 & -\delta_1 2^{-1} \\ \delta_1 2^{-1} & 1 \end{bmatrix} \begin{bmatrix} 1 & -\delta_0 \\ \delta_0 & 1 \end{bmatrix}, \tag{3.A.13}$$

and so the matrix product in (3.A.12) [or (3.A.10)] represents an efficient approximate factorization of the rotation operator in (3.A.6). The approximation gets better and better as n increases, and in the limit as $n \to \infty$ becomes exact.

The computational complexity of the CORDIC rotation algorithm may be described as follows. In Eq. (3.A.10) there are exactly $2n$ shifts (i.e., multiplications by 2^{-k}), and $2n + 2$ additions, plus two scalings by factor K_n. As well, only $n + 1$ bits of storage are needed to save the sequence $\{\delta_k\}$.

We conclude this section with a numerical example to show how to obtain the sequence $\{\delta_k\}$ via (3.A.1c).

Example 3.A.1   Suppose that we want to rotate a vector by an angle of $\theta = 20°$, and we decide that $n = 4$ gives sufficient accuracy for the application at hand. Via (3.A.1c) and Table 3.A.1, we have

$$\theta^{(0)} = 0°$$

$$\theta^{(1)} = \theta_0 = 45° \text{ as } \delta_0 = +1 \text{ since } \theta \geq \theta^{(0)} = 0°$$

$$\theta^{(2)} = \theta^{(1)} + \delta_1 \theta_1$$

$$= 45° - 26.565° \text{ as } \delta_1 = -1 \text{ since } \theta < \theta^{(1)}$$

$$= 18.435°$$

$$\theta^{(3)} = \theta^{(2)} + \delta_2 \theta_2$$

$$= 18.435° + 14.036° \text{ as } \delta_2 = +1 \text{ since } \theta \geq \theta^{(2)}$$

$$= 32.471°$$

$$\theta^{(4)} = \theta^{(3)} + \delta_3 \theta_3$$

$$= 32.471° - 7.1250° \text{ as } \delta_3 = -1 \text{ since } \theta < \theta^{(3)}$$

$$= 25.346°$$

$$\theta^{(5)} = \theta^{(4)} + \delta_4 \theta_4$$

$$= 25.346° - 3.5763° \text{ as } \delta_4 = -1 \text{ since } \theta < \theta^{(4)}$$

$$= 21.770° \approx \theta$$

Via (3.A.4) and $|\theta - \theta^{(n+1)}| \leq \theta_n$

$$|\epsilon_5| = 1.770° < \theta_4 = 3.5763°$$

so the error bound in Schelin's theorem is actually somewhat conservative, at least in this special case.

3.A.4   Computing Arctangents

The results in Section 3.A.3 can be modified to obtain a CORDIC algorithm for computing an approximation to $\theta = \tan^{-1}(y/x)$. The idea is to find the sequence $\{\delta_k | k = 0, 1, \ldots, n - 1, n\}$ to rotate the vector $[\ x \quad y\ ]^T = [\ x_0 \quad y_0\ ]^T$ to the vector $[x_n \quad y_n]^T$, where $y_n \approx 0$. More specifically, we would select δ_k so that $|y_{k+1}| < |y_k|$.

Let $\hat{\theta}$ denote the approximation to θ. The desired algorithm to compute $\hat{\theta}$ may be expressed as Pascal-like pseudocode:

```
θ̂ := 0;
x₀ := x; y₀ := y;
```

```
for k := 0 to n do begin
   if yk ≥ 0 then begin
      δk := −1;
      end
   else begin
      δk := +1;
      end ;
   θ̂ := −δkθk + θ̂;
   xk+1 := xk − δk2⁻ᵏyk;
   yk+1 := δk2⁻ᵏxk + yk;
   end ;
```

In this pseudocode

$$\begin{bmatrix} x_{k+1} \\ y_{k+1} \end{bmatrix} = \begin{bmatrix} 1 & -\delta_k 2^{-k} \\ \delta_k 2^{-k} & 1 \end{bmatrix} \begin{bmatrix} x_k \\ y_k \end{bmatrix},$$

and we see that for the manner in which sequence $\{\delta_k\}$ is constructed by the pseudocode, the inequality $|y_{k+1}| < |y_k|$ is satisfied. We choose n to achieve the desired accuracy of our estimate $\hat{\theta}$ of θ, specifically, $|\theta - \hat{\theta}| \leq \theta_n$.

3.A.5   Final Remarks

As an exercise, the reader should modify the previous results to determine a CORDIC method for computing $\cos\theta$, and $\sin\theta$. [*Hint*: Take a good look at (3.A.13).]

The CORDIC philosophy can be extended to the computation of hyperbolic trigonometric functions, logarithms[9] and other functions [4, 7]. It can also perform multiplication and division (see Table on p. 324 of Schelin [6]). As shown by Hu and Naganathan [9], the rate of convergence of the CORDIC method can be accelerated by a method similar to the Booth algorithm (see pp. 287–289 of Hamacher et al. [10]) for multiplication. However, this is at the expense of somewhat more complicated hardware structures. A roundoff error analysis of the CORDIC method has been performed by Hu [8]. We do not present these results in this book as they are quite involved. Hu claims to have fairly tight bounds on the errors, however. Fixed-point and floating-point schemes are both analyzed. A tutorial presentation of CORDIC-based VLSI architectures for digital signal processing applications appears in Hu [11]. Other papers on the CORDIC method are those by Timmermann et al. [12] and Lee and Lang [13] (which appeared in the *IEEE Transactions on Computers*, "Special Issue on Computer Arithmetic" of August 1992). An alternative summary of the CORDIC method may be found in Hwang [14]. Many of the ideas in Hu's paper [11] are applicable in a gate-array technology environment. Applications include the computation of discrete transforms (e.g., the DFT), digital filtering, adaptive filtering, Kalman filtering, the solution of special linear

---

[9]A clever alternative to the CORDIC approach for log calculations appears in Lo and Chen [15], and a method of computing square roots without division appears in Mikami et al. [16].

systems of equations (e.g., Toeplitz), deconvolution, and eigenvalue and singular value decompositions.

APPENDIX 3.B   MATHEMATICAL INDUCTION

The basic idea of mathematical induction is as follows. Assume that we are given a sequence of statements

$$S_0, S_1, \ldots, S_n, \ldots$$

and each S_i is true or it is false. To prove that all of the statements are true (i.e., to prove that S_n is true for all n) by induction: (1) prove that S_n is true for $n = 0$, and then (2) assume that S_n is true for any $n = k$ and then show that S_n is true for $n = k + 1$.

Example 3.B.1   Prove

$$S_n = \sum_{i=0}^{n} i^2 = \frac{n(n+1)(2n+1)}{6}, \quad n \geq 0.$$

Proof   We will use induction, but note that there are other methods (e.g., via z transforms). For $n = 0$, we obtain

$$S_0 = \sum_{i=0}^{0} i^2 = 0 \quad \text{and} \quad \frac{n(n+1)(2n+1)}{6}\bigg|_{n=0} = 0.$$

Thus, S_n is certainly true for $n = 0$.
Assume now that S_n is true for $n = k$ so that

$$S_k = \sum_{i=0}^{k} i^2 = \frac{k(k+1)(2k+1)}{6}. \tag{3.B.1}$$

We have

$$S_{k+1} = \sum_{i=0}^{k+1} i^2 = \sum_{i=0}^{k} i^2 + (k+1)^2 = S_k + (k+1)^2,$$

and so

$$S_k + (k+1)^2 = \frac{k(k+1)(2k+1)}{6} + (k+1)^2 = \frac{(k+1)(k+2)(2k+3)}{6}$$

$$= \frac{n(n+1)(2n+1)}{6}\bigg|_{n=k+1}$$

where we have used (3.B.1).

Therefore, S_n is true for $n = k + 1$ if S_n is true for $n = k$. Therefore, S_n is true for all $n \geq 0$ by induction.

APPENDIX 3.C   CATASTROPHIC CANCELLATION

The phenomenon of *catastrophic cancellation* is illustrated in the following output from a MATLAB implementation that ran on a Sun Microsystems Ultra 10 workstation using MATLAB version 6.0.0.88, release 12, in an attempt to compute $\exp(-20)$ using the Maclaurin series for $\exp(x)$ directly:

```
---------------------------------
term k              x^k/ k !
---------------------------------
   0                 1.000000000000000
   1               -20.000000000000000
   2               200.000000000000000
   3             -1333.333333333333258
   4              6666.666666666666970
   5            -26666.666666666667879
   6             88888.888888888890506
   7           -253968.253968253964558
   8            634920.634920634911396
   9          -1410934.744268077658489
  10           2821869.488536155316979
  11          -5130671.797338464297354
  12           8551119.662230772897601
  13         -13155568.711124267429113
  14          18793669.587320379912853
  15         -25058226.116427175700665
  16          31322782.645533967763186
  17         -36850332.524157606065273
  18          40944813.915730677545071
  19         -43099804.121821768581867
  20          43099804.121821768581867
  21         -41047432.496973112225533
  22          37315847.724521011114120
  23         -32448563.238713916391134
  24          27040469.365594934672117
  25         -21632375.492475949227810
  26          16640288.840366113930941
  27         -12326139.881752677261829
  28           8804385.629823340103030
  29          -6071990.089533339254558
  30           4047993.393022226169705
  31          -2611608.640659500379115
  32           1632255.400412187911570
  33           -989245.697219507652335
  34            581909.233658534009010
  35           -332519.562090590829030
  36            184733.090050328260986
  37            -99855.724351528784609
  38             52555.644395541465201
```

```
39          -26951.612510534087050
40           13475.806255267041706
41           -6573.564026959533294
42            3130.268584266443668
43           -1455.938876402997266
44             661.790398364998850
45            -294.129065939999407
46             127.882202582608457
47             -54.417958545790832
48              22.674149394079514
49              -9.254754854726333
50               3.701901941890533
51              -1.451726251721777
52               0.558356250662222
53              -0.210700471948008
54               0.078037211832596
55              -0.028377167939126
56               0.010134702835402
57              -0.003556036082597
58               0.001226219338827
59              -0.000415667572484
60               0.000138555857495
61              -0.000045428149998
62               0.000014654241935
63              -0.000004652140297
64               0.000001453793843
65              -0.000000447321182
66               0.000000135551873
67              -0.000000040463246
68               0.000000011900955
69              -0.000000003449552
70               0.000000000985586
71              -0.000000000277630
72               0.000000000077119
73              -0.000000000021129
74               0.000000000005710
75              -0.000000000001523
76               0.000000000000401
77              -0.000000000000104
78               0.000000000000027
79              -0.000000000000007
80               0.000000000000002
81              -0.000000000000000
82               0.000000000000000
83              -0.000000000000000
84               0.000000000000000
85              -0.000000000000000
86               0.000000000000000
87              -0.000000000000000
88               0.000000000000000
```

```
exp(-20) from sum of the above terms = 0.000000004173637

True value of exp(-20)                = 0.000000002061154
```

REFERENCES

1. E. Kreyszig, *Introductory Functional Analysis with Applications*, Wiley, New York, 1978.

2. W. Rudin, *Principles of Mathematical Analysis*, 3rd ed., McGraw-Hill, New York, 1976.

3. J. E. Volder, "The CORDIC Trigonometric Computing Technique," *IRE Trans. Electron. Comput.* **EC-8**, 330–334 (Sept. 1959).

4. J. S. Walther, "A Unified Algorithm for Elementary Functions," *AFIPS Conf. Proc.*, Vol. 38, 1971 Spring Joint Computer Conf., 379–385 May 18–20, 1971.

5. G. L. Haviland and A. A. Tuszynski, "A CORDIC Arithmetic Processor Chip," *IEEE Trans. Comput.* **C-29**, 68–79 (Feb. 1980).

6. C. W. Schelin, "Calculator Function Approximation," *Am. Math. Monthly* **90**, 317–325 (May 1983).

7. J.-M. Muller, "Discrete Basis and Computation of Elementary Functions," *IEEE Trans. Comput.* **C-34**, 857–862 (Sept. 1985).

8. Y. H. Hu, "The Quantization Effects of the CORDIC Algorithm," *IEEE Trans. Signal Process.* **40**, 834–844 (April 1992).

9. Y. H. Hu and S. Naganathan, "An Angle Recoding Method for CORDIC Algorithm Implementation," *IEEE Trans. Comput.* **42**, 99–102 (Jan. 1993).

10. V. C. Hamacher, Z. G. Vranesic, and S. G. Zaky, *Computer Organization*, 3rd ed., McGraw-Hill, New York, 1990.

11. Y. H. Hu, "CORDIC-Based VLSI Architectures for Digital Signal Processing," *IEEE Signal Process. Mag.* **9**, 16–35 (July 1992).

12. D. Timmermann, H. Hahn, and B. J. Hosticka, "Low Latency Time CORDIC Algorithms," *IEEE Trans. Comput.* **41**, 1010–1015 (Aug. 1992).

13. J. Lee and T. Lang, "Constant-Factor Redundant CORDIC for Angle Calculation and Rotation," *IEEE Trans. Comput.* **41**, 1016–1025 (Aug. 1992).

14. K. Hwang, *Computer Arithmetic: Principles, Architecture, and Design*, Wiley, New York, 1979.

15. H.-Y. Lo and J.-L. Chen, "A Hardwired Generalized Algorithm for Generating the Logarithm Base-k by Iteration," *IEEE Trans. Comput.* **C-36**, 1363–1367 (Nov. 1987).

16. N. Mikami, M. Kobayashi, and Y. Yokoyama, "A New DSP-Oriented Algorithm for Calculation of the Square Root Using a Nonlinear Digital Filter," *IEEE Trans. Signal Process.* **40**, 1663–1669 (July 1992).

17. G. G. Walter, *Wavelets and Other Orthogonal Systems with Applications*, CRC Press, Boca Raton, FL, 1994.

18. I. S. Gradshteyn and I. M. Ryzhik, in *Table of Integrals, Series and Products*, A. Jeffrey, ed., 5th ed., Academic Press, San Diego, CA, 1994.

19. L. Bers, *Calculus: Preliminary Edition*, Vol. 2, Holt, Rinehart, Winston, New York, 1967.

20. A. Leon-Garcia, *Probability and Random Processes for Electrical Engineering*, 2nd ed., Addison-Wesley, Reading, MA, 1994.

21. G. E. Forsythe, M. A. Malcolm, and C. B. Moler, *Computer Methods for Mathematical Computations*, Prentice-Hall, Englewood Cliffs, NJ, 1977.

22. M. R. Spiegel, *Theory and Problems of Advanced Calculus* (Schaum's Outline Series). Schaum (McGraw-Hill), New York, 1963.

23. A. Papoulis, *Signal Analysis*, McGraw-Hill, New York, 1977.
24. E. Kreyszig, *Advanced Engineering Mathematics*, 4th ed., Wiley, New York, 1979.
25. W. D. Lakin and D. A. Sanchez, *Topics in Ordinary Differential Equations*, Dover Publications, New York, 1970.

PROBLEMS

3.1. Prove the following theorem: Every convergent sequence in a metric space is a Cauchy sequence.

3.2. Let $f_n(x) = x^n$ for $n \in \{1, 2, 3, \ldots\} = \mathbf{N}$, and $f_n(x) \in C[0, 1]$ for all $n \in \mathbf{N}$.

(a) What is $f(x) = \lim_{n \to \infty} f_n(x)$ (for $x \in [0, 1]$)?

(b) Is $f(x) \in C[0, 1]$?

3.3. Sequence (x_n) is defined to be $x_n = (n + 1)/(n + 2)$ for $n \in \mathbf{Z}^+$. Clearly, if $X = [0, 1) \subset \mathbf{R}$, then $x_n \in X$ for all $n \in \mathbf{Z}^+$. Assume the metric for metric space X is $d(x, y) = |x - y|$ $(x, y \in X)$.

(a) What is $x = \lim_{n \to \infty} x_n$?

(b) Is X a complete space?

(c) Prove that (x_n) is Cauchy.

3.4. Recall Section 3.A.3 wherein the rotation operator was defined [Eq. (3.A.6)].

(a) Find an expression for angle θ such that for $y \neq 0$

$$\underbrace{\begin{bmatrix} \cos \theta & -\sin \theta \\ \sin \theta & \cos \theta \end{bmatrix}}_{=G(\theta)} \begin{bmatrix} x \\ y \end{bmatrix} = \begin{bmatrix} x' \\ 0 \end{bmatrix},$$

where x' is some arbitrary nonzero constant.

(b) Prove that $G^{-1}(\theta) = G^T(\theta)$ [i.e., the inverse of $G(\theta)$ is given by its transpose].

(c) Consider the matrix

$$A = \begin{bmatrix} 4 & 2 & 0 \\ 1 & 4 & 1 \\ 0 & 2 & 4 \end{bmatrix}.$$

Let $0_{n \times m}$ denote an array (matrix) of zeros with n rows, and m columns. Find $G(\theta_1)$, and $G(\theta_2)$ so that

$$\underbrace{\begin{bmatrix} 1 & 0_{1 \times 2} \\ 0_{2 \times 1} & G(\theta_2) \end{bmatrix} \begin{bmatrix} G(\theta_1) & 0_{2 \times 1} \\ 0_{1 \times 2} & 1 \end{bmatrix} A}_{=Q^T} = R,$$

where R is an upper triangular matrix (defined in Section 4.5, if you do not recall what this is).

(d) Find Q^{-T} (inverse of Q^T).

(*Comment:* The procedure illustrated by this problem is important in various applications such as solving least-squares approximations, and in finding the eigenvalues and eigenvectors of matrices.)

3.5. Review Appendix 3.A. Suppose that we wish to rotate a vector $[xy]^T \in \mathbf{R}^2$ through an angle $\theta = 25° \pm 1°$. Find n, and the required delta sequence (δ_k) to achieve this accuracy.

3.6. Review Appendix 3.A. A certain CORDIC routine has the following pseudocode description:

```
Input x and z (|z| ≤ 1);
y0 := 0; z0 := z;
for k := 0 to n − 1 do begin
   if zk < 0 then begin
      δk := −1;
   end
   else begin
      δk := +1;
   end;
   yk+1 := yk + δkx2−k;
   zk+1 := zk − δk2−k;
end;
```

The algorithm's output is y_n. What is y_n?

3.7. Suppose that

$$x_n(t) = \frac{t}{n+t}$$

for $n \in \mathbf{Z}^+$, and $t \in (0, 1) \subset \mathbf{R}$. Show that $(x_n(t))$ is uniformly convergent on $S = (0, 1)$.

3.8. Suppose that $u_n > 0$, and also that $\sum_{n=0}^{\infty} u_n$ converges. Prove that $\prod_{n=0}^{\infty} (1 + u_n)$ converges.

[*Hint:* Recall Lemma 2.1 (of Chapter 2).]

3.9. Prove that $\lim_{n \to \infty} x^n = 0$ if $|x| < 1$.

3.10. Prove that for $a, b, x \in \mathbf{R}$

$$\frac{1}{1 + |a - b|} \geq \frac{1}{1 + |a - x|} \frac{1}{1 + |b - x|}.$$

(*Comment:* This inequality often appears in the context of convergence proofs for certain series expansions.)

3.11. Consider the function

$$K_N(x) = \frac{2\pi}{N+1} \sum_{n=0}^{N} D_n(x)$$

[recall (3.24)]. Since $D_n(x)$ is 2π-periodic, $K_N(x)$ is also 2π-periodic. We may assume that $x \in [-\pi, \pi]$.

(a) Prove that

$$K_N(x) = \frac{1}{N+1} \frac{1 - \cos(N+1)x}{1 - \cos x}.$$

[*Hint:* Consider $\sum_{n=0}^{N} \sin\left((n + \frac{1}{2})x\right) = \text{Im}\left[\sum_{n=0}^{N} e^{j(n+\frac{1}{2})x}\right]$.]

(b) Prove that

$$K_N(x) \geq 0.$$

(c) Prove that

$$\frac{1}{2\pi} \int_{-\pi}^{\pi} K_N(x)\,dx = 1.$$

[*Comment:* The *partial sums* of the complex Fourier series expansion of the 2π-periodic function $f(x)$ (again $x \in [-\pi, \pi]$) are given by

$$f_N(x) = \sum_{n=-N}^{N} f_n e^{jnx},$$

where $f_n = \frac{1}{2\pi} \int_{-\pi}^{\pi} f(x)e^{-jnx}\,dx$. Define

$$\sigma_N(x) = \frac{1}{N+1} \sum_{n=0}^{N} f_n(x).$$

It can be shown that

$$\sigma_N(x) = \frac{1}{2\pi} \int_{-\pi}^{\pi} f(x - t)K_N(t)\,dt.$$

It is also possible to prove that $\sigma_N(x) \to f(x)$ uniformly on $[-\pi, \pi]$ if $f(x)$ is continuous. This is often called *Fejér's theorem*.]

3.12. Repeat the analysis of Example 3.12 for $f_L(0)$.

3.13. If $f(t) \in L^2(0, 2\pi)$ and $f(t) = \sum_{n \in \mathbf{Z}} f_n e^{jnt}$, show that

$$\frac{1}{2\pi} \int_0^{2\pi} |f(t)|^2\,dt = \sum_{n=-\infty}^{\infty} |f_n|^2.$$

This relates the energy/power of $f(t)$ to the energy/power of its Fourier series coefficients.

[*Comment*: For example, if $f(t)$ is one period of a 2π-periodic signal, then the power interpretation applies. In particular, suppose that $f(t) = i(t)$ a current waveform, and that this is the current into a resistor of resistance R; then the average power delivered to R is

$$P_{av} = \frac{1}{2\pi} \int_0^{2\pi} Ri^2(t)\,dt.]$$

3.14. Use the result of Example 1.20 to prove that

$$\sum_{n=0}^{\infty} \frac{(-1)^n}{2n+1} = \frac{\pi}{4}.$$

3.15. Prove that

$$\frac{1}{2\pi} \int_0^{\pi} \frac{\sin[(L+\frac{1}{2})t]}{\sin[\frac{1}{2}t]}\,dt = \frac{1}{2}.$$

3.16. Use mathematical induction to prove that

$$(1.1)^n \geq 1 + \frac{1}{10}n$$

for all $n \in \mathbf{N}$.

3.17. Use mathematical induction to prove that

$$(1+h)^n \geq 1 + nh$$

for all $n \in \mathbf{N}$, with $h \geq -1$. (This is *Bernoulli's inequality*.)

3.18. Use mathematical induction to prove that $4^n + 2$ is a multiple of 6 for all $n \in \mathbf{N}$.

3.19. Conjecture a formula for ($n \in \mathbf{N}$)

$$S_n = \sum_{k=1}^{n} \frac{1}{k(k+1)},$$

and prove it using mathematical induction.

3.20. Suppose that $f(x) = \tan x$. Use (3.55) to approximate $f(x)$ for all $x \in (-1, 1)$. Find an upper bound on $|e(x)|$, where $e(x)$ is the error of the approximation.

3.21. Given $f(x) = x^2 + 1$, find all $\xi \in (1, 2)$ such that

$$f^{(1)}(\xi) = \frac{f(2) - f(1)}{2 - 1}.$$

3.22. Use (3.65) to find an approximation to $\sqrt{1 + x}$ for $x \in (\frac{1}{4}, \frac{7}{4})$. Find an upper bound on the magnitude of $e(x)$.

3.23. Using a pocket calculator, compute $(0.97)^{1/3}$ for $n = 3$ using (3.82). Find upper and lower bounds on the error $e_{n+1}(x)$.

3.24. Using a pocket calculator, compute $[1.05]^{1/4}$ using (3.82). Choose $n = 2$ (i.e., quadratic approximation). Estimate the error involved in doing this using the error bound expression. Compare the bounds to the actual error.

3.25. Show that

$$\sum_{k=0}^{n} \binom{n}{k} = 2^n.$$

3.26. Show that

$$\frac{n}{j} \binom{n-1}{j-1} = \binom{n}{j}$$

for $j = 1, 2, \ldots, n - 1, n$.

3.27. Show that for $n \geq 2$

$$\sum_{k=1}^{n} k \binom{n}{k} = n2^{n-1}.$$

3.28. For

$$p_n(k) = \binom{n}{k} p^k (1 - p)^{n-k}, \, 0 \leq p \leq 1,$$

where $k = 0, 1, \ldots, n$, show that

$$p_n(k + 1) = \frac{(n - k)p}{(k + 1)(1 - p)} p_n(k).$$

[*Comment:* This recursive approach for finding $p_n(k)$ extends the range of n for which $p_n(k)$ may be computed before experiencing problems with numerical errors.]

3.29. Identity (3.103) was confirmed using an argument associated with Fig. 3.7. A somewhat different approach is the following. Confirm (3.103) by working with

$$\left[\frac{1}{\sqrt{2\pi}} \int_{-\infty}^{\infty} e^{-x^2/2} \, dx \right]^2 = \frac{1}{2\pi} \int_{-\infty}^{\infty} \int_{-\infty}^{\infty} e^{-(x^2 + y^2)/2} \, dx \, dy.$$

(*Hint:* Use the Cartesian to polar coordinate conversion $x = r \cos\theta$, $y = r \sin\theta$.)

3.30. Find the Maclaurin (infinite) series expansion for $f(x) = \sin^{-1} x$. What is the radius of convergence? [*Hint:* Theorem 3.7 and Eq. (3.82) are useful.]

3.31. From Eq. (3.85) for $x > -1$

$$\log_e(1 + x) = \sum_{k=1}^{n} \frac{(-1)^{k-1} x^k}{k} + r(x), \qquad (3.P.1)$$

where

$$|r(x)| \leq \begin{cases} \frac{1}{n+1} x^{n+1}, & x \geq 0 \\ \frac{|x|^{n+1}}{(1-|x|)(n+1)}, & -1 < x \leq 0 \end{cases} \qquad (3.P.2)$$

[Eq. (3.86)]. For what range of values of x does the series

$$\log_e(1 + x) = \sum_{k=1}^{\infty} \frac{(-1)^{k-1} x^k}{k}$$

converge? Explain using (3.P.2).

3.32. The following problems are easily worked with a pocket calculator.

(**a**) It can be shown that

$$\sin(x) = \sum_{k=0}^{n} \frac{(-1)^k}{(2k + 1)!} x^{2k+1} + e_{2n+3}(x), \qquad (3.P.3)$$

where

$$|e_{2n+3}(x)| \leq \frac{1}{(2n + 3)!} |x|^{2n+3}. \qquad (3.P.4)$$

Use (3.P.4) to compute $\sin(x)$ to 3 decimal places of accuracy for $x = 1.5$ radians. How large does n need to be?

(**b**) Use the approximation

$$\log_e(1 + x) \approx \sum_{n=1}^{N} \frac{(-1)^{n-1} x^n}{n}$$

to compute $\log_e(1.5)$ to three decimal places of accuracy. How large should N be to achieve this level of accuracy?

3.33. Assuming that

$$\left| \frac{2}{\pi} \int_0^\pi \frac{\sin x}{x} \, dx - 2 \sum_{r=0}^n \frac{\pi^{2r}(-1)^r}{(2r+1)(2r+1)!} \right| \leq \frac{2\pi^{2n+1}}{(2n+2)(2n+2)!}, \quad (3.P.5)$$

show that for suitable n

$$\frac{2}{\pi} \int_0^\pi \frac{\sin x}{x} \, dx > 1.17.$$

What is the smallest n needed? Justify Eq. (3.P.5).

3.34. Using integration by parts, find the asymptotic expansion of

$$c(x) = \int_x^\infty \cos t^2 \, dt.$$

3.35. Using integration by parts, find the asymptotic expansion of

$$s(x) = \int_x^\infty \sin t^2 \, dt.$$

3.36. Use MATLAB to plot (on the same graph) function $K_N(x)$ in Problem 3.11 for $N = 2, 4, 15$.

4 Linear Systems of Equations

4.1 INTRODUCTION

The necessity to solve linear systems of equations is commonplace in numerical computing. This chapter considers a few examples of how such problems arise (more examples will be seen in subsequent chapters) and the numerical problems that are frequently associated with attempts to solve them. We are particularly interested in the phenomenon of *ill conditioning*. We will largely concentrate on how the problem arises, what its effects are, and how to test for this problem.[1] In addition to this, we will also consider methods of solving linear systems other than the Gaussian elimination method that you most likely learned in an elementary linear algebra course.[2] More specifically, we consider LU and QR matrix factorization methods, and iterative methods of linear system solution. The concept of a singular value decomposition (SVD) is also introduced.

We will often employ the term "linear systems" instead of the longer phrase "linear systems of equations." However, the reader must be warned that the phrase "linear systems" can have a different meaning from our present usage. In signals and systems courses you will most likely see that a "linear system" is either a continuous-time (i.e., analog) or discrete-time (i.e., digital) dynamic system whose input/output (I/O) behavior satisfies superposition. However, such dynamic systems can be described in terms of linear systems of equations.

4.2 LEAST-SQUARES APPROXIMATION AND LINEAR SYSTEMS

Suppose that $f(x), g(x) \in L^2[0, 1]$, and that these functions are real-valued. Recalling Chapter 1, their inner product is therefore given by

$$\langle f, g \rangle = \int_0^1 f(x)g(x)\,dx. \tag{4.1}$$

---

[1]Methods employed to avoid ill-conditioned linear systems of equations will be mainly considered in a later chapter. These chiefly involve working with orthogonal basis sets.

[2]Review the Gaussian elimination procedure *now* if necessary. It is *mandatory* that you recall the basic matrix and vector operations and properties [$(AB)^T = B^T A^T$, etc.] from elementary linear algebra, too.

An Introduction to Numerical Analysis for Electrical and Computer Engineers, by C.J. Zarowski
ISBN 0-471-46737-5 © 2004 John Wiley & Sons, Inc.

Note that here we will assume that all members of $L^2[0, 1]$ are real-valued (for simplicity). Now assume that $\{\phi_n(x) | n \in \mathbf{Z}_N\}$ form a linearly independent set such that $\phi_n(x) \in L^2[0, 1]$ for all n. We wish to find the coefficients a_n such that

$$f(x) \approx \sum_{n=0}^{N-1} a_n \phi_n(x) \quad \text{for} \quad x \in [0, 1]. \tag{4.2}$$

A popular approach to finding a_n [1] is to choose them to minimize the functional

$$V(a) = ||f(x) - \sum_{n=0}^{N-1} a_n \phi_n(x)||^2, \tag{4.3}$$

where, for future convenience, we will treat a_n as belonging to the vector $a = [a_0 a_1 \cdots a_{N-2} a_{N-1}]^T \in \mathbf{R}^N$. Of course, in (4.3) $||f||^2 = \langle f, f \rangle$ via (4.1). We are at liberty to think of

$$e(x) = f(x) - \sum_{n=0}^{N-1} a_n \phi_n(x) \tag{4.4}$$

as the error between $f(x)$ and its approximation $\sum_n a_n \phi_n(x)$. So, our goal is to pick a to minimize $||e(x)||^2$, which, we have previously seen, may be interpreted as the energy of the error $e(x)$. This methodology of approximation is called *least-squares approximation*. The version of this that we are now considering is only one of a great many variations. We will see others later.

We may rewrite (4.3) as follows:

$$V(a) = ||f(x) - \sum_{n=0}^{N-1} a_n \phi_n(x)||^2$$

$$= \left\langle f - \sum_{n=0}^{N-1} a_n \phi_n, f - \sum_{k=0}^{N-1} a_k \phi_k \right\rangle$$

$$= \langle f, f \rangle - \left\langle f, \sum_{k=0}^{N-1} a_k \phi_k \right\rangle$$

$$- \left\langle \sum_{n=0}^{N-1} a_n \phi_n, f \right\rangle + \left\langle \sum_{n=0}^{N-1} a_n \phi_n, \sum_{k=0}^{N-1} a_k \phi_k \right\rangle$$

$$= ||f||^2 - \sum_{k=0}^{N-1} a_k \langle f, \phi_k \rangle - \sum_{n=0}^{N-1} a_n \langle \phi_n, f \rangle$$

$$+ \sum_{n=0}^{N-1} \sum_{k=0}^{N-1} a_k a_n \langle \phi_k, \phi_n \rangle$$

$$= ||f||^2 - 2 \sum_{k=0}^{N-1} a_k \langle f, \phi_k \rangle$$

$$+ \sum_{n=0}^{N-1} \sum_{k=0}^{N-1} a_k a_n \langle \phi_k, \phi_n \rangle. \tag{4.5}$$

Naturally, we have made much use of the inner product properties of Chapter 1. It is very useful to define

$$\rho = ||f||^2, \quad g_k = \langle f, \phi_k \rangle, \quad r_{n,k} = \langle \phi_n, \phi_k \rangle, \tag{4.6}$$

along with the vector

$$g = [g_0 g_1 \cdots g_{N-1}]^T \in \mathbf{R}^N, \tag{4.7a}$$

and the matrix

$$R = [r_{n,k}] \in \mathbf{R}^{N \times N}. \tag{4.7b}$$

We immediately observe that $R = R^T$ (i.e., R is *symmetric*). This is by virtue of the fact that $r_{n,k} = \langle \phi_n, \phi_k \rangle = \langle \phi_k, \phi_n \rangle = r_{k,n}$. Immediately we may rewrite (4.5) in order to obtain the *quadratic form*[3]

$$V(a) = a^T R a - 2a^T g + \rho. \tag{4.8}$$

The quadratic form occurs very widely in optimization and approximation problems, and so warrants considerable study. An expanded view of R is

$$R = \begin{bmatrix} \int_0^1 \phi_0^2(x)\,dx & \int_0^1 \phi_0(x)\phi_1(x)\,dx & \cdots & \int_0^1 \phi_0(x)\phi_{N-1}(x)\,dx \\ \int_0^1 \phi_0(x)\phi_1(x)\,dx & \int_0^1 \phi_1^2(x)\,dx & \cdots & \int_1^1 \phi_1(x)\phi_{N-1}(x)\,dx \\ \vdots & \vdots & & \vdots \\ \int_0^1 \phi_0(x)\phi_{N-1}(x)\,dx & \int_0^1 \phi_1(x)\phi_{N-1}(x)\,dx & \cdots & \int_0^1 \phi_{N-1}^2(x)\,dx \end{bmatrix}. \tag{4.9}$$

Note that the reader must get used to visualizing matrices and vectors in the general manner now being employed. The practical use of linear/matrix algebra demands this. Writing programs for anything involving matrix methods (which encompasses a great deal) is almost impossible without this ability. Moreover, modern software tools (e.g., MATLAB) assume that the user is skilled in this manner.

---

[3]The quadratic form is a generalization of the familiar quadratic $ax^2 + bx + c$, for which $x \in \mathbf{C}$ (if we are interested in the roots of a quadratic equation; otherwise we usually consider $x \in \mathbf{R}$) , and also $a, b, c \in \mathbf{R}$.

It is essential to us that R^{-1} exist (i.e., we need R to be *nonsingular*). Fortunately, this is always the case because $\{\phi_n | n \in \mathbf{Z}_N\}$ is an independent set. We shall prove this. If R is singular, then there is a set of coefficients α_i such that

$$\sum_{i=0}^{N-1} \alpha_i \langle \phi_i, \phi_j \rangle = 0 \tag{4.10}$$

for $j \in \mathbf{Z}_N$. This is equivalent to saying that the columns of R are linearly dependent. Now consider the function

$$\hat{f}(x) = \sum_{i=0}^{N-1} \alpha_i \phi_i(x)$$

so if (4.10) holds for all $j \in \mathbf{Z}_N$, then

$$\langle \hat{f}, \phi_j \rangle = \left\langle \sum_{i=0}^{N-1} \alpha_i \phi_i, \phi_j \right\rangle = \sum_{i=0}^{N-1} \alpha_i \langle \phi_i, \phi_j \rangle = 0$$

for all $j \in \mathbf{Z}_N$. Thus

$$\sum_{j=0}^{N-1} \alpha_j \langle \hat{f}, \phi_j \rangle = \sum_{j=0}^{N-1} \alpha_j \left[\sum_{i=0}^{N-1} \alpha_i \langle \phi_i, \phi_j \rangle \right] = 0,$$

implying that

$$\left\langle \sum_{i=0}^{N-1} \alpha_i \phi_i, \sum_{j=0}^{N-1} \alpha_j \phi_j \right\rangle = 0,$$

or in other words, $\langle \hat{f}, \hat{f} \rangle = ||\hat{f}||^2 = 0$, and so $\hat{f}(x) = 0$ for all $x \in [0, 1]$. This contradicts the assumption that $\{\phi_n | n \in \mathbf{Z}_N\}$ is a linearly independent set. So R^{-1} must exist.

From (4.3) it is clear that (via basic norm properties) $V(a) \geq 0$ for all $a \in \mathbf{R}^N$. If we now assume that $f(x) = 0$ (all $x \in [0, 1]$), we have $\rho = 0$, and $g = 0$ too. Thus, $a^T R a \geq 0$ for all a.

Definition 4.1: Positive Semidefinite Matrix, Positive Definite Matrix   Suppose that $A = A^T$ and that $A \in \mathbf{R}^{n \times n}$. Suppose that $x \in \mathbf{R}^n$. We say that A is *positive semidefinite (psd)* iff

$$x^T A x \geq 0$$

for all x. We say that A is *positive definite (pd)* iff

$$x^T A x > 0$$

for all $x \neq 0$.

If A is psd, we often symbolize this by writing $A \geq 0$, and if A is pd, we often symbolize this by writing $A > 0$. If a matrix is pd then it is clearly psd, but the converse is not necessarily true.

So far it is clear that $R \geq 0$. But in fact $R > 0$. This follows from the linear independence of the columns of R. If the columns of R were linearly dependent, then there would be an $a \neq 0$ such that $Ra = 0$, but we have already shown that R^{-1} exists, so it must be so that $Ra = 0$ iff $a = 0$. Immediately we conclude that R is positive definite.

Why is $R > 0$ so important? Recall that we may solve $ax^2 + bx + c = 0 \ (x \in \mathbf{C})$ by *completing the square*:

$$ax^2 + bx + c = a\left[x + \frac{b}{2a}\right]^2 + c - \frac{b^2}{4a}. \tag{4.11}$$

Now if $a > 0$, then $y(x) = ax^2 + bx + c$ has a unique minimum. Since $y^{(1)}(x) = 2ax + b = 0$ for $x = -\frac{b}{2a}$, this choice of x forces the first term of (4.11) (right-hand side of the equality) to zero, and we see that the minimum value of $y(x)$ is

$$y\left(-\frac{b}{2a}\right) = c - \frac{b^2}{4a} = \frac{4ac - b^2}{4a}. \tag{4.12}$$

Thus, completing the square makes the location of the minimum (if it exists), and the value of the minimum of a quadratic very obvious. For the same purpose we may complete the square of (4.8):

$$V(a) = [a - R^{-1}g]^T R[a - R^{-1}g] + \rho - g^T R^{-1}g. \tag{4.13}$$

It is quite easy to confirm that (4.13) gives (4.8) (so these equations must be equivalent):

$$[a - R^{-1}g]^T R[a - R^{-1}g] + \rho - g^T R^{-1}g$$
$$= [a^T - g^T(R^{-1})^T][Ra - g] + \rho - g^T R^{-1}g$$
$$= a^T Ra - g^T R^{-1} Ra - a^T g + g^T R^{-1}g + \rho - g^T R^{-1}g$$
$$= a^T Ra - g^T a - a^T g + \rho = a^T Ra - 2a^T g + \rho.$$

We have used $(R^{-1})^T = (R^T)^{-1} = R^{-1}$, and the fact that $a^T g = g^T a$. If vector $x = a - R^{-1}g$, then

$$[a - R^{-1}g]^T R[a - R^{-1}g] = x^T Rx.$$

So, because $R > 0$, it follows that $x^T Rx > 0$ for all $x \neq 0$. The last two terms of (4.13) do not depend on a. So we can minimize $V(a)$ only by minimizing the first term. $R > 0$ implies that this minimum must be for $x = 0$, implying $a = \hat{a}$, where

$$\hat{a} - R^{-1}g = 0,$$

or

$$R\hat{a} = g. \tag{4.14}$$

Thus

$$\hat{a} = \underset{a \in \mathbf{R}^N}{\operatorname{argmin}} V(a). \tag{4.15}$$

We see that to minimize $||e(x)||^2$ we must solve a linear system of equations, namely, Eq. (4.14). We remark that for $R > 0$, the minimum of $V(a)$ is at a unique location $\hat{a} \in \mathbf{R}^N$; that is, the minimum is unique.

In principle, solving least-squares approximation problems seems quite simple because we have systematic (and numerically reliable) methods to solve (4.14) (e.g., Gaussian elimination with partial pivoting). However, one apparent difficulty is the need to determine various integrals:

$$g_k = \int_0^1 f(x)\phi_k(x)\,dx, \, r_{n,k} = \int_0^1 \phi_n(x)\phi_k(x)\,dx.$$

Usually, the independent set $\{\phi_k | k \in \mathbf{Z}_N\}$ is chosen to make finding $r_{n,k}$ relatively straightforward. In fact, sometimes nice closed-form expressions exist. But numerical integration is generally needed to find g_k. Practically, this could involve applying series expansions such as considered in Chapter 3, or perhaps using quadratures such as will be considered in a later chapter. Other than this, there is a more serious problem. This is the problem that R might be *ill-conditioned*.

4.3 LEAST-SQUARES APPROXIMATION AND ILL-CONDITIONED LINEAR SYSTEMS

A popular choice for an independent set $\{\phi_k(x)\}$ would be

$$\phi_k(x) = x^k \quad \text{for} \quad x \in [0, 1], \, k \in \mathbf{Z}_N. \tag{4.16}$$

Certainly, these functions belong to the inner product space $L^2[0, 1]$. Thus, for $f(x) \in L^2[0, 1]$ an approximation to it is

$$\hat{f}(x) = \sum_{k=0}^{N-1} a_k x^k \in \mathbf{P}^{N-1}[0, 1], \tag{4.17}$$

and so we wish to *fit a degree $N - 1$ polynomial to* $f(x)$. Consequently

$$g_k = \int_0^1 x^k f(x)\,dx \tag{4.18a}$$

which is sometimes called the *kth moment*[4] of $f(x)$ on $[0, 1]$, and also

$$r_{n,k} = \int_0^1 \phi_n(x)\phi_k(x)\,dx = \int_0^1 x^{n+k}\,dx = \frac{1}{n+k+1} \tag{4.18b}$$

for $n, k \in \mathbf{Z}_N$.

----

[4]The concept of a moment is also central to probability theory.

For example, suppose that $N = 3$ (i.e., a quadratic fit); then

$$g = \left[\underbrace{\int_0^1 f(x)\,dx}_{=g_0} \quad \underbrace{\int_0^1 xf(x)\,dx}_{=g_1} \quad \underbrace{\int_0^1 x^2 f(x)\,dx}_{=g_2} \right]^T, \qquad (4.19a)$$

and

$$R = \begin{bmatrix} 1 & \frac{1}{2} & \frac{1}{3} \\ \frac{1}{2} & \frac{1}{3} & \frac{1}{4} \\ \frac{1}{3} & \frac{1}{4} & \frac{1}{5} \end{bmatrix} = \begin{bmatrix} r_{00} & r_{01} & r_{02} \\ r_{10} & r_{11} & r_{12} \\ r_{20} & r_{21} & r_{22} \end{bmatrix}, \qquad (4.19b)$$

and $\hat{a} = [\hat{a}_0 \ \hat{a}_1 \ \hat{a}_2]^T$, so we wish to solve

$$\begin{bmatrix} 1 & \frac{1}{2} & \frac{1}{3} \\ \frac{1}{2} & \frac{1}{3} & \frac{1}{4} \\ \frac{1}{3} & \frac{1}{4} & \frac{1}{5} \end{bmatrix} \begin{bmatrix} \hat{a}_0 \\ \hat{a}_1 \\ \hat{a}_2 \end{bmatrix} = \begin{bmatrix} \int_0^1 f(x)\,dx \\ \int_0^1 xf(x)\,dx \\ \int_0^1 x^2 f(x)\,dx \end{bmatrix}. \qquad (4.20)$$

We remark that R does not depend on the "data" $f(x)$, only the elements of g do. This is true in general, and it can be used to advantage. Specifically, if $f(x)$ changes frequently (i.e., we must work with different data), but the independent set does not change, then we need to invert R only once.

Matrix R in (4.19b) is a special case of the famous *Hilbert matrix* [2–4]. The general form of this matrix is (for any $N \in \mathbf{N}$)

$$R = \begin{bmatrix} 1 & \frac{1}{2} & \frac{1}{3} & \cdots & \frac{1}{N} \\ \frac{1}{2} & \frac{1}{3} & \frac{1}{4} & \cdots & \frac{1}{N+1} \\ \frac{1}{3} & \frac{1}{4} & \frac{1}{5} & \cdots & \frac{1}{N+2} \\ \vdots & \vdots & \vdots & & \vdots \\ \frac{1}{N} & \frac{1}{N+1} & \frac{1}{N+2} & \cdots & \frac{1}{2N-1} \end{bmatrix} \in \mathbf{R}^{N \times N}. \qquad (4.21)$$

Thus, (4.20) is a special case of a *Hilbert linear system of equations*. The matrix R in (4.21) seems "harmless," but it is actually a menace from a numerical computing standpoint. We now demonstrate this concept.

Suppose that our data are something very simple. Say that

$$f(x) = 1 \quad \text{for all} \quad x \in [0, 1].$$

In this case $g_k = \int_0^1 x^k \, dx = \frac{1}{k+1}$. Therefore, for any $N \in \mathbf{N}$, we are compelled to solve

$$
\begin{bmatrix}
1 & \dfrac{1}{2} & \dfrac{1}{3} & \cdots & \dfrac{1}{N} \\[2mm]
\dfrac{1}{2} & \dfrac{1}{3} & \dfrac{1}{4} & \cdots & \dfrac{1}{N+1} \\[2mm]
\dfrac{1}{3} & \dfrac{1}{4} & \dfrac{1}{5} & \cdots & \dfrac{1}{N+2} \\[2mm]
\vdots & \vdots & \vdots & & \vdots \\[2mm]
\dfrac{1}{N} & \dfrac{1}{N+1} & \dfrac{1}{N+2} & \cdots & \dfrac{1}{2N-1}
\end{bmatrix}
\begin{bmatrix}
\hat{a}_0 \\[1mm] \hat{a}_1 \\[1mm] \hat{a}_2 \\[1mm] \vdots \\[1mm] \hat{a}_{N-1}
\end{bmatrix}
=
\begin{bmatrix}
1 \\[1mm] \dfrac{1}{2} \\[1mm] \dfrac{1}{3} \\[1mm] \vdots \\[1mm] \dfrac{1}{N}
\end{bmatrix}.
\tag{4.22}
$$

A moment of thought reveals that solving (4.22) is trivial because g is the first column of R. Immediately, we see that

$$\hat{a} = [1\,00\cdots 00]^T. \tag{4.23}$$

(No other solution is possible since R^{-1} exists, implying that $R\hat{a} = g$ always possesses a unique solution.)

MATLAB implements Gaussian elimination (with partial pivoting) using the operator "\" to solve linear systems. For example, if we want x in $Ax = y$ for which A^{-1} exists, then $x = A\backslash y$. MATLAB also computes R using function "hilb"; that is, $R = \text{hilb}(N)$ will result in R being set to a $N \times N$ Hilbert matrix. Using the MATLAB "\" operator to solve for $\hat{a}$ in (4.22) gives the expected answer (4.23) for $N \le 50$ (at least). The computer-generated answers are correct to several decimal places. (Note that it is somewhat unusual to want to fit polynomials to data that are of such large degree.) So far, so good.

Now consider the results in Appendix 4.A. The MATLAB function "inv" may be used to compute the inverse of matrices. The appendix shows R^{-1} (computed via inv) for $N = 10, 11, 12$, and the MATLAB computed product RR^{-1} for these cases. Of course, $RR^{-1} = I$ (identity matrix) is expected in all cases. For the number of decimal places shown, we observe that $RR^{-1} \ne I$. Not only that, but the error $E = RR^{-1} - I$ rapidly becomes large with an increase in N. For $N = 12$, the error is substantial. In fact, the MATLAB function inv has built-in features to warn of trouble, and it does so for case $N = 12$. Since RR^{-1} is not being computed correctly, something has clearly gone wrong, and this has happened for rather small values of N. This is in striking contrast with the previous problem, where we wanted to compute $\hat{a}$ in (4.22). In this case, apparently, nothing went wrong.

We may consider changing our data to $f(x) = x^{N-1}$. In this case $g_k = 1/(N + k)$ for $k \in \mathbf{Z}_N$. The vector g in this case will be the last column of R. Thus, mathematically, $\hat{a} = [00 \cdots 001]^T$. If we use MATLAB "\\" to compute $\hat{a}$ for this problem we obtain the *computed* solutions:

$$\hat{\hat{a}} = [0.0000 \ \ 0.0000 \ \ 0.0000 \ \ 0.0000 \ \ \ldots$$

$$0.0002 \ \ -0.0006 \ \ 0.0013 \ \ -0.00170.0014 \ \ -0.0007 \ \ 1.0001]^T \ \ (N = 11),$$

$$\hat{\hat{a}} = [0.0000 \ \ 0.0000 \ \ 0.0000 \ \ -0.0002 \ \ \ldots \ \ 0.0015 \ \ -0.0067$$

$$0.0187 \ \ -0.0342 \ \ 0.0403 \ \ -0.0297 \ \ 0.0124 \ \ 0.9978]^T \ \ (N = 12).$$

The errors in the computed solutions $\hat{\hat{a}}$ here are much greater than those experienced in computing $\hat{a}$ in (4.23).

It turns out that the Hilbert matrix R is a classical example of an *ill-conditioned matrix* (with respect to the problem of solving linear systems of equations). The linear system in which it resides [i.e., Eq. (4.14)] is therefore an *ill-conditioned linear system*. In such systems the final answer (which is $\hat{a}$ here) can be exquisitely sensitive to very small perturbations (i.e., disturbances) in the inputs. The inputs in this case are the elements of R and g. From Chapter 2 we remember that R and g will not have an exact representation on the computer because of quantization errors. Additionally, as the computation proceeds rounding errors will cause further disturbances. The result is that in the end the final computed solution can deviate enormously from the correct mathematical solution. On the other hand, we have also shown that it is possible for the computed solution to be very close to the mathematically correct solution even in the presence of ill conditioning. Our problem then is to be able to detect when ill conditioning arises, and hence *might* pose a problem.

4.4   CONDITION NUMBERS

In the previous section there appears to be a problem involved in accurately computing the inverse of R (Hilbert matrix). This was attributed to the so-called ill conditioning of R. We begin here with some simpler lower-order examples that illustrate how the solution to a linear system $Ax = y$ can depend sensitively on A and y. This will lead us to develop a theory of *condition numbers* that warn us that the solution x might be inaccurately computed due to this sensitivity.

We will consider $Ax = y$ on the assumption that $A \in \mathbf{R}^{n \times n}$, and $x, y \in \mathbf{R}^n$. Initially we will assume $n = 2$, so

$$\begin{bmatrix} a_{00} & a_{01} \\ a_{10} & a_{11} \end{bmatrix} \begin{bmatrix} x_0 \\ x_1 \end{bmatrix} = \begin{bmatrix} y_0 \\ y_1 \end{bmatrix}. \tag{4.24}$$

In practice, we may be uncertain about the accuracy of the entries of A and y. Perhaps these entities originate from experimental data. So the entries may be subject to experimental errors. Additionally, as previously mentioned, the elements of A and y cannot normally be exactly represented on a computer because of the need to quantize their entries. Thus, we must consider the *perturbed* system

$$\left\{ \begin{bmatrix} a_{00} & a_{01} \\ a_{10} & a_{11} \end{bmatrix} + \underbrace{\begin{bmatrix} \delta a_{00} & \delta a_{01} \\ \delta a_{10} & \delta a_{11} \end{bmatrix}}_{=\delta A} \right\} \underbrace{\begin{bmatrix} \hat{x}_0 \\ \hat{x}_1 \end{bmatrix}}_{=\hat{x}} = \left\{ \begin{bmatrix} y_0 \\ y_1 \end{bmatrix} + \underbrace{\begin{bmatrix} \delta y_0 \\ \delta y_1 \end{bmatrix}}_{=\delta y} \right\}. \quad (4.25)$$

The perturbations are δA and δy. We will assume that these are "small." As you might expect, the practical definition of "small" will force us to define and work with suitable norms. This is dealt with below. We further assume that the computing machine we use to solve (4.25) is a "magical machine" that computes without rounding errors. Thus, any errors in the computed solution, denoted $\hat{x}$ here, can be due only to the perturbations δA and δy. It is our hope that $\hat{x} \approx x$. Unfortunately, this will not always be so, even for $n = 2$ with small perturbations.

Because n is small, we may obtain closed-form expressions for A^{-1}, x, $[A + \delta A]^{-1}$, and $\hat{x}$. More specifically

$$A^{-1} = \frac{1}{a_{00}a_{11} - a_{01}a_{10}} \begin{bmatrix} a_{11} & -a_{01} \\ -a_{10} & a_{00} \end{bmatrix}, \quad (4.26)$$

and

$$[A + \delta A]^{-1} = \frac{1}{(a_{00} + \delta a_{00})(a_{11} + \delta a_{11}) - (a_{01} + \delta a_{01})(a_{10} + \delta a_{10})}$$
$$\times \begin{bmatrix} a_{11} + \delta a_{11} & -(a_{01} + \delta a_{01}) \\ -(a_{10} + \delta a_{10}) & (a_{00} + \delta a_{00}) \end{bmatrix}. \quad (4.27)$$

The reader can confirm these by multiplying A^{-1} as given in (4.26) by A. The 2×2 identity matrix should be obtained. Using these formulas, we may consider the following example.

Example 4.1   Suppose that

$$A = \begin{bmatrix} 1 & -.01 \\ 2 & .01 \end{bmatrix}, \quad y = \begin{bmatrix} 2 \\ 1 \end{bmatrix}.$$

Nominally, the correct solution is

$$x = \begin{bmatrix} 1 \\ -100 \end{bmatrix}.$$

Let us consider different perturbation cases:

1. Suppose that

$$\delta A = \begin{bmatrix} 0 & 0 \\ 0 & .005 \end{bmatrix}, \quad \delta y = [0 \quad 0]^T.$$

In this case

$$\hat{x} = [1.1429 \quad -85.7143]^T.$$

2. Suppose that

$$\delta A = \begin{bmatrix} 0 & 0 \\ 0 & -.03 \end{bmatrix}, \quad \delta y = [0 \quad 0]^T.$$

for which

$$A + \delta A = \begin{bmatrix} 1 & -.01 \\ 2 & -.02 \end{bmatrix}.$$

This matrix is mathematically singular, so it does not possess an inverse. If MATLAB tries to compute $\hat{x}$ using (4.27), then we obtain

$$\hat{x} = 1.0 \times 10^{17} \times [-0.0865 \quad -8.6469]^T$$

and MATLAB issues a warning that the answer may not be correct. Obviously, this is truly a nonsense answer.

3. Suppose that

$$\delta A = \begin{bmatrix} 0 & 0 \\ 0 & -.02 \end{bmatrix}, \delta y = [0.10 \quad -0.05]^T.$$

In this case

$$\hat{x} = [-1.1500 \quad -325.0000]^T.$$

It is clear that small perturbations of A and y can lead to large errors in the computed value for x. These errors are not a result of accumulated rounding errors in the computational algorithm for solving the problem. For computations on a "nonmagical" (i.e., "real") computer, this should be at least intuitively plausible since our formulas for $\hat{x}$ are very simple in the sense of creating little opportunity for rounding error to grow (there are very few arithmetical operations involved). Thus, the errors $x - \hat{x}$ must be due entirely (or nearly so) to uncertainties in the original inputs. We conclude that the real problem is that the linear system we are solving is too sensitive to perturbations in the inputs. This naturally raises the question of how we may detect such sensitivity.

In view of this, we shall say that a matrix A is *ill-conditioned* if the solution x (in $Ax = y$) is very sensitive to perturbations on A and y. Otherwise, the matrix is said to be *well-conditioned*.

We will need to introduce appropriate norms in order to objectively measure the sizes of objects in our problem. However, before doing this we make a few observations that give additional insight into the nature of the problem. In Example 4.1 we note that the first column of A is big (in some sense), while the second column is small. The smallness of the second column makes A close to being singular. A similar observation may be made about Hilbert matrices. For a general $N \times N$ Hilbert matrix, the last two columns are given by

$$
\begin{bmatrix}
\dfrac{1}{N-1} & \dfrac{1}{N} \\[2mm]
\dfrac{1}{N} & \dfrac{1}{N+1} \\[2mm]
\vdots & \vdots \\[2mm]
\dfrac{1}{2N-2} & \dfrac{1}{2N-1}
\end{bmatrix} .
$$

For very large N, it is apparent that these two columns are almost linearly dependent; that is, one may be taken as close to being equal to the other. A simple numerical example is that $\frac{1}{100} \approx \frac{1}{101}$. Thus, at least at the outset, it seems that ill-conditioned matrices are close to being singular, and that this is the root cause of the sensitivity problem.

We now need to extend our treatment of the concept of norms from what we have seen in earlier chapters. Our main source is Golub and Van Loan [5], but similar information is to be found in Ref. 3 or 4 (or the references cited therein). A fairly rigorous treatment of matrix and vector norms can be found in Horn and Johnson [6].

Suppose again that $x \in \mathbf{R}^n$. The p-norm of x is defined to be

$$
||x||_p = \left[\sum_{k=0}^{n-1} |x_k|^p \right]^{1/p} , \tag{4.28}
$$

where $p \geq 1$. The most important special cases are, respectively, the 1-norm, 2-norm, and ∞-norm:

$$
||x||_1 = \sum_{k=0}^{n-1} |x_k|, \tag{4.29a}
$$

$$
||x||_2 = [x^T x]^{1/2}, \tag{4.29b}
$$

$$
||x||_\infty = \max_{0 \leq k \leq n-1} |x_k|. \tag{4.29c}
$$

The operation "max" means to select the biggest $|x_k|$. A *unit vector* with respect to norm $|| \cdot ||$ is a vector x such that $||x|| = 1$. Note that if x is a unit vector with respect to one norm, then it is not necessarily a unit vector with respect to another choice of norm. For example, suppose that $x = [\frac{\sqrt{3}}{2} \frac{1}{2}]^T$; then

$$||x||_2 = 1, ||x||_1 = \frac{\sqrt{3}+1}{2}, ||x||_\infty = \frac{\sqrt{3}}{2}.$$

The vector x is a unit vector under the 2-norm, but is not a unit vector under the 1-norm or the ∞-norm.

Norms have various properties that we will list without proof. Assume that $x, y \in \mathbf{R}^n$. For example, the *Hölder inequality* [recall (1.16) (in Chapter 1) for comparison] is

$$|x^T y| \le ||x||_p ||y||_q \tag{4.30}$$

for which $\frac{1}{p} + \frac{1}{q} = 1$. A special case is the *Cauchy–Schwarz inequality*

$$|x^T y| \le ||x||_2 ||y||_2, \tag{4.31}$$

which is a special instance of Theorem 1.1 (in Chapter 1). An important feature of norms is that they are *equivalent*. This means that if $|| \cdot ||_\alpha$ and $|| \cdot ||_\beta$ are norms on $\mathbf{R}^n$, then there are $c_1, c_2 > 0$ such that

$$c_1 ||x||_\alpha \le ||x||_\beta \le c_2 ||x||_\alpha \tag{4.32}$$

for all $x \in \mathbf{R}^n$. Some special instances of this are

$$||x||_2 \le ||x||_1 \le \sqrt{n} ||x||_2, \tag{4.33a}$$

$$||x||_\infty \le ||x||_2 \le \sqrt{n} ||x||_\infty, \tag{4.33b}$$

$$||x||_\infty \le ||x||_1 \le n ||x||_\infty. \tag{4.33c}$$

Equivalence is significant with respect to our problem in the following manner. When we define condition numbers below, we shall see that the specific value of the condition number depends in part on the choice of norm. However, equivalence says that if a matrix is ill-conditioned with respect to one type of norm, then it must be ill-conditioned with respect to any other type of norm. This can simplify analysis in practice because it allows us to compute the condition number using whatever norms are the easiest to work with. Equivalence can be useful in another respect. If we have a sequence of vectors in the space $\mathbf{R}^n$, then, if the sequence is Cauchy with respect to some chosen norm, it must be Cauchy with respect to any other choice of norm. This can simplify convergence analysis, again because we may pick the norm that is easiest to work with.

In Chapter 2 we considered absolute and relative error in the execution of floating-point operations. In this setting, operations were on scalars, and scalar solutions were generated. Now we must redefine absolute and relative error for vector quantities using the norms defined in the previous paragraph. Since $\hat{x} \in \mathbf{R}^n$

is the computed (i.e., approximate) solution to $x \in \mathbf{R}^n$ it is reasonable to define the *absolute error* to be

$$\epsilon_a = ||\hat{x} - x||, \tag{4.34a}$$

and the *relative error* is

$$\epsilon_r = \frac{||\hat{x} - x||}{||x||}. \tag{4.34b}$$

Of course, $x \neq 0$ is assumed here. The choice of norm is in principle arbitrary. However, if we use the ∞-norm, then the concept of relative error with respect to it can be made equivalent to a statement about the correct number of significant digits in $\hat{x}$:

$$\frac{||\hat{x} - x||_\infty}{||x||_\infty} \approx 10^{-d}. \tag{4.35}$$

In other words, the largest element of the computed solution $\hat{x}$ is correct to approximately d decimal digits. For example, suppose that $x = [1.256 - 2.554]^T$, and $\hat{x} = [1.251 - 2.887]^T$; then $\hat{x} - x = [-0.005 - 0.333]^T$, and so

$$||\hat{x} - x||_\infty = 0.333, \quad ||x||_\infty = 2.554,$$

so therefore $\epsilon_r = 0.1304 \approx 10^{-1}$. Thus, $\hat{x}$ has a largest element that is accurate to about one decimal digit, but the smallest element is observed to be correct to about three significant digits.

Matrices can have norms defined on them. We have remarked that ill conditioning seems to arise when a matrix is close to singular. Suitable matrix norms can allow us to measure how close a matrix is to being singular, and thus gives insight into its condition. Suppose that $A, B \in \mathbf{R}^{m \times n}$ (so A and B are not necessarily square matrices). $|| \cdot |||\mathbf{R}^{m \times n} \to \mathbf{R}$ is a *matrix norm*, provided the following axioms hold:

(MN1) $||A|| \geq 0$ for all A, and $||A|| = 0$ iff $A = 0$.
(MN2) $||A + B|| \leq ||A|| + ||B||$.
(MN3) $||\alpha A|| = |\alpha| \, ||A||$. Constant α is from the same field as the elements of the matrix A.

In the present context we usually consider $\alpha \in \mathbf{R}$. Extensions to complex-valued matrices and vectors are possible. The axioms above are essentially the same as for the norm in all other cases (see Definition 1.3 for comparison). The most common matrix norms are the *Frobenius norm*

$$||A||_F = \sqrt{\sum_{k=0}^{m-1} \sum_{l=0}^{n-1} |a_{k,l}|^2} \tag{4.36a}$$

and the p-norms

$$||A||_p = \sup_{x \neq 0} \frac{||Ax||_p}{||x||_p}. \tag{4.36b}$$

We see that in (4.36b) the matrix p-norm is dependent on the vector p-norm. Via (4.36b), we have

$$||Ax||_p \leq ||A||_p ||x||_p. \tag{4.36c}$$

We may regard A as an operator applied to x that yields output Ax. Equation (4.36c) gives an upper bound on the size of the output, as we know the size of A and the size of x as given by their respective p-norms. Also, since $A \in \mathbf{R}^{m \times n}$ it must be the case that $x \in \mathbf{R}^n$, but $y \in \mathbf{R}^m$. We observe that

$$||A||_p = \sup_{x \neq 0} \left|\left| A\left(\frac{x}{||x||_p}\right) \right|\right|_p = \max_{||x||_p = 1} ||Ax||_p. \tag{4.37}$$

This is an alternative means to compute the matrix p-norm: Evaluate $||Ax||_p$ at all points on the *unit sphere*, which is the set of vectors $\{x | ||x||_p = 1\}$, and then pick the largest value of $||Ax||_p$. Note that the term "sphere" is an extension of what we normally mean by a sphere. For the 2-norm in n dimensions, the unit sphere is clearly

$$||x||_2 = [x_0^2 + x_1^2 + \cdots + x_{n-1}^2]^{1/2} = 1. \tag{4.38}$$

This represents our intuitive (i.e., Euclidean) notion of a sphere. But, say, for the 1-norm the unit sphere is

$$||x||_1 = |x_0| + |x_1| + \cdots + |x_{n-1}| = 1. \tag{4.39}$$

Equations (4.38) and (4.39) specify very different looking surfaces in n-dimensional space. A suggested exercise is to sketch these spheres for $n = 2$.

As with vector norms, matrix norms have various properties. One property possessed by the matrix p-norms is called the *submultiplicative property*:

$$||AB||_p \leq ||A||_p ||B||_p \quad A \in \mathbf{R}^{m \times n}, \quad B \in \mathbf{R}^{n \times q}. \tag{4.40}$$

(The reader is warned that not all matrix norms possess this property; a counterexample appears on p. 57 of Golub and Van Loan [5]). A miscellany of other properties (including equivalences) is

$$||A||_2 \leq ||A||_F \leq \sqrt{n} ||A||_2, \tag{4.41a}$$

$$\max_{i,j} |a_{i,j}| \leq ||A||_2 \leq \sqrt{mn} \max_{i,j} |a_{i,j}|, \tag{4.41b}$$

$$||A||_1 = \max_{j \in \mathbf{Z}_n} \sum_{i=0}^{m-1} |a_{i,j}|, \tag{4.41c}$$

$$||A||_\infty = \max_{i \in \mathbf{Z}_m} \sum_{j=0}^{n-1} |a_{i,j}|, \tag{4.41d}$$

$$\frac{1}{\sqrt{n}} ||A||_\infty \le ||A||_2 \le \sqrt{m} ||A||_\infty, \tag{4.41e}$$

$$\frac{1}{\sqrt{m}} ||A||_1 \le ||A||_2 \le \sqrt{n} ||A||_1. \tag{4.41f}$$

The equivalences [e.g., (4.41a) and (4.41b)] have the same significance for matrices as the analogous equivalences for vectors seen in (4.32) and (4.33).

From (4.41c,d) we see that computing matrix 1-norms and ∞-norms is easy. However, computing matrix 2-norms is not easy. Consider (4.37) with $p = 2$:

$$||A||_2 = \max_{||x||_2=1} ||Ax||_2. \tag{4.42}$$

Let $R = A^T A \in \mathbf{R}^{n \times n}$ (no, R is not a Hilbert matrix here; we have "recycled" the symbol for another use), so then

$$||Ax||_2^2 = x^T A^T A x = x^T R x. \tag{4.43}$$

Now consider $n = 2$. Thus

$$x^T R x = [x_0 x_1] \begin{bmatrix} r_{00} & r_{01} \\ r_{10} & r_{11} \end{bmatrix} \begin{bmatrix} x_0 \\ x_1 \end{bmatrix}, \tag{4.44}$$

where $r_{01} = r_{10}$ because $R = R^T$. The vectors and matrix in (4.44) multiply out to become

$$x^T R x = r_{00} x_0^2 + 2 r_{01} x_0 x_1 + r_{11} x_1^2. \tag{4.45}$$

Since $||A||_2^2 = \max_{||x||_2=1} ||Ax||_2^2$, we may find $||A||_2^2$ by maximizing (4.45) subject to the equality constraint $||x||_2^2 = 1$, i.e., $x^T x = x_0^2 + x_1^2 = 1$. This problem may be solved by using *Lagrange multipliers* (considered somewhat more formally in Section 8.5). Thus, we must maximize

$$V(x) = x^T R x - \lambda [x^T x - 1], \tag{4.46}$$

where λ is the Lagrange multiplier. Since

$$V(x) = r_{00} x_0^2 + 2 r_{01} x_0 x_1 + r_{11} x_1^2 - \lambda [x_0^2 + x_1^2 - 1],$$

we have

$$\frac{\partial V(x)}{\partial x_0} = 2r_{00}x_0 + 2r_{01}x_1 - 2\lambda x_0 = 0,$$

$$\frac{\partial V(x)}{\partial x_1} = 2r_{01}x_0 + 2r_{11}x_1 - 2\lambda x_1 = 0,$$

and these equations may be rewritten in matrix form as

$$\begin{bmatrix} r_{00} & r_{01} \\ r_{10} & r_{11} \end{bmatrix} \begin{bmatrix} x_0 \\ x_1 \end{bmatrix} = \lambda \begin{bmatrix} x_0 \\ x_1 \end{bmatrix}. \tag{4.47}$$

In other words, $Rx = \lambda x$. Thus, the optimum choice of x is an *eigenvector* of $R = A^T A$. But which eigenvector is it?

First note that A^{-1} exists (by assumption), so $x^T Rx = x^T A^T Ax = (Ax)^T(Ax) > 0$ for all $x \neq 0$. Therefore, $R > 0$. Additionally, $R = R^T$, so all of the eigenvalues of R are real numbers.[5] Furthermore, because $R > 0$, all of its eigenvalues are positive. This follows if we consider $Rx = \lambda x$, and assume that $\lambda < 0$. In this case $x^T Rx = \lambda x^T x = \lambda ||x||_2^2 < 0$ for any $x \neq 0$. [If $\lambda = 0$, then $Rx = 0 \cdot x = 0$ implies that $x = 0$ (as R^{-1} exists), so $x^T Rx = 0$.] But this contradicts the assumption that $R > 0$, and so all of the eigenvalues of R must be positive. Now, since $||Ax||_2^2 = x^T A^T Ax = x^T Rx = x^T(\lambda x) = \lambda ||x||_2^2$, and since $||x||_2^2 = 1$, it must be the case that $||Ax||_2^2$ is biggest for the eigenvector of R corresponding to the biggest eigenvalue of R. If the eigenvalues of R are denoted λ_1 and λ_0 with $\lambda_1 \geq \lambda_0 > 0$, then finally we must have

$$||A||_2^2 = \lambda_1. \tag{4.48}$$

This argument can be generalized for all $n > 2$. If $R > 0$, we assume that all of its eigenvalues are distinct (this is not always true). If we denote them by $\lambda_0, \lambda_1, \ldots, \lambda_{n-1}$, then we may arrange them in decreasing order:

$$\lambda_{n-1} > \lambda_{n-2} > \cdots > \lambda_1 > \lambda_0 > 0. \tag{4.49}$$

Therefore, for $A \in \mathbf{R}^{m \times n}$

$$||A||_2^2 = \lambda_{n-1}. \tag{4.50}$$

The problem of computing the eigenvalues and eigenvectors of a matrix has its own special numerical difficulties. At this point we warn the reader that these problems must never be treated lightly.

---

[5] If A is a real-valued symmetric square matrix, then we may prove this claim as follows. Suppose that for eigenvector x of A, the eigenvalue is λ, that is, $Ax = \lambda x$. Now $((Ax)^*)^T = ((\lambda x)^*)^T$, and so $(x^*)^T A^T = \lambda^*(x^*)^T$. Therefore, $(x^*)^T A^T x = \lambda^*(x^*)^T x$. But $(x^*)^T A^T x = (x^*)^T Ax = \lambda(x^*)^T x$, so finally $\lambda^*(x^*)^T x = \lambda(x^*)^T x$, so we must have $\lambda = \lambda^*$. This can be true only if $\lambda \in \mathbf{R}$.

Example 4.2   Let $\det(A)$ denote the determinant of A. Suppose that $R = A^T A$, where

$$R = \begin{bmatrix} 1 & 0.5 \\ 0.5 & 1 \end{bmatrix}.$$

We will find $||A||_2$. Consider

$$\begin{bmatrix} 1 & 0.5 \\ 0.5 & 1 \end{bmatrix} \begin{bmatrix} x_0 \\ x_1 \end{bmatrix} = \lambda \begin{bmatrix} x_0 \\ x_1 \end{bmatrix}.$$

We must solve $\det(\lambda I - R) = 0$ for λ. [Recall that $\det(\lambda I - R)$ is the *characteristic polynomial* of R.] Thus

$$\det(\lambda I - R) = \det\left(\begin{bmatrix} \lambda - 1 & -0.5 \\ -0.5 & \lambda - 1 \end{bmatrix}\right) = (\lambda - 1)^2 - \frac{1}{4} = 0,$$

and $(\lambda - 1)^2 - \frac{1}{4} = \lambda^2 - 2\lambda + \frac{3}{4} = 0$, for

$$\lambda = \frac{-(-2) \pm \sqrt{(-2)^2 - 4 \cdot 1 \cdot \frac{3}{4}}}{2 \cdot 1} = \frac{2 \pm 1}{2} = \frac{1}{2}, \frac{3}{2}.$$

So, $\lambda_1 = \frac{3}{2}$, $\lambda_0 = \frac{1}{2}$. Thus, $||A||_2^2 = \lambda_1 = \frac{3}{2}$, and so finally

$$||A||_2 = \sqrt{\frac{3}{2}}.$$

(We do not need the eigenvectors of R to compute the 2-norm of A.)

We see that the essence of computing the 2-norm of matrix A is to find the zeros of the characteristic polynomial of $A^T A$. The problem of finding polynomial zeros is the subject of a later chapter. Again, this problem has its own special numerical difficulties that must never be treated lightly.

We now derive the condition number. Begin by assuming that $A \in \mathbf{R}^{n \times n}$, and that A^{-1} exists. The *error* between computed solution $\hat{x}$ to $Ax = y$ and x is

$$e = x - \hat{x}. \tag{4.51}$$

$Ax = y$, but $A\hat{x} \neq y$ in general. So we may define the *residual*

$$r = y - A\hat{x}. \tag{4.52}$$

We see that

$$Ae = Ax - A\hat{x} = y - A\hat{x} = r. \tag{4.53}$$

Thus, $e = A^{-1}r$. We observe that if $e = 0$, then $r = 0$, but if r is small, then e is not necessarily small because A^{-1} might be big, making $A^{-1}r$ big. In other words, *a small residual r does not guarantee that $\hat{x}$ is close to x*. Sometimes r is computed as a cursory check to see if $\hat{x}$ is "reasonable." The main advantage of r is that it may always be computed, whereas x is not known in advance and so e may never be computed exactly. Below it will be shown that considering r in combination with a condition number is a more reliable method of assessing how close $\hat{x}$ is to x.

Now, since $e = A^{-1}r$, we can say that $||e||_p = ||A^{-1}r||_p \leq ||A^{-1}||_p||r||_p$ [via (4.36c)]. Similarly, since $r = Ae$, we have $||r||_p = ||Ae||_p \leq ||A||_p||e||_p$. Thus

$$\frac{||r||_p}{||A||_p} \leq ||e||_p \leq ||A^{-1}||_p||r||_p. \tag{4.54}$$

Similarly, $x = A^{-1}y$, so immediately

$$\frac{||y||_p}{||A||_p} \leq ||x||_p \leq ||A^{-1}||_p||y||_p. \tag{4.55}$$

If $||x||_p \neq 0$, and $||y||_p \neq 0$, then taking reciprocals in (4.55) yields

$$\frac{1}{||A^{-1}||_p||y||_p} \leq \frac{1}{||x||_p} \leq \frac{||A||_p}{||y||_p}. \tag{4.56}$$

We may multiply corresponding terms in (4.56) and (4.54) to obtain

$$\frac{1}{||A^{-1}||_p||A||_p}\frac{||r||_p}{||y||_p} \leq \frac{||e||_p}{||x||_p} \leq ||A^{-1}||_p||A||_p\frac{||r||_p}{||y||_p}. \tag{4.57}$$

We recall from (4.34b) that $\epsilon_r = \frac{||x-\hat{x}||_p}{||x||_p} = \frac{||e||_p}{||x||_p}$, so

$$\frac{1}{||A^{-1}||_p||A||_p}\frac{||r||_p}{||y||_p} \leq \epsilon_r \leq ||A^{-1}||_p||A||_p\frac{||r||_p}{||y||_p}. \tag{4.58}$$

We call

$$\frac{||r||_p}{||y||_p} = \frac{||y - A\hat{x}||_p}{||y||_p} \tag{4.59}$$

the *relative residual*. We define

$$\kappa_p(A) = ||A||_p||A^{-1}||_p \tag{4.60}$$

to be the *condition number of A*. It is immediately apparent that $\kappa_p(A) \geq 1$ for any A and valid p. We see that ϵ_r is between $1/\kappa_p(A)$ and $\kappa_p(A)$ times the relative residual. In particular, if $\kappa_p(A) >> 1$ (i.e., if the condition number is very large), even if the relative residual is tiny, then ϵ_r *might* be large. On the other hand, if $\kappa_p(A)$ is close to unity, then ϵ_r will be small if the relative residual is small. In conclusion, *if $\kappa_p(A)$ is large, it is a warning (not a certainty) that small* *perturbations in A and y may cause $\hat{x}$ to differ greatly from x. Equivalently, if $\kappa_p(A)$ is large, then a small r does not imply that $\hat{x}$ is close to x.*

A rule of thumb in interpreting condition numbers is as follows [3, p. 229], and is more or less true regardless of p in (4.60). If $\kappa_p(A) \approx d \times 10^k$, where d is a decimal digit from one to nine, we can expect to lose (at worst) about k digits of accuracy. The reason that p does not matter too much is because we recall that matrix norms are equivalent. Therefore, for this rule of thumb to be useful, the working precision of the computing machine/software package must be known. For example, MATLAB computes to about 16 decimal digits of precision. Thus, $k \geq 16$ would give us concern that $\hat{x}$ is not close to x.

Example 4.3 Suppose that

$$A = \begin{bmatrix} 1 & 1 - \epsilon \\ 1 & 1 \end{bmatrix} \in \mathbf{R}^{2 \times 2}, |\epsilon| << 1.$$

We will determine an estimate of $\kappa_1(A)$. Clearly

$$A^{-1} = \frac{1}{\epsilon} \begin{bmatrix} 1 & -1 + \epsilon \\ -1 & 1 \end{bmatrix} = \begin{bmatrix} b_{00} & b_{01} \\ b_{10} & b_{11} \end{bmatrix}.$$

We have

$$\sum_{i=0}^{1} |a_{i,0}| = |a_{00}| + |a_{10}| = 2, \sum_{i=0}^{1} |a_{i,1}| = |a_{01}| + |a_{11}| = |1 - \epsilon| + 1,$$

so via (4.41c), $||A||_1 = \max\{2, |1 - \epsilon| + 1\} \approx 2$. Similarly

$$\sum_{i=0}^{1} |b_{i,0}| = |b_{00}| + |b_{10}| = \frac{2}{|\epsilon|}, \sum_{i=0}^{1} |b_{i,1}| = |b_{01}| + |b_{11}| = \frac{1}{|\epsilon|} + \left| \frac{-1 + \epsilon}{\epsilon} \right|,$$

so again via (4.41c) $||A^{-1}||_1 = \max \left\{ \frac{2}{|\epsilon|}, \left| 1 - \frac{1}{\epsilon} \right| + \frac{1}{|\epsilon|} \right\} \approx \frac{2}{|\epsilon|}$. Thus

$$\kappa_1(A) \approx \frac{4}{|\epsilon|}.$$

We observe that if $\epsilon = 0$, then A^{-1} does not exist, so our approximation to $\kappa_1(A)$ is a reasonable result because

$$\lim_{|\epsilon| \to 0} \kappa_1(A) = \infty.$$

We may wish to compute $\kappa_2(A) = ||A||_2 ||A^{-1}||_2$. We will suppose that $A \in \mathbf{R}^{n \times n}$ and that A^{-1} exists. But we recall that computing matrix 2-norms involves finding eigenvalues. More specifically, $||A||_2^2$ is the largest eigenvalue of $R = A^T A$

[recall (4.50)]. Suppose, as in (4.49), that λ_0 is the smallest eigenvalue of R for which the corresponding eigenvector is denoted by v, that is, $Rv = \lambda_0 v$. Then we observe that $R^{-1}v = \frac{1}{\lambda_0}v$. In other words, $1/\lambda_0$ is an eigenvalue of R^{-1}. By similar reasoning, $1/\lambda_k$ for $k \in \mathbf{Z}_n$ must all be eigenvalues of R^{-1}. Thus, $1/\lambda_0$ will be the biggest eigenvalue of R^{-1}. For present simplicity assume that A is a *normal matrix*. This means that $AA^T = A^TA = R$. The reader is cautioned that not all matrices A are normal. However, in this case we have $R^{-1} = A^{-1}A^{-T} = A^{-T}A^{-1}$. [Recall that $(A^{-1})^T = (A^T)^{-1} = A^{-T}$.] We have that $||A^{-1}||_2^2$ is the largest eigenvalue of $A^{-T}A^{-1}$, but $R^{-1} = A^{-T}A^{-1}$ since A is assumed normal. The largest eigenvalue of R^{-1} has been established to be $1/\lambda_0$, so it must be the case that for a normal matrix A (real-valued and invertible)

$$\kappa_2(A) = \sqrt{\frac{\lambda_{n-1}}{\lambda_0}}, \qquad (4.61)$$

that is, A is ill-conditioned if the ratio of the biggest to smallest eigenvalue of A^TA is large. In other words, a *large eigenvalue spread* is associated with matrix ill conditioning. It turns out that this conclusion holds even if A is not normal; that is, (4.61) is valid even if A is not normal. But we will not prove this. (The interested reader can see pp. 312 and 340 of Horn and Johnson [6] for more information.)

An obvious difficulty with condition numbers is that their exact calculation often seems to require knowledge of A^{-1}. Clearly this is problematic since computing A^{-1} accurately may not be easy or possible (because of ill conditioning). We seem to have a "chicken and egg" problem. This problem is often dealt with by using *condition number estimators*. This in turn generally involves placing bounds on condition numbers. But the subject of condition number estimation is not within the scope of this book. The interested reader might consult Higham [7] for further information on this subject if desired. There is some information on this matter in the treatise by Golub and Van Loan [5, pp. 128–130], which includes a pseudocode algorithm for ∞-norm condition number estimation of an upper triangular nonsingular matrix. We remark that $||A||_2$ is sometimes called the *spectral norm* of A, and is actually best computed using entities called *singular values* [5, p. 72]. This is because computing singular values avoids the necessity of computing A^{-1}, and can be done in a numerically reliable manner. Singular values will be discussed in more detail later.

We conclude this section with a remark about the Hilbert matrix R of Section 4.3. As discussed by Hill [3, p. 232], we have

$$\kappa_2(R) \propto e^{\alpha N}$$

for some $\alpha > 0$. (Recall that symbol $\propto$ means "proportional to.") Proving this is tough, and we will not attempt it. Thus, the condition number of R grows very rapidly with N and explains why the attempt to invert R in Appendix 4.A failed for so small a value of N.

4.5 *LU* DECOMPOSITION

In this section we will assume $A \in \mathbf{R}^{n \times n}$, and that A^{-1} exists. Many algorithms to solve $Ax = y$ work by factoring the matrix A in various ways. In this section we consider a Gaussian elimination approach to writing A as

$$A = LU, \tag{4.62}$$

where L is a nonsingular *lower triangular matrix*, and U is a nonsingular *upper triangular matrix*. This is the *LU decomposition (factorization)* of A. Naturally, $L, U \in \mathbf{R}^{n \times n}$, and $L = [l_{i,j}]$, $U = [u_{i,j}]$. Since these matrices are lower and upper triangular, respectively, it must be the case that

$$l_{i,j} = 0 \quad \text{for} \quad j > i \quad \text{and} \quad u_{i,j} = 0 \quad \text{for} \quad j < i. \tag{4.63}$$

For example, the following are (respectively) lower and upper triangular matrices:

$$L = \begin{bmatrix} 1 & 0 & 0 \\ 1 & 1 & 0 \\ 1 & 1 & 1 \end{bmatrix}, \qquad U = \begin{bmatrix} 1 & 2 & 3 \\ 0 & 4 & 5 \\ 0 & 0 & 6 \end{bmatrix}.$$

These matrices are clearly nonsingular since their determinants are 1 and 24, respectively. In fact, L is nonsingular iff $l_{i,i} \neq 0$ for all i, and U is nonsingular iff $u_{j,j} \neq 0$ for all j. We note that with A factored as in (4.62), the solution of $Ax = y$ becomes quite easy, but the details of this will be considered later. We now concentrate on finding the factors L, U.

We begin by defining a *Gauss transformation* matrix G_k such that

$$G_k x = \begin{bmatrix} 1 & \cdots & 0 & 0 & \cdots & 0 \\ \vdots & & \vdots & \vdots & & \vdots \\ 0 & \cdots & 1 & 0 & \cdots & 0 \\ 0 & \cdots & -\tau_k^k & 1 & \cdots & 0 \\ \vdots & & \vdots & \vdots & & \vdots \\ 0 & \cdots & -\tau_{n-1}^k & 0 & \cdots & 1 \end{bmatrix} \begin{bmatrix} x_0 \\ \vdots \\ x_{k-1} \\ x_k \\ \vdots \\ x_{n-1} \end{bmatrix} = \begin{bmatrix} x_0 \\ \vdots \\ x_{k-1} \\ 0 \\ \vdots \\ 0 \end{bmatrix} \tag{4.64}$$

for

$$\tau_i^k = \frac{x_i}{x_{k-1}}, i = k, \ldots, n - 1. \tag{4.65}$$

The superscript k on τ_j^k **does** *not* denote raising τ_j to a power; it is simply part of the name of the symbol. This naming convention is needed to account for the fact that there is a different set of τ values for every G_k. For this to work requires that $x_{k-1} \neq 0$. Equation (4.65) followed from considering the matrix–vector product in (4.64):

$$-\tau_k^k x_{k-1} + x_k = 0$$

$$-\tau_{k+1}^k x_{k-1} + x_{k+1} = 0$$

$$\vdots$$

$$-\tau_{n-1}^k x_{k-1} + x_{n-1} = 0.$$

We observe that G_k is "designed" to annihilate the last $n - k$ elements of vector x. We also see that G_k is lower triangular, and if it exists, always possesses an inverse because the main diagonal elements are all equal to unity. A lower triangular matrix where all of the main diagonal elements are equal to unity is called *unit lower triangular*. Similar terminology applies to upper triangular matrices. Define the *kth Gauss vector*

$$(\tau^k)^T = [\underbrace{0 \cdots 0}_{k \text{ zeros}} \tau_k^k \tau_{k+1}^k \cdots \tau_{n-1}^k]. \tag{4.66}$$

The kth unit vector is

$$e_k^T = [\underbrace{0 \cdots 0}_{k \text{ zeros}} 1 \underbrace{0 \cdots 0}_{n-k-1 \text{ zeros}}]. \tag{4.67}$$

If I is an $n \times n$ identity matrix, then

$$G_k = I - \tau^k e_{k-1}^T \tag{4.68}$$

for $k = 1, 2, \ldots, n - 1$. For example, if $n = 4$, we have

$$G_1 = \begin{bmatrix} 1 & 0 & 0 & 0 \\ -\tau_1^1 & 1 & 0 & 0 \\ -\tau_2^1 & 0 & 1 & 0 \\ -\tau_3^1 & 0 & 0 & 1 \end{bmatrix}, \quad G_2 = \begin{bmatrix} 1 & 0 & 0 & 0 \\ 0 & 1 & 0 & 0 \\ 0 & -\tau_2^2 & 1 & 0 \\ 0 & -\tau_3^2 & 0 & 1 \end{bmatrix},$$

$$G_3 = \begin{bmatrix} 1 & 0 & 0 & 0 \\ 0 & 1 & 0 & 0 \\ 0 & 0 & 1 & 0 \\ 0 & 0 & -\tau_3^3 & 1 \end{bmatrix}. \tag{4.69}$$

The Gauss transformation matrices may be applied to A, yielding an upper triangular matrix. This is illustrated by the following example.

Example 4.4 Suppose that

$$A = \begin{bmatrix} 1 & 2 & 3 & 4 \\ -1 & 1 & 2 & 1 \\ 0 & 2 & 1 & 3 \\ 0 & 0 & 1 & 1 \end{bmatrix} (= A^0).$$

We introduce matrices A^k, where $A^k = G_k A^{k-1}$ for $k = 1, 2, \ldots, n - 1$, and finally $U = A^{n-1}$. Once again, A^k is not the kth power of A, but rather denotes the kth matrix in a sequence of matrices. Now consider

$$
G_1 A^0 = \begin{bmatrix} 1 & 0 & 0 & 0 \\ 1 & 1 & 0 & 0 \\ 0 & 0 & 1 & 0 \\ 0 & 0 & 0 & 1 \end{bmatrix} \begin{bmatrix} 1 & 2 & 3 & 4 \\ -1 & 1 & 2 & 1 \\ 0 & 2 & 1 & 3 \\ 0 & 0 & 1 & 1 \end{bmatrix} = \begin{bmatrix} 1 & 2 & 3 & 4 \\ 0 & 3 & 5 & 5 \\ 0 & 2 & 1 & 3 \\ 0 & 0 & 1 & 1 \end{bmatrix} = A^1
$$

for which the τ_i^1 entries in the first column of G_1 depend on the first column of A^0 (i.e., of A) according to (4.65). Similarly

$$
G_2 A^1 = \begin{bmatrix} 1 & 0 & 0 & 0 \\ 0 & 1 & 0 & 0 \\ 0 & -\frac{2}{3} & 1 & 0 \\ 0 & 0 & 0 & 1 \end{bmatrix} \begin{bmatrix} 1 & 2 & 3 & 4 \\ 0 & 3 & 5 & 5 \\ 0 & 2 & 1 & 3 \\ 0 & 0 & 1 & 1 \end{bmatrix} = \begin{bmatrix} 1 & 2 & 3 & 4 \\ 0 & 3 & 5 & 5 \\ 0 & 0 & -\frac{7}{3} & -\frac{1}{3} \\ 0 & 0 & 1 & 1 \end{bmatrix} = A^2
$$

for which the τ_i^2 entries in the second column of G_2 depend on the second column of A^1, and also

$$
G_3 A^2 = \begin{bmatrix} 1 & 0 & 0 & 0 \\ 0 & 1 & 0 & 0 \\ 0 & 0 & 1 & 0 \\ 0 & 0 & \frac{3}{7} & 1 \end{bmatrix} \begin{bmatrix} 1 & 2 & 3 & 4 \\ 0 & 3 & 5 & 5 \\ 0 & 0 & -\frac{7}{3} & -\frac{1}{3} \\ 0 & 0 & 1 & 1 \end{bmatrix} = \begin{bmatrix} 1 & 2 & 3 & 4 \\ 0 & 3 & 5 & 5 \\ 0 & 0 & -\frac{7}{3} & -\frac{1}{3} \\ 0 & 0 & 0 & \frac{6}{7} \end{bmatrix} = U
$$

for which the τ_i^3 entries in the third column of G_3 depend on the third column of A^2. We see that U is indeed upper triangular, and it is also nonsingular. We also see that

$$
U = \underbrace{G_3 G_2 G_1}_{=L_1} A.
$$

Since the product of lower triangular matrices is a lower triangular matrix, it is the case that $L_1 = G_3 G_2 G_1$ is lower triangular. Thus

$$
A = L_1^{-1} U.
$$

Since the inverse (if it exists) of a lower triangular matrix is also a lower triangular matrix, we can define $L = L_1^{-1}$, and so $A = LU$. Thus

$$
L = G_1^{-1} G_2^{-1} G_3^{-1}.
$$

From this example it appears that we need to do much work in order to find G_k^{-1}. However, this is not the case. It turns out that

$$
G_k^{-1} = I + \tau^k e_{k-1}^T. \tag{4.70}
$$

This is easy to confirm. From (4.68) and (4.70)

$$G_k G_k^{-1} = [I - \tau^k e_{k-1}^T][I + \tau^k e_{k-1}^T] = I - \tau^k e_{k-1}^T + \tau^k e_{k-1}^T - \tau^k e_{k-1}^T \tau^k e_{k-1}^T$$
$$= I - \tau^k e_{k-1}^T \tau^k e_{k-1}^T.$$

But from (4.66) and (4.67) $e_{k-1}^T \tau^k = 0$, so finally $G_k G_k^{-1} = I$.

To obtain τ_i^k from (4.65), we see that we must divide by x_{k-1}. In our matrix factorization application of the Gauss transformation, we have seen (in Example 4.4) that x_{k-1} will be an element of A^k. These elements are called *pivots*. It is apparent that the factorization procedure cannot work if a pivot is zero. The occurrence of zero-valued pivots is a common situation. A simple example of a matrix that cannot be factored with our algorithm is

$$A = \begin{bmatrix} 0 & 1 \\ 1 & 0 \end{bmatrix}. \tag{4.71}$$

In this case

$$G_1 A = \begin{bmatrix} 1 & 0 \\ -\tau_1^1 & 1 \end{bmatrix} \begin{bmatrix} 0 & 1 \\ 1 & 0 \end{bmatrix} = \begin{bmatrix} 0 & 1 \\ 1 & -\tau_1^1 \end{bmatrix}, \tag{4.72}$$

and from (4.65)

$$\tau_1^1 = \frac{x_1}{x_0} = \frac{a_{10}}{a_{00}} = \frac{1}{0} \to \infty. \tag{4.73}$$

This result implies that not all matrices possess an LU factorization. Let $\det(A)$ denote the *determinant of A*. We may state a general condition for the existence of the LU factorization:

Theorem 4.1: Since $A = [a_{i,j}]_{i,j=0,\dots,n-1} \in \mathbf{R}^{n \times n}$ we define the *kth leading principle submatrix of A* to be $A_k = [a_{i,j}]_{i,j=0,\dots,k-1} \in \mathbf{R}^{k \times k}$ for $k = 1, 2, \dots, n$ (so that $A = A_n$, and $A_1 = [a_{00}] = a_{00}$). There exists a unit lower triangular matrix L and an upper triangular matrix U such that $A = LU$, provided that $\det(A_k) \neq 0$ for all $k = 1, 2 \dots, n$. Furthermore, with $U = [u_{i,j}] \in \mathbf{R}^{n \times n}$ we have $\det(A_k) = \prod_{i=0}^{k-1} u_{i,i}$.

The proof is given in Golub and Van Loan [5]. It will not be considered here. For A in (4.71), we see that $A_1 = [0] = 0$, so $\det(A_1) = 0$. Thus, even though A^{-1} exists, it does not possess an LU decomposition. It is also easy to verify that for

$$A = \begin{bmatrix} 1 & 4 & 1 \\ 2 & 8 & 1 \\ 0 & -1 & 1 \end{bmatrix},$$

although A^{-1} exists, again A does not possess an LU decomposition. In this case we have $\det(A_2) = \det\left(\begin{bmatrix} 1 & 4 \\ 2 & 8 \end{bmatrix}\right) = 0$. Theorem 4.1 leads to a test of positive definiteness according to the following theorem.

Theorem 4.2: Suppose $R \in \mathbf{R}^{n \times n}$ with $R = R^T$. Suppose that $R = LDL^T$, where L is unit lower triangular, and D is a diagonal matrix ($L = [l_{i,j}]_{i,j=0,\ldots,n-1}$, $D = [d_{i,j}]_{i,j=0,\ldots,n-1}$). If $d_{i,i} > 0$ for all $i \in \mathbf{Z}_n$, then $R > 0$.

Proof L is unit lower triangular, so for any $y \in \mathbf{R}^n$ there will be a unique $x \in \mathbf{R}^n$ such that

$$y = L^T x \quad (y^T = x^T L)$$

because L^{-1} exists. Thus, assuming $D > 0$

$$x^T R x = x^T L D L^T x = y^T D y = \sum_{i=0}^{n-1} y_i^2 d_{i,i} > 0$$

for all $y \neq 0$, since $d_{i,i} > 0$ for all $i \in \mathbf{Z}_n$. In fact, $\sum_{i=0}^{n-1} y_i^2 d_{i,i} = 0$ iff $y_i = 0$ for all $i \in \mathbf{Z}_n$. Consequently, $x^T R x > 0$ for all $x \neq 0$, and so immediately $R > 0$.

We relate D in Theorem 4.2 to U in Theorem 4.1 according to $U = DL^T$. If the LDL^T decomposition of a matrix R exists, then matrix D immediately tells us whether R is pd just by viewing the signs of the diagonal elements.

We may define (as in Example 4.4) $A^k = [a_{i,j}^k]$, where $k = 0, 1, \ldots, n-1$ and $A^0 = A$. Consequently

$$\tau_i^k = \frac{x_i}{x_{k-1}} = \frac{a_{i,k-1}^{k-1}}{a_{k-1,k-1}^{k-1}} \tag{4.74}$$

for $i = k, k+1, \ldots, n-1$. This follows because G_k contains τ_i^k, and as observed in the example above, τ_i^k depends on the column indexed $k-1$ in A^{k-1}. Thus, a pseudocode program for finding U can therefore be stated as follows:

```
A⁰ := A;
for k := 1 to n − 1 do begin
    for i := k to n − 1 do begin
        τᵏᵢ := aᵏ⁻¹ᵢ,ₖ₋₁/aᵏ⁻¹ₖ₋₁,ₖ₋₁; {This loop computes τᵏ}
    end;
    Aᵏ := GₖAᵏ⁻¹; { Gₖ contains τᵏᵢ via (4.64) }
end;
U := Aⁿ⁻¹;
```

We see that the pivots are $a_{k-1,k-1}^{k-1}$ for $k = 1, 2, \ldots, n-1$. Now

$$U = G_{n-1} G_{n-2} \cdots G_2 G_1 A.$$

so

$$A = \underbrace{G_1^{-1} G_2^{-1} \cdots G_{n-2}^{-1} G_{n-1}^{-1}}_{=L} U. \tag{4.75}$$

Consequently, from (4.70), we obtain

$$L = (I + \tau^1 e_0^T)(I + \tau^2 e_1^T) \cdots (I + \tau^{n-1} e_{n-2}^T) = I + \sum_{k=1}^{n-1} \tau^k e_{k-1}^T. \quad (4.76)$$

To confirm the last equality of (4.76), consider defining $L_m = G_1^{-1} \cdots G_m^{-1}$ for $m = 1, \ldots, n-1$. Assume that $L_m = I + \sum_{k=1}^{m} \tau^k e_{k-1}^T$, which is true for $m = 1$ because $L_1 = G_1^{-1} = I + \tau^1 e_0^T$. Consider $L_{m+1} = L_m G_{m+1}^{-1}$, so

$$L_{m+1} = \left(I + \sum_{k=1}^{m} \tau^k e_{k-1}^T \right) (I + \tau^{m+1} e_m^T)$$

$$= I + \sum_{k=1}^{m} \tau^k e_{k-1}^T + \tau^{m+1} e_m^T + \sum_{k=1}^{m} \tau^k e_{k-1}^T \tau^{m+1} e_m^T.$$

But $e_{k-1}^T \tau^{m+1} = 0$ for $k = 1, \ldots, m$ from (4.66) and (4.67). Thus

$$L_{m+1} = I + \sum_{k=1}^{m} \tau^k e_{k-1}^T + \tau^{m+1} e_m^T = I + \sum_{k=1}^{m+1} \tau^k e_{k-1}^T.$$

Therefore, (4.76) is valid by mathematical induction. (A simpler example of a proof by induction appears in Appendix 3.B.) Because of (4.76), the previous pseudocode implicitly computes L as well as U. Thus, if no zero-valued pivots are encountered, the algorithm will terminate, having provided us with both L and U. [As an exercise, the reader should use (4.76) to find L in Example 4.4 simply by looking at the appropriate entries of the matrices G_k; that is, do not use $L = G_1^{-1} G_2^{-1} G_3^{-1}$. Having found L by this means, confirm that $LU = A$.] We remark that (4.76) shows that L is *unit* lower triangular.

It is worth mentioning that certain classes of matrix are guaranteed to possess an *LU* decomposition. Suppose that $A \in \mathbf{R}^{n \times n}$ with $A = A^T$ and $A > 0$. Let $v = [\underbrace{v_0 \cdots v_{k-1}}_{=u^T} \ \underbrace{0 \cdots 0}_{n-k \text{ zeros}}]^T$; then, if $v \neq 0$, we have $v^T A v > 0$, but if A_k is the kth leading principle submatrix of A, then

$$v^T A v = u^T A_k u > 0$$

which holds for all $k = 1, 2, \ldots, n$. Consequently, $A_k > 0$ for all k, and so A_k^{-1} exists for all k. Since A_k^{-1} exists for all k, it follows that $\det(A_k) \neq 0$ for all k. The conditions of Theorem 4.1 are met, and so A possesses an *LU* decomposition. That is, *all real-valued, symmetric positive definite matrices possess an LU decomposition.*

We recall that the class of positive definite matrices is an important one since they have a direct association with least-squares approximation problems. This was demonstrated in Section 4.2.

How many *floating-point operations* (flops) are needed by the algorithm for finding the *LU* decomposition of a matrix? Answering this question gives us an indication of the *computational complexity* of the algorithm. Neglecting multiplication by zero or by one, to compute $A^k = G_k A^{k-1}$ requires $(n-k)(n-k+1)$ multiplications, and the same number of additions. This follows from considering the product $G_k A^{k-1}$ with the factors partitioned into submatrices according to

$$
G_k = \begin{bmatrix} I_k & 0 \\ T_k & I_{n-k} \end{bmatrix}, \qquad A^{k-1} = \begin{bmatrix} A_{00}^{k-1} & A_{01}^{k-1} \\ 0 & A_{11}^{k-1} \end{bmatrix}, \qquad (4.77)
$$

where I_k is a $k \times k$ identity matrix, T_k is $(n-k) \times k$ and is zero-valued except for its last column, which contains $-\tau^k$ [see (4.64)]. Similarly, A_{00}^{k-1} is $(k-1) \times (k-1)$, A_{01}^k is $(k-1) \times (n-k+1)$, and A_{11}^{k-1} is $(n-k+1) \times (n-k+1)$. From the pseudocode, we see that we need $\sum_{k=1}^{n-1}(n-k)$ division operations. Operation $A^k = G_k A^{k-1}$ is executed for $k = 1$ to $n-1$, so the total number of operations is:

$$
\sum_{k=1}^{n-1}(n-k)(n-k+1) \quad \text{multiplications}
$$

$$
\sum_{k=1}^{n-1}(n-k)(n-k+1) \quad \text{additions}
$$

$$
\sum_{k=1}^{n-1}(n-k) \qquad\qquad \text{divisions}
$$

We now recognize that

$$
\sum_{k=1}^{N} k = \frac{N(N+1)}{2}, \qquad \sum_{k=1}^{N} k^2 = \frac{N(N+1)(2N+1)}{6}, \qquad (4.78)
$$

where the second summation identity was proven in Appendix 3.B. The first summation identity may be proved in a similar manner. Therefore

$$
\sum_{k=1}^{n-1}(n-k)(n-k+1) = \sum_{k=1}^{n-1}[n^2 + n - (2n+1)k + k^2]
$$

$$
= (n-1)(n^2 + n) - (2n+1)\sum_{k=1}^{n-1} k + \sum_{k=1}^{n-1} k^2
$$

$$= \frac{1}{3}n^3 - \frac{1}{3}n, \tag{4.79a}$$

$$\sum_{k=1}^{n-1}(n-k) = n(n-1) - \sum_{k=1}^{n-1}k = \frac{1}{2}n^2 - \frac{1}{2}n. \tag{4.79b}$$

So-called asymptotic complexity measures are defined using

Definition 4.2: Big *O*   We say that $f(n) = O(g(n))$ if there is a $0 < c < \infty$, and an $N \in \mathbf{N}$ ($N < \infty$) such that

$$f(n) \leq cg(n)$$

for all $n > N$.

Our algorithm needs a total of $f(n) = 2(\frac{1}{3}n^3 - \frac{1}{3}n) + \frac{1}{2}n^2 - \frac{1}{2}n = \frac{2}{3}n^3 + \frac{1}{2}n^2 - \frac{7}{6}n$ flops. We may say that $O(n^3)$ operations (flops) are needed (so here $g(n) = n^3$). We may read $O(n^3)$ as "order *n*-cubed," so order *n*-cubed operations are needed. If one operation takes one unit of time on a computing machine we say the *asymptotic time complexity* of the algorithm is $O(n^3)$. Parameter *n* (matrix order) is the *size of the problem*. We might also say that the time complexity of the algorithm is *cubic in the size of the problem* since the number of operations $f(n)$ is a cubic polynomial in *n*. But we caution the reader about flop counting:

> "Flop counting is a necessarily crude approach to the measuring of program efficiency since it ignores subscripting, memory traffic, and the countless other overheads associated with program execution. We must not infer too much from a comparison of flops counts. ... Flop counting is just a 'quick and dirty' accounting method that captures only one of several dimensions of the efficiency issue."
>
> —Golub and Van Loan [5, p. 20]

Asymptotic complexity measures allow us to talk about algorithmic resource demands without getting bogged down in detailed expressions for computing time, memory requirements, and other variables. However, the comment by Golub and Van Loan above may clearly be extended to asymptotic measures.

Suppose that *A* is *LU*-factorable, and that we know *L* and *U*. Suppose that we wish to solve $Ax = y$. Thus

$$LUx = y, \tag{4.80}$$

and define $Ux = z$, so we begin by considering

$$Lz = y. \tag{4.81}$$

In expanded form this becomes

$$
\begin{bmatrix}
l_{00} & 0 & 0 & \cdots & 0 \\
l_{10} & l_{11} & 0 & \cdots & 0 \\
l_{20} & l_{21} & l_{22} & \cdots & 0 \\
\vdots & \vdots & \vdots & & \vdots \\
l_{n-1,0} & l_{n-1,1} & l_{n-1,2} & \cdots & l_{n-1,n-1}
\end{bmatrix}
\begin{bmatrix}
z_0 \\ z_1 \\ z_2 \\ \vdots \\ z_{n-1}
\end{bmatrix}
=
\begin{bmatrix}
y_0 \\ y_1 \\ y_2 \\ \vdots \\ y_{n-1}
\end{bmatrix}. \quad (4.82)
$$

Since L^{-1} exists, solving (4.81) is easy using *forward elimination* (*forward substitution*). Specifically, from (4.82)

$$
z_0 = \frac{y_0}{l_{0,0}}
$$

$$
z_1 = \frac{1}{l_{1,1}}[y_1 - z_0 l_{1,0}]
$$

$$
z_2 = \frac{1}{l_{2,2}}[y_2 - z_0 l_{2,0} - z_1 l_{2,1}]
$$

$$
\vdots
$$

$$
z_{n-1} = \frac{1}{l_{n-1,n-1}}\left[y_{n-1} - \sum_{k=0}^{n-2} z_k l_{n-1,k} \right].
$$

Thus, in general

$$
z_k = \frac{1}{l_{k,k}}\left[y_k - \sum_{i=0}^{k-1} z_i l_{k,i} \right] \quad (4.83)
$$

for $k = 1, 2, \ldots, n-1$ with $z_0 = y_0/l_{0,0}$. Since we now know z, we may solve $Ux = z$ by *backward substitution*. To see this, express the problem in expanded form:

$$
\begin{bmatrix}
u_{0,0} & u_{0,1} & \cdots & u_{0,n-2} & u_{0,n-1} \\
0 & u_{1,1} & \cdots & u_{1,n-2} & u_{1,n-1} \\
\vdots & \vdots & & \vdots & \vdots \\
0 & 0 & \cdots & u_{n-2,n-2} & u_{n-2,n-1} \\
0 & 0 & \cdots & 0 & u_{n-1,n-1}
\end{bmatrix}
\begin{bmatrix}
x_0 \\ x_1 \\ \vdots \\ x_{n-2} \\ x_{n-1}
\end{bmatrix}
=
\begin{bmatrix}
z_0 \\ z_1 \\ \vdots \\ z_{n-2} \\ z_{n-1}
\end{bmatrix}. \quad (4.84)
$$

From (4.84), we obtain

$$
x_{n-1} = \frac{z_{n-1}}{u_{n-1,n-1}}
$$

$$
x_{n-2} = \frac{1}{u_{n-2,n-2}}[z_{n-2} - x_{n-1} u_{n-2,n-1}]
$$

$$x_{n-3} = \frac{1}{u_{n-3,n-3}}[z_{n-3} - x_{n-1}u_{n-3,n-1} - x_{n-2}u_{n-3,n-2}]$$

$$\vdots$$

$$x_0 = \frac{1}{u_{0,0}}\left[z_0 - \sum_{k=1}^{n-1} x_k u_{0,k}\right].$$

In general

$$x_k = \frac{1}{u_{k,k}}\left[z_k - \sum_{i=k+1}^{n-1} x_i u_{k,i}\right] \tag{4.85}$$

for $k = n - 2, \ldots, 0$ with $x_{n-1} = z_{n-1}/u_{n-1,n-1}$. The forward-substitution and backward-substitution algorithms that we have just derived have an asymptotic time complexity of $O(n^2)$. The reader should confirm this as an exercise. This result suggests that most of the computational effort needed to solve for x in $Ax = y$ lies in the *LU* decomposition stage.

So far we have said nothing about the performance of our linear system solution method with respect to finite precision arithmetic effects (i.e., rounding error). Before considering this matter, we make a few remarks regarding the *stability* of our method. We have noted that the *LU* decomposition algorithm will fail if a zero-valued pivot is encountered. This can happen even if $Ax = y$ has a solution and A is well-conditioned. In other words, our algorithm is actually *unstable* since we can input numerically well-posed problems that cause it to fail. This does not necessarily mean that our algorithm should be totally rejected. For example, we have shown that positive definite matrices will never result in a zero-valued pivot. Furthermore, if $A > 0$, *and* it is well-conditioned, then it can be shown that an accurate answer will be provided by the algorithm despite its faults. Nonetheless, the problem of failure due to encountering a zero-valued pivot needs to be addressed. Also, what happens if a pivot is not exactly zero, but is close to zero? We might expect that this can result in a computed solution $\hat{x}$ that differs greatly from the mathematically exact solution x, especially where rounding error is involved, even if A is well-conditioned.

Recall (2.15) from Chapter 2,

$$fl[x \; op \; y] = (x \; op \; y)(1 + \epsilon) \tag{4.86}$$

for which $|\epsilon| \leq 2^{-t}$. If we store A in a floating-point machine, then, because of the necessity to quantize we are really storing the elements

$$[fl[A]]_{i,j} = fl[a_{i,j}] = a_{i,j}(1 + \epsilon_{i,j}) \tag{4.87}$$

with $|\epsilon_{i,j}| \leq 2^{-t}$. Suppose now that $A, B \in \mathbf{R}^{m \times n}$; we then define[6]

$$|A| = [|a_{i,j}|] \in \mathbf{R}^{m \times n}, \tag{4.88}$$

and by $B \leq A$, we mean $b_{i,j} \leq a_{i,j}$ for all i and j. So we may express (4.87) more compactly as

$$|fl[A] - A| \leq u|A|, \tag{4.89}$$

where $u = 2^{-t}$, since $|\epsilon_{i,j}| \leq 2^{-t}$.

Forsythe and Moler [4, pp. 104–105] show that the computed solution $\hat{z}$ to $Lz = y$ [recall (4.81)] as obtained by forward substitution is actually the *exact* solution to a perturbed lower triangular system

$$(L + \delta L)\hat{z} = y, \tag{4.90}$$

where δL is a lower triangular perturbation matrix, and where

$$|\delta L| \leq 1.01nu|L|. \tag{4.91}$$

A very similar bound exists for the problem of solving $Ux = z$ by backward-substitution. We will not derive these bounds, but will simply mention that the derivation involves working with a bound similar to (2.39) in Chapter 2. From (4.91) we have *relative perturbations* $\frac{|\delta l_{i,j}|}{|l_{i,j}|} \leq 1.01nu$. It is apparent that since u is typically quite tiny, unless n (matrix order) is quite huge, these relative perturbations will not be significant. In other words, forward substitution and backward substitution are very stable procedures that are quite resistant to the effects of rounding errors. Thus, any difficulties with our linear system solution procedure in terms of a rounding error likely involve only the LU factorization stage.

The rounding error analysis for our Gaussian elimination algorithm is even more involved than the effort required to obtain (4.91), so again we will content ourselves with citing the main result without proof. We cite Theorem 3.3.1 in Golub and Van Loan [5] as follows.

Theorem 4.3: Assume that A is an $n \times n$ matrix of floating-point numbers. If no zero-valued pivots are encountered during the execution of the Gaussian elimination algorithm for which A is the input, then the computed triangular factors (here denoted $\hat{L}$ and $\hat{U}$) satisfy

$$\hat{L}\hat{U} = A + \delta A \tag{4.92a}$$

such that

$$|\delta A| \leq 3(n-1)u(|A| + |\hat{L}||\hat{U}|) + O(u^2). \tag{4.92b}$$

---

[6]There is some danger in confusing this with the determinant. That is, some people use $|A|$ to denote the determinant of A. We will avoid this here by sticking with $\det(A)$ as the notation for determinant of A.

In this theorem the term $O(u^2)$ denotes a part of the error term dependent on u^2. This is quite small as $u^2 = 2^{-2t}$ (rounding assumed), and so may be practically disregarded. The term arises in the work of Golub and Van Loan [5] because those authors prefer to work with slightly looser bounding results than are to be found in the volume by Forsythe and Moler [4]. The bound in (4.92b) gives us cause for concern. The perturbation matrix δA may not be small. This is because $|\hat{L}||\hat{U}|$ can be quite large. An example of this would be

$$A = \begin{bmatrix} 1 & 4 & 1 \\ 2 & 8.001 & 1 \\ 0 & -1 & 1 \end{bmatrix},$$

for which

$$\hat{L} = \begin{bmatrix} 1 & 0 & 0 \\ 2 & 1 & 0 \\ 0 & -1000 & 1 \end{bmatrix}, \quad \hat{U} = \begin{bmatrix} 1 & 4 & 1 \\ 0 & 0.001 & -1 \\ 0 & 0 & -999 \end{bmatrix}.$$

This has happened because

$$A^1 = \begin{bmatrix} 1 & 4 & 1 \\ 0 & 0.001 & -1 \\ 0 & -1 & 1 \end{bmatrix},$$

which has $a_{1,1}^1 = 0.001$. This is a small pivot and is ultimately responsible for giving us "big" triangular factors. Clearly, the smaller the pivot the bigger the potential problem. Golub and Van Loan's [5] Theorem 3.3.2 (which we will not repeat here) goes on to demonstrate that the errors in the computed triangular factors can adversely affect the solution to $Ax = LUx = y$ as obtained by forward substitution and backward substitution. Thus, if we use the computed solutions $\hat{L}$ and $\hat{U}$ in $\hat{L}\hat{U}\hat{x} = y$, then the computed solution $\hat{x}$ may not be close to x.

How may our Gaussian elimination *LU* factorization algorithm be modified to make it more stable? The standard solution is to employ *partial pivoting*. We do not consider the method in detail here, but illustrate it with a simple example (Example 4.5). Essentially, before applying a Gauss transformation G_k, the rows of matrix A^{k-1} are permuted (i.e., exchanged) in such a manner as to make the pivot as large as possible while simultaneously ensuring that A^k is as close to being upper triangular as possible. Permutation operations have a matrix description, and such matrices may be denoted by P_k. We remark that $P_k^{-1} = P_k^T$.

Example 4.5 Suppose that

$$A = \begin{bmatrix} 1 & 4 & 1 \\ 2 & 8 & 1 \\ 0 & -1 & 1 \end{bmatrix} = A^0.$$

Thus

$$
G_1 P_1 A^0 = \begin{bmatrix} 1 & 0 & 0 \\ -\frac{1}{2} & 1 & 0 \\ 0 & 0 & 1 \end{bmatrix} \begin{bmatrix} 0 & 1 & 0 \\ 1 & 0 & 0 \\ 0 & 0 & 1 \end{bmatrix} \begin{bmatrix} 1 & 4 & 1 \\ 2 & 8 & 1 \\ 0 & -1 & 1 \end{bmatrix}
$$

$$
= \begin{bmatrix} 1 & 0 & 0 \\ -\frac{1}{2} & 1 & 0 \\ 0 & 0 & 1 \end{bmatrix} \begin{bmatrix} 2 & 8 & 1 \\ 1 & 4 & 1 \\ 0 & -1 & 1 \end{bmatrix} = \begin{bmatrix} 2 & 8 & 1 \\ 0 & 0 & \frac{1}{2} \\ 0 & -1 & 1 \end{bmatrix} = A^1,
$$

and

$$
G_2 P_2 A^1 = \begin{bmatrix} 1 & 0 & 0 \\ 0 & 1 & 0 \\ 0 & 0 & 1 \end{bmatrix} \begin{bmatrix} 1 & 0 & 0 \\ 0 & 0 & 1 \\ 0 & 1 & 0 \end{bmatrix} \begin{bmatrix} 2 & 8 & 1 \\ 0 & 0 & \frac{1}{2} \\ 0 & -1 & 1 \end{bmatrix}
$$

$$
= \begin{bmatrix} 1 & 0 & 0 \\ 0 & 1 & 0 \\ 0 & 0 & 1 \end{bmatrix} \begin{bmatrix} 2 & 8 & 1 \\ 0 & -1 & 1 \\ 0 & 0 & \frac{1}{2} \end{bmatrix} = A^2
$$

for which $U = A^2$. We see that P_2 interchanges rows 2 and 3 rather than 1 and 2 because to do otherwise would ruin the upper triangular structure we seek. It is apparent that

$$
G_2 P_2 G_1 P_1 A = U,
$$

so that

$$
A = \underbrace{P_1^{-1} G_1^{-1} P_2^{-1} G_2^{-1}}_{=L} U,
$$

for which

$$
L = \begin{bmatrix} \frac{1}{2} & 0 & 1 \\ 1 & 0 & 0 \\ 0 & 1 & 0 \end{bmatrix}.
$$

This matrix is manifestly not lower triangular. Thus, our use of partial pivoting to achieve algorithmic stability has been purchased at the expense of some loss of structure (although Theorem 3.4.1 in Ref. 5 shows how to recover much of what is lost.[7]) Also, permutations involve moving data around in the computer, and this is a potentially significant cost. But these prices are usually worth paying.

---

[7]In general, the Gaussian elimination with partial pivoting algorithm generates

$$
G_{n-1} P_{n-1} \cdots G_2 P_2 G_1 P_1 A = U,
$$

and it turns out that

$$
P_{n-1} \cdots P_2 P_1 A = LU
$$

for which L is unit lower triangular, and U is upper triangular. The expression for L in terms of the factors G_k is messy, and so we omit it. The interested reader can see pp. 112–113 of Ref. 5 for details.

It is worth mentioning that the need to trade off algorithm speed in favor of stability is common in numerical computing; that is, fast algorithms often have stability problems. Much of numerical computing is about creating the fastest possible stable algorithms. This is a notoriously challenging engineering problem.

A much more detailed account of Gaussian elimination with partial pivoting appears in Golub and Van Loan [5, pp. 108–116]. This matter will not be discussed further in this book.

4.6 LEAST-SQUARES PROBLEMS AND *QR* DECOMPOSITION

In this section we consider the *QR* decomposition of $A \in \mathbf{R}^{m \times n}$ for which $m \geq n$, and A is of full rank [i.e., rank $(A) = n$]. *Full rank* in this sense means that the columns of A are linearly independent. The *QR* decomposition of A is

$$A = QR, \tag{4.93}$$

where $Q \in \mathbf{R}^{m \times m}$ is an *orthogonal matrix* [i.e., $Q^T Q = QQ^T = I$ (identity matrix)], and $R \in \mathbf{R}^{m \times n}$ is upper triangular in the following sense:

$$R = \begin{bmatrix} r_{0,0} & r_{0,1} & \cdots & r_{0,n-1} \\ 0 & r_{1,1} & \cdots & r_{1,n-1} \\ \vdots & \vdots & & \vdots \\ 0 & 0 & \cdots & r_{n-1,n-1} \\ 0 & 0 & \cdots & 0 \\ \vdots & \vdots & & \vdots \\ 0 & 0 & \cdots & 0 \end{bmatrix} = \begin{bmatrix} \mathcal{R} \\ 0 \end{bmatrix}. \tag{4.94}$$

Here $\mathcal{R} \in \mathbf{R}^{n \times n}$ is a square upper triangular matrix and is nonsingular because A is full rank. The bottom block of zeros in R of (4.94) is $(m - n) \times n$.

It should be immediately apparent that the existence of a *QR* decomposition for A makes it quite easy to solve for x in $Ax = y$, if A^{-1} exists (which implies that in this special case A is square). Thus, $Ax = QRx = y$, and so $Rx = Q^T y$. The upper triangular linear system $Rx = Q^T y$ may be readily solved by backward substitution (recall the previous section).

The case where $m > n$ is important because it arises in *overdetermined least-squares approximation problems*. We illustrate with the following example based on a real-world problem.[8] Figure 4.1 is a plot of some simulated body core temperature

---

[8]This example is from the problem of estimating the circadian rhythm parameters of human patients who have sustained head injuries. The estimates are obtained by the suitable processing of various physiological data sets (e.g., body core temperature, heart rate, blood pressure). The nature of the injury has made the patients' rhythms deviate from the nominal 24-h cycle. Correct estimation of rhythm parameters can lead to improved clinical treatment because of improved timing in the administering of

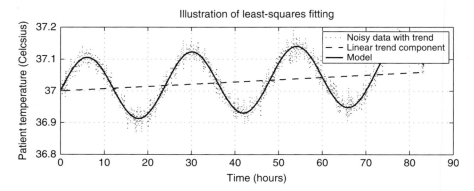

Figure 4.1 Simulated human patient temperature data to illustrate overdetermined least-squares model parameter estimation. Here we have $N = 1000$ samples f_n (the dots), for $T_s = 300$ (seconds), $T = 24$ (hours), $a = 2 \times 10^{-7} °C/s$, $b = 37°C$, and $c = 0.1°C$. The solution to (4.103) is $\hat{a} = 2.0582 \times 10^{-7} °C/s$, $\hat{b} = 36.9999°C$, $\hat{c} = 0.1012°C$.

measurements from a human patient (this is the noisy data with trend). The data has three components:

1. A sinusoidal component
2. Random noise.
3. A linear trend.

Our problem is to estimate the parameters of the sinusoid (i.e., the amplitude, period, and phase), which represents the patient's circadian rhythm. In other words, the noise and trend are undesirable and so are to be, in effect, removed from the desired sinusoidal signal component. Here we will content ourselves with estimating only the amplitude of the sinusoid. The problem of estimating the remaining parameters is tougher. Methods to estimate the remaining parameters will be considered later. (This is a nonlinear optimization problem.)

We assume the model for the data in Fig. 4.1 is the analog signal

$$f(t) = at + b + c \sin\left(\frac{2\pi}{T}t\right) + \eta(t). \tag{4.95}$$

Here the first two terms model the trend (assumed to be a straight line), the third term is the desired sinusoidal signal component, and $\eta(t)$ is a random noise component. We only possess samples of the signal $f_n = f(nT_s)$ (i.e., $t = nT_s$), for $n = 0, 1, \ldots, N - 1$, where T_s is the *sampling period* of the data collection system.

medication. We emphasize that the model in (4.95) is grossly oversimplified. Indeed, a better model is to replace term $at + b$ with subharmonic, and harmonic terms of $\sin\left(\frac{2\pi}{T}t\right)$. A harmonic term is one of frequency $\frac{2\pi}{T}n$, while a subharmonic has frequency $\frac{2\pi}{T}\frac{1}{n}$. Cosine terms should also be included in the improved model.

We assume that we know T which is the period of the patient's circadian rhythm. Our model also implicitly assumes knowledge of the phase of the sinusoid, too. These are very artificial assumptions since in practice these are the most important parameters we are trying to estimate, and they are never known in advance. However, our present circumstances demand simplification. Our estimate of f_n may be defined by

$$\hat{f}_n = aT_s n + b + c \sin\left(\frac{2\pi}{T} nT_s\right). \tag{4.96}$$

This is a sampled version of the analog model, except the noise term has been deleted.

We may estimate the unknown model parameters a, b, c by employing the same basic strategy we used in Section 4.2, specifically, a least-squares approach. Thus, defining $x = [a \quad b \quad c]^T$ (vector of unknown parameters), we strive to minimize

$$V(x) = \sum_{n=0}^{N-1} e_n^2 = \sum_{n=0}^{N-1} [f_n - \hat{f}_n]^2 \tag{4.97}$$

with respect to x. Using matrix/vector notation was very helpful in Section 4.2, and it remains so here. Define

$$v_n = \left[T_s n \quad 1 \quad \sin\left(\frac{2\pi}{T} T_s n\right)\right]^T. \tag{4.98}$$

Thus

$$e_n = f_n - v_n^T x. \tag{4.99}$$

We may define the *error vector* $e = [e_0 \ e_1 \ \cdots \ e_{N-1}]^T$, *data vector* $f = [f_0 \ f_1 \ \cdots \ f_{N-1}]^T$, and the *matrix of basis vectors*

$$A = \begin{bmatrix} v_0^T \\ v_1^T \\ \vdots \\ v_{N-1}^T \end{bmatrix} \in \mathbf{R}^{N \times 3}. \tag{4.100}$$

Consequently, via (4.99)

$$e = f - Ax. \tag{4.101}$$

Obviously, we would like to have $e = 0$, which implies the desire to solve $Ax = f$. If we have $N = 3$ and A^{-1} exists, then we may uniquely solve for x given any f. However, in practice, $N \gg 3$, so our linear system is overdetermined. Thus, no *unique* solution is possible. We have no option but to select x to minimize e in some sense. Once again, previous experience from Section 4.2 says least-squares is a viable choice. Thus, since $||e||_2^2 = e^T e = \sum_{n=0}^{N-1} e_n^2$, we consider

$$V(x) = e^T e = \underbrace{f^T f}_{=\rho} - 2x^T \underbrace{A^T f}_{=g} + x^T \underbrace{A^T A}_{=P} x \qquad (4.102)$$

[which is a more compact version of (4.97)]. This is yet another quadratic form [recall (4.8)]. We see that $P \in \mathbf{R}^{3 \times 3}$, and $g \in \mathbf{R}^3$. In our problem A is full rank so from the results in Section 4.2 we see that $P > 0$. Naturally, from the discussions of Sections 4.3 and 4.4, the conditioning of P is a concern. Here it turns out that because P is of low order (largely because we are interested only in estimating three parameters) it typically has a low condition number. However, as the order of P rises, the conditioning of P usually rapidly worsens; that is, ill conditioning tends to be a severe problem when the number of parameters to be estimated rises. From Section 4.2 we know that the optimum choice for x, denoted $\hat{x}$, is obtained by solving the linear system

$$P\hat{x} = g. \qquad (4.103)$$

The model curve of Fig. 4.1 (solid line) is the curve obtained using $\hat{x}$ in (4.96). Thus, since $\hat{x} = [\hat{a} \ \hat{b} \ \hat{c}]^T$, we plot $\hat{f}_n$ for $\hat{a}, \hat{b}, \hat{c}$ in place of a, b, c in (4.96). Equation (4.103) can be written as

$$A^T A \hat{x} = A^T f. \qquad (4.104)$$

This is just the overdetermined linear system $A\hat{x} = f$ multiplied on the left (i.e., premultiplied) by A^T. The system (4.104) is often referred to in the literature as the *normal equations*.

How is the previous applications example relevant to the problem of QR factorizing A as in Eq. (4.93)? To answer this, we need to consider the condition numbers of A, and of $P = A^T A$, and to see how orthogonal matrices Q facilitate the solution of overdetermined least-squares problems. We will then move on to the problem of how to practically compute the QR factorization of a full-rank matrix. We will consider the issue of conditioning first since this is a justification for considering QR factorization methods as opposed to the linear system solution methods of the previous section.

Singular values were mentioned in Section 4.4 as being relevant to the problem of computing spectral norms, and so of computing $\kappa_2(A)$. Now we need to consider the consequences of

Theorem 4.4: Singular Value Decomposition (SVD)   Suppose $A \in \mathbf{R}^{m \times n}$; then there exist orthogonal matrices

$$U = [u_0 u_1 \cdots u_{m-1}] \in \mathbf{R}^{m \times m}, \ V = [v_0 v_1 \cdots v_{n-1}] \in \mathbf{R}^{n \times n}$$

such that

$$\Sigma = U^T A V = \text{diag} \ (\sigma_0, \sigma_1, \ldots, \sigma_{p-1}) \in \mathbf{R}^{m \times n}, \ p = \min\{m, n\}, \qquad (4.105)$$

where $\sigma_0 \geq \sigma_1 \geq \ldots \geq \sigma_{p-1} \geq 0$.

An outline proof appears in Ref. 5 (p. 71) and is omitted here. The notation diag$(\sigma_0, \ldots, \sigma_{p-1})$ means a diagonal matrix with main diagonal elements $\sigma_0, \ldots, \sigma_{p-1}$. For example, if $m = 3, n = 2$, then $p = 2$, and

$$U^T A V = \begin{bmatrix} \sigma_0 & 0 \\ 0 & \sigma_1 \\ 0 & 0 \end{bmatrix},$$

but if $m = 2, n = 3$, then again $p = 2$, but now

$$U^T A V = \begin{bmatrix} \sigma_0 & 0 & 0 \\ 0 & \sigma_1 & 0 \end{bmatrix}.$$

The numbers σ_i are called *singular values*. Vector u_i is the *ith left singular vector*, and v_i is the *ith right singular vector*. The following notation is helpful:

$\sigma_i(A) = $ the ith singular value of A ($i \in \mathbf{Z}_p$).
$\sigma_{\max}(A) = $ the biggest singular value of A.
$\sigma_{\min}(A) = $ the smallest singular value of A.

We observe that because $AV = U\Sigma$, and $A^T U = V\Sigma^T$ we have, respectively

$$A v_i = \sigma_i u_i, \quad A^T u_i = \sigma_i v_i \tag{4.106}$$

for $i \in \mathbf{Z}_p$. Singular values give matrix 2-norms; as noted in the following theorem.

Theorem 4.5:
$$||A||_2 = \sigma_0 = \sigma_{\max}(A).$$

Proof Recall the result (4.37). From (4.105) $A = U\Sigma V^T$ so

$$||Ax||_2^2 = x^T A^T A x,$$

and

$$A^T A = V \Sigma^T \Sigma V^T = \sum_{i=0}^{p-1} \sigma_i^2 v_i v_i^T \in \mathbf{R}^{n \times n}. \tag{4.107}$$

For any $x \in \mathbf{R}^n$ there exist d_i such that

$$x = \sum_{i=0}^{n-1} d_i v_i \tag{4.108}$$

(because V is orthogonal so its column vectors form an orthogonal basis for $\mathbf{R}^n$). Thus

$$||x||_2^2 = x^T x = \sum_{i=0}^{n-1} \sum_{j=0}^{n-1} d_i d_j v_i^T v_j = \sum_{i=0}^{n-1} d_i^2 \tag{4.109}$$

(via $v_i^T v_j = \delta_{i-j}$). Now

$$x^T A^T A x = \sum_{i=0}^{p-1} \sigma_i^2 (x^T v_i)(v_i^T x) = \sum_{i=0}^{p-1} \sigma_i^2 \langle x, v_i \rangle^2, \tag{4.110}$$

but

$$\langle x, v_i \rangle = \left\langle \sum_j d_j v_j, v_i \right\rangle = \sum_j d_j \langle v_i, v_j \rangle = d_i. \tag{4.111}$$

Using (4.111) in (4.110), we obtain

$$||Ax||_2^2 = \sum_{i=0}^{n-1} \sigma_i^2 d_i^2 \tag{4.112}$$

for which it is understood that $\sigma_i^2 = 0$ for $i > p - 1$. We maximize $||Ax||_2^2$ subject to constraint $||x||_2^2 = 1$, which means employing Lagrange multipliers; that is, we maximize

$$L(d) = \sum_{i=0}^{n-1} \sigma_i^2 d_i^2 - \lambda \left(\sum_{i=0}^{n-1} d_i^2 - 1 \right), \tag{4.113}$$

where $d = [d_0 \ d_1 \ \cdots \ d_{n-1}]^T$. Thus

$$\frac{\partial L(d)}{\partial d_j} = 2\sigma_j^2 d_j - 2\lambda d_j = 0,$$

or

$$\sigma_j^2 d_j = \lambda d_j. \tag{4.114}$$

From (4.114) into (4.112)

$$||Ax||_2^2 = \lambda \sum_{i=0}^{n-1} d_i^2 = \lambda \tag{4.115}$$

for which we have used the fact that $||x||_2^2 = 1$ in (4.109). From (4.114) λ is the eigenvalue of a diagonal matrix containing σ_i^2. Consequently, λ is maximized for $\lambda = \sigma_0^2$. Therefore, $||A||_2 = \sigma_0$.

Suppose that

$$\sigma_0 \geq \cdots \geq \sigma_{r-1} > \sigma_r = \cdots = \sigma_{p-1} = 0, \tag{4.116}$$

then

$$\text{rank } (A) = r. \tag{4.117}$$

Thus, the SVD of A can tell us the rank of A. In our overdetermined least-squares problem we have $m \geq n$ and A is assumed to be of full-rank. This implies that $r = n$. Also, $p = n$. Thus, all singular values of a full-rank matrix are bigger than zero. Now suppose that A^{-1} exists. From (4.105) $A^{-1} = V\Sigma^{-1}U^T$. Immediately, $||A^{-1}||_2 = 1/\sigma_{\min}(A)$. Hence

$$\kappa_2(A) = ||A||_2||A^{-1}||_2 = \frac{\sigma_{\max}(A)}{\sigma_{\min}(A)}. \tag{4.118}$$

Thus, a large singular value spread is associated with matrix ill conditioning. [Recall (4.61) and the related discussion.] As remarked on p. 223 of Ref. 5, Eq. (4.118) can be extended to cover full-rank rectangular matrices with $m \geq n$:

$$A \in \mathbf{R}^{m \times n}, \text{rank } (A) = n \Rightarrow \kappa_2(A) = \frac{\sigma_{\max}(A)}{\sigma_{\min}(A)}. \tag{4.119}$$

This also holds for the transpose of A because $A^T = V\Sigma^T U^T$, so A^T has the same singular values as A. Thus, $\kappa_2(A^T) = \kappa_2(A)$. Golub and Van Loan [5, p. 225] claim (without formal proof) that $\kappa_2(A^T A) = [\kappa_2(A)]^2$. In other words, if the linear system $Ax = f$ is ill-conditioned, then $A^T A\hat{x} = A^T f$ is even more ill-conditioned. The condition number of the latter system is the square of that of the former system. More information on the conditioning of rectangular matrices is to be found in Appendix 4.B. This includes justification that $\kappa_2(A^T A) = [\kappa_2(A)]^2$.

A popular approach toward solving the normal equations $A^T A\hat{x} = A^T f$ is based on Cholesky decomposition

Theorem 4.6: Cholesky Decomposition   If $R \in \mathbf{R}^{n \times n}$ is symmetric and positive definite, then there exists a unique lower triangular matrix $L \in \mathbf{R}^{n \times n}$ with positive diagonal entries such that $R = LL^T$. This is the *Cholesky decomposition* (*factorization*) of R.

Algorithms to find this decomposition appear in Chapter 4 of Ref. 5. We do not consider them except to note that if they are used, then the computed solution to $A^T A\hat{x} = A^T f$, which we denote by $\hat{\hat{x}}$, may satisfy

$$\frac{||\hat{\hat{x}} - \hat{x}||_2}{||\hat{x}||_2} \approx u[\kappa_2(A)]^2, \tag{4.120}$$

where u is as in (4.89). Thus, this method of linear system solution is potentially highly susceptible to errors due to ill-conditioned problems. On the other hand, Cholesky approaches are computationally efficient in that they require about $n^3/3$ flops (Floating-point operations). Clearly, Gaussian elimination may be employed to solve the normal equations as well, but we recall that Gaussian elimination needed about $2n^3/3$ flops. Gaussian elimination is less efficient because it does not account for symmetry in matrix R. Note that these counts do not take into consideration the number of flops needed to determine $A^T A$ and $A^T f$, and do not account for the number of flops needed by the forward/backward substitution steps. However, the comparison between Cholesky decomposition and Gaussian elimination is reasonably fair because these other steps are essentially the same for both approaches.

Recall that $||e||_2^2 = ||Ax - f||_2^2$. Thus, for orthogonal matrix Q

$$||Q^T e||_2^2 = [Q^T e]^T Q^T e = e^T Q Q^T e = e^T e = ||e||_2^2. \tag{4.121}$$

Thus, the 2-norm is *invariant* to orthogonal transformations. This is one of the more important properties of 2-norms. Now consider

$$||e||_2^2 = ||Q^T Ax - Q^T f||_2^2. \tag{4.122}$$

Suppose that

$$Q^T f = \left[\begin{array}{c} f^u \\ f^l \end{array} \right] \tag{4.123}$$

for which $f^u \in \mathbf{R}^n$, and $f^l \in \mathbf{R}^{m-n}$. Thus, from (4.94) and $Q^T A = R$, we obtain

$$Q^T Ax - Q^T f = \left[\begin{array}{c} \mathcal{R}x - f^u \\ -f^l \end{array} \right], \tag{4.124}$$

implying that

$$||e||_2^2 = ||\mathcal{R}x - f^u||_2^2 + ||f^l||_2^2. \tag{4.125}$$

Immediately, we see that

$$\mathcal{R}\hat{x} = f^u. \tag{4.126}$$

The least-squares optimal solution $\hat{x}$ is therefore found by backward substitution. Equally clearly, we see that

$$\min_x ||e||_2^2 = ||f^l||_2^2 = \rho_{LS}^2. \tag{4.127}$$

This is the minimum error energy. Quantity ρ_{LS}^2 is also called the *minimum sum of squares*, and e is called the *residual* [5]. It is easy to verify that $\kappa_2(Q) = 1$ (Q is orthogonal). In other words, orthogonal matrices are perfectly conditioned. This means that the operation $Q^T A$ will not result in a matrix that is not as well conditioned as A. This in turn suggests that solving our least-squares problem using QR decomposition might be numerically more reliable than working with the normal equations. As explained on p. 230 of Ref. 5, this is not necessarily always true, but it is nevertheless a good reason to contemplate QR approaches to solving least-squares problems.[9]

---

[9]If the residual is big and the problem is ill-conditioned, then neither QR nor normal equation methods may give an accurate answer. However, QR approaches may be more accurate for small residuals in ill-conditioned problems than normal equation approaches.

How may we compute Q? There are three major approaches:

1. Gram–Schmidt algorithms
2. Givens rotation algorithms
3. Householder transformation algorithms

We will consider only Householder transformations.

We begin by a review of how vectors are projected onto vectors. Recall the law of cosines from trigonometry in reference to Fig. 4.2a. Assume that $x, y \in \mathbf{R}^n$. Suppose that $||x - y||_2 = a$, $||x||_2 = b$, and that $||y||_2 = c$. Therefore, where θ is the angle between x and y ($0 \le \theta \le \pi$ radians)

$$a^2 = b^2 + c^2 - 2bc \cos \theta, \tag{4.128}$$

or in terms of the vectors x and y, Eq. (4.128) becomes

$$||x - y||_2^2 = ||x||_2^2 + ||y||_2^2 - 2||x||_2||y||_2 \cos \theta. \tag{4.129}$$

In terms of inner products, this becomes

$$\langle x - y, x - y \rangle = \langle x, x \rangle + \langle y, y \rangle - 2[\langle x, x \rangle]^{1/2}[\langle y, y \rangle]^{1/2} \cos \theta,$$

which reduces to

$$\langle x, y \rangle = [\langle x, x \rangle]^{1/2}[\langle y, y \rangle]^{1/2} \cos \theta,$$

or

$$\langle x, y \rangle = ||x||_2||y||_2 \cos \theta. \tag{4.130}$$

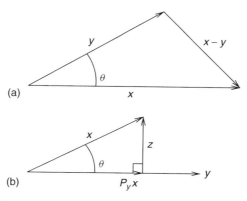

(a)

(b)

Figure 4.2   Illustration of the law of cosines (a) and the projection of vector x onto vector y (b).

Now consider Fig. 4.2b. Vector $P_y x$ is the *projection of x onto y*, where P_y denotes the *projection operator* that projects x onto y. It is immediately apparent that

$$||P_y x||_2 = ||x||_2 \cos \theta. \tag{4.131}$$

This is the Euclidean length of $P_y x$. The unit vector in the direction of y is $y/||y||_2$. Therefore

$$P_y x = \frac{||x||_2 \cos \theta}{||y||_2} y. \tag{4.132}$$

But from (4.130) this becomes

$$P_y x = \frac{\langle x, y \rangle}{||y||_2^2} y. \tag{4.133}$$

Since $\langle x, y \rangle = x^T y = y^T x$, we see that

$$P_y x = \frac{y^T x}{||y||_2^2} y = \left[\frac{1}{||y||_2^2} y y^T \right] x. \tag{4.134}$$

In (4.134) $y y^T \in \mathbf{R}^{n \times n}$, so the operator P_y has the matrix representation

$$P_y = \frac{1}{||y||_2^2} y y^T. \tag{4.135}$$

In Fig. 4.2b we see that $z = x - P_y x$, and that

$$z = (I - P_y) x = \left[I - \frac{1}{||y||_2^2} y y^T \right] x, \tag{4.136}$$

which is the component of x that is orthogonal to y. We observe that

$$P_y^2 = \frac{1}{||y||_2^4} y y^T y y^T = \frac{y ||y||_2^2 y^T}{||y||_2^4} = \frac{1}{||y||_2^2} y y^T = P_y. \tag{4.137}$$

If $A^2 = A$, we say that matrix A is *idempotent*. Thus, projection operators are idempotent. Also, $P_y^T = P_y$ so projection operators are also symmetric.

In Fig. 4.3, $x, y, z \in \mathbf{R}^n$, and $y^T z = 0$. Define the *Householder transformation matrix*

$$H = I - 2 \frac{y y^T}{||y||_2^2}. \tag{4.138}$$

We see that $H = I - 2 P_y$ [via (4.135)]. Hence Hx is as shown in Fig. 4.3; that is, the Householder transformation finds the reflection of vector x with respect to vector z, and $z \perp y$ (Definition 1.6 of Chapter 1). Recall the unit vector $e_i \in \mathbf{R}^n$

$$e_i = [\underbrace{0 \cdots 0}_{i \text{ zeros}} 1 0 \cdots 0]^T,$$

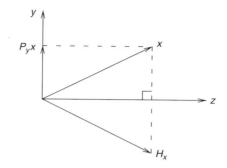

Figure 4.3   Geometric interpretation of the Householder transformation operator H. Note that $z^T y = 0$.

so $e_0 = [10 \cdots 0]^T$. Suppose that we want $Hx = ae_0$ for some $a \in \mathbf{R}$ with $a \neq 0$; that is, we wish to design H to annihilate all elements of x except for the top element. Let $y = x + \alpha e_0$; then

$$y^T x = (x^T + \alpha e_0^T)x = x^T x + \alpha x_0 \tag{4.139a}$$

(as $x = [x_0 \ x_1 \ \cdots \ x_{n-1}]^T$), and

$$\|y\|_2^2 = (x^T + \alpha e_0^T)(x + \alpha e_0) = x^T x + 2\alpha x_0 + \alpha^2. \tag{4.139b}$$

Therefore

$$Hx = x - 2\frac{yy^T}{\|y\|_2^2}x = x - 2\frac{y^T x}{\|y\|_2^2}y$$

so from (4.139), this becomes

$$Hx = x - 2\frac{(x^T x + \alpha x_0)(x + \alpha e_0)}{\|y\|_2^2}$$

$$= x - 2\frac{(x^T x + \alpha x_0)x + \alpha(y^T x)e_0}{\|y\|_2^2}$$

$$= \left[1 - 2\frac{x^T x + \alpha x_0}{x^T x + 2\alpha x_0 + \alpha^2}\right]x - 2\alpha\frac{y^T x}{\|y\|_2^2}e_0. \tag{4.140}$$

To force the first term to zero, we require

$$x^T x + 2\alpha x_0 + \alpha^2 - 2(x^T x + \alpha x_0) = 0,$$

which implies that $\alpha^2 = x^T x$, or in other words, we need

$$\alpha = \pm\|x\|_2. \tag{4.141}$$

Consequently, we select $y = x \pm ||x||_2 e_0$. In this case

$$
Hx = -2\alpha \frac{y^T x}{||y||_2^2} e_0 = -2\alpha \frac{x^T x + \alpha x_0}{x^T x + 2\alpha x_0 + \alpha^2} e_0
$$

$$
= -2\alpha \frac{\alpha^2 + \alpha x_0}{2\alpha^2 + 2\alpha x_0} e_0 = -\alpha e_0 = \mp ||x||_2 e_0, \tag{4.142}
$$

so $Hx = a e_0$ for $a = -\alpha$ if $y = x + \alpha e_0$ with $\alpha = \pm ||x||_2$.

Example 4.6 Suppose $x = [4 \quad 3 \quad 0]^T$, so $||x||_2 = 5$. Choose $\alpha = 5$. Thus $y = [9 \quad 3 \quad 0]^T$, and

$$
H = I - 2\frac{yy^T}{y^T y} = \frac{1}{45}\begin{bmatrix} -36 & -27 & 0 \\ -27 & 36 & 0 \\ 0 & 0 & 45 \end{bmatrix}.
$$

We see that

$$
Hx = \frac{1}{45}\begin{bmatrix} -36 & -27 & 0 \\ -27 & 36 & 0 \\ 0 & 0 & 45 \end{bmatrix}\begin{bmatrix} 4 \\ 3 \\ 0 \end{bmatrix} = \frac{1}{45}\begin{bmatrix} -225 \\ 0 \\ 0 \end{bmatrix} = \begin{bmatrix} -5 \\ 0 \\ 0 \end{bmatrix},
$$

so $Hx = -\alpha e_0$.

The Householder transformation is designed to annihilate elements of vectors. But in contrast with the Gauss transformations of Section 4.5, *Householder matrices are orthogonal*. To see this observe that

$$
H^T H = \left[I - 2\frac{yy^T}{y^T y} \right]\left[I - 2\frac{yy^T}{y^T y} \right]
$$

$$
= I - 4\frac{yy^T}{y^T y} + 4\frac{yy^T yy^T}{[y^T y]^2}
$$

$$
= I - 4\frac{yy^T}{y^T y} + 4\frac{yy^T}{y^T y} = I.
$$

Thus, no matter how we select y, $\kappa_2(H) = 1$. Householder matrices are therefore perfectly conditioned.

To obtain R in (4.94), we define

$$
\tilde{H}_k = \begin{bmatrix} I_{k-1} & 0 \\ 0 & H_k \end{bmatrix} \in \mathbf{R}^{m \times m}, \tag{4.143}
$$

where $k = 1, 2, \ldots, n$, I_{k-1} is an order $k - 1$ identity matrix, H_k is an order $m - k + 1$ Householder transformation matrix. We design H_k to annihilate elements k to $m - 1$ of column $k - 1$ in A^{k-1}, where $A^0 = A$, and

$$A^k = \tilde{H}_k A^{k-1}, \tag{4.144}$$

so $A^n = R$ (in (4.94)). Much as in Section 4.5, we have $A^k = [a_{i,j}^k] \in \mathbf{R}^{m \times n}$, and we assume $m > n$.

Example 4.7   Suppose

$$A = A^0 = \begin{bmatrix} a_{00}^0 & a_{01}^0 & a_{02}^0 \\ a_{10}^0 & a_{11}^0 & a_{12}^0 \\ a_{20}^0 & a_{21}^0 & a_{22}^0 \\ a_{30}^0 & a_{31}^0 & a_{32}^0 \end{bmatrix} \in \mathbf{R}^{4 \times 3},$$

and so therefore

$$A^1 = \tilde{H}_1 A^0 = \begin{bmatrix} \times & \times & \times & \times \\ \times & \times & \times & \times \\ \times & \times & \times & \times \\ \times & \times & \times & \times \end{bmatrix} \begin{bmatrix} a_{00}^0 & a_{01}^0 & a_{02}^0 \\ a_{10}^0 & a_{11}^0 & a_{12}^0 \\ a_{20}^0 & a_{21}^0 & a_{22}^0 \\ a_{30}^0 & a_{31}^0 & a_{32}^0 \end{bmatrix} = \begin{bmatrix} a_{00}^1 & a_{01}^1 & a_{02}^1 \\ 0 & a_{11}^1 & a_{12}^1 \\ 0 & a_{21}^1 & a_{22}^1 \\ 0 & a_{31}^1 & a_{32}^1 \end{bmatrix},$$

$$A^2 = \tilde{H}_2 A^1 = \begin{bmatrix} 1 & 0 & 0 & 0 \\ 0 & \times & \times & \times \\ 0 & \times & \times & \times \\ 0 & \times & \times & \times \end{bmatrix} \begin{bmatrix} a_{00}^1 & a_{01}^1 & a_{02}^1 \\ 0 & a_{11}^1 & a_{12}^1 \\ 0 & a_{21}^1 & a_{22}^1 \\ 0 & a_{31}^1 & a_{32}^1 \end{bmatrix} = \begin{bmatrix} a_{00}^1 & a_{01}^1 & a_{02}^1 \\ 0 & a_{11}^2 & a_{12}^2 \\ 0 & 0 & a_{22}^2 \\ 0 & 0 & a_{32}^2 \end{bmatrix},$$

and

$$A^3 = \tilde{H}_3 A^2 = \begin{bmatrix} 1 & 0 & 0 & 0 \\ 0 & 1 & 0 & 0 \\ 0 & 0 & \times & \times \\ 0 & 0 & \times & \times \end{bmatrix} \begin{bmatrix} a_{00}^1 & a_{01}^1 & a_{02}^1 \\ 0 & a_{11}^2 & a_{12}^2 \\ 0 & 0 & a_{22}^2 \\ 0 & 0 & a_{32}^2 \end{bmatrix}$$

$$= \begin{bmatrix} a_{00}^1 & a_{01}^1 & a_{02}^1 \\ 0 & a_{11}^2 & a_{12}^2 \\ 0 & 0 & a_{22}^3 \\ 0 & 0 & 0 \end{bmatrix} = R.$$

The $\times$ signs denote the Householder matrix elements that are not specified. This example is intended only to show the general pattern of elements in the matrices.

Define

$$x^k = \left[a_{k-1,k-1}^{k-1} \ a_{k,k-1}^{k-1} \ \cdots \ a_{m-1,k-1}^{k-1}\right]^T \in \mathbf{R}^{m-k+1}, \tag{4.145}$$

so if $x^k = \left[x_0^k \ x_1^k \ \cdots \ x_{m-k}^k\right]^T$ then $x_i^k = a_{i+k-1,k-1}^{k-1}$, and so

$$H_k = I_{m-k+1} - 2\frac{y^k(y^k)^T}{(y^k)^T y^k}, \tag{4.146}$$

where $y^k = x^k \pm ||x^k||_2 e_0^k$, and $e_0^k = [1 \ 0 \ \cdots \ 0]^T \in \mathbf{R}^{m-k+1}$. A pseudocode analogous to that for Gaussian elimination (recall Section 4.5) is as follows:

```
A⁰ := A;
for k := 1 to n do begin
    for i := 0 to m − k do begin
        xᵏᵢ := aᵏ⁻¹ᵢ₊ₖ₋₁,ₖ₋₁;  {This loop makes xᵏ}
        end;
    yᵏ := xᵏ + sign(xᵏ₀)||xᵏ||₂eᵏ₀;
    Aᵏ := H̃ₖAᵏ⁻¹;  { H̃ₖ contains Hₖ via (4.146) }
    end;
R := Aⁿ;
```

From (4.143) $\tilde{H}_k^T \tilde{H}_k = I_m$ because $H_k^T H_k = I_{m-k+1}$, and of course $I_{k-1}^T I_{k-1} = I_{k-1}$; that is, $\tilde{H}_k$ is orthogonal for all k. Since

$$R = A^n = \tilde{H}_n \tilde{H}_{n-1} \cdots \tilde{H}_2 \tilde{H}_1 A, \tag{4.147}$$

we have

$$A = \underbrace{\tilde{H}_1^T \tilde{H}_2^T \cdots \tilde{H}_n^T}_{=Q} R. \tag{4.148}$$

Thus, the pseudocode above implicitly computes Q because it creates the orthogonal factors $\tilde{H}_k$.

In the pseudocode we see that

$$y^k = x^k + \text{sign}(x_0^k) \ ||x^k||_2 e_0^k. \tag{4.149}$$

Recall from (4.141) that $\alpha = \pm ||x||_2$, so we must choose the sign of α. It is best that $\alpha = \text{sign}(x_0)||x||_2$, where $\text{sign}(x_0) = +1$ for $x_0 \geq 0$, and $\text{sign}(x_0) = -1$ if $x_0 < 0$. This turns out to ensure that H remains as close as possible to perfect orthogonality in the face of rounding errors. Because $||x||_2$ might be very large or very small, there is a risk of overflow or underflow in the computation of $||x||_2$. Thus, it is often better to compute y from $x/||x||_\infty$. This works because scaling x does not mathematically alter H (which may be confirmed as an exercise). Typically, $m >> n$ (e.g., in the example of Fig. 4.1 we had $m = N = 1000$, while $n = 3$), so, since $\tilde{H}_k \in \mathbf{R}^{m \times m}$, we rarely can accumulate and store the elements of $\tilde{H}_k$ for all k as too much memory is needed for such a task. Instead, it is much better to observe that (for example) if $H \in \mathbf{R}^{m \times m}$, $A \in \mathbf{R}^{m \times n}$ then, as $y \in \mathbf{R}^m$, we have

$$HA = \left[I - 2\frac{yy^T}{y^T y}\right] A = A - \frac{2}{y^T y} y (A^T y)^T. \tag{4.150}$$

From (4.150) $A^T y \in \mathbf{R}^n$, which has jth element

$$[A^T y]_j = \sum_{k=0}^{m-1} a_{k,j} y_k \tag{4.151}$$

for $j = 0, 1, \ldots, n - 1$. If $\beta = 2/y^T y$, then, from (4.150) and (4.151), we have

$$[HA]_{i,j} = a_{i,j} - \beta y_i \left[\sum_{k=0}^{m-1} a_{k,j} y_k\right] \tag{4.152}$$

for $i = 0, 1, \ldots, m - 1$, and $j = 0, 1, \ldots, n - 1$. A pseudocode program that implements this is as follows:

```
β := 2/yᵀy;
for j = 0 to n − 1 do begin
  s := ∑ᵐ⁻¹ₖ₌₀ aₖ,ⱼyₖ;
  s := βs;
  for i := 0 to m − 1 do begin
    aᵢ,ⱼ := aᵢ,ⱼ − syᵢ;
    end;
  end;
```

This program is written to overwrite matrix A with matrix HA. This reduces computer system memory requirements. Recall (4.123), where we see that $Q^T f$ must be computed so that f^u can be found. Knowledge of f^u is essential to compute $\hat{x}$ via (4.126). As in the problem of computing $\tilde{H}_k A^{k-1}$, we do not wish to accumulate and save the factors $\tilde{H}_k$ in

$$Q^T f = \tilde{H}_n \tilde{H}_{n-1} \cdots \tilde{H}_1 f. \tag{4.153}$$

Instead, $Q^T f$ would be computed using an algorithm similar to that suggested by (4.152).

All the suggestions in the previous paragraph are needed in a practical implementation of the Householder transformation matrix method for *QR* factorization. As noted in Ref. 5, the rounding error performance of the practical Householder *QR* factorization algorithm is quite good. It is stated [5] as well that the number of flops needed by the Householder method for finding $\hat{x}$ is greater than that needed by

Cholesky factorization. Somewhat simplistically, the Cholesky method is computationally more efficient than the Householder method, but the Householder method is less susceptible to ill conditioning and to rounding errors than is the Cholesky method. More or less, there is therefore a tradeoff between speed and accuracy involved in selecting between these competing methods for solving the overdetermined least-squares problem. The Householder approach is also claimed [5] to require more memory than the Cholesky approach.

4.7 ITERATIVE METHODS FOR LINEAR SYSTEMS

Matrix $A \in \mathbf{R}^{n \times n}$ is said to be *sparse* if most of its n^2 elements are zero-valued. Such matrices can arise in various applications, such as in the numerical solution of partial differential equations (PDEs). Sections 4.5 and 4.6 have presented such *direct methods* as the LU and QR decompositions (factorizations) of A in order to solve $Ax = b$ (assuming that A is nonsingular). However, these procedures do not in themselves take advantage of any structure that may be possessed by A such as sparsity. Thus, they are not necessarily computationally efficient procedures. Therefore, in the present section, we consider *iterative methods* to determine $x \in \mathbf{R}^n$ in $Ax = b$. In this section, whenever we consider $Ax = b$, we will always assume that A^{-1} exists. Iterative methods work by creating a Cauchy sequence of vectors $(x^{(k)})$ that converges to x.[10] Iterative methods may be particularly advantageous when A is not only sparse, but is also large (i.e., large n). This is because direct methods often require the considerable movement of data around the computing machine memory system, and this can slow the computation down substantially. But a properly conceived and implemented iterative method can alleviate this problem.

Our presentation of iterative methods here is based largely on the work of Quarteroni et al. [8, Chapter 4]. We use much of the same notation as that in Ref. 8. But it is a condensed presentation as this section is intended only to convey the main ideas about iterative linear system solvers.

In Section 4.4 matrix and vector norms were considered in order to characterize the sizes of errors in the numerical estimate of x in $Ax = b$ due to perturbations of A, and b. We will need to consider such norms here. As noted above, our goal here is to derive a methodology to generate vector sequence $(x^{(k)})$[11] such that

$$\lim_{k \to \infty} x^{(k)} = x, \tag{4.154}$$

where $x = [x_0 \ x_1 \ \cdots \ x_{n-1}]^T \in \mathbf{R}^n$ satisfies $Ax = b$ and $x^{(k)} = [x_0^{(k)} \ x_1^{(k)} \ \cdots \ x_{n-1}^{(k)}]^T \in \mathbf{R}^n$. The basic idea is to find an *operator* T such that $x^{(k+1)} = Tx^{(k)} (= T(x^{(k)}))$, for $k = 0, 1, 2, \ldots$. Because $(x^{(k)})$ is designed to be Cauchy (recall

---

[10]As such, we will be revisiting ideas first seen in Section 3.2.

[11]Note that the "(k)" in $x^{(k)}$ does not denote the raising of x to a power or the taking of the kth derivative, but rather is part of the name of the vector. Similar notation applies to matrices. So, A^k is the kth power of A, but $A^{(k)}$ is not.

Section 3.2) for any $\epsilon > 0$, there will be an $m \in \mathbf{Z}^+$ such that $||x^{(m)} - x|| < \epsilon$ [recall that $d(x^{(k)}, x) = ||x^{(k)} - x||$]. The operator T is defined according to

$$x^{(k+1)} = Bx^{(k)} + f, \qquad (4.155)$$

where $x^{(0)} \in \mathbf{R}^n$ is the starting value (*initial guess* about the solution x), $B \in \mathbf{R}^{n \times n}$ is called the *iteration matrix*, and $f \in \mathbf{R}^n$ is derived from A and b in $Ax = b$. Since we want (4.154) to hold, from (4.155) we seek B and f such that $x = Bx + f$, or $A^{-1}b = BA^{-1}b + f$ (using $Ax = b$, implying $x = A^{-1}b$), so

$$f = (I - B)A^{-1}b. \qquad (4.156)$$

The *error vector at step k* is defined to be

$$e^{(k)} = x^{(k)} - x, \qquad (4.157)$$

and naturally we want $\lim_{k \to \infty} e^{(k)} = 0$. Convergence would be in some suitably selected norm.

As matters now stand, there is no guarantee that (4.154) will hold. We achieve convergence only by the proper selection of B, and for matrices A possessing suitable properties (considered below). Before we can consider these matters we require certain basic results involving matrix norms.

Definition 4.3: Spectral Radius   Let $s(A)$ denote the set of eigenvalues of matrix $A \in \mathbf{R}^{n \times n}$. The *spectral radius* of A is

$$\rho(A) = \max_{\lambda \in s(A)} |\lambda|.$$

An important property possessed by $\rho(A)$ is as follows.

Property 4.1   If $A \in \mathbf{R}^{n \times n}$ with $\epsilon > 0$, then there is a norm denoted $|| \cdot ||_\epsilon$ (i.e., a norm perhaps dependent on ϵ) satisfying the *consistency condition* (4.36c), and such that

$$||A||_\epsilon \leq \rho(A) + \epsilon.$$

Proof   See Isaacson and Keller [9].

This is just a formal way of saying that there is always a matrix norm that is arbitrarily close to the spectral radius of A

$$\rho(A) = \inf_{||\cdot||} ||A|| \qquad (4.158)$$

with the infimum (defined in Section 1.3) taken over all possible norms that satisfy (4.36c). We say that the sequence of matrices $(A^{(k)})$ [with $A^{(k)} \in \mathbf{R}^{n \times n}$] *converges* to $A \in \mathbf{R}^{n \times n}$ iff

$$\lim_{k \to \infty} ||A^{(k)} - A|| = 0. \qquad (4.159)$$

The norm in (4.159) is arbitrary because of norm equivalence (recall discussion on this idea in Section 4.4).

Theorem 4.7: Let $A \in \mathbf{R}^{n \times n}$; then

$$\lim_{k \to \infty} A^k = 0 \Leftrightarrow \rho(A) < 1. \tag{4.160}$$

As well, the matrix geometric series $\sum_{k=0}^{\infty} A^k$ converges iff $\rho(A) < 1$. In this instance

$$\sum_{k=0}^{\infty} A^k = (I - A)^{-1}. \tag{4.161}$$

So, if $\rho(A) < 1$, then matrix $I - A$ is invertible, and also

$$\frac{1}{1 + ||A||} \leq ||(I - A)^{-1}|| \leq \frac{1}{1 - ||A||}, \tag{4.162}$$

where $|| \cdot ||$ here is an *induced matrix norm* (i.e., (4.36b) holds) such that $||A|| < 1$.

Proof   We begin by showing (4.160) holds. Let $\rho(A) < 1$ so there must be an $\epsilon > 0$ such that $\rho(A) < 1 - \epsilon$, and from Property 4.1 there is a consistent matrix norm $|| \cdot ||$ such that

$$||A|| \leq \rho(A) + \epsilon < 1.$$

Because [recall (4.40)] of $||A^k|| \leq ||A||^k < 1$, and the definition of convergence, as $k \to \infty$, we have $A^k \to 0 \in \mathbf{R}^{n \times n}$. Conversely, assume that $\lim_{k \to \infty} A^k = 0$, and let λ be any eigenvalue of A. For eigenvector x ($\neq 0$) of A associated with eigenvalue λ, we have $A^k x = \lambda^k x$, and so $\lim_{k \to \infty} \lambda^k = 0$. Thus, $|\lambda| < 1$, and hence $\rho(A) < 1$. Now consider (4.161). If λ is an eigenvalue of A, then $1 - \lambda$ is an eigenvalue of $I - A$. We observe that

$$(I - A)(I + A + A^2 + \cdots + A^{n-1} + A^n) = I - A^{n+1}. \tag{4.163}$$

Since $\rho(A) < 1$, $I - A$ has an inverse, and letting $n \to \infty$ in (4.163) yields

$$(I - A) \sum_{k=0}^{\infty} A^k = I$$

so that (4.161) holds.

Now, because matrix norm $|| \cdot ||$ satisfies (4.36b), we must have $||I|| = 1$. Thus

$$1 = ||I|| \leq ||I - A|| \, ||(I - A)^{-1}|| \leq (1 + ||A||)||(I - A)^{-1}||,$$

which gives the first inequality in (4.162). Since $I = (I - A) + A$, we have

$$(I - A)^{-1} = I + A(I - A)^{-1}$$

so that

$$||(I - A)^{-1}|| \leq 1 + ||A|| \, ||(I - A)^{-1}||.$$

Condition $||A|| < 1$ implies that this yields the second inequality in (4.162).

We mention that in Theorem 4.7 an induced matrix norm exists to give $||A|| < 1$ because of Property 4.1 (recall that $(A^{(k)})$ is convergent, giving $\rho(A) < 1$). Theorem 4.7 now leads us to the following theorem.

Theorem 4.8: Suppose that $f \in \mathbf{R}^n$ satisfies (4.156); then $(x^{(k)})$ converges to x satisfying $Ax = b$ for any $x^{(0)}$ iff $\rho(B) < 1$.

Proof From (4.155)–(4.157), we have

$$
\begin{aligned}
e^{(k+1)} = x^{(k+1)} - x &= Bx^{(k)} + f - x = Bx^{(k)} + (I - B)A^{-1}b - x \\
&= Be^{(k)} + Bx + (I - B)A^{-1}b - x \\
&= Be^{(k)} + Bx + x - Bx - x \\
&= Be^{(k)}.
\end{aligned}
$$

Immediately, we see that

$$e^{(k)} = B^k e^{(0)} \tag{4.164}$$

for $k \in \mathbf{Z}^+$. From Theorem 4.7

$$\lim_{k \to \infty} B^k e^{(0)} = 0$$

for all $e^{(0)} \in \mathbf{R}^n$ iff $\rho(B) < 1$.

On the other hand, suppose $\rho(B) \geq 1$; then there is at least one eigenvalue λ of B such that $|\lambda| \geq 1$. Let $e^{(0)}$ be the eigenvector associated with λ, so $Be^{(0)} = \lambda e^{(0)}$, implying that $e^{(k)} = \lambda^k e^{(0)}$. But this implies that $e^{(k)} \not\to 0$ as $k \to \infty$ since $|\lambda| \geq 1$.

This theorem gives a general condition on B so that iterative procedure (4.155) converges. Theorem 4.9 (below) will say more. However, our problem now is to find B. From (4.158), and Theorem 4.7 a sufficient condition for convergence is that $||B|| < 1$, for any matrix norm.

A general approach to constructing iterative methods is to use the *additive splitting* of the matrix A according to

$$A = P - N, \tag{4.165}$$

where $P, N \in \mathbf{R}^{n \times n}$ are suitable matrices, and P^{-1} exists. Matrix P is sometimes called a *preconditioning matrix*, or *preconditioner* (for reasons we will not consider here, but that are explained in Ref. 8). To be specific, we rewrite (4.155) as

$$x^{(k+1)} = P^{-1} N x^{(k)} + P^{-1} b,$$

that is, for $k \in \mathbf{Z}^+$

$$P x^{(k+1)} = N x^{(k)} + b, \tag{4.166}$$

so that $f = P^{-1} b$, and $B = P^{-1} N$. Alternatively

$$x^{(k+1)} = x^{(k)} + P^{-1} \underbrace{[b - A x^{(k)}]}_{=r^{(k)}}, \tag{4.167}$$

where $r^{(k)}$ is the *residual vector at step* k. From (4.167) we see that to obtain $x^{(k+1)}$ requires us to solve a linear system of equations involving P. Clearly, for this approach to be worth the trouble, P must be nonsingular, and be easy to invert as well in order to save on computations.

We will now make the additional assumption that the main diagonal elements of A are nonzero (i.e., $a_{i,i} \neq 0$ for all $i \in \mathbf{Z}_n$). All the iterative methods we consider in this section will assume this. In this case we may express $Ax = b$ in the equivalent form

$$x_i = \frac{1}{a_{ii}} \left[b_i - \sum_{\substack{j=0 \\ j \neq i}}^{n-1} a_{ij} x_j \right] \tag{4.168}$$

for $i = 0, 1, \ldots, n-1$.

The expression (4.168) immediately leads to, for any initial guess $x^{(0)}$, the *Jacobi method*, which is defined by the iterations

$$x_i^{(k+1)} = \frac{1}{a_{ii}} \left[b_i - \sum_{\substack{j=0 \\ j \neq i}}^{n-1} a_{ij} x_j^{(k)} \right] \tag{4.169}$$

for $i = 0, 1, \ldots, n-1$. It is easy to show that this algorithm implements the splitting

$$P = D, N = D - A = L + U, \tag{4.170}$$

where $D = \mathrm{diag}(a_{0,0}, a_{1,1}, \ldots, a_{n-1,n-1})$ (i.e., diagonal matrix that is the main diagonal elements of A), L is the lower triangular matrix such that $l_{ij} = -a_{ij}$ if $i > j$, and $l_{ij} = 0$ if $i \leq j$, and U is the upper triangular matrix such that $u_{ij} = -a_{ij}$ if $j > i$, and $u_{ij} = 0$ if $j \leq i$. Here the iteration matrix B is given by

$$B = B_J = P^{-1} N = D^{-1}(L + U) = I - D^{-1} A. \tag{4.171}$$

The Jacobi method generalizes according to

$$x_i^{(k+1)} = \frac{\omega}{a_{ii}} \left[b_i - \sum_{\substack{j=0 \\ j \neq i}}^{n-1} a_{ij} x_j^{(k)} \right] + (1 - \omega) x_i^{(k)}, \tag{4.172}$$

where $i = 0, 1, \ldots, n - 1$, and ω is the *relaxation parameter*. Relaxation parameters are introduced into iterative procedures in order to control convergence rates. The algorithm (4.172) is called the *Jacobi overrelaxation* (JOR) *method*. In this algorithm the iteration matrix B takes on the form

$$B = B_J(\omega) = \omega B_J + (1 - \omega) I, \tag{4.173}$$

and (4.172) can be expressed in the form (4.167) according to

$$x^{(k+1)} = x^{(k)} + \omega D^{-1} r^{(k)}. \tag{4.174}$$

The JOR method satisfies (4.156) provided that $\omega \neq 0$. The method is easily seen to reduce to the Jacobi method when $\omega = 1$.

An alternative to the Jacobi method is the *Gauss–Seidel method*. This is defined as

$$x_i^{(k+1)} = \frac{1}{a_{ii}} \left[b_i - \sum_{j=0}^{i-1} a_{ij} x_j^{(k+1)} - \sum_{j=i+1}^{n-1} a_{ij} x_j^{(k)} \right], \tag{4.175}$$

where $i = 0, 1, \ldots, n - 1$. In matrix form (4.175) can be expressed as

$$Dx^{(k+1)} = b + Lx^{(k+1)} + Ux^{(k)}, \tag{4.176}$$

where D, L, and U are the same matrices as those associated with the Jacobi method. In the Gauss–Seidel method we implement the splitting

$$P = D - L, \quad N = U \tag{4.177}$$

with the iteration matrix

$$B = B_{GS} = (D - L)^{-1} U. \tag{4.178}$$

As there is an overrelaxation method for the Jacobi approach, the same idea applies for the Gauss–Seidel case. The *Gauss–Seidel successive overrelaxation* (SOR) *method* is defined to be

$$x_i^{(k+1)} = \frac{\omega}{a_{ii}} \left[b_i - \sum_{j=0}^{i-1} a_{ij} x_j^{(k+1)} - \sum_{j=i+1}^{n-1} a_{ij} x_j^{(k)} \right] + (1 - \omega) x_i^{(k)}, \tag{4.179}$$

again for $i = 0, 1, \ldots, n - 1$. In matrix form this procedure can be expressed as

$$Dx^{(k+1)} = \omega[b + Lx^{(k+1)} + Ux^{(k)}] + (1 - \omega)Dx^{(k)}$$

or

$$[I - \omega D^{-1}L]x^{(k+1)} = \omega D^{-1}b + [(1 - \omega)I + \omega D^{-1}U]x^{(k)}, \qquad (4.180)$$

for which the iteration matrix is now

$$B = B_{GS}(\omega) = [I - \omega D^{-1}L]^{-1}[(1 - \omega)I + \omega D^{-1}U]. \qquad (4.181)$$

We see from (4.180) (on multiplying both sides by D) that

$$[D - \omega L]x^{(k+1)} = \omega b + [(1 - \omega)D + \omega U]x^{(k)},$$

so from the fact that $A = D - (L + U)$ [recall (4.170)], this may be rearranged as

$$x^{(k+1)} = x^{(k)} + \left[\frac{1}{\omega}D - L\right]^{-1} r^{(k)} \qquad (4.182)$$

$(r^{(k)} = b - Ax^{(k)})$, which is the form (4.167). Condition (4.156) holds if $\omega \neq 0$. The case $\omega = 1$ corresponds to the Gauss–Seidel method in (4.175). If $\omega \in (0, 1)$, the technique is often called an *underrelaxation* method, while for $\omega \in (1, \infty)$ it is an *overrelaxation* method.

We will now summarize, largely without proof, results concerning the convergence of $(x^{(k)})$ to x for sequences generated by the previous iterative algorithms. We observe that every iteration in any of the proposed methods needs (in the worst case, assuming that A is not sparse) $O(n^2)$ arithmetic operations. The total number of iterations is m, and is needed to achieve desired accuracy $||x^{(m)} - x|| < \epsilon$, and so in turn the total number of arithmetic operations needed is $O(mn^2)$. Gaussian elimination needs $O(n^3)$ operations to solve $Ax = b$, so the iterative methods are worthwhile computationally only if m is sufficiently small. If m is about the same size as n, then little advantage can be expected from iterative methods. On the other hand, if A is sparse, perhaps possessing only $O(n)$ nonzero elements, then the iterative methods require only $O(mn)$ operations to achieve $||x^{(m)} - x|| < \epsilon$. We need to give conditions on A so that $x^{(k)} \to x$, and also to say something about the number of iterations needed to achieve convergence to desired accuracy.

Let us begin with the following definition.

Definition 4.4: Diagonal Dominance   Matrix $A \in \mathbf{R}^{n \times n}$ is *diagonally dominant* if

$$|a_{i,i}| > \sum_{\substack{j=0 \\ j \neq i}}^{n-1} |a_{ij}| \qquad (4.183)$$

for $i = 0, 1, \ldots, n - 1$.

We mention here that Definition 4.4 is a bit different from Definition 6.2 (Chapter 6), where diagonal dominance concepts appear in the context of spline interpolation problems. It can be shown that if A in $Ax = b$ is diagonally dominant according to Definition 4.4, then the Jacobi and Gauss-Seidel methods both converge. Proof for the Jacobi method appears in Theorem 4.2 of Ref. 8, while the Gauss–Seidel case is proved by Axelsson [10].

If $A = A^T$, and $A > 0$ both the Jacobi and Gauss–Seidel methods will converge. A proof for the Gauss–Seidel case appears in Golub and Van Loan [5, Theorem 10.1.2]. The Jacobi case is considered in Ref. 8. Convergence results exist for the overrelaxation methods JOR and SOR. For example, if $A = A^T$ with $A > 0$, the SOR method is convergent iff $0 < \omega < 2$ [8]. Naturally, we wish to select ω so that convergence occurs as rapidly as possible (i.e., m in $||x^{(m)} - x|| < \epsilon$ is minimal). However, the problem of selecting the optimal value for ω is well beyond the scope of this book.

We recall that our iterative procedures have the general form in (4.155), where it is intended that $x = Bx + f$. We may regard $y = Tx = Bx + f$ as a mapping $T|\mathbf{R}^n \rightarrow \mathbf{R}^n$. On linear vector space $\mathbf{R}^n$ we may define the metric

$$d(x, y) = \max_{j \in \mathbf{Z}_n} |x_j - y_j| \tag{4.184}$$

(recall the properties of metrics from Chapter 1). Space $(\mathbf{R}^n, d)$ is a complete metric space [11, p. 308]. From Kreyszig [11] we have the following theorem.

Theorem 4.9: If the linear system $x = Bx + f$ is such that

$$\sum_{j=0}^{n-1} |b_{ij}| < 1$$

for $i = 0, 1, \ldots, n - 1$ then solution x is unique. The solution can be obtained as the limit of the vector sequence $(x^{(k)})$ for $k = 0, 1, 2, \ldots$ ($x^{(0)}$ is arbitrary), where

$$x^{(k+1)} = Bx^{(k)} + f,$$

and where for $\alpha = \max_{i \in \mathbf{Z}_n} \sum_{j=0}^{n-1} |b_{ij}|$, we have the *error bounds*

$$d(x^{(m)}, x) \leq \frac{\alpha}{1 - \alpha} d(x^{(m-1)}, x^{(m)}) \leq \frac{\alpha^m}{1 - \alpha} d(x^{(0)}, x^{(1)}). \tag{4.185}$$

Proof We will give only an outline proof. This theorem is really just a special instance of the *contraction theorem*, which appears and is proved in Chapter 7 (see Theorem 7.3 and Corollary 7.1).

The essence of the proof is to consider the fact that

$$d(Tx, Ty) = \max_{i \in \mathbf{Z}_n} \left| \sum_{j=0}^{n-1} b_{ij}(x_j - y_j) \right|$$

$$\leq \max_{j \in \mathbf{Z}_n} |x_j - y_j| \max_{i \in \mathbf{Z}_n} \sum_{j=0}^{n-1} |b_{ij}|$$

$$= d(x, y) \max_{i \in \mathbf{Z}_n} \sum_{j=0}^{n-1} |b_{ij}|,$$

so $d(Tx, Ty) \leq \alpha d(x, y)$, if we define

$$\alpha = \max_{i \in \mathbf{Z}_n} \sum_{j=0}^{n-1} |b_{ij}| = ||B||_\infty$$

[recall (4.41d)].

In this theorem we see that if $\alpha < 1$, then $x^{(k)} \to x$. In this case $d(Tx, Ty) < d(x, y)$ for all $x, y \in \mathbf{R}^n$. Such a mapping T is called a *contraction mapping* (or *contractive mapping*). We see that contraction mappings have the effect of moving points in a space closer together. The error bounds stated in (4.185) give us an idea about the number of iterations m needed to achieve $||x^{(m)} - x|| < \epsilon$ ($\epsilon > 0$). We emphasize that condition $\alpha < 1$ is sufficient for convergence, so $(x^{(k)})$ may converge to x even if this condition is violated. It is also noteworthy that convergence will be fast if α is small, that is, if $||B||_\infty$ is small. The result in Theorem 4.8 certainly suggests convergence ought to be fast if $\rho(B)$ is small.

Example 4.8   We shall consider the application of SOR to the problem of solving $Ax = b$, where

$$A = \begin{bmatrix} 4 & 1 & 0 & 0 \\ 1 & 4 & 1 & 0 \\ 0 & 1 & 4 & 1 \\ 0 & 0 & 1 & 4 \end{bmatrix}, \quad b = \begin{bmatrix} 1 \\ 2 \\ 3 \\ 4 \end{bmatrix}.$$

We shall assume that $x^{(0)} = [0000]^T$. Note that SOR is not the best way to solve this problem. A better approach is to be found in Section 6.5 (Chapter 6). This example is for illustration only. However, it is easy to confirm that

$$x = [0.1627 \quad 0.3493 \quad 0.4402 \quad 0.8900]^T.$$

Recall that the SOR iterations are specified by (4.179). However, we have not discussed how to terminate the iterative process. A popular choice is to recall that $r^{(k)} = b - Ax^{(k)}$ [see (4.167)], and to stop the iterations when for $k = m$

$$\frac{||r^{(m)}||}{||r^{(0)}||} \leq \tau \tag{4.186}$$

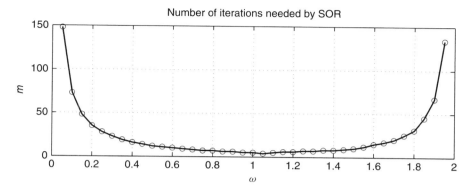

Figure 4.4   Plot of the number of iterations m needed by SOR as a function of ω for the parameters of Example 4.8 in order to satisfy the stopping condition $||r^{(m)}||_\infty/||r^{(0)}||_\infty \le \tau$.

for some $\tau > 0$ (a small value). For our present purposes $|| \cdot ||$ shall be the norm in (4.29c), which is compatible with the needs of Theorem 4.9. We shall choose $\tau = 0.001$.

We observe that A is diagonally dominant, so convergence is certainly expected for $\omega = 1$. In fact, $A > 0$ so convergence of the SOR method can be expected for all $\omega \in (0, 2)$. Figure 4.4 plots the m that achieves (4.186) versus ω, and we see that there is an optimal choice for ω that is somewhat larger than $\omega = 1$. In this case though the optimal choice does not lead to much of an improvement over choice $\omega = 1$.

For our problem [recalling (4.170)], we have

$$D = \begin{bmatrix} 4 & 0 & 0 & 0 \\ 0 & 4 & 0 & 0 \\ 0 & 0 & 4 & 0 \\ 0 & 0 & 0 & 4 \end{bmatrix}, L = \begin{bmatrix} 0 & 0 & 0 & 0 \\ -1 & 0 & 0 & 0 \\ 0 & -1 & 0 & 0 \\ 0 & 0 & -1 & 0 \end{bmatrix},$$

$$U = \begin{bmatrix} 0 & -1 & 0 & 0 \\ 0 & 0 & -1 & 0 \\ 0 & 0 & 0 & -1 \\ 0 & 0 & 0 & 0 \end{bmatrix}.$$

From (4.178)

$$B_{GS} = \begin{bmatrix} 0.0000 & -0.2500 & 0.0000 & 0.0000 \\ 0.0000 & 0.0625 & -0.2500 & 0.0000 \\ 0.0000 & -0.0156 & 0.0625 & -0.2500 \\ 0.0000 & 0.0039 & -0.0156 & 0.0625 \end{bmatrix}.$$

We therefore find that $||B_{GS}||_\infty = 0.3281$. It is possible to show (preferably using MATLAB or some other software tool that is good with eigenproblems) that

$\rho(B_{GS}) = 0.1636$. Given (4.185) in Theorem 4.9 we therefore expect fast convergence for our problem since α is fairly small. In fact

$$||x^{(m)} - x||_\infty \le \frac{||B_{GS}||_\infty^m}{1 - ||B_{GS}||_\infty}||(D - L)^{-1}b||_\infty \qquad (4.187)$$

[using $x^{(0)} = 0$, $x^{(1)} = (D - L)^{-1}b$]. For the stopping criterion of (4.186) we obtained (recalling that $\omega = 1$, and $\tau = 0.001$) $m = 5$ with

$$x^{(5)} = [0.1630 \quad 0.3490 \quad 0.4403 \quad 0.8899]^T$$

so that

$$||x^{(5)} - x||_\infty = 3.6455 \times 10^{-4}.$$

The right-hand side of (4.187) evaluates to

$$\frac{||B_{GS}||_\infty^m}{1 - ||B_{GS}||_\infty}||(D - L)^{-1}b||_\infty = 4.7523 \times 10^{-3}.$$

Thus, (4.187) certainly holds true.

4.8   FINAL REMARKS

We have seen that inaccurate solutions to linear systems of equations can arise when the linear system is ill-conditioned. Condition numbers warn us if this is a potential problem. However, even if a problem is well-conditioned, an inaccurate solution may arise if the algorithm applied to solve it is unstable. In the case of problems arising out of algorithm instability, we naturally replace the unstable algorithm with a stable one (e.g., Gaussian elimination may need to be replaced by Gaussian elimination with partial pivoting). In the case of an ill-conditioned problem, we may try to improve the accuracy of the solution by either

1. Using an algorithm that does not worsen the conditioning of the underlying problem (e.g., choosing QR factorization in preference to Cholesky factorization)
2. Reformulating the problem so that it is better conditioned

We have not considered the second alternative in this chapter. This will be done in Chapter 5.

APPENDIX 4.A   HILBERT MATRIX INVERSES

Consider the following MATLAB code:

```
R = hilb(10);
inv(R)
```

```
ans =

   1.0e+12 *

 Columns 1 through 7

     0.0000   -0.0000    0.0000   -0.0000    0.0000   -0.0000    0.0000
    -0.0000    0.0000   -0.0000    0.0000   -0.0002    0.0005   -0.0008
     0.0000   -0.0000    0.0001   -0.0010    0.0043   -0.0112    0.0178
    -0.0000    0.0000   -0.0010    0.0082   -0.0379    0.1010   -0.1616
     0.0000   -0.0002    0.0043   -0.0379    0.1767   -0.4772    0.7712
    -0.0000    0.0005   -0.0112    0.1010   -0.4772    1.3014   -2.1208
     0.0000   -0.0008    0.0178   -0.1616    0.7712   -2.1208    3.4803
    -0.0000    0.0008   -0.0166    0.1529   -0.7358    2.0376   -3.3636
     0.0000   -0.0004    0.0085   -0.0788    0.3820   -1.0643    1.7659
    -0.0000    0.0001   -0.0018    0.0171   -0.0832    0.2330   -0.3883

 Columns 8 through 10

    -0.0000    0.0000   -0.0000
     0.0008   -0.0004    0.0001
    -0.0166    0.0085   -0.0018
     0.1529   -0.0788    0.0171
    -0.7358    0.3820   -0.0832
     2.0376   -1.0643    0.2330
    -3.3636    1.7659   -0.3883
     3.2675   -1.7231    0.3804
    -1.7231    0.9122   -0.2021
     0.3804   -0.2021    0.0449

R*inv(R)

ans =

 Columns 1 through 7

     1.0000    0.0000    0.0000   -0.0000    0.0001    0.0001   -0.0001
    -0.0000    1.0000    0.0000   -0.0000    0.0001    0.0001   -0.0002
    -0.0000    0.0000    1.0000   -0.0000    0.0001    0.0000   -0.0001
    -0.0000    0.0000    0.0000    1.0000    0.0000    0.0000   -0.0000
    -0.0000    0.0000    0.0000   -0.0000    1.0000   -0.0000   -0.0000
    -0.0000    0.0000    0.0000   -0.0000    0.0000    1.0000   -0.0000
    -0.0000    0.0000    0.0000   -0.0000    0.0000    0.0000    0.9999
    -0.0000    0.0000    0.0000   -0.0000    0.0000    0.0001   -0.0000
    -0.0000    0.0000    0.0000   -0.0000    0.0000    0.0000   -0.0001
    -0.0000    0.0000    0.0000   -0.0000    0.0000    0.0000   -0.0000

 Columns 8 through 10

    -0.0000   -0.0001    0.0000
     0.0001   -0.0001    0.0000
     0.0000   -0.0001    0.0000
    -0.0000   -0.0000    0.0000
```

```
    0.0000   -0.0000    0.0000
   -0.0000   -0.0000    0.0000
    0.0000   -0.0000    0.0000
    1.0000   -0.0000    0.0000
    0.0000    1.0000    0.0000
   -0.0000   -0.0000    1.0000
```

```
R = hilb(11);
inv(R)
```

ans =

   1.0e+14 *

  Columns 1 through 7

```
    0.0000   -0.0000    0.0000   -0.0000    0.0000   -0.0000    0.0000
   -0.0000    0.0000   -0.0000    0.0000   -0.0000    0.0000   -0.0000
    0.0000   -0.0000    0.0000   -0.0000    0.0002   -0.0006    0.0012
   -0.0000    0.0000   -0.0000    0.0003   -0.0019    0.0064   -0.0137
    0.0000   -0.0000    0.0002   -0.0019    0.0110   -0.0381    0.0817
   -0.0000    0.0000   -0.0006    0.0064   -0.0381    0.1329   -0.2877
    0.0000   -0.0000    0.0012   -0.0137    0.0817   -0.2877    0.6270
   -0.0000    0.0001   -0.0016    0.0183   -0.1101    0.3902   -0.8555
    0.0000   -0.0000    0.0013   -0.0149    0.0905   -0.3227    0.7111
   -0.0000    0.0000   -0.0006    0.0068   -0.0415    0.1487   -0.3292
    0.0000   -0.0000    0.0001   -0.0013    0.0081   -0.0293    0.0651
```

  Columns 8 through 11

```
   -0.0000    0.0000   -0.0000    0.0000
    0.0001   -0.0000    0.0000   -0.0000
   -0.0016    0.0013   -0.0006    0.0001
    0.0183   -0.0149    0.0068   -0.0013
   -0.1101    0.0905   -0.0415    0.0081
    0.3902   -0.3227    0.1487   -0.0293
   -0.8555    0.7111   -0.3292    0.0651
    1.1733   -0.9796    0.4553   -0.0903
   -0.9796    0.8212   -0.3830    0.0762
    0.4553   -0.3830    0.1792   -0.0357
   -0.0903    0.0762   -0.0357    0.0071
```

```
R*inv(R)
```

ans =

  Columns 1 through 7

```
    0.9997   -0.0009    0.0022    0.0028   -0.0164    0.0558   -0.1229
   -0.0002    0.9992    0.0020    0.0023   -0.0132    0.0454   -0.1029
   -0.0002   -0.0007    1.0018    0.0019   -0.0112    0.0385   -0.0844
   -0.0002   -0.0006    0.0016    1.0017   -0.0097    0.0331   -0.0736
   -0.0002   -0.0006    0.0015    0.0015    0.9915    0.0285   -0.0638
   -0.0002   -0.0005    0.0014    0.0013   -0.0076    1.0258   -0.0581
```

```
-0.0002   -0.0005    0.0013    0.0012   -0.0070    0.0234    0.9468
-0.0001   -0.0005    0.0012    0.0011   -0.0063    0.0216   -0.0474
-0.0001   -0.0004    0.0011    0.0010   -0.0059    0.0201   -0.0448
-0.0001   -0.0004    0.0010    0.0009   -0.0053    0.0187   -0.0406
-0.0001   -0.0004    0.0010    0.0009   -0.0052    0.0179   -0.0395
```

Columns 8 through 11

```
  0.1665   -0.1405    0.0652   -0.0091
  0.1351   -0.1165    0.0530   -0.0071
  0.1125   -0.0973    0.0452   -0.0058
  0.0964   -0.0844    0.0385   -0.0047
  0.0858   -0.0739    0.0341   -0.0041
  0.0745   -0.0661    0.0300   -0.0037
  0.0696   -0.0592    0.0280   -0.0033
  1.0635   -0.0547    0.0251   -0.0029
  0.0581    0.9495    0.0235   -0.0028
  0.0536   -0.0458    1.0213   -0.0024
  0.0527   -0.0445    0.0207    0.9976
```

```
R = hilb(12);
inv(R)
```
Warning: Matrix is close to singular or badly scaled.
        Results may be inaccurate. RCOND = 2.632091e-17.

ans =

   1.0e+15 *

Columns 1 through 7

```
  0.0000   -0.0000    0.0000   -0.0000    0.0000   -0.0000    0.0000
 -0.0000    0.0000   -0.0000    0.0000   -0.0000    0.0000   -0.0000
  0.0000   -0.0000    0.0000   -0.0000    0.0001   -0.0002    0.0006
 -0.0000    0.0000   -0.0000    0.0001   -0.0008    0.0032   -0.0086
  0.0000   -0.0000    0.0001   -0.0008    0.0054   -0.0229    0.0624
 -0.0000    0.0000   -0.0002    0.0032   -0.0229    0.0990   -0.2720
  0.0000   -0.0000    0.0006   -0.0086    0.0624   -0.2720    0.7528
 -0.0000    0.0000   -0.0011    0.0151   -0.1107    0.4863   -1.3545
  0.0000   -0.0000    0.0013   -0.0173    0.1276   -0.5640    1.5794
 -0.0000    0.0000   -0.0009    0.0124   -0.0920    0.4090   -1.1511
  0.0000   -0.0000    0.0004   -0.0050    0.0377   -0.1686    0.4765
 -0.0000    0.0000   -0.0001    0.0009   -0.0067    0.0301   -0.0855
```

Columns 8 through 12

```
 -0.0000    0.0000   -0.0000    0.0000   -0.0000
  0.0000   -0.0000    0.0000   -0.0000    0.0000
 -0.0011    0.0013   -0.0009    0.0004   -0.0001
  0.0151   -0.0173    0.0124   -0.0050    0.0009
 -0.1107    0.1275   -0.0920    0.0377   -0.0067
  0.4863   -0.5639    0.4090   -0.1686    0.0301
 -1.3544    1.5793   -1.1510    0.4765   -0.0855
  2.4505   -2.8712    2.1015   -0.8732    0.1572
```

```
-2.8713    3.3786   -2.4821    1.0348   -0.1869
 2.1016   -2.4822    1.8297   -0.7651    0.1385
-0.8732    1.0348   -0.7651    0.3208   -0.0582
 0.1572   -0.1869    0.1386   -0.0582    0.0106
```

R*inv(R)
Warning: Matrix is close to singular or badly scaled.
          Results may be inaccurate. RCOND = 2.632091e-17.

ans =

Columns 1 through 7

```
1.0126   -0.0066   -0.0401    0.0075    0.1532   -0.8140    2.2383
0.0113    0.9943   -0.0361    0.0100    0.1162   -0.6265    1.7168
0.0103   -0.0050    0.9673    0.0106    0.0952   -0.5300    1.4834
0.0094   -0.0045   -0.0299    1.0104    0.0797   -0.4573    1.2725
0.0087   -0.0041   -0.0275    0.0104    1.0703   -0.4038    1.0986
0.0081   -0.0037   -0.0255    0.0102    0.0621    0.6416    0.9971
0.0075   -0.0034   -0.0237    0.0099    0.0554   -0.3245    1.9062
0.0071   -0.0032   -0.0222    0.0095    0.0495   -0.2944    0.8232
0.0066   -0.0030   -0.0209    0.0093    0.0439   -0.2686    0.7246
0.0063   -0.0028   -0.0197    0.0087    0.0424   -0.2532    0.7002
0.0059   -0.0026   -0.0187    0.0087    0.0372   -0.2307    0.6309
0.0057   -0.0024   -0.0177    0.0081    0.0358   -0.2168    0.6064
```

Columns 8 through 12

```
-4.0762    4.7656   -3.5039    1.4385   -0.2183
-3.1582    3.7754   -2.7520    1.1123   -0.1649
-2.6250    3.1055   -2.3301    0.9219   -0.1390
-2.2676    2.6602   -1.9922    0.7905   -0.1163
-1.9863    2.4023   -1.7139    0.7104   -0.0992
-1.7969    2.1094   -1.5430    0.6289   -0.0897
-1.6133    1.9258   -1.4043    0.5581   -0.0779
-0.4658    1.7598   -1.2734    0.5146   -0.0715
-1.3047    2.5762   -1.1445    0.4629   -0.0651
-1.2793    1.5098   -0.1055    0.4424   -0.0619
-1.1387    1.3438   -0.9873    1.3955   -0.0529
-1.1025    1.2998   -0.9395    0.3809    0.9474
```

diary off

The MATLAB rcond function (which gave the number RCOND above) needs some explanation. A useful reference on this is Hill [3, pp. 229–230]. It is based on a condition number estimator in the old FORTRAN codes known as "LINPACK". It is based on 1-norms. rcond(A) will give the *reciprocal* of the 1-norm condition number of A. If A is well-conditioned, then rcond(A) will be close to unity (i.e., close to one), and will be very tiny if A is ill-conditioned. The rule of thumb involved in interpreting an rcond output is "if rcond(A) $\approx d \times 10^{-k}$, where d is a digit from 1 to 9, then the elements of xcomp can usually be expected to have k fewer significant digits of accuracy than the elements of A" [3]. Here xcomp is simply the computed solution to $Ax = y$; that is, in the notation of the present set of notes, $\hat{x}$ =xcomp. MATLAB does arithmetic with about 16 decimal digits [3, p. 228], so in the preceding example of a Hilbert matrix inversion problem for $N = 12$, since RCOND is about 10^{-17}, we have lost about 17 digits in computing R^{-1}. Of course, this loss is catastrophic for our problem.

APPENDIX 4.B   SVD AND LEAST SQUARES

From Theorem 4.4, $A = U\Sigma V^T$, so this expands into the summation

$$A = \sum_{i=0}^{p-1} \sigma_i u_i v_i^T. \qquad (4.A.1)$$

But if $r = \text{rank}\ (A)$, then (4.A.1) reduces to

$$A = \sum_{i=0}^{r-1} \sigma_i u_i v_i^T. \qquad (4.A.2)$$

In the following theorem $\rho_{LS}^2 = ||f^l||_2^2 = ||A\hat{x} - f||_2^2$ [see (4.127)], and $\hat{x}$ is the least-squares optimal solution to $Ax = f$.

Theorem 4.B.1: Let A be represented as in (4.A.2) with $A \in \mathbf{R}^{m \times n}$ and $m \geq n$. If $f \in \mathbf{R}^m$ then

$$\hat{x} = \sum_{i=0}^{r-1} \left(\frac{u_i^T f}{\sigma_i} \right) v_i, \qquad (4.A.3)$$

$$\rho_{LS}^2 = \sum_{i=r}^{m-1} (u_i^T f)^2. \qquad (4.A.4)$$

Proof   For all $x \in \mathbf{R}^n$, using the invariance of the 2-norm to orthogonal transformations, and the fact that $VV^T = I_n$ ($n \times n$ identity matrix)

$$||Ax - f||_2^2 = ||U^T AV(V^T x) - U^T f||_2^2 = ||\Sigma \alpha - U^T f||_2^2, \qquad (4.A.5)$$

where $\alpha = V^T x$, so as $\alpha = [\alpha_0 \cdots \alpha_{n-1}]^T$ we have $\alpha_i = v_i^T x$. Equation (4.A.5) expands as

$$||Ax - f||_2^2 = \alpha^T \Sigma^T \Sigma \alpha - 2\alpha^T \Sigma^T U^T f + f^T UU^T f, \qquad (4.A.6)$$

which further expands as

$$\|Ax - f\|_2^2 = \sum_{i=0}^{r-1} \sigma_i^2 \alpha_i^2 - 2 \sum_{i=0}^{r-1} \alpha_i \sigma_i u_i^T f + \sum_{i=0}^{m-1} [u_i^T f]^2$$

$$= \sum_{i=0}^{r-1} [\sigma_i^2 \alpha_i^2 - 2\alpha_i \sigma_i u_i^T f + [u_i^T f]^2] + \sum_{i=r}^{m-1} [u_i^T f]^2$$

$$= \sum_{i=0}^{r-1} [\sigma_i \alpha_i - u_i^T f]^2 + \sum_{i=r}^{m-1} [u_i^T f]^2. \qquad (4.A.7)$$

To minimize this we must have $\sigma_i \alpha_i - u_i^T f = 0$, and so

$$\alpha_i = \frac{1}{\sigma_i} u_i^T f \qquad (4.A.8)$$

for $i \in \mathbf{Z}_r$. As $\alpha = V^T x$, we have $x = V\alpha$, so if we set $\alpha_r = \cdots = \alpha_{n-1} = 0$, then from Eq. (4.A.8), we obtain

$$\hat{x} = \sum_{i=0}^{r-1} \alpha_i v_i = \sum_{i=0}^{r-1} \frac{u_i^T f}{\sigma_i} v_i,$$

which is (4.A.3). For this choice of $\hat{x}$ from (4.A.7)

$$\|A\hat{x} - f\|_2^2 = \sum_{i=r}^{m-1} [u_i^T f]^2 = \rho_{LS}^2,$$

which is (4.A.4).

Define $A^+ = V\Sigma^+ U^T$ (again $A \in \mathbf{R}^{m \times n}$ with $m \geq n$), where

$$\Sigma^+ = \mathrm{diag}(\sigma_0^{-1}, \ldots, \sigma_{r-1}^{-1}, 0, \ldots, 0) \in \mathbf{R}^{n \times m}. \qquad (4.A.9)$$

We observe that

$$A^+ f = V\Sigma^+ U^T f = \sum_{i=0}^{r-1} \frac{u_i^T f}{\sigma_i} v_i = \hat{x}. \qquad (4.A.10)$$

We call A^+ the *pseudoinverse* of A. We have established that if rank $(A) = n$, then $A^T A\hat{x} = A^T f$, so $\hat{x} = (A^T A)^{-1} A^T f$, which implies that in this case $A^+ = (A^T A)^{-1} A^T$. If $A \in \mathbf{R}^{n \times n}$ and A^{-1} exists, then $A^+ = A^{-1}$.

If A^{-1} exists (i.e., $m = n$ and rank $(A) = n$), then we recall that $\kappa_2(A) = \|A\|_2 \|A^{-1}\|_2$. If $A \in \mathbf{R}^{m \times n}$ and $m \geq n$, then we extend this definition to

$$\kappa_2(A) = \|A\|_2 \|A^+\|_2. \qquad (4.A.11)$$

We have established that $||A||_2 = \sigma_0$ so since $A^+ = V\Sigma^+ U^T$ we have $||A^+||_2 = 1/\sigma_{r-1}$. Consequently

$$\kappa_2(A) = \frac{\sigma_0}{\sigma_{r-1}} \tag{4.A.12}$$

which provides a somewhat better justification of (4.119), because if rank $(A) = n$ then (4.A.12) is $\kappa_2(A) = \sigma_0/\sigma_{n-1} = \sigma_{max}(A)/\sigma_{min}(A)$ [which is (4.119)]. From (4.A.11), $\kappa_2(A^T A) = ||A^T A||_2 ||(A^T A)^+||_2$. With $A = U\Sigma V^T$, and $A^T = V\Sigma^T U^T$ we have

$$A^T A = V\Sigma^T \Sigma V^T,$$

and

$$(A^T A)^+ = V(\Sigma^T \Sigma)^+ V^T.$$

Thus, $||A^T A||_2 = \sigma_0^2$, and $||(A^T A)^+||_2 = \sigma_{n-1}^{-2}$ (rank $(A) = n$). Thus

$$\kappa_2(A^T A) = \frac{\sigma_0^2}{\sigma_{n-1}^2} = [\kappa_2(A)]^2.$$

The condition number definition $\kappa_p(A)$ in (4.60) was fully justified because of (4.58). An analogous justification exists for (4.A.11), but is much more difficult to derive, and this is why we do not consider it in this book.

REFERENCES

1. J. R. Rice, *The Approximation of Functions*, Vol. I: *Linear Theory*, Addison-Wesley, Reading, MA, 1964.
2. M.-D. Choi, "Tricks or Treats with the Hilbert Matrix," *Am. Math. Monthly* **90**(5), 301–312 (May 1983).
3. D. R. Hill, *Experiments in Computational Matrix Algebra* (C. B. Moler, consulting ed.), Random House, New York, 1988.
4. G. E. Forsythe and C. B. Moler, *Computer Solution of Linear Algebraic Systems*, Prentice-Hall, Englewood Cliffs, NJ, 1967.
5. G. H. Golub and C. F. Van Loan, *Matrix Computations*, 2nd ed., Johns Hopkins Univ. Press, Baltimore, MD, 1989.
6. R. A. Horn and C. R. Johnson, *Matrix Analysis*, Cambridge Univ. Press, Cambridge, MA, 1985.
7. N. J. Higham, *Accuracy and Stability of Numerical Algorithms*, SIAM, Philadelphia, PA, 1996.
8. A. Quarteroni, R. Sacco, and F. Saleri, *Numerical Mathematics* (Texts in Applied Mathematics series, Vol. 37), Springer-Verlag, New York, 2000.
9. E. Isaacson and H. B. Keller, *Analysis of Numerical Methods*, Wiley, New York, 1966.
10. O. Axelsson, *Iterative Solution Methods*, Cambridge Univ. Press, New York, 1994.
11. E. Kreyszig, *Introductory Functional Analysis with Applications*, Wiley, New York, 1978.

PROBLEMS

4.1. Function $f(x) \in L^2[0, 1]$ is to be approximated according to

$$f(x) \approx a_0 x + \frac{a_1}{x + c}$$

using least squares, where $c \in \mathbf{R}$ is some fixed parameter. This involves solving the linear system

$$\underbrace{\begin{bmatrix} \frac{1}{3} & 1 - c \log_e\left(1 + \frac{1}{c}\right) \\ 1 - c \log_e\left(1 + \frac{1}{c}\right) & \frac{1}{c(c+1)} \end{bmatrix}}_{=R} \underbrace{\begin{bmatrix} \hat{a}_0 \\ \hat{a}_1 \end{bmatrix}}_{=\hat{a}} = \underbrace{\begin{bmatrix} \int_0^1 x f(x)\, dx \\ \int_0^1 \frac{f(x)}{x+c}\, dx \end{bmatrix}}_{=g},$$

where $\hat{a}$ is the vector from $\mathbf{R}^2$ that minimizes the energy $V(a)$ [Eq. (4.8)].

(a) Suppose that $f(x) = x + \frac{1}{x+1}$. Find $\hat{a}$ for $c = 1$. For this special case it is possible to know the answer in advance without solving the linear system above. However, this problem requires you to solve the system. [*Hint:* It helps to recall that

$$\begin{bmatrix} a & b \\ c & d \end{bmatrix}^{-1} = \frac{1}{ad - bc} \begin{bmatrix} d & -b \\ -c & a \end{bmatrix}.$$

(b) Derive R [which is a special case of (4.9)].

4.2. Suppose that

$$A = \begin{bmatrix} 1 & 2 \\ -1 & -5 \end{bmatrix}.$$

Find $||A||_\infty$, $||A||_1$, and $||A||_2$.

4.3. Suppose that

$$A = \begin{bmatrix} 1 & 0 \\ 2 & \epsilon \end{bmatrix} \in \mathbf{R}^{2 \times 2}, \epsilon \geq 0,$$

and that A^{-1} exists. Find $\kappa_\infty(A)$ if ϵ is small.

4.4. Suppose that

$$A = \begin{bmatrix} 1 & 1 \\ 0 & \epsilon \end{bmatrix} \in \mathbf{R}^{2 \times 2},$$

and assume $\epsilon > 0$ (so that A^{-1} always exists). Find $\kappa_2(A) = ||A||_2||A^{-1}||_2$. What happens to condition number $\kappa_2(A)$ if $\epsilon \to 0$?

4.5. Let $A(\epsilon), B(\epsilon) \in \mathbf{R}^{n \times n}$. For example, $A(\epsilon) = [a_{ij}(\epsilon)]$ so element $a_{ij}(\epsilon)$ of $A(\epsilon)$ depends on the parameter $\epsilon \in \mathbf{R}$.

(a) Prove that

$$\frac{d}{d\epsilon}[A(\epsilon)B(\epsilon)] = A(\epsilon)\frac{dB(\epsilon)}{d\epsilon} + \frac{dA(\epsilon)}{d\epsilon}B(\epsilon),$$

where $dA(\epsilon)/dt = [da_{ij}(\epsilon)/d\epsilon]$, and $dB(\epsilon)/dt = [db_{ij}(\epsilon)/d\epsilon]$.

(b) Prove that

$$\frac{d}{d\epsilon}A^{-1}(\epsilon) = -A^{-1}(\epsilon)\left[\frac{dA(\epsilon)}{d\epsilon}\right]A^{-1}(\epsilon).$$

[*Hint:* Consider $\frac{d}{d\epsilon}[A(\epsilon)A^{-1}(\epsilon)] = \frac{d}{d\epsilon}I = 0$, and use (a).]

4.6. This problem is an alternative derivation of $\kappa(A)$. Suppose that $\epsilon \in \mathbf{R}$, A, $F \in \mathbf{R}^{n \times n}$, and $x(\epsilon)$, y, $f \in \mathbf{R}^n$. Consider the perturbed linear system of equations

$$(A + \epsilon F)x(\epsilon) = y + \epsilon f, \tag{4.P.1}$$

where $Ax = y$, so $x(0) = x$ is the correct solution. Clearly, ϵF models the errors in A, while ϵf models the errors in y. From (4.P.1), we obtain

$$x(\epsilon) = [A + \epsilon F]^{-1}(y + \epsilon f). \tag{4.P.2}$$

The Taylor series expansion for $x(\epsilon)$ about $\epsilon = 0$ is

$$x(\epsilon) = x(0) + \frac{dx(0)}{d\epsilon}\epsilon + O(\epsilon^2), \tag{4.P.3}$$

where $O(\epsilon^2)$ denotes terms in the expansion containing ϵ^k for $k \geq 2$.

Use (4.P.3), results from the previous problem, and basic matrix–vector norm properties (Section 4.4) to derive the bound in

$$\frac{||x(\epsilon) - x||}{||x||} \leq \epsilon \underbrace{||A||\,||A^{-1}||}_{=\kappa(A)}\left\{\frac{||f||}{||y||} + \frac{||F||}{||A||}\right\}$$

[*Comment:* The *relative error in A* is $\rho_A = \epsilon||F||/||A||$, and the *relative error in y* is $\rho_y = \epsilon||f||/||y||$, so more concisely $||x(\epsilon) - x||/||x|| \leq \kappa(A)(\rho_A + \rho_y)$.]

4.7. This problem follows the example relating to Eq. (4.95). An analog signal $f(t)$ is modeled according to

$$f(t) = \sum_{j=0}^{p-1} a_j t^j + \eta(t),$$

where $a_j \in \mathbf{R}$ for all $j \in \mathbf{Z}_p$, and $\eta(t)$ is some random noise term. We only possess samples of the signal; that is, we only have the finite-length sequence

(f_n) defined by $f_n = f(nT_s)$ for $n \in \mathbf{Z}_N$, where $T_s > 0$ is the sampling period of the data acquisition system that gave us the samples. Our estimate of f_n is therefore

$$\hat{f}_n = \sum_{j=0}^{p-1} a_j T_s^j n^j.$$

With $a = [a_0 \ a_1 \ \cdots \ a_{p-2} \ a_{p-1}]^T \in \mathbf{R}^p$, find

$$V(a) = \sum_{n=0}^{N-1} [f_n - \hat{f}_n]^2$$

in the form of Eq. (4.102). This implies that you must specify ρ, g, A, and P.

4.8. Sampled data f_n ($n = 0, 1, \ldots, N-1$) is modeled according to

$$\hat{f}_n = a + bn + C \sin(\theta n + \phi).$$

Recall $\sin(A + B) = \sin A \cos B + \cos A \sin B$. Also recall that

$$V(\theta) = \sum_{n=0}^{N-1} [f_n - \hat{f}_n]^2 = f^T f - f^T A(\theta)[A^T(\theta)A(\theta)]^{-1}A^T(\theta)f,$$

where $f = [f_0 f_1 \cdots f_{N-1}]^T \in \mathbf{R}^N$. Give a detailed expression for $A(\theta)$.

4.9. Prove that the product of two lower triangular matrices is a lower triangular matrix.

4.10. Prove that the product of two upper triangular matrices is an upper triangular matrix.

4.11. Find the LU decomposition of the matrix

$$A = \begin{bmatrix} 2 & -1 & 0 \\ -1 & 2 & -1 \\ 0 & -1 & 2 \end{bmatrix}$$

using Gauss transformations as recommended in Section 4.5. Use $A = LU$ to rewrite the factorization of A as $A = LDL^T$, where L is unit lower triangular, and D is a diagonal matrix. Is $A > 0$? Why?

4.12. Use Gaussian elimination to LU factorize the matrix

$$A = \begin{bmatrix} 4 & -1 & \frac{1}{2} \\ -1 & 4 & -1 \\ \frac{1}{2} & -1 & 4 \end{bmatrix}.$$

Is $A > 0$? Why?

4.13. (a) Consider $A \in \mathbf{R}^{n \times n}$, and $A = A^T$. Suppose that the leading principal submatrices of A all have positive determinants. Prove that $A > 0$.

(b) Is

$$\begin{bmatrix} 5 & -3 & 0 \\ -3 & 5 & 1 \\ 0 & 1 & 5 \end{bmatrix} > 0?$$

Why?

4.14. The vector space $\mathbf{R}^n$ is an inner product space with inner product $\langle x, y \rangle = x^T y$ for all $x, y \in \mathbf{R}^n$. Suppose that $A \in \mathbf{R}^{n \times n}$, $A = A^T$, and also that $A > 0$. Prove that $\langle x, y \rangle = x^T A y$ is also an inner product on the vector space $\mathbf{R}^n$.

4.15. In the quadratic form

$$V(x) = f^T f - 2x^T A^T f + x^T A^T A x,$$

we assume $x \in \mathbf{R}^n$, and $A^T A > 0$. Prove that

$$V(\hat{x}) = f^T f - f^T A [A^T A]^{-1} A^T f,$$

where $\hat{x}$ is the vector that minimizes $V(x)$.

4.16. Suppose that $A \in \mathbf{R}^{N \times M}$ with $M \leq N$, and rank $(A) = M$. Suppose also that $\langle x, y \rangle = x^T y$. Consider $P = A[A^T A]^{-1} A^T$, $P_\perp = I - P$ (I is the $N \times N$ identity matrix). Prove that for all $x \in \mathbf{R}^N$ we have $\langle Px, P_\perp x \rangle = 0$. (*Comment*: Matrices P and $P_\perp$ are examples of *orthogonal projection operators*.)

4.17. (a) Write a MATLAB function for forward substitution (solving $Lz = y$). Write a MATLAB function for backward substitution (solving $Ux = z$). Test your functions out on the following matrices and vectors:

$$L = \begin{bmatrix} 1 & 0 & 0 & 0 \\ -1 & 1 & 0 & 0 \\ 0 & \frac{2}{3} & 1 & 0 \\ 0 & 0 & -\frac{3}{7} & 1 \end{bmatrix}, \quad U = \begin{bmatrix} 1 & 2 & 3 & 4 \\ 0 & 3 & 5 & 5 \\ 0 & 0 & -\frac{7}{3} & -\frac{1}{3} \\ 0 & 0 & 0 & \frac{6}{7} \end{bmatrix},$$

$$y = \begin{bmatrix} 1 & 1 & -1 & -1 \end{bmatrix}^T, z = \begin{bmatrix} 1 & 2 & -\frac{7}{3} & -2 \end{bmatrix}^T.$$

(b) Write a MATLAB function to implement the LU decomposition algorithm based on Gauss transformations considered in Section 4.5. Test your function out on the following A matrices:

$$A = \begin{bmatrix} 1 & 2 & 3 & 4 \\ -1 & 1 & 2 & 1 \\ 0 & 2 & 1 & 3 \\ 0 & 0 & 1 & 1 \end{bmatrix}, \quad A = \begin{bmatrix} 2 & -1 & 0 \\ -1 & 2 & -1 \\ 0 & -1 & 2 \end{bmatrix}.$$

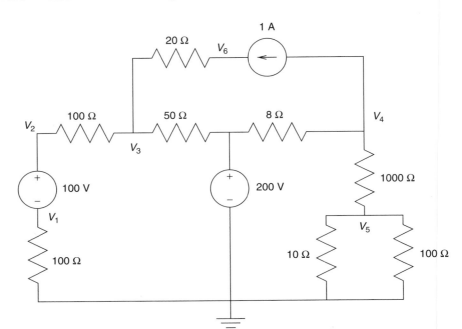

Figure 4.P.1   The DC electric circuit of Problem 4.18.

4.18. Consider the DC electric circuit in Fig. 4.P.1.

Write the node equations for the node voltages $V_1, V_2, \ldots, V_6$ as shown. These may be loaded into a vector

$$v = [V_1 V_2 V_3 V_4 V_5 V_6]^T$$

such that the node equations have the form

$$Gv = y. \tag{4.P.4}$$

Use the Gaussian elimination, forward substitution, and backward substitution MATLAB functions from the previous problem to solve the linear system (4.P.4).

4.19. Let $I_n \in \mathbf{R}^{n \times n}$ be the order n identity matrix, and define

$$T_n = \begin{bmatrix} a & b & 0 & \cdots & 0 & 0 \\ b & a & b & \cdots & 0 & 0 \\ 0 & b & a & \cdots & 0 & 0 \\ \vdots & \vdots & \vdots & & \vdots & \vdots \\ 0 & 0 & 0 & \cdots & a & b \\ 0 & 0 & 0 & \cdots & b & a \end{bmatrix} \in \mathbf{R}^{n \times n}.$$

Matrix T_n is *tridiagonal*. It is also an example of a symmetric matrix that is *Toeplitz* (defined in a later problem). The characteristic polynomial of T_n is $p_n(\lambda) = \det(\lambda I_n - T_n)$. Show that

$$p_n(\lambda) = (\lambda - a)p_{n-1}(\lambda) - b^2 p_{n-2}(\lambda) \qquad (4.P.5)$$

for $n = 3, 4, 5, \ldots$. Find $p_1(\lambda)$, and $p_2(\lambda)$ [initial conditions for the polynomial recursion in (4.P.5)].

4.20. Consider the following If: $T \in \mathbf{C}^{n \times n}$ is *Toeplitz* if it has the form $T = [t_{i-j}]_{i,j=0,1,\ldots,n-1}$. Thus, for example, if $n = 3$ we have

$$T = \begin{bmatrix} t_0 & t_{-1} & t_{-2} \\ t_1 & t_0 & t_{-1} \\ t_2 & t_1 & t_0 \end{bmatrix}.$$

Observe that in a Toeplitz matrix all of the elements on any given diagonal are equal to each other. A *symmetric Toeplitz matrix* has the form $T = [t_{|i-j|}]$ since $T = T^T$ implies that $t_{-i} = t_i$ for all i. Let $x_n = [x_{n,0} \ x_{n,1} \ \cdots \ x_{n,n-2} \ x_{n,n-1}]^T \in \mathbf{C}^n$. Let $J_n \in \mathbf{C}^{n \times n}$ be the $n \times n$ *exchange matrix* (also called the *contra-identity matrix*) which is defined as the matrix yielding $\hat{x}_n = J_n x_n = [x_{n,n-1} x_{n,n-2} \cdots x_{n,1} x_{n,0}]^T$. We see that J_n simply reverses the order of the elements of x_n. An immediate consequence of this is that $J_n^2 x_n = x_n$ (i.e., $J_n^2 = I_n$). What is J_3? (Write this matrix out completely.) Suppose that T_n is a symmetric Toeplitz matrix.

(a) Show that (noting that $\hat{\tau}_n = J_n \tau_n$)

$$T_{n+1} = \begin{bmatrix} T_n & \tau_n \\ \tau_n^T & t_0 \end{bmatrix} = \begin{bmatrix} t_0 & \hat{\tau}_n^T \\ \hat{\tau}_n & T_n \end{bmatrix}$$

(*nesting property*). What is τ_n?

(b) Show that

$$J_n T_n J_n = T_n$$

(*persymmetry property*).

(*Comment:* Toeplitz matrices have an important role to play in digital signal processing. For example, they appear in problems in spectral analysis, and in voice compression algorithms.)

4.21. This problem is about a computationally efficient method to solve the linear system

$$R_n a_n = \sigma_n^2 e_0, \qquad (4.P.6)$$

where $R_n \in \mathbf{R}^{n \times n}$ is symmetric and Toeplitz (recall Problem 4.20). All of the leading principle submatrices (recall the definition in Theorem 4.1) of R_n are nonsingular. Also, $e_0 = [1 \ 0 \ 0 \ \cdots \ 0 \ 0]^T \in \mathbf{R}^n$, $a_n = [1 \ a_{n,1} \ \cdots \ a_{n,n-2} \ a_{n,n-1}]^T$, and $\sigma_n^2 \in \mathbf{R}$ is unknown as well. Define $e_n = J_n e_0$. Clearly, $a_{n,0} = 1$ (all n).

(a) Prove that

$$R_n \hat{a}_n = \sigma_n^2 \hat{e}_0 = \sigma_n^2 e_n, \qquad (4.P.7)$$

where $J_n a_n = \hat{a}_n$.

(b) Observe that $R_n [a_n \ \hat{a}_n] = \sigma_n^2 [e_0 \ e_n]$. Augmented matrix $[a_n \ \hat{a}_n]$ is $n \times 2$, and so $[e_0 \ e_n]$ is $n \times 2$ as well. Prove that

$$R_{n+1} \begin{bmatrix} a_n & 0 \\ 0 & \hat{a}_n \end{bmatrix} = \begin{bmatrix} \sigma_n^2 e_0 & \eta_n \\ \eta_n & \sigma_n^2 e_n \end{bmatrix}. \qquad (4.P.8)$$

What is η_n? (That is, find a simple expression for it.)

(c) We wish to obtain

$$R_{n+1} [a_{n+1} \ \hat{a}_{n+1}] = \sigma_{n+1}^2 [e_0 \ e_{n+1}] \qquad (4.P.9)$$

from a manipulation of (4.P.8). To this end, find a formula for parameter $K_n \in \mathbf{R}$ in

$$R_n \begin{bmatrix} a_n & 0 \\ 0 & \hat{a}_n \end{bmatrix} \begin{bmatrix} 1 & K_n \\ K_n & 1 \end{bmatrix} = \begin{bmatrix} \sigma_n^2 e_0 & \eta_n \\ \eta_n & \sigma_n^2 e_n \end{bmatrix} \begin{bmatrix} 1 & K_n \\ K_n & 1 \end{bmatrix}$$

such that (4.P.9) is obtained. This implies that we obtain the vector recursions

$$a_{n+1} = \begin{bmatrix} a_n \\ 0 \end{bmatrix} + K_n \begin{bmatrix} 0 \\ \hat{a}_n \end{bmatrix}$$

and

$$\hat{a}_{n+1} = K_n \begin{bmatrix} a_n \\ 0 \end{bmatrix} + \begin{bmatrix} 0 \\ \hat{a}_n \end{bmatrix}.$$

Find the initial condition $a_1 \in \mathbf{R}^{1 \times 1}$. What is σ_1^2?

(d) Prove that

$$\sigma_{n+1}^2 = \sigma_n^2 (1 - K_n^2).$$

(e) Summarize the algorithm obtained in the previous steps in the form of pseudocode. The resulting algorithm is often called the *Levinson–Durbin algorithm*.

(f) Count the number of arithmetic operations needed to implement the Levinson–Durbin algorithm in (e). Compare this number to the number of arithmetic operations needed by the general LU decomposition algorithm presented in Section 4.5.

(g) Write a MATLAB function to implement the Levinson-Durbin algorithm. Test your algorithm out on the matrix

$$R = \begin{bmatrix} 2 & 1 & 0 \\ 1 & 2 & 1 \\ 0 & 1 & 2 \end{bmatrix}.$$

(*Hint:* You will need the properties in Problem 4.20. The parameters K_n are called *reflection coefficients*, and are connected to certain problems in electrical transmission line theory.)

4.22. This problem is about proving that solving (4.P.6) yields the LDL^T decomposition of R_n^{-1}. (Of course, R_n is real-valued, symmetric, and Toeplitz.) Observe that (via the nesting property for Toeplitz matrices)

$$R_n \underbrace{\begin{bmatrix} 1 & 0 & \cdots & 0 & 0 \\ a_{n,1} & 1 & \cdots & 0 & 0 \\ a_{n,2} & a_{n-1,1} & \cdots & 0 & 0 \\ \vdots & \vdots & & \vdots & \vdots \\ a_{n,n-2} & a_{n-1,n-3} & \cdots & 1 & 0 \\ a_{n,n-1} & a_{n-1,n-2} & \cdots & a_{2,1} & 1 \end{bmatrix}}_{=L_n} = \underbrace{\begin{bmatrix} \sigma_n^2 & \times & \cdots & \times & \times \\ 0 & \sigma_{n-1}^2 & \cdots & \times & \times \\ 0 & 0 & \cdots & \times & \times \\ \vdots & \vdots & & \vdots & \vdots \\ 0 & 0 & \cdots & \sigma_2^2 & \times \\ 0 & 0 & \cdots & 0 & \sigma_1^2 \end{bmatrix}}_{=\tilde{U}_n},$$

(4.P.10)

where $\times$ denotes "don't care" entries; thus, the particular value of such an entry is of no interest to us. Use (4.P.10) to prove that

$$L_n^T R_n L_n = \tilde{D}_n,$$
(4.P.11)

where $\tilde{D}_n = \text{diag}\,(\sigma_n^2, \sigma_{n-1}^2, \ldots, \sigma_2^2, \sigma_1^2)$ (diagonal matrix; see the comments following Theorem 4.4). Use (4.P.11) to prove that

$$R_n^{-1} = L_n D_n L_n^T.$$
(4.P.12)

What is D_n? [*Hint:* Using (4.P.10), and $R_n = R_n^T$ we note that $L_n^T R_n L_n$ can be expressed in two distinct but equivalent ways. This observation is used to establish (4.P.11).]

4.23. A matrix T_n is said to be *strongly nonsingular* (*strongly regular*) if all of its leading principle submatrices are nonsingular (recall the definition in Theorem 4.1). Suppose that $x_n = [x_{n,0}\ x_{n,1}\ \cdots\ x_{n,n-2}\ x_{n,n-1}]^T \in \mathbf{C}^n$, and define

$Z_n \in \mathbf{C}^{n \times n}$ according to

$$Z_n x_n = [0 \; x_{n,0} \; x_{n,1} \; \cdots \; x_{n,n-2}]^T \in \mathbf{C}^n.$$

Thus, Z_n shifts the elements of any column vector down by one position. The top position is filled in with a zero, while the last element $x_{n,n-1}$ is lost. Assume that T_n is real-valued, symmetric and Toeplitz. Consider

$$T_n - Z_n T_n Z_n^T = X_n.$$

(a) Find X_n.

(b) If T_n is strongly nonsingular then what is rank (X_n)? (Be careful. There may be separate cases to consider.)

(c) Use δ_k (Krönecker delta) to specify Z_n; that is, $Z_n = [z_{i,j}]$, so what is $z_{i,j}$ in terms of the Krönecker delta? (*Hint:* For example, the identity matrix can be described as $I = [\delta_{i-j}]$.)

4.24. Suppose that $R \in \mathbf{R}^{N \times N}$ is strongly nonsingular (see Problem 4.23 for the definition of this term), and that $R = R^T$. Thus, there exists the factorization

$$R = L_N D_N L_N^T, \tag{4.P.13}$$

where L_N is a unit lower triangular matrix, and D_N is a diagonal matrix. Let

$$L_N = [l_0 \; l_1 \; \cdots \; l_{N-2} \; l_{N-1}],$$

so l_i is column i of L_N. Let

$$D_N = \text{diag}(d_0, d_1, \ldots, d_{N-1}).$$

Thus, via (4.P.13), we have

$$R = \sum_{k=0}^{N-1} d_k l_k l_k^T. \tag{4.P.14}$$

Consider the following algorithm:

```
R₀ := R;
for n := 0 to N − 1 do begin
    dₙ := eₙᵀRₙeₙ;
    lₙ := dₙ⁻¹Rₙeₙ;
    Rₙ₊₁ := Rₙ − dₙlₙlₙᵀ;
end;
```

As usual, we have the unit vector

$$e_i = [\underbrace{0\ 0\ \cdots\ 0}_{i\ \text{zeros}}\ 1\ 0\ \cdots\ 0]^T \in \mathbf{R}^N.$$

The algorithm above is the *Jacobi procedure (algorithm)* for computing the *Cholesky factorization* (recall Theorem 4.6) of R.

Is $R > 0$ a necessary condition for the algorithm to work? Explain. Test the Jacobi procedure out on

$$R = \begin{bmatrix} 2 & 1 & 0 \\ 1 & 2 & 1 \\ 0 & 1 & 2 \end{bmatrix}.$$

Is $R > 0$? Justify your answer. How many arithmetic operations are needed to implement the Jacobi procedure? How does this compare with the Gaussian elimination method for general LU factorization considered in Section 4.5?

4.25. Suppose that A^{-1} exists, and that $I + V^T A^{-1} U$ is also nonsingular. Of course, I is the identity matrix.

(a) Prove the *Sherman–Morrison–Woodbury formula*

$$[A + U V^T]^{-1} = A^{-1} - A^{-1} U [I + V^T A^{-1} U]^{-1} V^T A^{-1}.$$

(b) Prove that if $U = u \in \mathbf{C}^n$ and $V = v \in \mathbf{C}^n$, then

$$[A + u v^T]^{-1} = A^{-1} - \frac{A^{-1} u v^T A^{-1}}{1 + v^T A^{-1} u}.$$

(*Comment:* These identities can be used to develop adaptive filtering algorithms, for example.)

4.26. Suppose that

$$A = \begin{bmatrix} a & b \\ 0 & c \end{bmatrix}.$$

Find conditions on a, b, and c to ensure that $A > 0$.

4.27. Suppose that $A \in \mathbf{R}^{n \times n}$, and that A is not necessarily symmetric. We still say that $A > 0$ iff $x^T A x > 0$ for all $x \neq 0$ ($x \in \mathbf{R}^n$). Show that $A > 0$ iff $B = \frac{1}{2}(A + A^T) > 0$. Matrix B is often called the *symmetric part of A*. (*Note:* In this book, unless stated to the contrary, a pd matrix is always assumed to be symmetric.)

4.28. Prove that for $A \in \mathbf{R}^{n \times n}$

$$||A||_\infty = \max_{0 \leq i \leq n-1} \sum_{j=0}^{n-1} |a_{i,j}|.$$

4.29. Derive Eq. (4.128) (the law of cosines).

4.30. Consider $A_n \in \mathbf{R}^{n \times n}$ such that

$$A_n = \begin{bmatrix} 1 & -1 & -1 & \cdots & -1 & -1 \\ 0 & 1 & -1 & \cdots & -1 & -1 \\ 0 & 0 & 1 & \cdots & -1 & -1 \\ \vdots & \vdots & \vdots & & \vdots & \vdots \\ 0 & 0 & 0 & \cdots & 1 & -1 \\ 0 & 0 & 0 & \cdots & 0 & 1 \end{bmatrix},$$

$$A_n^{-1} = \begin{bmatrix} 1 & 1 & 2 & \cdots & 2^{n-3} & 2^{n-2} \\ 0 & 1 & 1 & \cdots & 2^{n-4} & 2^{n-3} \\ 0 & 0 & 1 & \cdots & 2^{n-5} & 2^{n-4} \\ \vdots & \vdots & \vdots & & \vdots & \vdots \\ 0 & 0 & 0 & \cdots & 1 & 1 \\ 0 & 0 & 0 & \cdots & 0 & 1 \end{bmatrix}.$$

Show that $\kappa_\infty(A_n) = n2^{n-1}$. Clearly, $\det(A_n) = 1$. Consider

$$D_n = \operatorname{diag}(10^{-1}, 10^{-1}, \ldots, 10^{-1}) \in \mathbf{R}^n.$$

What is $\kappa_p(D_n)$? Clearly, $\det(D_n) = 10^{-n}$. What do these two cases say about the relationship between $\det(A)$, and $\kappa(A)$ ($A \in \mathbf{R}^{n \times n}$, and is nonsingular) in general?

4.31. Recall Problem 1.7 (in Chapter 1). Suppose $a = [a_0 a_1 \cdots a_n]^T$, $b = [b_0 b_1 \cdots b_m]^T$, and $c = [c_0 c_1 \cdots c_{n+m}]^T$, where

$$c_l = \sum_{k=0}^{n} a_k b_{l-k}.$$

Find matrix A such that $c = Ab$, and find matrix B such that $c = Ba$. What are the sizes of matrices A, and B? (*Hint: A* and *B* will be *rectangular* Toeplitz matrices. This problem demonstrates the close association between Toeplitz matrices and the convolution operation, and so partially explains the central importance of Toeplitz matrices in digital signal processing.)

4.32. Matrix $P \in \mathbf{R}^{n \times n}$ is a *permutation matrix* if it possesses exactly one one per row and column, and zeros everywhere else. Such a matrix simply reorders the elements in a vector. For example

$$
\begin{bmatrix}
0 & 1 & 0 & 0 \\
0 & 0 & 1 & 0 \\
1 & 0 & 0 & 0 \\
0 & 0 & 0 & 1
\end{bmatrix}
\begin{bmatrix}
x_0 \\
x_1 \\
x_2 \\
x_3
\end{bmatrix}
=
\begin{bmatrix}
x_1 \\
x_2 \\
x_0 \\
x_3
\end{bmatrix}.
$$

Show that $P^{-1} = P^T$ (i.e., P is an orthogonal matrix).

4.33. Find c and s in

$$
\begin{bmatrix}
c & -s \\
s & c
\end{bmatrix}
\begin{bmatrix}
x_0 \\
x_1
\end{bmatrix}
=
\begin{bmatrix}
\sqrt{x_0^2 + x_1^2} \\
0
\end{bmatrix},
$$

where $s^2 + c^2 = 1$.

4.34. Consider

$$
A =
\begin{bmatrix}
2 & 1 & 0 & 0 \\
1 & 2 & 1 & 0 \\
0 & 1 & 2 & 1 \\
0 & 0 & 1 & 2
\end{bmatrix}.
$$

Use Householder transformation matrices to find the QR factorization of matrix A.

4.35. Consider

$$
A =
\begin{bmatrix}
5 & 2 & 1 & 0 \\
2 & 5 & 2 & 1 \\
1 & 2 & 5 & 2 \\
0 & 1 & 2 & 5
\end{bmatrix}.
$$

Using Householder transformation matrices, find orthogonal matrices Q_0, and Q_1 such that

$$
Q_1 Q_0 A =
\begin{bmatrix}
h_{00} & h_{01} & h_{02} & h_{03} \\
h_{10} & h_{11} & h_{12} & h_{13} \\
0 & h_{21} & h_{22} & h_{23} \\
0 & 0 & h_{32} & h_{33}
\end{bmatrix}
= H.
$$

[*Comment:* Matrix H is *upper Hessenberg*. This problem is an illustration of a process that is important in finding matrix eigenvalues (Chapter 11).]

4.36. Write a MATLAB function to implement the Householder QR factorization algorithm as given by the pseudocode in Section 4.6 [between Eqs. (4.146), and (4.147)]. The function must output the separate factors $\tilde{H}_k^T$ that make up Q, in addition to the factors Q and R. Test your function out on the matrix A in Problem 4.34.

4.37. Prove Eq. (4.117), and establish that

$$\text{rank } (A) = \text{rank } (A^T).$$

4.38. If $A \in \mathbf{C}^{n \times n}$, then the *trace of* A is given by

$$\text{Tr}(A) = \sum_{k=0}^{n-1} a_{k,k}$$

(i.e., is the sum of the main diagonal elements of matrix A). Prove that

$$||A||_F^2 = \text{Tr}(AA^H)$$

[recall (4.36a)].

4.39. Suppose that $A, B \in \mathbf{R}^{n \times n}$, and $Q \in \mathbf{R}^{n \times n}$ is orthogonal; then, if $A = QBQ^T$, prove that

$$\text{Tr}(A) = \text{Tr}(B).$$

4.40. Recall Theorem 4.4. Prove that for $A \in \mathbf{R}^{n \times n}$

$$||A||_F^2 = \sum_{k=0}^{p-1} \sigma_k^2.$$

(*Hint:* Use the results from Problems 4.38 and 4.39.)

4.41. Consider Theorem 4.9. Is $\alpha < 1$ a *necessary* condition for convergence? Explain.

4.42. Suppose that $A \in \mathbf{R}^{n \times n}$ is pd and prove that the JOR method (Section 4.7) converges- if

$$0 < \omega < \frac{2}{\rho(D^{-1}A)}.$$

4.43. Repeat the analysis made in Example 4.8, but instead use the matrix

$$A = \begin{bmatrix} 4 & 2 & 1 & 0 \\ 1 & 4 & 1 & 0 \\ 0 & 1 & 4 & 1 \\ 0 & 0 & 2 & 4 \end{bmatrix}.$$

This will involve writing and running a suitable MATLAB function. Find the optimum choice for ω to an accuracy of ± 0.02.

5 Orthogonal Polynomials

5.1 INTRODUCTION

Orthogonal polynomials arise in highly diverse settings. They can be solutions to special classes of differential equations that arise in mathematical physics problems. They are vital in the design of analog and digital filters. They arise in numerical integration methods, and they have a considerable role to play in solving least-squares and uniform approximation problems.

Therefore, in this chapter we begin by considering some of the properties shared by all orthogonal polynomials. We then consider the special cases of Chebyshev, Hermite, and Legendre polynomials. Additionally, we consider the application of orthogonal polynomials to least-squares and uniform approximation problems. However, we emphasize the case of least-squares approximation, which was first considered in some depth in Chapter 4. The approach to least-squares problems taken here alleviates some of the concerns about ill-conditioning that were noted in Chapter 4.

5.2 GENERAL PROPERTIES OF ORTHOGONAL POLYNOMIALS

We are interested here in the inner product space $L^2(D)$, where D is the domain of definition of the functions in the space. Typically, $D = [a, b]$, $D = \mathbf{R}$, or $D = [0, \infty)$. We shall, as in Chapter 4, assume that all members of $L^2(D)$ are real-valued to simplify matters. So far, our inner product has been

$$\langle f, g \rangle = \int_D f(x)g(x)\,dx, \tag{5.1}$$

but now we consider the *weighted inner product*

$$\langle f, g \rangle = \int_D w(x)f(x)g(x)\,dx, \tag{5.2}$$

where $w(x) \geq 0$ ($x \in D$) is the *weighting function*. This includes (5.1) as a special case for which $w(x) = 1$ for all $x \in D$.

An Introduction to Numerical Analysis for Electrical and Computer Engineers, by C.J. Zarowski
ISBN 0-471-46737-5 © 2004 John Wiley & Sons, Inc.

Our goal is to consider polynomials

$$\phi_n(x) = \sum_{k=0}^{n} \phi_{n,k} x^k, \, x \in D \tag{5.3}$$

of degree n such that

$$\langle \phi_n, \phi_m \rangle = \delta_{n-m} \tag{5.4}$$

for all $n, m \in \mathbf{Z}^+$. The inner product is that of (5.2). Our polynomials $\{\phi_n(x) | n \in \mathbf{Z}^+\}$. are *orthogonal polynomials on D with respect to weighting function $w(x)$.* Changing D and/or $w(x)$ will generate very different polynomials, and we will consider important special cases later. However, all orthogonal polynomials possess certain features in common with each other regardless of the choice of D or $w(x)$. We shall consider a few of these in this section.

If $p(x)$ is a polynomial of degree n, then we may write $\deg(p(x)) = n$.

Theorem 5.1: Any three consecutive orthogonal polynomials are related by the *three-term recurrence formula (relation)*

$$\phi_{n+1}(x) = (A_n x + B_n)\phi_n(x) + C_n \phi_{n-1}(x), \tag{5.5}$$

where

$$A_n = \frac{\phi_{n+1,n+1}}{\phi_{n,n}}, \qquad B_n = \frac{\phi_{n+1,n+1}}{\phi_{n,n}} \left(\frac{\phi_{n+1,n}}{\phi_{n+1,n+1}} - \frac{\phi_{n,n-1}}{\phi_{n,n}} \right),$$

$$C_n = -\frac{\phi_{n+1,n+1}\phi_{n-1,n-1}}{\phi_{n,n}^2}. \tag{5.6}$$

Proof   Our proof is a somewhat expanded version of that in Isaacson and Keller [1, pp. 204–205].

Observe that for $A_n = \phi_{n+1,n+1}/\phi_{n,n}$, we have

$$q_n(x) = \phi_{n+1}(x) - A_n x \phi_n(x) = \sum_{k=0}^{n+1} \phi_{n+1,k} x^k - A_n \sum_{k=0}^{n} \phi_{n,k} x^{k+1}$$

$$= \sum_{k=0}^{n+1} \phi_{n+1,k} x^k - A_n \sum_{j=1}^{n+1} \phi_{n,j-1} x^j$$

$$= \sum_{k=0}^{n} \phi_{n+1,k} x^k - A_n \sum_{j=1}^{n} \phi_{n,j-1} x^j + [\phi_{n+1,n+1} - A_n \phi_{n,n}] x^{n+1}$$

$$= \sum_{k=0}^{n} \phi_{n+1,k} x^k - A_n \sum_{j=1}^{n} \phi_{n,j-1} x^j,$$

which is a polynomial of degree at most n, i.e., $\deg(q_n(x)) \leq n$. Thus, for suitable α_k

$$q_n(x) = \sum_{k=0}^{n} \alpha_k \phi_k(x),$$

so because $\langle \phi_k, \phi_j \rangle = \delta_{k-j}$, we have $\alpha_j = \langle q_n, \phi_j \rangle$. In addition

$$\phi_{n+1}(x) - A_n x \phi_n(x) = \alpha_n \phi_n(x) + \alpha_{n-1} \phi_{n-1}(x) + \sum_{k=0}^{n-2} \alpha_k \phi_k(x). \tag{5.7}$$

Now $\deg(x\phi_j(x)) \leq n - 1$ for $j \leq n - 2$. Thus, there are β_k such that

$$x\phi_j(x) = \sum_{k=0}^{n-1} \beta_k \phi_k(x)$$

so $\langle \phi_r, x\phi_j \rangle = 0$ if $r > n - 1$, or

$$\langle \phi_r, x\phi_j \rangle = 0 \tag{5.8}$$

for $j = 0, 1, \ldots, n - 2$. From (5.7) via (5.8), we obtain

$$\langle \phi_{n+1} - A_n x \phi_n, \phi_k \rangle = \langle \phi_{n+1}, \phi_k \rangle - A_n \langle \phi_n, x\phi_k \rangle = 0$$

for $k = 0, 1, \ldots, n - 2$. This is the inner product of $\phi_k(x)$ with the left-hand side of (5.7). For the right-hand side of (5.7)

$$0 = \left\langle \alpha_n \phi_n + \alpha_{n-1}\phi_{n-1} + \sum_{j=0}^{n-2} \alpha_j \phi_j, \phi_k \right\rangle$$

$$= \alpha_n \langle \phi_n, \phi_k \rangle + \alpha_{n-1} \langle \phi_{n-1}, \phi_k \rangle + \sum_{j=0}^{n-2} \alpha_j \langle \phi_j, \phi_k \rangle$$

$$= \sum_{j=0}^{n-2} \alpha_j \langle \phi_j, \phi_k \rangle$$

again for $k = 0, \ldots, n - 2$. We can only have $\sum_{j=0}^{n-2} \alpha_j \langle \phi_j, \phi_k \rangle = 0$ if $\alpha_j = 0$ for $j = 0, 1, \ldots, n - 2$. Thus, (5.7) reduces to

$$\phi_{n+1}(x) - A_n x \phi_n(x) = \alpha_n \phi_n(x) + \alpha_{n-1} \phi_{n-1}(x)$$

or

$$\phi_{n+1}(x) = (A_n x + \alpha_n)\phi_n(x) + \alpha_{n-1}\phi_{n-1}(x), \tag{5.9}$$

which has the form of (5.5). We now need to verify that $\alpha_n = B_n$, $\alpha_{n-1} = C_n$ as in (5.6). From (5.9)

$$\phi_n(x) = A_{n-1}x\phi_{n-1}(x) + \alpha_{n-1}\phi_{n-1}(x) + \alpha_{n-2}\phi_{n-2}(x)$$

or

$$x\phi_{n-1}(x) = \frac{1}{A_{n-1}}\phi_n(x) + \underbrace{\left[-\frac{1}{A_{n-1}}(\alpha_{n-1}\phi_{n-1}(x) + \alpha_{n-2}\phi_{n-2}(x)) \right]}_{=p_{n-1}(x)} \qquad (5.10)$$

for which $\deg(p_{n-1}(x)) \le n - 1$. Thus, from (5.9)

$$\langle \phi_{n+1}, \phi_{n-1} \rangle = A_n \langle x\phi_n, \phi_{n-1} \rangle + \alpha_n \langle \phi_n, \phi_{n-1} \rangle + \alpha_{n-1} \langle \phi_{n-1}, \phi_{n-1} \rangle$$

so that

$$\alpha_{n-1} = -A_n \langle x\phi_n, \phi_{n-1} \rangle = -A_n \langle \phi_n, x\phi_{n-1} \rangle,$$

and via (5.10)

$$\alpha_{n-1} = -A_n \left\langle \phi_n, \frac{1}{A_{n-1}}\phi_n + p_{n-1} \right\rangle,$$

which becomes

$$\alpha_{n-1} = -\frac{A_n}{A_{n-1}} \langle \phi_n, \phi_n \rangle = -\frac{A_n}{A_{n-1}}$$

or

$$C_n = \alpha_{n-1} = -\frac{\phi_{n+1,n+1}}{\phi_{n,n}} \frac{\phi_{n-1,n-1}}{\phi_{n,n}},$$

which is the expression for C_n in (5.6). Expanding (5.5)

$$\sum_{k=0}^{n+1} \phi_{n+1,k}x^k = A_n \sum_{k=1}^{n+1} \phi_{n,k-1}x^k + B_n \sum_{k=0}^{n} \phi_{n,k}x^k + C_n \sum_{k=0}^{n-1} \phi_{n-1,k}x^k,$$

where, on comparing the terms for $k = n$, we see that

$$\phi_{n+1,n} = A_n\phi_{n,n-1} + B_n\phi_{n,n}$$

(since $\phi_{n-1,n} = 0$). Therefore

$$B_n = \frac{1}{\phi_{n,n}}[\phi_{n+1,n} - A_n\phi_{n,n-1}] = \frac{1}{\phi_{n,n}}\left[\phi_{n+1,n} - \frac{\phi_{n+1,n+1}}{\phi_{n,n}}\phi_{n,n-1} \right]$$

$$= \frac{\phi_{n+1,n+1}}{\phi_{n,n}}\left[\frac{\phi_{n+1,n}}{\phi_{n+1,n+1}} - \frac{\phi_{n,n-1}}{\phi_{n,n}} \right],$$

which is the expression for B_n in (5.6).

Theorem 5.2: Orthogonal polynomials satisfy the *Christoffel–Darboux formula* (*relation*)

$$(x - y) \sum_{j=0}^{n} \phi_j(x)\phi_j(y) = \frac{\phi_{n,n}}{\phi_{n+1,n+1}}[\phi_{n+1}(x)\phi_n(y) - \phi_{n+1}(y)\phi_n(x)]. \quad (5.11)$$

Proof The proof is an expanded version of that from Isaacson and Keller [1, p. 205].

From (5.5), we obtain

$$\phi_n(y)\phi_{n+1}(x) = (A_n x + B_n)\phi_n(x)\phi_n(y) + C_n\phi_{n-1}(x)\phi_n(y), \quad (5.12)$$

so reversing the roles of x and y gives

$$\phi_n(x)\phi_{n+1}(y) = (A_n y + B_n)\phi_n(y)\phi_n(x) + C_n\phi_{n-1}(y)\phi_n(x). \quad (5.13)$$

Subtracting (5.13) from (5.12) yields

$$\phi_n(y)\phi_{n+1}(x) - \phi_n(x)\phi_{n+1}(y) = A_n(x - y)\phi_n(x)\phi_n(y)$$
$$+ C_n(\phi_{n-1}(x)\phi_n(y) - \phi_n(x)\phi_{n-1}(y)). \quad (5.14)$$

We note that

$$\frac{C_n}{A_n} = -\frac{\phi_{n-1,n-1}}{\phi_{n,n}} = -\frac{1}{A_{n-1}}, \quad (5.15)$$

so (5.14) may be rewritten as

$$(x - y)\phi_n(x)\phi_n(y) = A_n^{-1}(\phi_{n+1}(x)\phi_n(y) - \phi_{n+1}(y)\phi_n(x))$$
$$- A_{n-1}^{-1}(\phi_n(x)\phi_{n-1}(y) - \phi_{n-1}(x)\phi_n(y)).$$

Now consider

$$(x - y)\sum_{j=0}^{n}\phi_j(x)\phi_j(y) = \sum_{j=0}^{n}\left\{A_j^{-1}[\phi_{j+1}(x)\phi_j(y) - \phi_{j+1}(y)\phi_j(x)]\right.$$
$$\left. - A_{j-1}^{-1}[\phi_j(x)\phi_{j-1}(y) - \phi_{j-1}(x)\phi_j(y)]\right\}. \quad (5.16)$$

Since $A_j = \phi_{j+1,j+1}/\phi_{j,j}$, we have $A_{-1}^{-1} = \phi_{-1,-1}/\phi_{0,0} = 0$ because $\phi_{-1,-1} = 0$. Taking this into account, the summation on the right-hand side of (5.16) reduces due to the cancellation of terms,[1] and so (5.16) finally becomes

$$(x - y)\sum_{j=0}^{n}\phi_j(x)\phi_j(y) = \frac{\phi_{n,n}}{\phi_{n+1,n+1}}[\phi_{n+1}(x)\phi_n(y) - \phi_{n+1}(y)\phi_n(x)],$$

which is (5.11).

---

[1] The cancellation process seen here is similar to the one that occurred in the derivation of the Dirichlet kernel identity (3.24). This mathematical technique seems to be a recurring theme in analysis.

The following corollary to Theorem 5.2 is from Hildebrand [2, p. 342]. However, there appears to be no proof in Ref. 2, so one is provided here.

Corollary 5.1 With $\phi_k^{(1)}(x) = d\phi_k(x)/dx$ we have

$$\sum_{j=0}^{n} \phi_j^2(x) = \frac{\phi_{n,n}}{\phi_{n+1,n+1}} [\phi_{n+1}^{(1)}(x)\phi_n(x) - \phi_n^{(1)}(x)\phi_{n+1}(x)]. \qquad (5.17)$$

Proof Since

$$\phi_{n+1}(x)\phi_n(y) - \phi_{n+1}(y)\phi_n(x) = [\phi_{n+1}(x) - \phi_{n+1}(y)]\phi_n(y)$$
$$- [\phi_n(x) - \phi_n(y)]\phi_{n+1}(y)$$

via (5.11)

$$\sum_{j=0}^{n} \phi_j(x)\phi_j(y)$$

$$= \frac{\phi_{n,n}}{\phi_{n+1,n+1}} \left[\frac{\phi_{n+1}(x) - \phi_{n+1}(y)}{x - y}\phi_n(y) - \frac{\phi_n(x) - \phi_n(y)}{x - y}\phi_{n+1}(y) \right]. \qquad (5.18)$$

Letting $y \to x$ in (5.18) immediately yields (5.17). This is so from the definition of the derivative, and the fact that all polynomials may be differentiated any number of times.

The three-term recurrence relation is a vital practical method for orthogonal polynomial generation. However, an alternative approach is the following.

If $\deg(q_{r-1}(x)) = r - 1$ ($q_{r-1}(x)$ is an *arbitrary* degree $r - 1$ polynomial), then

$$\langle \phi_r, q_{r-1} \rangle = \int_D w(x)\phi_r(x)q_{r-1}(x)\,dx = 0. \qquad (5.19)$$

Define

$$w(x)\phi_r(x) = \frac{d^r G_r(x)}{dx^r} = G_r^{(r)}(x) \qquad (5.20)$$

so that (5.19) becomes

$$\int_D G_r^{(r)}(x)q_{r-1}(x)\,dx = 0. \qquad (5.21)$$

If we repeatedly integrate by parts (using $q_{r-1}^{(r)}(x) = 0$ for all x), we obtain

$$\int_D G_r^{(r)}(x)q_{r-1}(x)\,dx = G_r^{(r-1)}(x)q_{r-1}(x)|_D - \int_D G_r^{(r-1)}(x)q_{r-1}^{(1)}(x)\,dx$$

$$= G_r^{(r-1)}(x)q_{r-1}(x)|_D - G_r^{(r-2)}(x)q_{r-1}^{(1)}(x)|_D$$

$$+ \int_D G_r^{(r-2)}(x)q_{r-1}^{(2)}(x)\,dx$$

$$= G_r^{(r-1)}(x)q_{r-1}(x)|_D - G_r^{(r-2)}(x)q_{r-1}^{(1)}(x)|_D$$

$$+ G_r^{(r-3)}(x)q_{r-1}^{(2)}(x)|_D + \int_D G_r^{(r-3)}(x)q_{r-1}^{(3)}(x)\,dx$$

and so on. Finally

$$[G_r^{(r-1)}q_{r-1} - G_r^{(r-2)}q_{r-1}^{(1)} + G_r^{(r-3)}q_{r-1}^{(2)} - \cdots + (-1)^{r-1}G_r q_{r-1}^{(r-1)}]_D = 0, \tag{5.22a}$$

or alternatively

$$\left[\sum_{k=1}^{r}(-1)^{k-1}G_r^{(r-k)}q_{r-1}^{(k-1)}\right]_D = 0. \tag{5.22b}$$

Since from (5.20)

$$\phi_r(x) = \frac{1}{w(x)}\frac{d^r G_r(x)}{dx^r}, \tag{5.23}$$

which is a polynomial of degree r, it must be the case that $G_r(x)$ satisfies the differential equation

$$\frac{d^{r+1}}{dx^{r+1}}\left[\frac{1}{w(x)}\frac{d^r G_r(x)}{dx^r}\right] = 0 \tag{5.24}$$

for $x \in D$. Recalling $D = [a, b]$ and allowing $a \to -\infty$, $b \to \infty$ (so we may have $D = [a, \infty)$, or $D = \mathbf{R}$), Eq. (5.22a) must be satisfied for *any values* of $q_{r-1}^{(k)}(a)$ and $q_{r-1}^{(k)}(b)$ ($k = 0, 1, \ldots, r - 1$). This implies that we have the *boundary conditions*

$$G_r^{(k)}(a) = 0, \, G_r^{(k)}(b) = 0 \tag{5.25}$$

for $k = 0, 1, \ldots, r - 1$. These restrict the solution to (5.24). It can be shown [6] that (5.24) has a nontrivial[2] solution for all $r \in \mathbf{Z}^+$ assuming that $w(x) \geq 0$ for all $x \in D$, and that the moments $\int_D x^k w(x)\,dx$ exist for all $k \in \mathbf{Z}^+$. Proving this is difficult, and so will not be considered. The expression in (5.23) may be called a *Rodrigues formula* for $\phi_r(x)$ (although it should be said that this terminology usually arises only in the context of Legendre polynomials).

The Christoffel–Darboux formulas arise in various settings. For example, they are relevant to the problem of designing Savitzky–Golay smoothing digital filters [3], and they arise in numerical integration methods [2, 4]. Another way in which the Christoffel–Darboux formula of (5.11) can make an appearance is as follows.

---

[2]The trivial solution for a differential equation is the identically zero function.

We may wish to approximate $f(x) \in L^2(D)$ as

$$f(x) \approx \sum_{j=0}^{n} a_j \phi_j(x), \tag{5.26}$$

so the *residual* is

$$r_n(x) = f(x) - \sum_{j=0}^{n} a_j \phi_j(x). \tag{5.27}$$

Adopting the now familiar least-squares approach, we select a_j to minimize functional $V(a) = ||r_n||^2 = \langle r_n, r_n \rangle$

$$V(a) = \int_D w(x) r_n^2(x) \, dx = \int_D w(x) \left[f(x) - \sum_{j=0}^{n} a_j \phi_j(x) \right]^2 dx, \tag{5.28}$$

where $a = [a_0 \ a_1 \ \cdots \ a_n]^T \in \mathbf{R}^{n+1}$. Of course, this expands to become

$$V(a) = \int_D w(x) f^2(x) \, dx - 2 \sum_{j=0}^{n} a_j \int_D w(x) f(x) \phi_j(x) \, dx$$

$$+ \sum_{j=0}^{n} \sum_{k=0}^{n} a_j a_k \int_D w(x) \phi_j(x) \phi_k(x) \, dx,$$

so if $g = [g_0 \ g_1 \ \cdots \ g_n]^T$ with $g_j = \int_D w(x) f(x) \phi_j(x) \, dx$, and $R = [r_{i,j}] \in \mathbf{R}^{(n+1) \times (n+1)}$ with $r_{i,j} = \int_D w(x) \phi_i(x) \phi_j(x) \, dx$, and $\rho = \int_D w(x) f^2(x) \, dx$, then

$$V(a) = \rho - 2a^T g + a^T R a. \tag{5.29}$$

But $r_{i,j} = \langle \phi_i, \phi_j \rangle = \delta_{i-j}$ [via (5.4)], so we have $R = I$ (identity matrix). Immediately, one of the advantages of working with orthogonal polynomials is that R is perfectly conditioned (contrast this with the Hilbert matrix of Chapter 4). This in itself is a powerful incentive to consider working with orthogonal polynomial expansions. Another nice consequence is that the optimal solution $\hat{a}$ satisfies

$$\hat{a} = g, \tag{5.30}$$

that is

$$\hat{a}_j = \int_D w(x) f(x) \phi_j(x) \, dx = \langle f, \phi_j \rangle, \tag{5.31}$$

where $j \in \mathbf{Z}_{n+1}$. If $a = \hat{a}$ in (5.27) then we have the *optimal residual*

$$\hat{r}_n(x) = f(x) - \sum_{j=0}^{n} \hat{a}_j \phi_j(x). \tag{5.32}$$

We may substitute (5.31) into (5.32) to obtain

$$\hat{r}_n(x) = f(x) - \sum_{j=0}^{n} \left[\int_D w(y) f(y) \phi_j(y) \, dy \right] \phi_j(x)$$

$$= f(x) - \int_D f(y) w(y) \left[\sum_{j=0}^{n} \phi_j(x) \phi_j(y) \right] dy. \qquad (5.33)$$

For convenience we define the *kernel function*[3]

$$K_n(x, y) = \sum_{j=0}^{n} \phi_j(x) \phi_j(y), \qquad (5.34)$$

so that (5.33) becomes

$$\hat{r}_n(x) = f(x) - \int_D w(y) f(y) K_n(x, y) \, dy. \qquad (5.35)$$

Clearly, $K_n(x, y)$ in (5.34) has the alternative formula given by (5.11). Now consider

$$\int_D f(x) w(y) K_n(x, y) \, dy = f(x) \int_D w(y) \sum_{j=0}^{n} \phi_j(x) \phi_j(y) \, dy$$

$$= f(x) \sum_{j=0}^{n} \phi_j(x) \int_D w(y) \phi_j(y) \, dy$$

$$= f(x) \sum_{j=0}^{n} \phi_j(x) \langle 1, \phi_j \rangle$$

$$= f(x) \sum_{j=0}^{n} \phi_j(x) \left\langle \frac{\phi_0(x)}{\phi_{0,0}}, \phi_j(x) \right\rangle$$

$$= f(x) \frac{\phi_0(x)}{\phi_{0,0}} = f(x)$$

because $\phi_0(x) = \phi_{0,0}$ for $x \in D$, and $\langle \phi_0, \phi_j \rangle = \delta_j$. Thus, (5.35) becomes

$$\hat{r}_n(x) = \int_D f(x) w(y) K_n(x, y) \, dy - \int_D f(y) w(y) K_n(x, y) \, dy$$

$$= \int_D w(y) K_n(x, y) [f(x) - f(y)] \, dy. \qquad (5.36)$$

---

[3] Recall the Dirichlet kernel of Chapter 3, which we now see is really just a special instance of the general idea considered here.

The optimal residual (optimal error) $\hat{r}_n(x)$ presumably gets smaller in some sense as $n \to \infty$. Clearly, insight into how it behaves in the limit as $n \to \infty$ can be provided by (5.36). The behavior certainly depends on $f(x)$, $w(x)$, and the kernel $K_n(x, y)$. Intuitively, the summation expression for $K_n(x, y)$ in (5.34) is likely to be less convenient to work with than the alternative expression we obtain from (5.11).

Some basic results on polynomial approximations are as follows.

Theorem 5.3: Weierstrass' Approximation Theorem   Let $f(x)$ be continuous on the closed interval $[a, b]$. For any $\epsilon > 0$ there exists an integer $N = N(\epsilon)$, and a polynomial $p_N(x) \in \mathbf{P}^N[a, b]$ $(\deg(p_N(x)) \le N)$ such that

$$|f(x) - p_N(x)| \le \epsilon$$

for all $x \in [a, b]$.

Various proofs of this result exist in the literature (e.g., see Rice [5, p. 121] and Isaacson and Keller [1, pp. 183–186]), but we omit them here. We see that the convergence of $p_N(x)$ to $f(x)$ is uniform as $N \to \infty$ (recall Definition 3.4 in Chapter 3). Theorem 5.3 states that any function continuous on an interval may be approximated with a polynomial to arbitrary accuracy. Of course, a large degree polynomial may be needed to achieve a particular level of accuracy depending upon $f(x)$. We also remark that Weierstrass' theorem is an *existence theorem*. It claims the existence of a polynomial that uniformly approximates a continuous function on $[a, b]$, but it does not tell us how to find the polynomial. Some information about the convergence behavior of a least-squares approximation is provided by the following theorem.

Theorem 5.4: Let $D = [a, b]$, $w(x) = 1$ for all $x \in D$. Let $f(x)$ be continuous on D, and let

$$q_n(x) = \sum_{j=0}^{n} \hat{a}_j \phi_j(x)$$

(least-squares polynomial approximation to $f(x)$ on D), so $\hat{a}_j = \langle f, \phi_j \rangle = \int_a^b f(x)\phi_j(x)\,dx$. Then

$$\lim_{n\to\infty} V(\hat{a}) = \lim_{n\to\infty} \int_a^b [f(x) - q_n(x)]^2\,dx = 0,$$

and we have *Parseval's equality*

$$\int_a^b f^2(x)\,dx = \sum_{j=0}^{\infty} \hat{a}_j^2. \tag{5.37}$$

Proof   We use proof by contradiction. Assume that $\lim_{n \to \infty} V(\hat{a}) = \delta > 0$. Pick any $\epsilon > 0$ such that $\epsilon^2 = \frac{1}{2(b-a)}\delta$. By Theorem 5.3 (Weierstrass' theorem) there is a polynomial $p_m(x)$ such that $|f(x) - p_m(x)| \le \epsilon$ for $x \in D$. Thus

$$\int_a^b [f(x) - p_m(x)]^2 \, dx \le \epsilon^2 [b - a] = \frac{1}{2}\delta.$$

Now via (5.29) and (5.30)

$$V(\hat{a}) = \rho - \hat{a}^T \hat{a},$$

that is

$$V(\hat{a}) = \int_a^b f^2(x) \, dx - \sum_{j=0}^n \hat{a}_j^2,$$

and $V(a) \ge 0$ for all $a \in \mathbf{R}^{n+1}$. So we have *Bessel's inequality*

$$V(\hat{a}) = \int_a^b f^2(x) \, dx - \sum_{j=0}^n \hat{a}_j^2 \ge 0, \tag{5.38}$$

and $V(\hat{a})$ must be a nonincreasing function of n. Hence the least-squares approximation of degree m, say, $q_m(x)$, satisfies

$$\frac{1}{2}\delta \ge \int_a^b [f(x) - q_m(x)]^2 \, dx \ge \delta$$

which is a contradiction unless $\delta = 0$. Since $\delta = 0$, we must have (5.37).

We observe that Parseval's equality of (5.37) relates the energy of $f(x)$ to the energy of the coefficients $\hat{a}_j$. The convergence behavior described by Theorem 5.4 is sometimes referred to as *convergence in the mean* [1, pp. 197–198]. This theorem does not say what happens if $f(x)$ is not continuous on $[a, b]$. The result (5.36) is independent regardless of whether $f(x)$ is continuous. It is a potentially more powerful result for this reason. It turns out that the convergence of the least-squares polynomial sequence $(q_n(x))$ to $f(x)$ is pointwise in general but, depending on $\phi_j(x)$ and $f(x)$, the convergence can sometimes be uniform. For uniform convergence, $f(x)$ must be sufficiently smooth. The pointwise convergence of the orthogonal polynomial series when $f(x)$ has a discontinuity implies that the Gibbs phenomenon can be expected. (Recall this phenomenon in the context of the Fourier series expansion as seen in Section 3.4.)

We have seen polynomial approximation to functions in Chapter 3. There we saw that the Taylor formula is a polynomial approximation to a function with a number of derivatives equal to the degree of the Taylor polynomial. This approximation technique is obviously limited to functions that are sufficiently differentiable. But our present polynomial approximation methodology has no such limitation.

We remark that Theorem 5.4 suggests (but does not prove) that if $f(x) \in L^2[a, b]$, then

$$f(x) = \sum_{j=0}^{\infty} \langle f, \phi_j \rangle \phi_j(x), \tag{5.39}$$

where

$$\langle f, \phi_j \rangle = \int_a^b f(x)\phi_j(x)\,dx. \tag{5.40}$$

This has the basic form of the Fourier series expansion that was first seen in Chapter 1. For this reason (5.39) is sometimes called a *generalized Fourier series expansion*, although in the most general form of this idea the orthogonal functions are not necessarily always polynomials. The idea of a generalized Fourier series expansion can be extended to domains such as $D = [0, \infty)$, or $D = \mathbf{R}$, and to any weighting functions that lead to solutions to (5.24).

The next three sections consider the Chebyshev, Hermite, and Legendre polynomials as examples of how to apply the core theory of this section.

5.3   CHEBYSHEV POLYNOMIALS

Suppose that $D = [a, b]$ and recall (5.28),

$$V(a) = \int_a^b w(x)r_n^2(x)\,dx, \tag{5.41}$$

which is the (weighted) energy of the approximation error (residual) in (5.27):

$$r_n(x) = f(x) - \sum_{j=0}^{n} a_j \phi_j(x). \tag{5.42}$$

The weighting function $w(x)$ is often selected to give more or less weight to errors in different places on the interval $[a, b]$. This is intended to achieve a degree of control over error behavior. If $w(x) = c > 0$ for $x \in [a, b]$, then equal importance is given to errors across the interval. This choice (with $c = 1$) gives rise to the Legendre polynomials, and will be considered later. If we wish to give more weight to errors at the ends of the interval, then a popular instance of this is for $D = [-1, 1]$ with weighting function

$$w(x) = \frac{1}{\sqrt{1 - x^2}}. \tag{5.43}$$

This choice leads to the famous *Chebyshev polynomials of the first kind.* The reader will most likely see these applied to problems in analog and digital filter design in subsequent courses. For now, we concentrate on their basic theory.

The following lemma and ideas expressed in it are pivotal in understanding the Chebyshev polynomials and how they are constructed.

Lemma 5.1:   If $n \in \mathbf{Z}^+$, then

$$\cos n\theta = \sum_{k=0}^{n} \beta_{n,k} \cos^k \theta \tag{5.44}$$

for suitable $\beta_{n,k} \in \mathbf{R}$.

Proof   First of all recall the trigonometric identities

$$\cos(m+1)\theta = \cos m\theta \cos \theta - \sin m\theta \sin \theta$$

$$\cos(m-1)\theta = \cos m\theta \cos \theta + \sin m\theta \sin \theta$$

(which follow from the more basic identity $\cos(a+b) = \cos a \cos b - \sin a \sin b$). From the sum of these identities

$$\cos(m+1)\theta = 2\cos m\theta \cos \theta - \cos(m-1)\theta. \tag{5.45}$$

For $n=1$ in (5.44), we have $\beta_{1,0} = 0$, $\beta_{1,1} = 1$, and for $n=0$, we have $\beta_{0,0} = 1$. These will form initial conditions in a recursion that we now derive using mathematical induction.

Assume that (5.44) is valid both for $n = m$ and for $n = m - 1$, and so

$$\cos m\theta = \sum_{k=0}^{m} \beta_{m,k} \cos^k \theta$$

$$\cos(m-1)\theta = \sum_{k=0}^{m-1} \beta_{m-1,k} \cos^k \theta,$$

and so via (5.45)

$$\cos(m+1)\theta = 2\cos \theta \sum_{k=0}^{m} \beta_{m,k} \cos^k \theta - \sum_{k=0}^{m-1} \beta_{m-1,k} \cos^k \theta$$

$$= 2\sum_{k=0}^{m} \beta_{m,k} \cos^{k+1} \theta - \sum_{k=0}^{m-1} \beta_{m-1,k} \cos^k \theta$$

$$= 2\sum_{r=1}^{m+1} \beta_{m,r-1} \cos^r \theta - \sum_{r=0}^{m-1} \beta_{m-1,r} \cos^r \theta$$

$$= \sum_{k=0}^{m+1} [2\beta_{m,k-1} - \beta_{m-1,k}] \cos^k \theta \, (\beta_{m,k} = 0 \text{ for } k < 0, \text{ and } k > m)$$

$$= \sum_{k=0}^{m+1} \beta_{m+1,k} \cos^k \theta,$$

which is to say that

$$\beta_{m+1,k} = 2\beta_{m,k-1} - \beta_{m-1,k}. \tag{5.46}$$

This is the desired three-term recursion for the coefficients in (5.44). As a consequence of this result, Eq. (5.44) is valid for $n = m + 1$ and thus is valid for all $n \geq 0$ by mathematical induction.

This lemma states that $\cos n\theta$ may be expressed as a polynomial of degree n in $\cos \theta$. Equation (5.46) along with the *initial conditions*

$$\beta_{0,0} = 1, \quad \beta_{1,0} = 0, \quad \beta_{1,1} = 1 \tag{5.47}$$

tells us how to find the polynomial coefficients. For example, from (5.46)

$$\beta_{2,k} = 2\beta_{1,k-1} - \beta_{0,k}$$

for $k = 0, 1, 2$. [In general, we evaluate (5.46) for $k = 0, 1, \ldots, m + 1$.] Therefore

$$\beta_{2,0} = 2\beta_{1,-1} - \beta_{0,0} = -1,$$
$$\beta_{2,1} = 2\beta_{1,0} - \beta_{0,1} = 0,$$
$$\beta_{2,2} = 2\beta_{1,1} - \beta_{0,2} = 2$$

which implies that $\cos 2\theta = -1 + 2 \cos^2 \theta$. This is certainly true as it is a well-known trigonometric identity.

Lemma 5.1 possesses a converse. Any polynomial in $\cos \theta$ of degree n can be expressed as a linear combination of members from the set $\{\cos k\theta \,|\, k = 0, 1, \ldots, n\}$.

We now have enough information to derive the Chebyshev polynomials. Recalling (5.19) we need a polynomial $\phi_r(x)$ of degree r such that

$$\int_{-1}^{1} \frac{1}{\sqrt{1 - x^2}} \phi_r(x) q_{r-1}(x) \, dx = 0, \tag{5.48}$$

where $q_{r-1}(x)$ is an arbitrary polynomial such that $\deg(q_{r-1}(x)) \leq r - 1$. Let us change variables according to

$$x = \cos \theta. \tag{5.49}$$

Then $dx = -\sin\theta\,d\theta$, and $x \in [-1, 1]$ maps to $\theta \in [0, \pi]$. Thus

$$\int_{-1}^{1} \frac{1}{\sqrt{1-x^2}} \phi_r(x) q_{r-1}(x)\,dx = -\int_{\pi}^{0} \frac{1}{\sqrt{1-\cos^2\theta}} \phi_r(\cos\theta) q_{r-1}(\cos\theta) \sin\theta\,d\theta$$

$$= \int_{0}^{\pi} \phi_r(\cos\theta) q_{r-1}(\cos\theta)\,d\theta = 0. \qquad (5.50)$$

Because of Lemma 5.1 (and the above mentioned converse to it)

$$\int_{0}^{\pi} \phi_r(\cos\theta) \cos k\theta\,d\theta = 0 \qquad (5.51)$$

for $k = 0, 1, \ldots, r - 1$. Consider $\phi_r(\cos\theta) = C_r \cos r\theta$, then

$$C_r \int_{0}^{\pi} \cos r\theta \cos k\theta\,d\theta = \frac{1}{2} C_r \int_{0}^{\pi} [\cos(r+k)\theta + \cos(r-k)\theta]\,d\theta$$

$$= \frac{1}{2} C_r \left[\frac{1}{r+k} \sin(r+k)\theta + \frac{1}{r-k} \sin(r-k)\theta \right]_{0}^{\pi} = 0$$

for $k = 0, 1, \ldots, r - 1$. Thus, we may indeed choose

$$\phi_r(x) = C_r \cos[r \cos^{-1} x]. \qquad (5.52)$$

Constant C_r is selected to *normalize* the polynomial according to user requirements. Perhaps the most common choice is simply to set $C_r = 1$ for all $r \in \mathbf{Z}^+$. In this case we set $T_r(x) = \phi_r(x)$:

$$T_r(x) = \cos[r \cos^{-1} x]. \qquad (5.53)$$

These are the *Chebyshev polynomials of the first kind*.

By construction, if $r \neq k$, then

$$\langle T_r, T_k \rangle = \int_{-1}^{1} \frac{1}{\sqrt{1-x^2}} T_r(x) T_k(x)\,dx = 0.$$

Consider $r = k$

$$\|T_r\|^2 = \int_{-1}^{1} \frac{1}{\sqrt{1-x^2}} T_r^2(x)\,dx$$

$$= \int_{-1}^{1} \frac{1}{\sqrt{1-x^2}} \cos^2[r \cos^{-1} x]\,dx, \qquad (5.54)$$

and apply (5.49). Thus

$$\|T_r\|^2 = \int_{0}^{\pi} \cos^2 r\theta\,d\theta = \begin{cases} \pi, & r = 0 \\ \frac{1}{2}\pi, & r > 0 \end{cases}. \qquad (5.55)$$

We have claimed $\phi_r(x)$ in (5.52), and hence $T_r(x)$ in (5.53), are *polynomials*. This is not immediately obvious, given that the expressions are in terms of trigonometric functions. We will now confirm that $T_r(x)$ is indeed a polynomial in x for all $r \in \mathbf{Z}^+$. Our approach follows the proof of Lemma 5.1.

First, we observe that

$$T_0(x) = \cos(0 \cdot \cos^{-1} x) = 1, \quad T_1(x) = \cos(1 \cdot \cos^{-1} x) = x. \quad (5.56)$$

Clearly, these are polynomials in x. Once again (see the proof of Lemma 5.1)

$$\begin{aligned}
T_{n+1}(x) &= \cos(n+1)\theta \, (\theta = \cos^{-1} x) \\
&= \cos n\theta \cos \theta - (\cos(n-1)\theta - \cos n\theta \cos \theta) \\
&= 2 \cos n\theta \cos \theta - \cos(n-1)\theta \\
&= 2x T_n(x) - T_{n-1}(x),
\end{aligned}$$

that is

$$T_{n+1}(x) = 2x T_n(x) - T_{n-1}(x) \quad (5.57)$$

for $n \in \mathbf{N}$. The initial conditions are expressed in (5.56). We see that this is a special case of (5.5) (Theorem 5.1). It is immediately clear from (5.57) and (5.56) that $T_n(x)$ is a polynomial in x for all $n \in \mathbf{Z}^+$.

Some plots of Chebyshev polynomials of the first kind appear in Fig. 5.1. We remark that on the interval $x \in [-1, 1]$ the "ripples" of the polynomials are of the same height. This fact makes this class of polynomials quite useful in certain *uniform approximation problems*. It is one of the main properties that makes this class of polynomials useful in filter design. Some indication of how the Chebyshev polynomials relate to uniform approximation problems appears in Section 5.7.

Example 5.1   This example is about how to program the recursion specified in (5.57). Define

$$T_n(x) = \sum_{j=0}^{n} T_{n,j} x^j$$

so that $T_{n,j}$ is the coefficient of x^j. This is the same kind of notation as in (5.3) for $\phi_n(x)$. In this setting it is quite important to note that $T_{n,j} = 0$ for $j < 0$, and for $j > n$. From (5.57), we obtain

$$\sum_{j=0}^{n+1} T_{n+1,j} x^j = 2x \sum_{j=0}^{n} T_{n,j} x^j - \sum_{j=0}^{n-1} T_{n-1,j} x^j,$$

and so

$$\sum_{j=0}^{n+1} T_{n+1,j} x^j = 2 \sum_{j=0}^{n} T_{n,j} x^{j+1} - \sum_{j=0}^{n-1} T_{n-1,j} x^j.$$

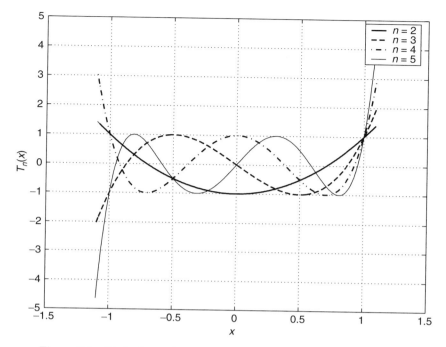

Figure 5.1 Chebyshev polynomials of the first kind of degrees $2, 3, 4, 5$.

Changing the variable in the second summation according to $k = j + 1$ (so $j = k - 1$) yields

$$\sum_{k=0}^{n+1} T_{n+1,k} x^k = 2 \sum_{k=1}^{n+1} T_{n,k-1} x^k - \sum_{k=0}^{n-1} T_{n-1,k} x^k,$$

and modifying the limits on the summations on the right-hand side yields

$$\sum_{k=0}^{n+1} T_{n+1,k} x^k = 2 \sum_{k=0}^{n+1} T_{n,k-1} x^k - \sum_{k=0}^{n+1} T_{n-1,k} x^k.$$

We emphasize that this is permitted because we recall that $T_{n,-1} = T_{n-1,n} = T_{n-1,n+1} = 0$ for all $n \geq 1$. Comparing like powers of x gives us the recursion

$$T_{n+1,k} = 2T_{n,k-1} - T_{n-1,k}$$

for $k = 0, 1, \ldots, n + 1$ with $n = 1, 2, 3, \ldots$. Since $T_0(x) = 1$, we have $T_{0,0} = 1$, and since $T_1(x) = x$, we have $T_{1,0} = 0$, $T_{1,1} = 1$, which are the *initial conditions* for the recursion. Therefore, a pseudocode program to compute $T_n(x)$ is

```
T_{0,0} := 1;
T_{1,0} := 0; T_{1,1} := 1;
for n := 1 to N − 1 do begin
    for k := 0 to n + 1 do begin
```

$$T_{n+1,k} := 2T_{n,k-1} - T_{n-1,k};$$
$$\text{end;}$$
$$\text{end;}$$

This computes $T_2(x), \ldots, T_N(x)$ (so we need $N \geq 2$).

We remark that the recursion in Example 5.1 may be implemented using integer arithmetic, and so there will be no rounding or quantization errors involved in computing $T_n(x)$. However, there is the risk of computing machine overflow.

Example 5.2   This example is about changing representations. Specifically, how might we express

$$p_n(x) = \sum_{j=0}^{n} p_{n,j} x^j \in \mathbf{P}^n[-1, 1]$$

in terms of the Chebyshev polynomials of the first kind? In other words, we wish to determine the series coefficients in

$$p_n(x) = \sum_{j=0}^{n} \alpha_j T_j(x).$$

We will consider this problem only for $n = 4$.

Therefore, we begin by noting that (via (5.56) and (5.57)) $T_0(x) = 1$, $T_1(x) = x$, $T_2(x) = 2x^2 - 1$, $T_3(x) = 4x^3 - 3x$, and $T_4(x) = 8x^4 - 8x^2 + 1$. Consequently

$$\sum_{j=0}^{4} \alpha_j T_j(x) = \alpha_0 + \alpha_1 x + 2\alpha_2 x^2 - \alpha_2$$

$$+ 4\alpha_3 x^3 - 3\alpha_3 x + 8\alpha_4 x^4 - 8\alpha_4 x^2 + \alpha_4$$

$$= (\alpha_0 - \alpha_2 + \alpha_4) + (\alpha_1 - 3\alpha_3)x + (2\alpha_2 - 8\alpha_4)x^2 + 4\alpha_3 x^3 + 8\alpha_4 x^4$$

implying that (on comparing like powers of x)

$$p_{4,0} = \alpha_0 - \alpha_2 + \alpha_4, \ p_{4,1} = \alpha_1 - 3\alpha_3,$$

$$p_{4,2} = 2\alpha_2 - 8\alpha_4, \ p_{4,3} = 4\alpha_3, \ p_{4,4} = 8\alpha_4.$$

This may be more conveniently expressed in matrix form:

$$\underbrace{\begin{bmatrix} 1 & 0 & -1 & 0 & 1 \\ 0 & 1 & 0 & -3 & 0 \\ 0 & 0 & 2 & 0 & -8 \\ 0 & 0 & 0 & 4 & 0 \\ 0 & 0 & 0 & 0 & 8 \end{bmatrix}}_{=U} \underbrace{\begin{bmatrix} \alpha_0 \\ \alpha_1 \\ \alpha_2 \\ \alpha_3 \\ \alpha_4 \end{bmatrix}}_{=\alpha} = \underbrace{\begin{bmatrix} p_{4,0} \\ p_{4,1} \\ p_{4,2} \\ p_{4,3} \\ p_{4,4} \end{bmatrix}}_{=p}.$$

The upper triangular system $U\alpha = p$ is certainly easy to solve using the backward substitution algorithm presented in Chapter 4 (see Section 4.5). We note that the elements of matrix U are the coefficients $\{T_{n,j}\}$ as might be obtained from the algorithm in Example 5.1.

Can you guess, on the basis of Example 5.2, what matrix U will be for any n ?

5.4 HERMITE POLYNOMIALS

Now let us consider $D = \mathbf{R}$ with weighting function

$$w(x) = e^{-\alpha^2 x^2}. \tag{5.58}$$

This is essentially the Gaussian pulse from Chapter 3. Recalling (5.23), we have

$$\phi_r(x) = e^{\alpha^2 x^2} \frac{d^r G_r(x)}{dx^r}, \tag{5.59}$$

where $G_r(x)$ satisfies the differential equation

$$\frac{d^{r+1}}{dx^{r+1}}\left[e^{\alpha^2 x^2} \frac{d^r G_r(x)}{dx^r}\right] = 0 \tag{5.60}$$

[recall (5.24)]. From (5.25) $G_r(x)$ and $G_r^{(k)}(x)$ (for $k = 1, 2, \ldots, r - 1$) must tend to zero as $x \to \pm\infty$. We may consider the trial solution

$$G_r(x) = C_r e^{-\alpha^2 x^2}. \tag{5.61}$$

The kth derivative of this is of the form

$$C_r e^{-\alpha^2 x^2} \times \text{(polynomial of degree } k),$$

so (5.61) satisfies both (5.60) and the required boundary conditions. Therefore

$$\phi_r(x) = C_r e^{\alpha^2 x^2} \frac{d^r}{dx^r}(e^{-\alpha^2 x^2}). \tag{5.62}$$

It is common practice to define the Hermite polynomials to be $\phi_r(x)$ for $C_r = (-1)^r$ with either $\alpha^2 = 1$ or $\alpha^2 = \frac{1}{2}$. We shall select $\alpha^2 = 1$, and so our *Hermite polynomials* are

$$H_r(x) = (-1)^r e^{x^2} \frac{d^r}{dx^r}(e^{-x^2}). \tag{5.63}$$

By construction, for $k \neq r$

$$\int_{-\infty}^{\infty} e^{-\alpha^2 x^2} \phi_r(x)\phi_k(x)\, dx = 0. \tag{5.64}$$

For the case where $k = r$, the following result is helpful.

It must be the case that

$$\phi_r(x) = \sum_{k=0}^{r} \phi_{r,k} x^k$$

so that

$$||\phi_r||^2 = \int_D w(x)\phi_r^2(x)\,dx = \int_D w(x)\phi_r(x)\left[\sum_{k=0}^{r} \phi_{r,k} x^k\right]dx,$$

but $\int_D w(x)\phi_r(x)x^i\,dx = 0$ for $i = 0, 1, \ldots, r-1$ [special case of (5.19)], and so

$$||\phi_r||^2 = \phi_{r,r}\int_D w(x)\phi_r(x)x^r\,dx = \phi_{r,r}\int_D x^r G_r^{(r)}(x)\,dx \qquad (5.65)$$

[via (5.20)]. We may integrate (5.65) by parts r times, and apply (5.25) to obtain

$$||\phi_r||^2 = (-1)^r r!\phi_{r,r}\int_D G_r(x)\,dx. \qquad (5.66)$$

So, for our present problem with $k = r$, we obtain

$$||\phi_r||^2 = \int_{-\infty}^{\infty} e^{-\alpha^2 x^2}\phi_r^2(x)\,dx = (-1)^r r!\phi_{r,r}\int_{-\infty}^{\infty} C_r e^{-\alpha^2 x^2}\,dx. \qquad (5.67)$$

Now (if $y = \alpha x$ with $\alpha > 0$)

$$\int_{-\infty}^{\infty} e^{-\alpha^2 x^2}\,dx = 2\int_0^{\infty} e^{-\alpha^2 x^2}\,dx = \frac{2}{\alpha}\int_0^{\infty} e^{-y^2}\,dy = \frac{1}{\alpha}\sqrt{\pi} \qquad (5.68)$$

via (3.103) in Chapter 3. Consequently, (5.67) becomes

$$||\phi_r||^2 = (-1)^r r!\phi_{r,r} C_r \frac{\sqrt{\pi}}{\alpha}. \qquad (5.69)$$

With $C_r = (-1)^r$ and $\alpha = 1$, we recall $H_r(x) = \phi_r(x)$, so (5.69) becomes

$$||H_r||^2 = r!H_{r,r}\sqrt{\pi}. \qquad (5.70)$$

We need an expression for $H_{r,r}$.
   We know that for suitable $p_{n,k}$

$$\frac{d^n}{dx^n}e^{-x^2} = \underbrace{\left[\sum_{k=0}^{n} p_{n,k} x^k\right]}_{=p_n(x)} e^{-x^2}. \qquad (5.71)$$

Now consider

$$\frac{d^{n+1}}{dx^{n+1}}e^{-x^2} = \frac{d}{dx}\left\{\left[\sum_{k=0}^{n} p_{n,k}x^k\right]e^{-x^2}\right\}$$

$$= -2x\left[\sum_{k=0}^{n} p_{n,k}x^k\right]e^{-x^2} + \left[\sum_{k=1}^{n} kp_{n,k}x^{k-1}\right]e^{-x^2}$$

$$= \left[-2\sum_{k=0}^{n+1} p_{n,k-1} + \sum_{k=0}^{n+1}(k+1)p_{n,k+1}\right]x^k e^{-x^2}$$

$$= \sum_{k=0}^{n+1}\underbrace{[-2p_{n,k-1} + (k+1)p_{n,k+1}]}_{=p_{n+1,k}}x^k e^{-x^2},$$

so we have the recurrence relation

$$p_{n+1,k} = -2p_{n,k-1} + (k+1)p_{n,k+1}. \tag{5.72}$$

From (5.71) and (5.63) $H_n(x) = (-1)^n p_n(x)$. From (5.72)

$$p_{n+1,n+1} = -2p_{n,n} + (n+2)p_{n,n+2} = -2p_{n,n},$$

and so

$$H_{n+1,n+1} = (-1)^{n+1}p_{n+1,n+1} = -(-1)^n(-2p_{n,n}) = 2H_{n,n}. \tag{5.73}$$

Because $H_0(x) = 1$ [via (5.63)], $H_{0,0} = 1$, and immediately we have $H_{n,n} = 2^n$ [solution to the difference equation (5.73)]. Therefore

$$\|H_r\|^2 = 2^r r!\sqrt{\pi}, \tag{5.74}$$

thus, $\int_{-\infty}^{\infty} e^{-x^2} H_r^2(x)\,dx = 2^r r!\sqrt{\pi}$.

A three-term recurrence relation for $H_n(x)$ is needed. Define the *generating function*

$$S(x,t) = \exp[-t^2 + 2xt] = \exp[x^2 - (t-x)^2]. \tag{5.75}$$

Observe that

$$\frac{\partial^n}{\partial t^n}S(x,t) = \exp[x^2]\frac{\partial^n}{\partial t^n}\exp[-(t-x)^2] = (-1)^n\exp[x^2]\frac{\partial^n}{\partial x^n}\exp[-(t-x)^2]. \tag{5.76}$$

Because of the second equality in (5.76), we have

$$S^{(n)}(x,0) = \frac{\partial^n}{\partial t^n}S(x,0) = (-1)^n\exp[x^2]\frac{d^n}{dx^n}\exp[-x^2] = H_n(x). \tag{5.77}$$

The Maclaurin expansion of $S(x, t)$ about $t = 0$ is [recall (3.75)]

$$S(x, t) = \sum_{n=0}^{\infty} \frac{S^{(n)}(x, 0)}{n!} t^n,$$

so via (5.77), this becomes

$$S(x, t) = \sum_{n=0}^{\infty} \frac{H_n(x)}{n!} t^n. \tag{5.78}$$

Now, from (5.75) and (5.78), we have

$$\frac{\partial S}{\partial x} = 2te^{-t^2+2tx} = \sum_{n=0}^{\infty} \frac{2t^{n+1}}{n!} H_n(x), \tag{5.79a}$$

and also

$$\frac{\partial S}{\partial x} = \sum_{n=0}^{\infty} \frac{t^n}{n!} \frac{\partial H_n(x)}{\partial x} = \sum_{n=-1}^{\infty} \frac{t^{n+1}}{(n+1)!} \frac{dH_{n+1}(x)}{dx}. \tag{5.79b}$$

Comparing the like powers of t in (5.79) yields

$$\frac{1}{(n+1)!} \frac{dH_{n+1}(x)}{dx} = \frac{2}{n!} H_n(x),$$

which implies that

$$\frac{dH_{n+1}(x)}{dx} = 2(n+1)H_n(x) \tag{5.80}$$

for $n \in \mathbf{Z}^+$. We may also consider

$$\frac{\partial S}{\partial t} = (-2t + 2x)e^{-t^2+2tx} = \sum_{n=0}^{\infty} \frac{(-2t + 2x)}{n!} t^n H_n(x) \tag{5.81a}$$

and

$$\frac{\partial S}{\partial t} = \sum_{n=1}^{\infty} \frac{t^{n-1}}{(n-1)!} H_n(x) = \sum_{n=0}^{\infty} \frac{t^n}{n!} H_{n+1}(x). \tag{5.81b}$$

From (5.81a)

$$\sum_{n=0}^{\infty} \frac{(-2t + 2x)}{n!} t^n H_n(x) = -2 \sum_{n=0}^{\infty} \frac{t^n}{(n-1)!} H_{n-1}(x) + 2x \sum_{n=0}^{\infty} \frac{t^n}{n!} H_n(x). \tag{5.82}$$

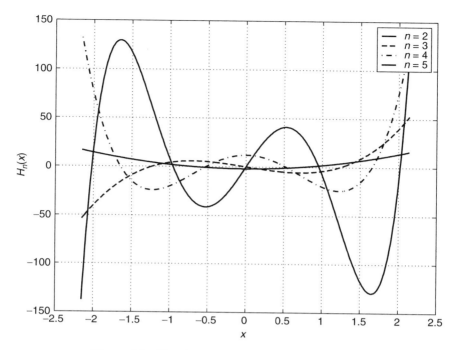

Figure 5.2   Hermite polynomials of degrees 2, 3, 4, 5.

Comparing like powers of t in (5.81b) and (5.82) yields

$$\frac{1}{n!} H_{n+1}(x) = -\frac{2}{(n-1)!} H_{n-1}(x) + 2x \frac{1}{n!} H_n(x)$$

or

$$H_{n+1}(x) = 2x H_n(x) - 2n H_{n-1}(x). \tag{5.83}$$

This holds for all $n \in \mathbf{N}$. Equation (5.83) is another special case of (5.5).

The Hermite polynomials are relevant to quantum mechanics (quantum harmonic oscillator problem), and they arise in signal processing as well (often in connection with the uncertainty principle for signals). A few Hermite polynomials are plotted in Fig. 5.2.

5.5   LEGENDRE POLYNOMIALS

We consider $D = [-1, 1]$ with uniform weighting function

$$w(x) = 1 \quad \text{for all} \quad x \in D. \tag{5.84}$$

In this case (5.24) becomes

$$\frac{d^{r+1}}{dx^{r+1}} \left[\frac{d^r G_r(x)}{dx^r} \right] = 0$$

or

$$\frac{d^{2r+1}}{dx^{2r+1}} G_r(x) = 0. \tag{5.85}$$

The boundary conditions of (5.25) become

$$G_r^{(k)}(\pm 1) = 0 \tag{5.86}$$

for $k \in \mathbf{Z}_r$. Consequently, the solution to (5.85) is given by

$$G_r(x) = C_r(x^2 - 1)^r. \tag{5.87}$$

Thus, from (5.23)

$$\phi_r(x) = C_r \frac{d^r}{dx^r} (x^2 - 1)^r. \tag{5.88}$$

The *Legendre polynomials* use $C_r = \frac{1}{2^r r!}$, and are denoted by

$$P_r(x) = \frac{1}{2^r r!} \frac{d^r}{dx^r} (x^2 - 1)^r, \tag{5.89}$$

which is the *Rodrigues formula* for $P_r(x)$. By construction, for $k \neq r$, we must have

$$\int_{-1}^{1} P_r(x) P_k(x) \, dx = 0. \tag{5.90}$$

From (5.66) and (5.87), we obtain

$$||P_r||^2 = \frac{(-1)^r}{2^r} P_{r,r} \int_{-1}^{1} (x^2 - 1)^r \, dx. \tag{5.91}$$

Recalling the binomial theorem

$$(a + b)^r = \sum_{k=0}^{r} \binom{r}{k} a^k b^{r-k}, \tag{5.92}$$

we see that

$$\frac{d^r}{dx^r}(x^2 - 1)^r = \frac{d^r}{dx^r} \sum_{k=0}^{r} \binom{r}{k} (-1)^{r-k} x^{2k}$$

$$= \frac{d^r}{dx^r} \left[x^{2r} + \sum_{k=0}^{r-1} \binom{r}{k} (-1)^{r-k} x^{2k} \right]$$

$$= 2r(2r - 1) \cdots (r + 1)x^r + \frac{d^r}{dx^r} \left[\sum_{k=0}^{r-1} \binom{r}{k} (-1)^{r-k} x^{2k} \right],$$

implying that (recall (5.89))

$$P_{r,r} = \frac{2r(2r - 1) \cdots (r + 1)}{2^r r!} = \frac{(2r)!}{2^r [r!]^2}. \tag{5.93}$$

But we need to evaluate the integral in (5.91) as well. With the change of variable $x = \sin \theta$, we have

$$I_r = \int_{-1}^{1} (x^2 - 1)^r \, dx = (-1)^r \int_{-\pi/2}^{\pi/2} \cos^{2r+1} \theta \, d\theta. \tag{5.94}$$

We may integrate by parts

$$\int_{-\pi/2}^{\pi/2} \cos^{2r+1} \theta \, d\theta = \int_{-\pi/2}^{\pi/2} \cos \theta \cos^{2r} \theta \, d\theta$$

$$= \cos^{2r} \theta \sin \theta |_{-\pi/2}^{\pi/2} + 2r \int_{-\pi/2}^{\pi/2} \cos^{2r-1} \theta \sin^2 \theta \, d\theta$$

(in $\int u \, dv = uv - \int v \, du$, we let $u = \cos^{2r} \theta$, and $dv = \cos \theta \, d\theta$), which becomes (on using the identity $\sin^2 \theta = 1 - \cos^2 \theta$)

$$\int_{-\pi/2}^{\pi/2} \cos^{2r+1} \theta \, d\theta = 2r \int_{-\pi/2}^{\pi/2} \cos^{2r-1} \theta \, d\theta - 2r \int_{-\pi/2}^{\pi/2} \cos^{2r+1} \theta \, d\theta$$

for $r \geq 1$. Therefore

$$(2r + 1) \int_{-\pi/2}^{\pi/2} \cos^{2r+1} \theta \, d\theta = 2r \int_{-\pi/2}^{\pi/2} \cos^{2r-1} \theta \, d\theta. \tag{5.95}$$

Now, since [via (5.94)]

$$I_{r-1} = -(-1)^r \int_{-\pi/2}^{\pi/2} \cos^{2r-1} \theta \, d\theta,$$

Eq. (5.95) becomes

$$(2r + 1)(-1)^r I_r = -2r(-1)^r I_{r-1},$$

or more simply

$$I_r = -\frac{2r}{2r + 1} I_{r-1}. \tag{5.96}$$

This holds for $r \geq 1$ with initial condition $I_0 = 2$. The solution to the difference equation is

$$I_r = (-1)^r \frac{2^{2r+1}[r!]^2}{(2r + 1)!}. \tag{5.97}$$

This can be confirmed by direct substitution of (5.97) into (5.96). Consequently, if we combine (5.91), (5.93), and (5.97), then

$$||P_r||^2 = \frac{(-1)^2}{2^r} \cdot \frac{(2r)!}{2^r[r!]^2} \cdot \frac{(-1)^r 2^{2r+1}[r!]^2}{(2r + 1)!},$$

which simplifies to

$$||P_r||^2 = \frac{2}{2r + 1}. \tag{5.98}$$

A closed-form expression for $P_n(x)$ is possible using (5.92) in (5.89). Specifically, consider

$$P_n(x) = \frac{1}{2^n n!} \frac{d^n}{dx^n}(x^2 - 1)^n = \frac{1}{2^n n!} \frac{d^n}{dx^n} \left[\sum_{k=0}^{M} \binom{n}{k} (-1)^k x^{2n-2k} \right], \tag{5.99}$$

where $M = n/2$ (n even), or $M = (n - 1)/2$ (n odd). We observe that

$$\frac{d^n}{dx^n} x^{2n-2k} = (2n - 2k)(2n - 2k - 1) \cdots (n - 2k + 1)x^{n-2k}. \tag{5.100}$$

Now

$$(2n - 2k)! = (2n - 2k)(2n - 2k - 1) \cdots (n - 2k + 1) \underbrace{(n - 2k) \cdots 2 \cdot 1}_{=(n-2k)!},$$

so (5.100) becomes

$$\frac{d^n}{dx^n} x^{2n-2k} = \frac{(2n - 2k)!}{(n - 2k)!} x^{n-2k}. \tag{5.101}$$

Thus, (5.99) reduces to

$$P_n(x) = \frac{1}{2^n n!} \sum_{k=0}^{M} \binom{n}{k} (-1)^k \frac{(2n - 2k)!}{(n - 2k)!} x^{n-2k},$$

or alternatively

$$P_n(x) = \sum_{k=0}^{M} (-1)^k \frac{(2n-2k)!}{2^n k!(n-k)!(n-2k)!} x^{n-2k}. \tag{5.102}$$

Consider $f(x) = \frac{1}{\sqrt{1-x}}$ so that $f^{(k)}(x) = \frac{(2k-1)(2k-3)\cdots 3\cdot 1}{2^k}(1-x)^{-(2k+1)/2}$ (for $k \geq 1$). Define $(2k-1)!! = (2k-1)(2k-3)\cdots 3\cdot 1$, and define $(-1)!! = 1$. As usual, define $0! = 1$. We note that $(2n)! = 2^n n!(2n-1)!!$ which may be seen by considering

$$(2n)! = (2n)(2n-1)(2n-2)(2n-3)\cdots 3\cdot 2\cdot 1$$

$$= [(2n)(2n-2)\cdots 4\cdot 2][(2n-1)(2n-3)\cdots 3\cdot 1] = 2^n n!(2n-1)!!. \tag{5.103}$$

Consequently, the Maclaurin expansion for $f(x)$ is given by

$$f(x) = \frac{1}{\sqrt{1-x}} = \sum_{k=0}^{\infty} \frac{f^{(k)}(0)}{k!} x^k = \sum_{k=0}^{\infty} \frac{(2k)!}{2^{2k}[k!]^2} x^k. \tag{5.104}$$

The *ratio test* [7, p. 709] confirms that this series converges if $|x| < 1$. Using (5.102) and (5.104), it is possible to show that

$$S(x,t) = \frac{1}{\sqrt{1-2xt+t^2}} = \sum_{n=0}^{\infty} P_n(x)t^n, \tag{5.105}$$

so this is the *generating function* for the Legendre polynomials $P_n(x)$. We observe that

$$\frac{\partial S}{\partial t} = \frac{x-t}{[1-2xt+t^2]^{3/2}} = \frac{x-t}{1-2xt+t^2} S. \tag{5.106}$$

Also

$$\frac{\partial S}{\partial t} = \sum_{n=0}^{\infty} n P_n(x)t^{n-1}. \tag{5.107}$$

Equating (5.107) and (5.106), we have

$$\frac{x-t}{1-2xt+t^2} \sum_{n=0}^{\infty} P_n(x)t^n = \sum_{n=0}^{\infty} n P_n(x)t^{n-1},$$

which becomes

$$x \sum_{n=0}^{\infty} P_n(x)t^n - \sum_{n=0}^{\infty} P_n(x)t^{n+1} = \sum_{n=0}^{\infty} n P_n(x)t^{n-1} - 2x \sum_{n=0}^{\infty} n P_n(x)t^n$$

$$+ \sum_{n=0}^{\infty} n P_n(x)t^{n+1},$$

and if $P_{-1}(x) = 0$, then this becomes

$$x \sum_{n=0}^{\infty} P_n(x)t^n - \sum_{n=0}^{\infty} P_{n-1}(x)t^n = \sum_{n=0}^{\infty}(n+1)P_{n+1}(x)t^n - 2x \sum_{n=0}^{\infty} nP_n(x)t^n$$
$$+ \sum_{n=0}^{\infty}(n-1)P_{n-1}(x)t^n,$$

so on comparing like powers of t in this expression, we obtain

$$xP_n(x) - P_{n-1}(x) = (n+1)P_{n+1}(x) - 2xnP_n(x) + (n-1)P_{n-1}(x),$$

which finally yields

$$(n+1)P_{n+1}(x) = (2n+1)xP_n(x) - nP_{n-1}(x)$$

or (for $n \geq 1$)

$$P_{n+1}(x) = \frac{2n+1}{n+1}xP_n(x) - \frac{n}{n+1}P_{n-1}(x), \tag{5.108}$$

which is the three-term recurrence relation for the Legendre polynomials, and hence is yet another special case of (5.5).

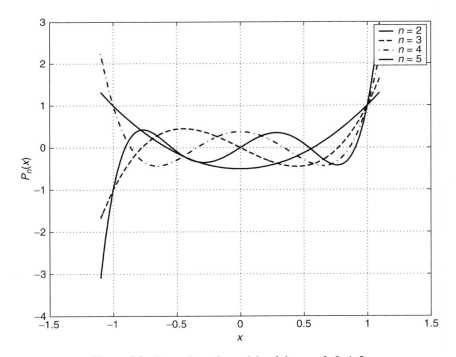

Figure 5.3 Legendre polynomials of degrees 2, 3, 4, 5.

The Legendre polynomials arise in potential problems in electromagnetics. Modeling the scattering of electromagnetic radiation by particles involves working with these polynomials. Legendre polynomials appear in quantum mechanics as part of the solution to Schrödinger's equation for the hydrogen atom. Some Legendre polynomials appear in Fig. 5.3.

5.6 AN EXAMPLE OF ORTHOGONAL POLYNOMIAL LEAST-SQUARES APPROXIMATION

Chebyshev polynomials of the first kind, and Legendre polynomials are both orthogonal on $[-1, 1]$, but their weighting functions are different. We shall illustrate the approximation behavior of these polynomials through an example wherein we wish to approximate

$$f(x) = \begin{cases} 0, & -1 \leq x < 0 \\ 1, & 0 \leq x \leq 1 \end{cases} \tag{5.109}$$

by both of these polynomial types. We will work with fifth-degree least-squares polynomial approximations in both cases.

We consider Legendre polynomial approximation first. We must therefore normalize the polynomials so that our basis functions have unit norm. Thus, our approximation will be

$$\hat{f}(x) = \sum_{k=0}^{5} \hat{a}_k \phi_k(x), \tag{5.110}$$

where

$$\phi_k(x) = \frac{1}{||P_k||} P_k(x),$$

and

$$\hat{a}_k = \langle f, \phi_k \rangle = \int_{-1}^{1} \frac{1}{||P_k||} f(x) P_k(x) \, dx,$$

for which we see that

$$\hat{f}(x) = \sum_{k=0}^{5} \underbrace{\left[\int_{-1}^{1} f(x) P_k(x) \, dx \right]}_{=\hat{\alpha}_k} \frac{P_k(x)}{||P_k||^2}. \tag{5.111}$$

We have [e.g., via (5.102)]

$$P_0(x) = 1, \ P_1(x) = x, \ P_2(x) = \frac{1}{2}[3x^2 - 1],$$

$$P_3(x) = \frac{1}{2}[5x^3 - 3x], \ P_4(x) = \frac{1}{8}[35x^4 - 30x^2 + 3],$$

$$P_5(x) = \frac{1}{8}[63x^5 - 70x^3 + 15x].$$

The squared norms [via (5.98)] are

$$||P_0||^2 = 2, \qquad ||P_1||^2 = \frac{2}{3}, \qquad ||P_2||^2 = \frac{2}{5}, \qquad ||P_3||^2 = \frac{2}{7},$$

$$||P_4||^2 = \frac{2}{9}, \qquad ||P_5||^2 = \frac{2}{11}.$$

By direct calculation, $\hat{\alpha}_k = \int_0^1 P_k(x)\,dx$ becomes

$$\hat{\alpha}_0 = \int_0^1 1 \cdot dx = 1, \hat{\alpha}_1 = \int_0^1 x\,dx = \frac{1}{2},$$

$$\hat{\alpha}_2 = \frac{1}{2}\int_0^1 [3x^2 - 1]\,dx = \frac{3}{2}\left[\frac{1}{3}x^3\right]_0^1 - \frac{1}{2}[x]_0^1 = 0,$$

$$\hat{\alpha}_3 = \frac{1}{2}\int_0^1 [5x^3 - 3x]\,dx = \frac{5}{2}\left[\frac{1}{4}x^4\right]_0^1 - \frac{3}{2}\left[\frac{1}{2}x^2\right]_0^1 = -\frac{1}{8},$$

$$\hat{\alpha}_4 = \frac{1}{8}\int_0^1 [35x^4 - 30x^2 + 3]\,dx = \frac{35}{8}\left[\frac{1}{5}x^5\right]_0^1 - \frac{30}{8}\left[\frac{1}{3}x^3\right]_0^1 + \frac{3}{8}[x]_0^1 = 0,$$

$$\hat{\alpha}_5 = \frac{1}{8}\int_0^1 [63x^5 - 70x^3 + 15x]\,dx$$

$$= \frac{63}{8}\left[\frac{1}{6}x^6\right]_0^1 - \frac{70}{8}\left[\frac{1}{4}x^4\right]_0^1 + \frac{15}{8}\left[\frac{1}{2}x^2\right]_0^1 = \frac{1}{16}.$$

The substitution of these (and the squared norms $||P_k||^2$) into (5.111) yields the least-squares Legendre polynomial approximation

$$\hat{f}(x) = \frac{1}{2}P_0(x) + \frac{3}{4}P_1(x) - \frac{7}{16}P_3(x) + \frac{11}{32}P_5(x). \tag{5.112}$$

We observe that $\hat{f}(0) = \frac{1}{2}$, $\hat{f}(1) = \frac{37}{32} = 1.15625$, and $\hat{f}(-1) = -\frac{5}{32} = -0.15625$.
Now we consider the Chebyshev polynomial approximation. In this case

$$\hat{f}(x) = \sum_{k=0}^{5} \hat{b}_k \phi_k(x), \tag{5.113}$$

where

$$\phi_k(x) = \frac{1}{||T_k||} T_k(x),$$

and

$$\hat{b}_k = \langle f, \phi_k \rangle = \int_{-1}^{1} \frac{1}{\sqrt{1 - x^2}} \frac{f(x)T_k(x)}{||T_k||}\,dx,$$

from which we see that

$$\hat{f}(x) = \sum_{k=0}^{5} \underbrace{\left[\int_{-1}^{1} \frac{1}{\sqrt{1-x^2}} f(x) T_k(x) \, dx\right]}_{=\hat{\beta}_k} \frac{1}{||T_k||^2} T_k(x). \qquad (5.114)$$

We have [e.g., via (5.57)] the polynomials

$$T_0(x) = 1, \qquad T_1(x) = x, \qquad T_2(x) = 2x^2 - 1,$$

$$T_3(x) = 4x^3 - 3x, \qquad T_4(x) = 8x^4 - 8x^2 + 1,$$

$$T_5(x) = 16x^5 - 20x^3 + 5x.$$

The squared norms [via (5.55)] are given by

$$||T_0||^2 = \pi, \; ||T_k||^2 = \frac{\pi}{2} \quad (k \geq 1).$$

By direct calculation $\hat{\beta}_k = \int_0^1 \frac{1}{\sqrt{1-x^2}} T_k(x) \, dx$ becomes (using $x = \cos\theta$, and $T_k(\cos\theta) = \cos(k\theta)$ [recall (5.53)] $\hat{\beta}_k = \int_0^{\pi/2} \cos(k\theta) \, d\theta$, and hence

$$\hat{\beta}_0 = \int_0^{\pi/2} 1 \cdot d\theta = \frac{\pi}{2},$$

$$\hat{\beta}_1 = \int_0^{\pi/2} \cos\theta \, d\theta = [\sin\theta]_0^{\pi/2} = 1,$$

$$\hat{\beta}_2 = \int_0^{\pi/2} \cos(2\theta) \, d\theta = \left[\frac{1}{2}\sin(2\theta)\right]_0^{\pi/2} = 0,$$

$$\hat{\beta}_3 = \int_0^{\pi/2} \cos(3\theta) \, d\theta = \left[\frac{1}{3}\sin(3\theta)\right]_0^{\pi/2} = -\frac{1}{3},$$

$$\hat{\beta}_4 = \int_0^{\pi/2} \cos(4\theta) \, d\theta = \left[\frac{1}{4}\sin(4\theta)\right]_0^{\pi/2} = 0,$$

$$\hat{\beta}_5 = \int_0^{\pi/2} \cos(5\theta) \, d\theta = \left[\frac{1}{5}\sin(5\theta)\right]_0^{\pi/2} = \frac{1}{5}.$$

Substituting these (and the squared norms $||T_k||^2$) into (5.114) yields the least-squares Chebyshev polynomial approximation

$$\hat{f}(x) = \frac{1}{\pi}\left[\frac{\pi}{2}T_0(x) + 2T_1(x) - \frac{2}{3}T_3(x) + \frac{2}{5}T_5(x)\right]. \qquad (5.115)$$

We observe that $\hat{f}(0) = \frac{1}{2}$, $\hat{f}(1) = 1.051737$, and $\hat{f}(-1) = -0.051737$.

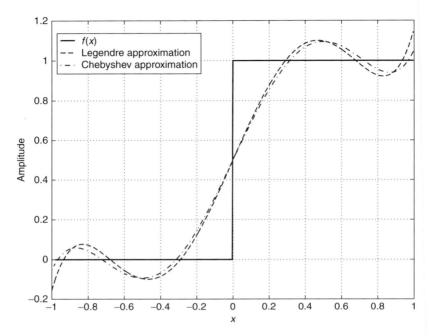

Figure 5.4 Plots of $f(x)$ from (5.109), the Legendre approximation (5.112), and the Chebyshev approximation (5.115). The least-squares approximations are fifth-degree polynomials in both cases.

Plots of $f(x)$ from (5.109) and the approximations in (5.112) and (5.115) appear in Fig. 5.4. Both polynomial approximations are fifth-degree polynomials, and yet they look fairly different. This is so because the weighting functions are different. The Chebyshev approximation is better than the Legendre approximation near $x = \pm 1$ because greater weight is given to errors near the ends of the interval $[-1, 1]$.

5.7 UNIFORM APPROXIMATION

The subject of uniform approximation is distinctly more complicated than that of least-squares approximation. So, we will not devote too much space to it in this book. We will concentrate on a relatively simple illustration of the main idea.

Recall the space $C[a, b]$ from Chapter 1. We recall that it is a normed space with the norm

$$||x|| = \sup_{t \in [a,b]} |x(t)| \tag{5.116}$$

for any $x(t) \in C[a, b]$. In fact, further recalling Chapter 3, this space happens to be a complete metric space for which the metric induced by (5.116) is

$$d(x, y) = \sup_{t \in [a,b]} |x(t) - y(t)|, \tag{5.117}$$

where $x(t), y(t) \in C[a, b]$. In *uniform approximation* we (for example) may wish to approximate $f(t) \in C[a, b]$ with $x(t) \in \mathbf{P}^n[a, b] \subset C[a, b]$ such that $||f - x||$ is minimized. The norm (5.116) is sometimes called the *Chebyshev norm*, and so our problem is sometimes called the *Chebyshev approximation problem*. The error $e(t) = f(t) - x(t)$ has the norm

$$||e|| = \sup_{t \in [a,b]} |f(t) - x(t)| \qquad (5.118)$$

and is the maximum deviation between $f(t)$ and $x(t)$ on $[a, b]$. We wish to find $x(t)$ to minimize this. Consequently, uniform approximation is sometimes also called *minimax approximation* as we wish to *mini*mize the *max*imum deviation between $f(t)$ and its approximation $x(t)$. We remark that because $f(t)$ is continuous (by definition) it will have a well-defined maximum on $[a, b]$ (although this maximum need not be at a unique location). We are therefore at liberty to replace "sup" by "max" in (5.116), (5.117) and (5.118) if we wish.

Suppose that $y_j(t) \in C[a, b]$ for $j = 0, 1, \ldots, n - 1$ and the set of functions $\{y_j(t) | j \in \mathbf{Z}_n\}$ are an independent set. This set generates an n-dimensional *sub-space*[4] of $C[a, b]$ that may be denoted by $Y = \left\{ \sum_{j=0}^{n-1} \alpha_j y_j(t) | \alpha_j \in \mathbf{R} \right\}$. From Kreyszig [8, p. 337] consider

Definition 5.1: Haar Condition   Subspace $Y \subset C[a, b]$ satisfies the *Haar condition* if every $y \in Y$ ($y \neq 0$) has at most $n - 1$ zeros in $[a, b]$, where $n = \dim(Y)$ (dimension of the subspace Y).

We may select $y_j(t) = t^j$ for $j \in \mathbf{Z}_n$ so any $y(t) \in Y$ has the form $y(t) = \sum_{j=0}^{n-1} \alpha_j t^j$ and thus is a polynomial of degree at most $n - 1$. A degree $n - 1$ polynomial has $n - 1$ zeros, and so such a subspace Y satisfies the Haar condition.

Definition 5.2: Alternating Set   Let $x \in C[a, b]$, and $y \in Y$, where Y is any subspace of $C[a, b]$. A set of points $t_0, \ldots, t_k$ in $[a, b]$ such that $t_0 < t_1 < \cdots < t_k$ is called an *alternating set* for $x - y$ if $x(t_j) - y(t_j)$ has alternately the values $+||x - y||$, and $-||x - y||$ at consecutive points t_j.

Thus, suppose $x(t_j) - y(t_j) = +||x - y||$, then $x(t_{j+1}) - y(t_{j+1}) = -||x - y||$, but if instead $x(t_j) - y(t_j) = -||x - y||$, then $x(t_{j+1}) - y(t_{j+1}) = +||x - y||$. The norm is, of course, that in (5.116).

Lemma 5.2: Best Approximation   Let Y be any subspace of $C[a, b]$ that satisfies the Haar condition. Given $f \in C[a, b]$, let $y \in Y$ be such that for $f - y$ there exists an alternating set of $n + 1$ points, where $n = \dim(Y)$, then y is the best uniform approximation of f out of Y.

The proof is omitted, but may be found in Kreyszig [8, pp. 345–346].

---

[4]A subspace of a vector space X is a nonempty subset Y of X such that for any $y_1, y_2 \in Y$, and all scalars a, b from the field of the vector space we have $ay_1 + by_2 \in Y$.

Consider the particular case of $C[-1, 1]$ with $f(t) \in C[-1, 1]$ such that for a given $n \in \mathbf{N}$

$$f(t) = t^n. \tag{5.119}$$

Now $Y = \left\{ \sum_{j=0}^{n-1} \alpha_j t^j \,|\, \alpha_j \in \mathbf{R} \right\}$, so $y_j(t) = t^j$ for $j \in \mathbf{Z}_n$. We wish to select α_j such that the error $e = f - y$

$$e(t) = f(t) - \sum_{j=0}^{n-1} \alpha_j t^j \tag{5.120}$$

is minimized with respect to the Chebyshev norm (5.116); that is, select α_j to minimize $||e||$. Clearly, $\dim(Y) = n$. According to Lemma 5.2 $||e||$ is minimized if $e(t)$ in (5.120) has an alternating set of $n + 1$ points.

Recall Lemma 5.1 (Section 5.3), which stated [see (5.44)] that

$$\cos n\theta = \sum_{k=0}^{n} \beta_{n,k} \cos^k \theta. \tag{5.121}$$

From (5.46) $\beta_{n+1,n+1} = 2\beta_{n,n}$, and $\beta_{0,0} = 1$. Consequently, $\beta_{n,n} = 2^{n-1}$ (for $n \geq 1$). Thus, (5.121) can be rewritten as

$$\cos n\theta = 2^{n-1} \cos^n \theta + \sum_{j=0}^{n-1} \beta_{n,j} \cos^j \theta \tag{5.122}$$

($n \geq 1$). Suppose that $t = \cos\theta$, so $\theta \in [0, \pi]$ maps to $t \in [-1, 1]$ and from (5.122)

$$\cos[n \cos^{-1} t] = 2^{n-1} t^n + \sum_{j=0}^{n-1} \beta_{n,j} t^j. \tag{5.123}$$

We observe that $\cos n\theta$ has an alternating set of $n + 1$ points $\theta_0, \theta_1, \ldots, \theta_n$ on $[0, \pi]$ for which $\cos n\theta_k = \pm 1$ (clearly $||\cos n\theta|| = 1$). For example, if $n = 1$, then

$$\theta_0 = 0, \qquad \theta_1 = \pi.$$

If $n = 2$, then

$$\theta_0 = 0, \qquad \theta_1 = \pi/2, \qquad \theta_2 = \pi,$$

and if $n = 3$, then

$$\theta_0 = 0, \qquad \theta_1 = \frac{\pi}{3}, \qquad \theta_2 = \frac{2\pi}{3}, \qquad \theta_3 = \pi.$$

In general, $\theta_k = \frac{k}{n}\pi$ for $k = 0, 1, \ldots, n$. Thus, if $t_k = \cos\theta_k$ then $t_k = \cos\left(\frac{k}{n}\pi\right)$. We may rewrite (5.123) as

$$\frac{1}{2^{n-1}} \cos[n \cos^{-1} t] = t^n + \frac{1}{2^{n-1}} \sum_{j=0}^{n-1} \beta_{n,j} t^j. \tag{5.124}$$

This is identical to $e(t)$ in (5.120) if we set $\beta_{n,j} = -2^{n-1}\alpha_j$. In other words, if we choose

$$e(t) = \frac{1}{2^{n-1}} \cos[n \cos^{-1} t], \qquad (5.125)$$

then $||e||$ is minimized since we know that e has an alternating set $t = t_k = \cos\left(\frac{k}{n}\pi\right)$, $k = 0, 1, \ldots, n$, and $t_k \in [-1, 1]$. We recall from Section 5.3 that $T_k(t) = \cos[k \cos^{-1} t]$ is the kth-degree Chebyshev polynomial of the first kind [see (5.53)]. So, $e(t) = T_n(t)/2^{n-1}$. Knowing this, we may readily determine the optimal coefficients α_j in (5.120) if we so desire.

Thus, the Chebyshev polynomials of the first kind determine the best degree $n - 1$ polynomial uniform approximation to $f(t) = t^n$ on the interval $t \in [-1, 1]$.

REFERENCES

1. E. Isaacson and H. B. Keller, *Analysis of Numerical Methods*, Wiley, New York, 1966.

2. F. B. Hildebrand, *Introduction to Numerical Analysis*, 2nd ed., McGraw-Hill, New York, 1974.

3. J. S. Lim and A. V. Oppenheim, eds., *Advanced Topics in Signal Processing*. Prentice-Hall, Englewood Cliffs, NJ, 1988.

4. P. J. Davis and P. Rabinowitz, *Numerical Integration*, Blaisdell, Waltham, MA, 1967.

5. J. R. Rice, *The Approximation of Functions*, Vol. I: *Linear Theory*, Addison-Wesley, Reading, MA, 1964.

6. G. Szegö, *Orthogonal Polynomials*, 3rd ed., American Mathematical Society, 1967.

7. L. Bers, *Calculus: Preliminary Edition*, Vol. 2, Holt, Rinehart, Winston, New York, 1967.

8. E. Kreyszig, *Introductory Functional Analysis with Applications*, Wiley, New York, 1978.

PROBLEMS

5.1. Suppose that $\{\phi_k(x)|k \in \mathbf{Z}_N\}$ are orthogonal polynomials on the interval $D = [a, b] \subset \mathbf{R}$ with respect to some weighting function $w(x) \geq 0$. Show that

$$\sum_{k=0}^{N-1} a_k\phi_k(x) = 0$$

holds for all $x \in [a, b]$ iff $a_k = 0$ for all $k \in \mathbf{Z}_N$ [i.e., prove that $\{\phi_k(x)|k \in \mathbf{Z}_N\}$ is an independent set].

5.2. Verify (5.11) by direct calculation for $\phi_k(x) = T_k(x)/||T_k||$ with $n = 2$; that is, find $\phi_0(x)$, $\phi_1(x)$, $\phi_2(x)$, and $\phi_3(x)$, and verify that the left- and right-hand sides of (5.11) are equal to each other.

5.3. Suppose that $\{\phi_n(x)|n \in \mathbf{Z}^+\}$ are orthogonal polynomials on the interval $[a, b]$ with respect to some weighting function $w(x) \geq 0$. Prove the following theorem: The roots x_j $(j = 1, 2, \ldots, n)$ of $\phi_n(x) = 0$ $(n \in \mathbf{N})$ are all real-valued, simple, and $a < x_j < b$ for all j.

5.4. Recall that $T_j(x)$ is the degree j Chebyshev polynomial of the first kind. Recall from Lemma 5.1 that $\cos(n\theta) = \sum_{k=0}^{n} \beta_{n,k} \cos^k \theta$. For $x \in [-1, 1]$ with $k \in \mathbf{Z}^+$, we can find coefficients a_j such that

$$x^k = \sum_{j=0}^{k} a_j T_j(x).$$

Prove that

$$a_j = \frac{1}{||T_j||^2} \sum_{r=0}^{j} \beta_{j,r} \int_0^{\pi} \cos^{k+r} \theta \, d\theta.$$

Is this the best way to find the coefficients a_j ? If not, specify an alternative approach.

5.5. This problem is about the converse to Lemma 5.1. Suppose that we have $p(\cos \theta) = \sum_{k=0}^{3} a_{3,k} \cos^k \theta$. We wish to find coefficients $b_{3,j}$ such that

$$p(\cos \theta) = \sum_{j=0}^{3} b_{3,j} \cos(j\theta).$$

Use Lemma 5.1 to show that $a = Ub$, where

$$a = [a_{3,0} \ a_{3,1} \ a_{3,2} \ a_{3,3}]^T, b = [b_{3,0} \ b_{3,1} \ b_{3,2} \ b_{3,3}]^T$$

and U is an upper triangular matrix containing the coefficients $\beta_{m,k}$ for $m = 0, 1, 2, 3$, and $k = 0, 1 \ldots, m$. Of course, we may use back-substitution to solve $Ub = a$ for b if this were desired.

5.6. Suppose that for $n \in \mathbf{Z}^+$ we are given $p(\cos \theta) = \sum_{k=0}^{n} a_{n,k} \cos^k \theta$. Show how to find the coefficients $b_{n,j}$ such that

$$p(\cos \theta) = \sum_{j=0}^{n} b_{n,j} \cos(j\theta),$$

that is, generalize the previous problem from $n = 3$ to any n.

5.7. Recall Section 5.6, where

$$f(x) = \begin{cases} 0, & -1 \leq x < 0 \\ 1, & 0 \leq x \leq 1 \end{cases}$$

was approximated by

$$\hat{f}(x) = \sum_{k=0}^{n} \hat{b}_k \phi_k(x), \quad \phi_k(x) = \frac{1}{||T_k||} T_k(x)$$

for $n = 5$. Find a general expression for $\hat{b}_k$ for all $k \in \mathbf{Z}^+$. Use the resulting expansion

$$f(x) = \sum_{k=0}^{\infty} \frac{\hat{b}_k}{||T_k||} T_k(x)$$

to prove that

$$\frac{\pi}{4} = \sum_{n=1}^{\infty} \frac{1}{2n-1}(-1)^{n-1}.$$

5.8. Suppose that

$$f(x) = \begin{cases} 0, & -1 \le x < 0 \\ x, & 0 \le x \le 1 \end{cases},$$

then find $\hat{b}_k$ in

$$f(x) = \sum_{k=0}^{\infty} \hat{b}_k \frac{1}{||T_k||} T_k(x).$$

5.9. Do the following:

(a) Solve the polynomial equation $T_k(x) = 0$ for all $k > 0$, and so find all the zeros of the polynomials $T_k(x)$.

(b) Show that $T_n(x)$ satisfies the differential equation

$$(1 - x^2)T_n^{(2)}(x) - x T_n^{(1)}(x) + n^2 T_n(x) = 0.$$

Recall that $T_n^{(r)}(x) = d^r T_n(x)/dx^r$.

5.10. The three-term recurrence relation for the *Chebyshev polynomials of the second kind* is

$$U_{r+1}(x) = 2x U_r(x) - U_{r-1}(x), \tag{5.P.1}$$

where $r \ge 1$, and where the initial conditions are

$$U_0(x) = 1, \quad U_1(x) = 2x. \tag{5.P.2}$$

We remark that (5.P.1) is identical to (5.57). Specifically, the recursion for Chebyshev polynomials of both kinds is the same, except that the initial conditions in (5.P.2) are not the same for both. [From (5.56) $T_0(x) = 1$, but $T_1(x) = x$.] Since $\deg(U_r) = r$, we have (for $r \geq 0$)

$$U_r(x) = \sum_{j=0}^{r} U_{r,j} x^j. \tag{5.P.3}$$

[Recall the notation for $\phi_r(x)$ in (5.3).] From the polynomial recursion in (5.P.1) we may obtain

$$U_{r+1,j} = 2U_{r,j-1} - U_{r-1,j}. \tag{5.P.4a}$$

This expression holds for $r \geq 1$, and for $j = 0, 1, \ldots, r, r + 1$. From (5.P.2) the initial conditions for (5.P.4a) are

$$U_{0,0} = 1, U_{1,0} = 0, U_{1,1} = 2. \tag{5.P.4b}$$

Write a MATLAB function that uses (5.P.4) to generate the Chebyshev polynomials of the second kind for $r = 0, 1, \ldots, N$. Test your program out for $N = 8$. Program output should be in the form of a table that is written to a file. The tabular format should be something like

```
degree   0    1    2    3    4    5    6    7    8
  0      1    0    0    0    0    0    0    0    0
  1      0    2    0    0    0    0    0    0    0
  2     -1    0    4    0    0    0    0    0    0
  3      0   -4    0    8    0    0    0    0    0

etc.
```

5.11. The Chebyshev polynomials of the second kind use

$$D = [-1, 1], w(x) = \sqrt{1 - x^2}. \tag{5.P.5}$$

Denote these polynomials as the set $\{\phi_n(x) | n \in \mathbf{Z}^+\}$. (This notation applies if the polynomials are normalized to possess unity-valued norms.) Of course, in our function space, we use the inner product

$$\langle f, g \rangle = \int_{-1}^{1} \sqrt{1 - x^2} f(x) g(x) \, dx. \tag{5.P.6}$$

Derive the polynomials $\{\phi_n(x) | n \in \mathbf{Z}^+\}$. (*Hint:* The process is much the same as the derivation of the Chebyshev polynomials of the first kind presented in Section 5.3.) Therefore, begin by considering $q_{r-1}(x)$ which is any polynomial of degree not more than $r - 1$. Thus, for suitable $c_k \in \mathbf{R}$, we must have $q_{r-1}(x) = \sum_{j=0}^{r-1} c_j \phi_j(x)$, and $\langle \phi_r, q_{r-1} \rangle = \sum_{j=0}^{r-1} c_j \langle \phi_r, \phi_j \rangle = 0$ because $\langle \phi_r, \phi_j \rangle = 0$ for $j = 0, 1, \ldots, r-1$. In expanded form

$$\langle \phi_r, q_{r-1} \rangle = \int_{-1}^{1} \sqrt{1 - x^2} \phi_r(x) q_{r-1}(x) \, dx = 0. \tag{5.P.7}$$

Use the change of variable $x = \cos\theta$ (so $dx = -\sin\theta \, d\theta$) to reduce (5.P.7) to

$$\int_0^{\pi} \sin^2\theta \cos(k\theta) \phi_r(\cos\theta) \, d\theta = 0 \tag{5.P.8}$$

for $k = 0, 1, 2, \ldots, r-1$, where Lemma 5.1 and its converse have also been employed. Next consider the candidate

$$\phi_r(\cos\theta) = C_r \frac{\sin(r+1)\theta}{\sin\theta}, \tag{5.P.9}$$

and verify that this satisfies (5.P.8). [Hence (5.P.9) satisfies (5.P.7).] Show that (5.P.9) becomes

$$\phi_r(x) = C_r \frac{\sin[(r+1)\cos^{-1}x]}{\sin\cos^{-1}x}.$$

Prove that for $C_r = 1$ (all $r \in \mathbf{Z}^+$)

$$\phi_{r+1}(x) = 2x\phi_r(x) - \phi_{r-1}(x).$$

In this case we normally use the notation $U_r(x) = \phi_r(x)$. Verify that $U_0(x) = 1$, and that $U_1(x) = 2x$. Prove that $\|U_n\|^2 = \frac{\pi}{2}$ for $n \in \mathbf{Z}^+$. Finally, of course, $\phi_n(x) = U_n(x)/\|U_n\|$.

5.12. Write a MATLAB function to produce plots of $U_k(x)$ for $k = 2, 3, 4, 5$ similar to Fig. 5.1.

5.13. For $\phi_k(x) = U_k(x)/\|U_k\|$, we have

$$\phi_k(x) = \sqrt{\frac{2}{\pi}} U_k(x).$$

Since $\langle \phi_k, \phi_j \rangle = \delta_{k-j}$, for any $f(x) \in L^2[-1, 1]$, we have the series expansion

$$f(x) = \sum_{k=0}^{\infty} \hat{a}_k \phi_k(x),$$

where

$$\hat{a}_k = \langle f, \phi_k \rangle = \frac{1}{\|U_k\|} \int_{-1}^{1} \sqrt{1 - x^2} f(x) U_k(x) \, dx.$$

Suppose that we work with the following function:

$$f(x) = \begin{cases} 0, & -1 \le x < 0 \\ 1, & 0 \le x \le 1 \end{cases}.$$

(a) Find a nice general formula for the elements of the sequence $(\hat{a}_k)$ $(k \in \mathbf{Z}^+)$.

(b) Use MATLAB to plot the approximation

$$\hat{f}_2(x) = \sum_{k=0}^{5} \hat{a}_k \phi_k(x)$$

on the same graph as that of $f(x)$ (i.e., create a plot similar to that of Fig. 5.4). Suppose that $\psi_k(x) = T_k(x)/\|T_k\|$; then another approximation to $f(x)$ is given by

$$\hat{f}_1(x) = \sum_{k=0}^{5} \hat{b}_k \psi_k(x).$$

Plot $\hat{f}_1(x)$ on the same graph as $\hat{f}_2(x)$ and $f(x)$.

(c) Compare the accuracy of the approximations $\hat{f}_1(x)$ and $\hat{f}_2(x)$ to $f(x)$ near the endpoints $x = \pm 1$. Which approximation is better near these endpoints ? Explain why if you can.

5.14. Prove the following:

(a) $T_n(x)$, and $T_{n-1}(x)$ have no zeros in common.

(b) Between any two neighboring zeros of $T_n(x)$, there is precisely one zero of $T_{n-1}(x)$. This is called the *interleaving of zeros* property.

(*Comment:* The interleaving of zeros property is possessed by all orthogonal polynomials. When this property is combined with ideas from later chapters, it can be used to provide algorithms to find the zeros of orthogonal polynomials in general.)

5.15. (a) Show that we can write

$$T_n(x) = \cos[n \cos^{-1} x] = \cosh[n \cosh^{-1} x].$$

[*Hint:* Note that $\cos x = \frac{1}{2}(e^{jx} + e^{-jx})$, $\cosh x = \frac{1}{2}(e^x + e^{-x})$, so that $\cos x = \cosh(jx)$.]

(b) Prove that $T_{2n}(x) = T_n(2x^2 - 1)$. [*Hint:* $\cos(2x) = 2\cos^2 x - 1$.]

5.16. Use Eq. (5.63) in the following problems:

(a) Show that

$$\frac{d}{dx} H_r(x) = 2r H_{r-1}(x).$$

(b) Show that

$$\frac{d}{dx}\left[e^{-x^2}\frac{d}{dx}H_r(x)\right] = -2re^{-x^2}H_r(x).$$

(c) From the preceding confirm that $H_r(x)$ satisfies the *Hermite differential equation*

$$H_r^{(2)}(x) - 2xH_r^{(1)}(x) + 2rH_r(x) = 0.$$

5.17. Find $H_k(x)$ for $k = 0, 1, 2, 3, 4, 5$ (i.e., find the first six Hermite polynomials) using (5.83).

5.18. Using (5.63) prove that

$$H_k(x) = n!\sum_{j=0}^{N}(-1)^j\frac{2^{n-2j}}{j!(n-2j)!}x^{n-2j},$$

where $N = n/2$ (n is even), $N = (n-1)/2$ (n is odd).

5.19. Suppose that $P_k(x)$ is the Legendre polynomial of degree k. Recall Eq. (5.90). Find constants α and β such that for $Q_k(x) = P_k(\alpha x + \beta)$ we have (for $k \neq r$)

$$\int_a^b Q_r(x)Q_k(x)\,dx = 0.$$

[*Comment:* This linear transformation of variable allows us to least-squares approximate $f(x)$ using Legendre polynomial series on any interval $[a, b]$.]

5.20. Recall Eq. (5.105):

$$\frac{1}{\sqrt{1 - 2xt + t^2}} = \sum_{n=0}^{\infty}P_n(x)t^n.$$

Verify the terms $n = 0, 1, 2, 3$.

[*Hint:* Recall Eq. (3.82) (From Chapter 3).]

5.21. The distance between two points A and B in $\mathbf{R}^3$ is r, while the distance from A to the origin O is r_1, and the distance from B to the origin O is r_2. The angle between the vector $\overrightarrow{OA}$, and vector $\overrightarrow{OB}$ is θ.

(a) Show that

$$\frac{1}{r} = \frac{1}{\sqrt{r_1^2 + r_2^2 - 2r_1r_2\cos\theta}}.$$

[*Hint:* Recall the law of cosines (Section 4.6).]

(b) Show that

$$\frac{1}{r} = \frac{1}{r_2} \sum_{n=0}^{\infty} P_n(\cos\theta) \left(\frac{r_1}{r_2}\right)^n,$$

where, of course, $P_n(x)$ is the Legendre polynomial of degree n.

(*Comment:* This result is important in electromagnetic potential theory.)

5.22. Recall Section 5.7. Write a MATLAB function to plot on the same graph both $f(t) = t^n$ and $e(t)$ [in (5.125)] for each of $n = 2, 3, 4, 5$. You must generate four separate plots, one for each instance of n.

5.23. The set of points $C = \{e^{j\theta} | \theta \in \mathbf{R}, j = \sqrt{-1}\}$ is the *unit circle* of the complex plane. If $R(e^{j\theta}) \geq 0$ for all $\theta \in \mathbf{R}$, then we may define an inner product on a suitable space of functions that are defined on C

$$\langle F, G \rangle = \frac{1}{2\pi} \int_{-\pi}^{\pi} R(e^{j\theta}) F^*(e^{j\theta}) G(e^{j\theta}) \, d\theta, \tag{5.P.10}$$

where, in general, $F(e^{j\theta}), G(e^{j\theta}) \in \mathbf{C}$. Since $e^{j\theta}$ is 2π-periodic in θ, the integration limits in (5.P.10) are $-\pi$ to π, but another standard choice is from 0 to 2π. The function $R(e^{j\theta})$ is a weighting function for the inner product in (5.P.10). For $R(e^{j\theta})$, there will be a real-valued sequence (r_k) such that

$$R(e^{j\theta}) = \sum_{k=-\infty}^{\infty} r_k e^{-jk\theta},$$

and we also have $r_{-k} = r_k$ [i.e., (r_k) is a *symmetric sequence*]. For $F(e^{j\theta})$ and $G(e^{j\theta})$, we have real-valued sequences (f_k), and (g_k) such that

$$F(e^{j\theta}) = \sum_{k=-\infty}^{\infty} f_k e^{-jk\theta}, G(e^{j\theta}) = \sum_{k=-\infty}^{\infty} g_k e^{-jk\theta}.$$

(a) Show that $\langle e^{-jn\theta}, e^{-jm\theta} \rangle = r_{n-m}$.
(b) Show that $\langle e^{-jn\theta} F(e^{j\theta}), e^{-jn\theta} G(e^{j\theta}) \rangle = \langle F(e^{j\theta}), G(e^{j\theta}) \rangle$.
(c) Show that $\langle F(e^{j\theta}), G(e^{j\theta}) \rangle = \langle 1, F(e^{-j\theta}) G(e^{j\theta}) \rangle$.

[*Comment:* The unit circle C is of central importance in the theory of stability of linear time-invariant (LTI) discrete-time systems, and so appears in the subjects of digital control and digital signal processing.]

5.24. Recall Problem 4.21 from and the previous problem (5.23). Given $a_n(z) = \sum_{k=0}^{n-1} a_{n,k} z^{-k}$ (and $z = e^{j\theta}$), show that

$$\langle a_n(z), z^{-k} \rangle = \sigma_n^2 \delta_k$$

for $k = 0, 1, \ldots, n - 1$.

[*Comment*: This result ultimately leads to an alternative derivation of the Levinson–Durbin algorithm, and suggests that this algorithm actually generates a sequence of orthogonal polynomials on the unit circle C. These polynomials are in the indeterminate z^{-1} (instead of x).]

5.25. It is possible to construct orthogonal polynomials on *discrete sets*. This problem is about a particular example of this. Suppose that

$$\phi_r[n] = \sum_{j=0}^{r} \phi_{r,j} n^j,$$

where $n \in [-L, U] \subset \mathbf{Z}$, $U \geq L \geq 0$ and $\phi_{r,r} \neq 0$, so that $\deg(\phi_r) = r$, and let us also assume that $\phi_{r,j} \in \mathbf{R}$ for all r, and j. We say that $\{\phi_r[n]|n \in \mathbf{Z}^+\}$ is an orthogonal set if $\langle \phi_k[n], \phi_m[n] \rangle = ||\phi_k||^2 \delta_{k-m}$, where the inner product is defined by

$$\langle f[n], g[n] \rangle = \sum_{n=-L}^{U} w[n] f[n] g[n], \tag{5.P.11}$$

and $w[n] \geq 0$ for all $n \in [-L, U]$ is a *weighting sequence* for our inner product space. (Of course, $||f||^2 = \langle f[n], f[n] \rangle$.) Suppose that $L = U = M$; then, for $w[n] = 1$ (all $n \in [-M, M]$), it can be shown (with much effort) that the *Gram polynomials* are given by the three-term recurrence relation

$$p_{k+1}[n] = \frac{2(2k+1)}{(k+1)(2M-k)} n p_k[n] - \frac{k}{k+1} \frac{2M+k+1}{2M-k} p_{k-1}[n], \tag{5.P.12}$$

where $p_0[n] = 1$, and $p_1[n] = n/M$.

(a) Use (5.P.12) to find $p_k[n]$ for $k = 2, 3, 4$, where $M = 2$.

(b) Use (5.P.12) to find $p_k[n]$ for $k = 2, 3, 4$, where $M = 3$.

(c) Use (5.P.11) to find $||p_k||^2$ for $k = 2, 3, 4$, where $M = 2$.

(d) Use (5.P.11) to find $||p_k||^2$ for $k = 2, 3, 4$, where $M = 3$.

(*Comment:* The uniform weighting function $w[n] = 1$ makes the Gram polynomials the discrete version of the Legendre polynomials. The Gram polynomials were actually invented by Chebyshev.)

5.26. Integration by parts is clearly quite important in analysis (e.g., recall Section 3.6). To derive the Gram polynomials (previous problem) makes use of *summation by parts*.

Suppose that $v[n]$, $u[n]$, and $f[n]$ are defined on $\mathbf{Z}$. Define the *forward difference operator* Δ according to

$$\Delta f[n] = f[n+1] - f[n]$$

[for any sequence $(f[n])$]. Prove the expression for summation by parts, which is

$$\sum_{n=-L}^{U} u[n]\Delta v[n] = u[n]v[n]\Big|_{-L}^{U+1} - \sum_{n=-L}^{U} v[n+1]\Delta u[n].$$

[*Hint*: Show that

$$u[n]\Delta v[n] = \Delta u[n]v[n] - v[n+1]\Delta u[n],$$

and then consider using the identity

$$\sum_{n=-L}^{U} \Delta f[n] = \sum_{n=-L}^{U} (f[n+1] - f[n]) = f[n]\Big|_{-L}^{U+1}.$$

Of course, $f[n]\Big|_{A}^{B} = f[B] - f[A]$.]

6 Interpolation

6.1 INTRODUCTION

Suppose that we have the data $\{(t_k, x(t_k)) | k \in \mathbf{Z}_{n+1}\}$, perhaps obtained experimentally. An example of this appeared in Section 4.6 . In this case we assumed that $t_k = kT_s$ for which $x(t_k) = x(t)|_{t=kT_s}$ are the samples of some analog signal. In this example these time samples were of simulated (and highly oversimplified) physiological data for human patients (e.g., blood pressure, heart rate, body core temperature). Our problem involved assuming a model for the data; thus, assuming $x(t)$ is explained by a particular mathematical function with certain unknown parameters to be estimated on the basis of the model and the data. In other words, we estimate $x(t)$ with $\hat{x}(t, \alpha)$, where α is the vector of unknown parameters (the model parameters to be estimated), and we chose α to minimize the error

$$e(t_k) = x(t_k) - \hat{x}(t_k, \alpha), \quad k \in \mathbf{Z}_{n+1} \tag{6.1}$$

according to some criterion. We have emphasized choosing α to minimize

$$V(\alpha) = \sum_{k=0}^{n} e^2(t_k). \tag{6.2}$$

This was the least-squares approach. However, the idea of choosing α to minimize $\max_{k \in \mathbf{Z}_{n+1}} |e(t_k)|$ is an alternative suggested by Section 5.7. Other choices are possible. However, no matter what choice we make, in all cases $\hat{x}(t_k, \alpha)$ is not necessarily exactly equal to $x(t_k)$ except perhaps by chance. The problem of finding $\hat{x}(t, \alpha)$ to minimize $e(t)$ in this manner is often called *curve fitting*. It is to be distinguished from *interpolation*, which may be defined as follows.

Usually we assume $t_0 < t_1 < \cdots < t_{n-1} < t_n$ with $t_0 = a$, and $t_n = b$ so that $t_k \in [a, b] \subset \mathbf{R}$. To *interpolate* the data $\{(t_k, x(t_k)) | k \in \mathbf{Z}_{n+1}\}$, we seek a function $p(t)$ such that $t \in [a, b]$, and

$$p(t_k) = x(t_k) \tag{6.3}$$

for all $k \in \mathbf{Z}_{n+1}$. We might know something about the properties of $x(t)$ for $t \neq t_k$ on interval $[a, b]$, and so we might select $p(t)$ to possess similar properties.

An Introduction to Numerical Analysis for Electrical and Computer Engineers, by C.J. Zarowski
ISBN 0-471-46737-5 © 2004 John Wiley & Sons, Inc.

However, we emphasize that the *interpolating function p(t) exactly matches x(t)* at the given sample points $t_k, k \in \mathbf{Z}_{n+1}$.

Curve fitting is used when the data are uncertain because of the corrupting effects of measurement errors, random noise, or interference. Interpolation is appropriate when the data are accurately or exactly known.

Interpolation is quite important in digital signal processing. For example, bandlimited signals may need to be interpolated in order to change sampling rates. Interpolation is vital in numerical integration methods, as will be seen later. In this application the integrand is typically known or can be readily found at some finite set of points with significant accuracy. Interpolation at these points leads to a function (usually a polynomial) that can be easily integrated, and so provides a useful approximation to the given integral.

This chapter discusses interpolation with polynomials only. In principle it is possible to interpolate using other functions (rational functions, trigonometric functions, etc.) But these other approaches are usually more involved, and so will not be considered in this book.

6.2   LAGRANGE INTERPOLATION

This chapter presents polynomial interpolation in three different forms. The first form, which might be called the *direct form*, involves obtaining the interpolating polynomial by the direct solution of a particular linear system of equations (Vandermonde system). The second form is an alternative called *Lagrange interpolation* and is considered in this section along with the direct form. The third form is called *Newton interpolation*, and is considered in the next section. All three approaches give the same polynomial but expressed in different mathematical forms each possessing particular advantages and disadvantages. No one form is useful in all applications, and this is why we must consider them all.

As in Section 6.1, we consider the data set $\{(t_k, x(t_k))|k \in \mathbf{Z}_{n+1}\}$, and we will let $x_k = x(t_k)$. We wish, as already noted, that our interpolating function be a polynomial of degree n:

$$p_n(t) = \sum_{j=0}^{n} p_{n,j} t^j. \qquad (6.4)$$

For example, if $n = 1$ (*linear interpolation*), then

$$x_k = \sum_{j=0}^{1} p_{1,j} t_k^j,$$

$$x_{k+1} = \sum_{j=0}^{1} p_{1,j} t_{k+1}^j,$$

or

$$p_{1,0} + p_{1,1}t_k = x_k,$$

$$p_{1,0} + p_{1,1}t_{k+1} = x_{k+1},$$

or in matrix form this becomes

$$\begin{bmatrix} 1 & t_k \\ 1 & t_{k+1} \end{bmatrix} \begin{bmatrix} p_{1,0} \\ p_{1,1} \end{bmatrix} = \begin{bmatrix} x_k \\ x_{k+1} \end{bmatrix}.$$

This has a unique solution, provided $t_k \neq t_{k+1}$ because in this instance the matrix has determinant $\det\left(\begin{bmatrix} 1 & t_k \\ 1 & t_{k+1} \end{bmatrix}\right) = t_{k+1} - t_k$. Polynomial $p_1(t) = p_{1,0} + p_{1,1}t$ linearly interpolates the points (t_k, x_k), and (t_{k+1}, x_{k+1}). As another example, if $n = 2$ (*quadratic interpolation*), then we have

$$x_k = \sum_{j=0}^{2} p_{2,j} t_k^j,$$

$$x_{k+1} = \sum_{j=0}^{2} p_{2,j} t_{k+1}^j,$$

$$x_{k+2} = \sum_{j=0}^{2} p_{2,j} t_{k+2}^j,$$

or in matrix form

$$\begin{bmatrix} 1 & t_k & t_k^2 \\ 1 & t_{k+1} & t_{k+1}^2 \\ 1 & t_{k+2} & t_{k+2}^2 \end{bmatrix} \begin{bmatrix} p_{2,0} \\ p_{2,1} \\ p_{2,2} \end{bmatrix} = \begin{bmatrix} x_k \\ x_{k+1} \\ x_{k+2} \end{bmatrix}. \tag{6.5}$$

The matrix in (6.5) has the determinant $(t_k - t_{k+1})(t_{k+1} - t_{k+2})(t_{k+2} - t_k)$, which will not be zero if t_k, t_{k+1} and t_{k+2} are all distinct. The polynomial $p_2(t) = p_{2,0} + p_{2,1}t + p_{2,2}t^2$ interpolates the points (t_k, x_k), (t_{k+1}, x_{k+1}), and (t_{k+2}, x_{k+2}). In general, for arbitrary n we have the linear system

$$\underbrace{\begin{bmatrix} 1 & t_k & \cdots & t_k^{n-1} & t_k^n \\ 1 & t_{k+1} & \cdots & t_{k+1}^{n-1} & t_{k+1}^n \\ \vdots & \vdots & & \vdots & \vdots \\ 1 & t_{k+n-1} & \cdots & t_{k+n-1}^{n-1} & t_{k+n-1}^n \\ 1 & t_{k+n} & \cdots & t_{k+n}^{n-1} & t_{k+n}^n \end{bmatrix}}_{=A} \underbrace{\begin{bmatrix} p_{n,0} \\ p_{n,1} \\ \vdots \\ p_{n,n-1} \\ p_{n,n} \end{bmatrix}}_{=p} = \underbrace{\begin{bmatrix} x_k \\ x_{k+1} \\ \vdots \\ x_{k+n-1} \\ x_{k+n} \end{bmatrix}}_{=x}. \tag{6.6}$$

Matrix A is called a *Vandermonde matrix*, and the linear system (6.6) is a *Vandermonde linear system of equations*. The solution to (6.6) (if it exists) gives the *direct form* of the interpolating polynomial stated in (6.4). For convenience we will let $k = 0$. If we let $t = t_0$, then

$$A = A(t) = \begin{bmatrix} 1 & t & \cdots & t^{n-1} & t^n \\ 1 & t_1 & \cdots & t_1^{n-1} & t_1^n \\ \vdots & \vdots & & \vdots & \vdots \\ 1 & t_{n-1} & \cdots & t_{n-1}^{n-1} & t_{n-1}^n \\ 1 & t_n & \cdots & t_n^{n-1} & t_n^n \end{bmatrix}. \tag{6.7}$$

Let $D(t) = \det(A(t))$, and we see that $D(t)$ is a polynomial in the indeterminate t of degree n. Therefore, $D(t) = 0$ is an equation with exactly n roots (via the *fundamental theorem of algebra*). But $D(t_k) = 0$ for $k = 1, \ldots, n$ since the rows of $A(t)$ are dependent for any $t = t_k$. So $t_1, t_2, \ldots, t_n$ are the only possible roots of $D(t) = 0$. Therefore, if $t_0, t_1, \ldots, t_n$ are all distinct, then we must have $\det(A(t_0)) \neq 0$. Hence A in (6.6) will always possess an inverse if $t_{k+i} \neq t_{k+j}$ for $i \neq j$.

For small values of n (e.g., $n = 1$, or $n = 2$), the direct solution of (6.6) can be a useful method of polynomial interpolation. However, it is known that Vandermonde matrices can be very ill-conditioned even for relatively small n (e.g., $n \geq 10$ or so). This is particularly likely to happen for the common case of *equispaced data*, i.e., the case where $t_k = t_0 + hk$ for $k = 0, 1, \ldots, n$ with $h > 0$, as mentioned in Hill [1, p. 233]. Thus, we much prefer to avoid interpolating by the direct solution of (6.6) when $n \gg 2$. Also, direct solution of (6.6) is computationally inefficient, unless one contemplates the use of *fast algorithms for Vandermonde system solution* in Golub and Van Loan [2]. These are significantly faster than Gaussian elimination as they possess asymptotic time complexities of only $O(n^2)$ versus the $O(n^3)$ complexity of Gaussian elimination approaches.

We remark that so far we have proved the *existence* of a polynomial of degree n that interpolates the data $\{(t_k, x_k) | k \in \mathbf{Z}_{n+1}\}$, provided $t_k \neq t_j$ $(j \neq k)$. The polynomial also happens to be *unique*, a fact readily apparent from the uniqueness of the solution to (6.6) assuming the existence condition is met. So, if we can find (by **any** method) any polynomial of degree $\leq n$ that interpolates the given data, then this is the only possible interpolating polynomial for the data.

Since we disdain the idea of solving (6.6) directly we seek alternative methods to obtain $p_n(t)$. In this regard better approach to polynomial interpolation is an often *Lagrange interpolation*, which works as follows.

Again assume that we wish to interpolate the data set $\{(t_k, x_k) | k \in \mathbf{Z}_{n+1}\}$. Suppose that we possess polynomials (called *Lagrange polynomials*) $L_j(t)$ with the property

$$L_j(t_k) = \begin{cases} 0, & j \neq k \\ 1, & j = k \end{cases} = \delta_{j-k}. \tag{6.8}$$

Then the interpolating polynomial for the data set is

$$p_n(t) = x_0 L_0(t) + x_1 L_1(t) + \cdots + x_n L_n(t) = \sum_{j=0}^{n} x_j L_j(t). \qquad (6.9)$$

We observe that

$$p_n(t_k) = \sum_{j=0}^{n} x_j L_j(t_k) = \sum_{j=0}^{n} x_j \delta_{j-k} = x_k \qquad (6.10)$$

for $k \in \mathbf{Z}_{n+1}$, so $p_n(t)$ in (6.9) does indeed interpolate the data set, and via uniqueness, if we were to write $p_n(t)$ in the direct form $p_n(t) = \sum_{j=0}^{n} p_{n,j} t^j$, then the polynomial coefficients $p_{n,j}$ would satisfy (6.6). We may see that for $j \in \mathbf{Z}_{n+1}$ the Lagrange polynomials are given by

$$L_j(t) = \prod_{\substack{i=0 \\ i \neq j}}^{n} \frac{t - t_i}{t_j - t_i}. \qquad (6.11)$$

Equation (6.9) is called the *Lagrange form of the interpolating polynomial.*

Example 6.1   Consider data set $\{(t_0, x_0), (t_1, x_1), (t_2, x_2)\}$ so $n = 2$. Therefore

$$L_0(t) = \frac{(t - t_1)(t - t_2)}{(t_0 - t_1)(t_0 - t_2)},$$

$$L_1(t) = \frac{(t - t_0)(t - t_2)}{(t_1 - t_0)(t_1 - t_2)},$$

$$L_2(t) = \frac{(t - t_0)(t - t_1)}{(t_2 - t_0)(t_2 - t_1)}.$$

It is not difficult to see that $p_2(t) = x_0 L_0(t) + x_1 L_1(t) + x_2 L_2(t)$. For example, $p_2(t_0) = x_0 L_0(t_0) + x_1 L_1(t_0) + x_2 L_2(t_0) = x_0$.

Suppose that the data set has the specific values $\{(0, 1), (1, 2), (2, 3)\}$, and so (6.6) becomes

$$\begin{bmatrix} 1 & 0 & 0 \\ 1 & 1 & 1 \\ 1 & 2 & 4 \end{bmatrix} \begin{bmatrix} p_{2,0} \\ p_{2,1} \\ p_{2,2} \end{bmatrix} = \begin{bmatrix} 1 \\ 2 \\ 3 \end{bmatrix}.$$

This has solution $p_2(t) = t + 1$ (i.e., $p_{2,2} = 0$). We see that the interpolating polynomial $p_n(t)$ need not have degree exactly equal to n, but can be of lower degree. We also see that

$$L_0(t) = \frac{(t - 1)(t - 2)}{(0 - 1)(0 - 2)} = \frac{1}{2}(t - 1)(t - 2) = \frac{1}{2}(t^2 - 3t + 2),$$

$$L_1(t) = \frac{(t-0)(t-2)}{(1-0)(1-2)} = -t(t-2) = -(t^2 - 2t),$$

$$L_2(t) = \frac{(t-0)(t-1)}{(2-0)(2-1)} = \frac{1}{2}t(t-1) = \frac{1}{2}(t^2 - t).$$

Observe that

$$x_0 L_0(t) + x_1 L_1(t) + x_2 L_2(t)$$

$$= 1 \cdot \frac{1}{2}(t^2 - 3t + 2) + 2 \cdot (-1)(t^2 - 2t) + 3 \cdot \frac{1}{2}(t^2 - t)$$

$$= \left(\frac{1}{2}t^2 - 2t + \frac{3}{2}t^2\right) + \left(-\frac{3}{2}t + 4t - \frac{3}{2}t\right) + \left(\frac{1}{2} \cdot 2\right) = t + 1,$$

which is $p_2(t)$. As expected, the Lagrange form of the interpolating polynomial, and the solution to the Vandermonde system are the same polynomial.

We remark that once $p_n(t)$ is found, we often wish to evaluate $p_n(t)$ for $t \neq t_k$ ($k \in \mathbf{Z}_{n+1}$). Suppose that $p_n(t)$ is known in direct form; then this should be done using *Horner's rule*:

$$p_n(t) = p_{n,0} + t[p_{n,1} + t[p_{n,2} + t[p_{n,3} + \cdots + t[p_{n,n-1} + tp_{n,n}]] \cdots]]]. \quad (6.12)$$

For example, if $n = 3$, then

$$p_3(t) = p_{3,0} + t[p_{3,1} + t[p_{3,2} + tp_{3,3}]]. \quad (6.13)$$

To evaluate this requires only 6 flops (3 floating-point multiplications, and 3 floating-point additions). Evaluating $p_3(t) = p_{3,0} + p_{3,1}t + p_{3,2}t^2 + p_{3,3}t^3$ *directly* needs 9 flops (6 floating-point multiplications and 3 floating-point additions). Thus, Horner's rule is more efficient from a computational standpoint. Using Horner's rule may be described as evaluating the polynomial from the inside out.

While Lagrange interpolation represents a simple way to solve (6.6) the form of the solution is (6.9), and so requires more effort to evaluate at $t \neq t_k$ than the direct form, the latter of which is easily evaluated using Horner's rule as noted in the previous paragraph. Another objection to Lagrange interpolation is that if we were to add elements to our data set, then the calculations for the current data set would have to be discarded, and this would force us to begin again. It is possible to overcome this inefficiency using *Newton's divided-difference method* which leads to the *Newton form of the interpolating polynomial*. This may be seen in Hildebrand [3, Chapter 2]. We will consider this methodology in the next section.

When we evaluate $p_n(t)$ for $t \neq t_k$, but with $t \in [a, b] = [t_0, t_n]$, then this is interpolation. If we wish to evaluate $p_n(t)$ for $t < a$ or $t > b$, then this is referred to as *extrapolation*. To do this is highly risky. Indeed, even if we constrain t to satisfy $t \in [a, b]$, the results can be poor. We illustrate with a famous example of

Runge's which is described in Forsythe et al. [4, pp. 69–70]. It is very possible that for $f(t) \in C[a, b]$, we have

$$\lim_{n \to \infty} \sup_{t \in [a,b]} |f(t) - p_n(t)| = \infty.$$

Runge's specific example is for $t \in [-5, 5]$ with

$$f(t) = \frac{1}{1 + t^2},$$

where he showed that for any t satisfying $3.64 \le |t| < 5$, then

$$\lim_{n \to \infty} \sup |f(t) - p_n(t)| = \infty.$$

This divergence with respect to the Chebyshev norm [recall Section 5.7 (of Chapter 5) for the definition of this term] is called *Runge's phenomenon*. A fairly detailed account of Runge's example appears in Isaacson and Keller [5, pp. 275–279].

We close this section with mention of the approximation error $f(t) - p_n(t)$. Suppose that $f(t) \in C^{n+1}[a, b]$, and it is understood that $f^{(n+1)}(t)$ is continuous as well; that is, $f(t)$ has continuous derivatives $f^{(k)}(t)$ for $k = 1, 2, \ldots, n + 1$. It can be shown that for suitable $\xi = \xi(t)$

$$e_n(t) = f(t) - p_n(t) = \frac{1}{(n + 1)!} f^{(n+1)}(\xi) \prod_{i=0}^{n} (t - t_i), \qquad (6.14)$$

where $\xi \in [a, b]$. If we know $f^{(n+1)}(t)$, then clearly (6.14) may yield useful bounds on $|e_n(t)|$. Of course, if we know nothing about $f(t)$, then (6.14) is useless. Equation (6.14) is useless if $f(t)$ is not sufficiently differentiable. A derivation of (6.14) is given by Hildebrand [3, pp. 81–83], but we omit it here. However, it will follow from results to be considered in the next section.

6.3 NEWTON INTERPOLATION

Define

$$x[t_0, t_1] = \frac{x(t_1) - x(t_0)}{t_1 - t_0} = \frac{x_1 - x_0}{t_1 - t_0}. \qquad (6.15)$$

This is called the *first divided difference of $x(t)$ relative to t_1 and t_0*. We see that $x[t_0, t_1] = x[t_1, t_0]$. We may linearly interpolate $x(t)$ for $t \in [t_0, t_1]$ according to

$$x(t) \approx x(t_0) + \frac{t - t_0}{t_1 - t_0}[x(t_1) - x(t_0)] = x(t_0) + (t - t_0)x[t_0, t_1]. \qquad (6.16)$$

It is convenient to define $p_0(t) = x(t_0)$ and $p_1(t) = x(t_0) + (t - t_0)x[t_0, t_1]$. This is notation consistent with Section 6.2. In fact, $p_1(t)$ agrees with the solution to

$$\begin{bmatrix} 1 & t_0 \\ 1 & t_1 \end{bmatrix} \begin{bmatrix} p_{1,0} \\ p_{1,1} \end{bmatrix} = \begin{bmatrix} x_0 \\ x_1 \end{bmatrix}, \tag{6.17}$$

as we expect.

Unless $x(t)$ is truly linear the *secant slope* $x[t_0, t_1]$ will depend on the *abscissas* t_0 and t_1. If $x(t)$ is a second-degree polynomial, then $x[t_1, t]$ will itself be a linear function of t for a given t_1. Consequently, the ratio

$$x[t_0, t_1, t_2] = \frac{x[t_1, t_2] - x[t_0, t_1]}{t_2 - t_0} \tag{6.18}$$

will be independent of t_0, t_1, and t_2. [This ratio is the *second divided difference of $x(t)$ with respect to t_0, t_1, and t_2.*] To see that this claim is true consider $x(t) = a_0 + a_1 t + a_2 t^2$, so then

$$x[t_1, t] = a_1 + a_2(t + t_1). \tag{6.19}$$

Therefore

$$x[t_1, t_2] - x[t_0, t_1] = x[t_1, t_2] - x[t_1, t_0] = a_2(t_2 - t_0)$$

so that $x[t_0, t_1, t_2] = a_2$. We also note from (6.18) that

$$x[t_0, t_1, t_2] = \frac{1}{t_2 - t_0}\left[\frac{x_2 - x_1}{t_2 - t_1} - \frac{x_1 - x_0}{t_1 - t_0}\right], \tag{6.20a}$$

$$x[t_1, t_0, t_2] = \frac{1}{t_2 - t_1}\left[\frac{x_2 - x_0}{t_2 - t_0} - \frac{x_0 - x_1}{t_0 - t_1}\right], \tag{6.20b}$$

for which we may rewrite these in *symmetric form*

$$x[t_0, t_1, t_2] = \frac{x_0}{(t_0 - t_1)(t_0 - t_2)} + \frac{x_1}{(t_1 - t_0)(t_1 - t_2)} + \frac{x_2}{(t_2 - t_0)(t_2 - t_1)}$$

$$= x[t_1, t_0, t_2], \tag{6.21}$$

that is, $x[t_0, t_1, t_2] = x[t_1, t_0, t_2]$. We see from (6.16) that

$$\frac{x(t) - x(t_0)}{t - t_0} \approx x[t_0, t_1],$$

that is

$$x[t_0, t] \approx x[t_0, t_1], \tag{6.22}$$

and from this we consider the difference

$$x[t_0, t] - x[t_0, t_1] = x[t_0, t] - x[t_1, t_0] = (t - t_1)x[t_0, t_1, t] \qquad (6.23)$$

via (6.18), and the symmetry property $x[t_0, t_1, t] = x[t_1, t_0, t]$. Since $x(t)$ is assumed to be quadratic, we may replace the approximation of (6.16) with the identity

$$x(t) = x(t_0) + (t - t_0)x[t_0, t]$$

$$= \underbrace{x(t_0) + (t - t_0)x[t_0, t_1]}_{=p_1(t)} + (t - t_0)(t - t_1)x[t_0, t_1, t] \qquad (6.24)$$

via (6.23). The first equality of (6.24) may be verified by direct calculation using $x(t) = a_0 + a_1 t + a_2 t^2$, and (6.19). We see that if $p_1(t)$ approximates $x(t)$, then the error involved is [from (6.24)]

$$e(t) = x(t) - p_1(t) = (t - t_0)(t - t_1)x[t_0, t_1, t]. \qquad (6.25)$$

These results generalize to $x(t)$ a polynomial of higher degree.

We may recursively define the divided differences of orders $0, 1, \ldots, k - 1, k$ according to

$$x[t_0] = x(t_0) = x_0,$$

$$x[t_0, t_1] = \frac{x(t_1) - x(t_0)}{t_1 - t_0} = \frac{x_1 - x_0}{t_1 - t_0},$$

$$x[t_0, t_1, t_2] = \frac{x[t_1, t_2] - x[t_0, t_1]}{t_2 - t_0},$$

$$\vdots$$

$$x[t_0, \ldots, t_k] = \frac{x[t_1, \ldots, t_k] - x[t_0, \ldots, t_{k-1}]}{t_k - t_0}. \qquad (6.26)$$

We have established the symmetry $x[t_0, t_1] = x[t_1, t_0]$ (case $k = 1$), and also $x[t_0, t_1, t_2] = x[t_1, t_0, t_2]$ (case $k = 2$). For $k = 2$, symmetry of this kind can also be deduced from the symmetric form (6.21). It seems reasonable [from (6.21)] that in general

$$x[t_0, \ldots, t_k]$$

$$= \frac{x_0}{(t_0 - t_1) \cdots (t_0 - t_k)} + \frac{x_1}{(t_1 - t_0) \cdots (t_1 - t_k)} + \cdots + \frac{x_k}{(t_k - t_0) \cdots (t_k - t_{k-1})}$$

$$= \sum_{j=0}^{k} \frac{1}{\prod_{\substack{i=0 \\ i \neq j}}^{k}(t_j - t_i)} x_j. \qquad (6.27)$$

It is convenient to define the coefficient of x_j as

$$\alpha_j^k = \frac{1}{\prod_{\substack{i=0 \\ i \neq j}}^{k} (t_j - t_i)} \tag{6.28}$$

for $j = 0, 1, \ldots, k$. Thus, $x[t_0, \ldots, t_k] = \sum_{j=0}^{k} \alpha_j^k x_j$. We may prove (6.27) formally by mathematical induction. We outline the detailed approach as follows. Suppose that it is true for $k = r$; that is, assume that

$$x[t_0, \ldots, t_r] = \sum_{j=0}^{r} \frac{1}{\prod_{\substack{i=0 \\ i \neq j}}^{r} (t_j - t_i)} x_j, \tag{6.29}$$

and consider [from definition (6.26)]

$$x[t_0, \ldots, t_{r+1}] = \frac{1}{t_{r+1} - t_0} \{x[t_1, \ldots, t_{r+1}] - x[t_0, \ldots, t_r]\}. \tag{6.30}$$

For (6.30)

$$x[t_1, \ldots, t_{r+1}] = \frac{x_1}{(t_1 - t_2) \cdots (t_1 - t_{r+1})}$$
$$+ \frac{x_2}{(t_2 - t_1) \cdots (t_2 - t_{r+1})} + \cdots + \frac{x_{r+1}}{(t_{r+1} - t_1) \cdots (t_{r+1} - t_r)}, \tag{6.31a}$$

and

$$x[t_0, \ldots, t_r] = \frac{x_0}{(t_0 - t_1) \cdots (t_0 - t_r)}$$
$$+ \frac{x_1}{(t_1 - t_0) \cdots (t_1 - t_r)} + \cdots + \frac{x_r}{(t_r - t_0) \cdots (t_r - t_{r-1})}. \tag{6.31b}$$

If we substitute (6.31) into (6.30), we see that, for example, for the terms involving only x_1

$$\frac{1}{t_{r+1} - t_0} \left\{ \frac{x_1}{(t_1 - t_2)(t_1 - t_3) \cdots (t_1 - t_r)(t_1 - t_{r+1})} \right.$$
$$\left. - \frac{x_1}{(t_1 - t_0)(t_1 - t_2) \cdots (t_1 - t_{r-1})(t_1 - t_r)} \right\}$$
$$= \frac{x_1}{t_{r+1} - t_0} \frac{1}{(t_1 - t_2) \cdots (t_1 - t_r)} \left\{ \frac{1}{t_1 - t_{r+1}} - \frac{1}{t_1 - t_0} \right\}$$
$$= \frac{x_1}{(t_1 - t_0)(t_1 - t_2) \cdots (t_1 - t_r)(t_1 - t_{r+1})} = \alpha_1^{r+1} x_1.$$

The same holds for all remaining terms in x_j for $j = 0, 1, \ldots, r + 1$. Hence (6.27) is valid by induction for all $k \geq 1$. Because of (6.27), the ordering of the arguments in $x[t_0, \ldots, t_k]$ is *irrelevant*. Consequently, $x[t_0, \ldots, t_k]$ can be expressed as the difference between two divided differences of order $k - 1$, having any $k - 1$ of their k arguments in common, divided by the difference between those arguments that are not in common. For example

$$x[t_0, t_1, t_2, t_3] = \frac{x[t_1, t_2, t_3] - x[t_0, t_1, t_2]}{t_3 - t_0} = \frac{x[t_0, t_2, t_3] - x[t_1, t_2, t_3]}{t_0 - t_1}.$$

What happens if two arguments of a divided difference become equal? The situation is reminiscent of Corollary 5.1 (in Chapter 5). For example, suppose $t_1 = t + \epsilon$; then

$$x[t, t_1] = \frac{x(t_1) - x(t)}{t_1 - t} = \frac{x(t + \epsilon) - x(t)}{\epsilon},$$

implying that

$$x[t, t] = \lim_{\epsilon \to 0} \frac{x(t + \epsilon) - x(t)}{\epsilon} = \frac{dx(t)}{dt}$$

so that

$$x[t, t] = \frac{dx(t)}{dt}. \tag{6.32}$$

This assumes that $x(t)$ is differentiable. By similar reasoning

$$\frac{d}{dt} x[t_0, \ldots, t_k, t] = x[t_0, \ldots, t_k, t, t] \tag{6.33}$$

(assuming that $t_0, \ldots, t_k$ are constants). Suppose $u_1, \ldots, u_n$ are differentiable functions of t, then it turns out that

$$\frac{d}{dt} x[t_0, \ldots, t_k, u_1, \ldots, u_n] = \sum_{j=1}^{n} x[t_0, \ldots, t_k, u_1, \ldots, u_n, u_j] \frac{du_j}{dt}. \tag{6.34}$$

Therefore, if $u_1 = u_2 = \cdots = u_n = t$, then, from (6.34)

$$\frac{d}{dt} x[t_0, \ldots, t_k, \underbrace{t, \ldots, t}_{n}] = n x[t_0, \ldots, t_k, \underbrace{t, \ldots, t}_{n+1}]. \tag{6.35}$$

Using (6.33), and (6.35) it may be shown that

$$\frac{d^r}{dt^r} x[t_0, \ldots, t_k, t] = r! \, x[t_0, \ldots, t_k, \underbrace{t, \ldots, t}_{r+1}]. \tag{6.36}$$

Of course, this assumes that $x(t)$ is sufficiently differentiable.

Equation (6.24) is just a special case of something more general. We note that if $x(t)$ is not a quadratic, then (6.24) is only an approximation for $t \notin \{t_0, t_1, t_2\}$, and so

$$x(t) \approx x(t_0) + (t - t_0)x[t_0, t_1] + (t - t_0)(t - t_1)x[t_0, t_1, t_2] = p_2(t). \quad (6.37)$$

It is easy to verify by direct evaluation that $p_2(t_i) = x(t_i)$ for $i \in \{0, 1, 2\}$. Equation (6.37) is the *second-degree interpolation formula*, while (6.16) is the *first-degree interpolation formula*. We may generalize (6.24) [and hence (6.37)] to higher degrees by using (6.26); that is

$$x(t) = x[t_0] + (t - t_0)x[t_0, t],$$

$$x[t_0, t] = x[t_0, t_1] + (t - t_1)x[t_0, t_1, t],$$

$$x[t_0, t_1, t] = x[t_0, t_1, t_2] + (t - t_2)x[t_0, t_1, t_2, t],$$

$$\vdots$$

$$x[t_0, \ldots, t_{n-1}, t] = x[t_0, \ldots, t_n] + (t - t_n)x[t_0, \ldots, t_n, t], \quad (6.38)$$

where the last equation follows from

$$
\begin{aligned}
x[t_0, \ldots, t_n, t] &= \frac{x[t_1, \ldots, t_n, t] - x[t_0, \ldots, t_n]}{t - t_0} \\
&= \frac{x[t_0, \ldots, t_{n-1}, t] - x[t_0, \ldots, t_n]}{t - t_n}
\end{aligned}
\quad (6.39)
$$

(the second equality follows by exchanging t_0 and t_n). If the second relation of (6.38) is substituted into the first, we obtain

$$x(t) = x[t_0] + (t - t_0)x[t_0, t_1] + (t - t_0)(t - t_1)x[t_0, t_1, t], \quad (6.40)$$

which is just (6.24) again. If we substitute the third relation of (6.38) into (6.40), we obtain

$$x(t) = x[t_0] + (t - t_0)x[t_0, t_1] + (t - t_0)(t - t_1)x[t_0, t_1, t_2]$$
$$+ (t - t_0)(t - t_1)(t - t_2)x[t_0, t_1, t_2, t]. \quad (6.41)$$

This leads to the *third-degree interpolation formula*

$$p_3(t) = x[t_0] + (t - t_0)x[t_0, t_1] + (t - t_0)(t - t_1)x[t_0, t_1, t_2]$$
$$+ (t - t_0)(t - t_1)(t - t_2)x[t_0, t_1, t_2, t_3].$$

Continuing in this fashion, we obtain

$$x(t) = x[t_0] + (t - t_0)x[t_0, t_1] + (t - t_0)(t - t_1)x[t_0, t_1, t_2]$$

$$+ \cdots + (t - t_0)(t - t_1) \cdots (t - t_{n-1})x[t_0, t_1 \ldots, t_n] + e(t), \qquad (6.42\text{a})$$

where

$$e(t) = (t - t_0)(t - t_1) \cdots (t - t_n)x[t_0, t_1, \ldots, t_n, t], \qquad (6.42\text{b})$$

and we define

$$p_n(t) = x[t_0] + (t - t_0)x[t_0, t_1]$$

$$+ (t - t_0)(t - t_1)x[t_0, t_1, t_2] + \cdots + (t - t_0) \cdots (t - t_{n-1})x[t_0, \ldots, t_n],$$

$$(6.42\text{c})$$

which is the *nth-degree interpolating formula*, and is clearly a polynomial of degree
n. So $e(t)$ is the error involved in interpolating $x(t)$ using polynomial $p_n(t)$. It is
the case that $p_n(t_k) = x(t_k)$ for $k = 0, 1, \ldots, n$. Equation (6.42a) is the *Newton
interpolating formula with divided differences*. If $x(t)$ is a polynomial of degree n
(or less), then $e(t) = 0$ (all t). This is more formally justified later.

Example 6.2   Consider $e(t)$ for $n = 2$. This requires [via (6.27)]

$$x[t_0, t_1, t_2, t] = \frac{x_0}{(t_0 - t_1)(t_0 - t_2)(t_0 - t)} + \frac{x_1}{(t_1 - t_0)(t_1 - t_2)(t_1 - t)}$$

$$+ \frac{x_2}{(t_2 - t_0)(t_2 - t_1)(t_2 - t)} + \frac{x(t)}{(t - t_0)(t - t_1)(t - t_2)}$$

Thus

$$e(t) = (t - t_0)(t - t_1)(t - t_2)x[t_0, t_1, t_2, t]$$

$$= \underbrace{-\frac{(t - t_1)(t - t_2)x_0}{(t_0 - t_1)(t_0 - t_2)} - \frac{(t - t_0)(t - t_2)x_1}{(t_1 - t_0)(t_1 - t_2)} - \frac{(t - t_0)(t - t_1)x_2}{(t_2 - t_0)(t_2 - t_1)}}_{=-p_2(t)} + x(t).$$

$$(6.43)$$

The form of $p_2(t)$ in (6.43) is that of $p_2(t)$ in Example 6.1:

$$p_2(t) = x_0 L_0(t) + x_1 L_1(t) + x_2 L_2(t).$$

If $x(t) = a_0 + a_1 t + a_2 t^2$, then clearly $e(t) = 0$ for all t. In fact, $p_{2,k} = a_k$.

Example 6.3   Suppose that we wish to interpolate $x(t) = e^t$ given that $t_k = kh$
with $h = 0.1$, and $k = 0, 1, 2, 3$. We are told that

$$x_0 = 1.000000, \qquad x_1 = 1.105171, \qquad x_2 = 1.221403, \qquad x_3 = 1.349859.$$

We will consider $n = 2$ (i.e., quadratic interpolation). The task is aided if we construct the *divided-difference table*:

$t_0 = 0$ $x(t_0) = 1.000000$

$x[t_0, t_1] = 1.05171$

$t_1 = 0.1$ $x(t_1) = 1.105171$ $x[t_0, t_1, t_2] = 0.55305$

$x[t_1, t_2] = 1.16232$

$t_2 = 0.2$ $x(t_2) = 1.221403$ $x[t_1, t_2, t_3] = 0.61120$

$x[t_2, t_3] = 1.28456$

$t_3 = 0.3$ $x(t_3) = 1.349859$

For $t \in [0, 0.2]$ we consider [from (6.37)]

$$x(t) \approx x(t_0) + (t - t_0)x[t_0, t_1] + (t - t_0)(t - t_1)x[t_0, t_1, t_2],$$

which for the data we are given becomes

$$x(t) \approx 1.000000 + 1.051710t + 0.55305t(t - 0.1) = p_2^a(t). \tag{6.44a}$$

If $t = 0.11$, then $p_2^a(0.11) = 1.116296$, while it turns out that $x(0.11) = 1.116278$, so $e(0.11) = x(0.11) - p_2^a(0.11) = -0.000018$. For $t \in [0.1, 0.3]$ we might consider

$$x(t) \approx x(t_1) + (t - t_1)x[t_1, t_2] + (t - t_1)(t - t_2)x[t_1, t_2, t_3],$$

which for the given data becomes

$$x(t) \approx 1.105171 + 1.16232(t - 0.1) + 0.61120(t - 0.1)(t - 0.2) = p_2^b(t). \tag{6.44b}$$

We observe that to calculate $p_2^b(t)$ does not require discarding all the results needed to determine $p_2^a(t)$. We do not need to begin again as with Lagrangian interpolation since, for example

$$x[t_0, t_1, t_2] = \frac{x[t_1, t_2] - x[t_0, t_1]}{t_2 - t_0}, \qquad x[t_1, t_2, t_3] = \frac{x[t_2, t_3] - x[t_1, t_2]}{t_3 - t_1},$$

and both of these divided differences require $x[t_1, t_2]$. If we wanted to use cubic interpolation then the table is very easily augmented to include $x[t_0, t_1, t_2, t_3]$. Thus, updating the interpolating polynomial due to the addition of more data to the table, or of increasing n, may proceed more efficiently than if we were to employ Lagrange interpolation.

We also observe that both $p_2^a(t)$ and $p_2^b(t)$ may be used to interpolate $x(t)$ for $t \in [0.1, 0.2]$. Which polynomial should be chosen? Ideally, we would select the one for which $e(t)$ is the smallest. Practically, this means seeking bounds for $|e(t)|$ and choosing the interpolating polynomial with the best error bound.

We have shown that if $x(t)$ is approximated by $p_n(t)$, then the error has the form

$$e(t) = \pi(t)x[t_0, \ldots, t_n, t] \tag{6.45a}$$

[recall (6.42b)], where

$$\pi(t) = (t - t_0)(t - t_1) \cdots (t - t_n), \tag{6.45b}$$

which is an degree $n + 1$ polynomial. The form of the error in (6.45a) can be useful in analyzing the accuracy of numerical integration and numerical differentiation procedures, but another form of the error can be found.

We note that $e(t) = x(t) - p_n(t)$ [recall (6.42)], so both $x(t) - p_n(t)$ and $\pi(t)$ vanish at $t_0, t_1, \ldots, t_n$. Consider the linear combination

$$X(t) = x(t) - p_n(t) - \kappa \pi(t). \tag{6.46}$$

We wish to select κ so that $X(\bar{t}) = 0$, where $\bar{t} \neq t_k$ for $k \in \mathbf{Z}_{n+1}$. Such a κ exists because $\pi(t)$ vanishes only at $t_0, t_1, \ldots, t_n$. Let $a = \min\{t_0, \ldots, t_n, \bar{t}\}$, $b = \max\{t_0, \ldots, t_n, \bar{t}\}$, and define the interval $\bar{I} = [a, b]$. By construction, $X(t)$ vanishes at least $n + 2$ times on $\bar{I}$. *Rolle's theorem* from calculus states that $X^{(k)}(t) = d^k X(t)/dt^k$ vanishes at least $n + 2 - k$ times inside $\bar{I}$. Specifically, $X^{(n+1)}(t)$ vanishes at least once inside $\bar{I}$. Let this point be called $\bar{\xi}$. Therefore, from (6.46), we obtain

$$x^{(n+1)}(\bar{\xi}) - p_n^{(n+1)}(\bar{\xi}) - \kappa \pi^{(n+1)}(\bar{\xi}) = 0. \tag{6.47}$$

But $p_n(t)$ is a polynomial of degree n, so $p^{(n+1)}(t) = 0$ for all $t \in \bar{I}$. From (6.45b) $\pi^{(n+1)}(t) = (n + 1)!$, so finally (6.47) reduces to

$$\kappa = \frac{1}{(n+1)!} x^{(n+1)}(\bar{\xi}). \tag{6.48}$$

From $X(\bar{t}) = 0$ in (6.46), and using (6.48), we find that

$$e(\bar{t}) = x(\bar{t}) - p_n(\bar{t}) = \frac{1}{(n+1)!} x^{(n+1)}(\bar{\xi})\pi(\bar{t}) \tag{6.49}$$

for some $\bar{\xi} \in \bar{I}$. If we were to let $\bar{t} = t_k$ for any $k \in \mathbf{Z}_{n+1}$, then both sides of (6.49) vanish even in this previously excluded case. This allows us to write

$$e(t) = \frac{1}{(n+1)!} x^{(n+1)}(\xi(t))\pi(t) \tag{6.50}$$

for some $\xi = \xi(t) \in I$ $(I = [a, b]$ with $a = \min\{t_0, \ldots, t_n, t\}$, and $b = \max\{t_0, \ldots, t_n, t\})$. If $x^{(n+1)}(t)$ is continuous for $t \in I$, then $x^{(n+1)}(t)$ is bounded on I, so there is an $M_{n+1} > 0$ such that

$$x^{(n+1)}(\xi) \leq M_{n+1}, \tag{6.51}$$

and hence

$$|e(t)| \leq \frac{M_{n+1}}{(n+1)!}|\pi(t)| \tag{6.52}$$

for all $t \in I$. It is to be emphasized that for this to hold $x^{(n+1)}(t)$ must exist, and we normally require it to be continuous, too. Equations (6.50) and (6.45a) are equivalent, and thus

$$\pi(t)x[t_0, \ldots, t_n, t] = \frac{1}{(n+1)!}x^{(n+1)}(\xi)\pi(t),$$

or in other words

$$x[t_0, \ldots, t_n, t] = \frac{1}{(n+1)!}x^{(n+1)}(\xi) \tag{6.53}$$

for some $\xi \in I$, whenever $x^{(n+1)}(t)$ exists in I. In particular, if $x(t)$ is a polynomial of degree n or less, then (6.53) yields $x[t_0, \ldots t_n, t] = 0$, hence $e(t) = 0$ for all t (a fact mentioned earlier).

Example 6.4   Recall Example 6.3. We considered $x(t) = e^t$ with $t_0 = 0$, $t_1 = 0.1$, $t_2 = 0.2$, and we found that

$$p_2^a(t) = 1.000000 + 1.051710t + 0.553050t(t - 0.1)$$

for which $p_2^a(0.11) = 1.116296$, and $x(0.11) = 1.116278$. The exact error is

$$e(0.11) = x(0.11) - p_2^a(0.11) = -0.000018.$$

We will compare this with the bound we obtain from (6.52). Since $n = 2$, $x^{(3)}(t) = e^t$, and $I = [0, 0.2]$, so

$$M_3 = e^{0.2} = 1.221403$$

(which was given data), and $\pi(t) = t(t - 0.1)(t - 0.2)$, so $|\pi(0.11)| = 9.9 \times 10^{-5}$. Consequently, from (6.52), we have

$$|e(0.11)| \leq \frac{1.221403}{3!}9.9 \times 10^{-5} = 0.000020.$$

The actual error certainly agrees with this bound.

We end this section by observing that (6.14) immediately follows from (6.50).

6.4   HERMITE INTERPOLATION

In the previous sections polynomial interpolation methods matched the polynomial only to the value of the function $f(x)$ at various points $x = x_k \in [a, b] \subset \mathbf{R}$. In this section we consider *Hermite interpolation* where the interpolating polynomial also matches the first derivatives $f^{(1)}(x)$ at $x = x_k$. This interpolation technique is important in the development of higher order numerical integration methods as will be seen in Chapter 9.

The following theorem is the main result, and is essentially Theorem 3.9 from Burden and Faires [6].

Theorem 6.1: Hermite Interpolation   Suppose that $f(x) \in C^1[a, b]$, and that $x_0, x_1, \ldots, x_n \in [a, b]$ are distinct, then the unique polynomial of degree (at most) $2n + 1$ denoted by $p_{2n+1}(x)$, and such that

$$p_{2n+1}(x_j) = f(x_j), \quad p_{2n+1}^{(1)}(x_j) = f^{(1)}(x_j) \tag{6.54}$$

($j \in \mathbf{Z}_{n+1}$) is given by

$$p_{2n+1}(x) = \sum_{k=0}^{n} h_k(x) f(x_k) + \sum_{k=0}^{n} \hat{h}_k(x) f^{(1)}(x_k), \tag{6.55}$$

where

$$h_k(x) = [1 - 2L_k^{(1)}(x_k)(x - x_k)][L_k(x)]^2, \tag{6.56}$$

and

$$\hat{h}_k(x) = (x - x_k)[L_k(x)]^2 \tag{6.57}$$

such that [recall (6.11)]

$$L_k(x) = \prod_{\substack{i=0 \\ i \neq k}}^{n} \frac{x - x_i}{x_k - x_i}. \tag{6.58}$$

Proof   To show (6.54) for $x_0, x_1, \ldots, x_n$, we require that $h_k(x)$, and $\hat{h}_k(x)$ in (6.56) and (6.57) satisfy the conditions

$$h_k(x_j) = \delta_{j-k}, \quad h_k^{(1)}(x_j) = 0 \tag{6.59a}$$

and

$$\hat{h}_k(x_j) = 0, \quad \hat{h}_k^{(1)}(x_j) = \delta_{j-k}. \tag{6.59b}$$

Assuming that these conditions hold, we may confirm (6.54) as follows. Via (6.55)

$$p_{2n+1}(x_j) = \sum_{k=0}^{n} h_k(x_j) f(x_k) + \sum_{k=0}^{n} \hat{h}_k(x_j) f^{(1)}(x_k),$$

and via (6.59), this becomes

$$p_{2n+1}(x_j) = \sum_{k=0}^{n} \delta_{j-k} f(x_k) + \sum_{k=0}^{n} 0 \cdot f^{(1)}(x_k) = f(x_j).$$

This confirms the first case in (6.54). Similarly, via (6.59)

$$p_{2n+1}^{(1)}(x_j) = \sum_{k=0}^{n} h_k^{(1)}(x_j) f(x_k) + \sum_{k=0}^{n} \hat{h}_k^{(1)}(x_j) f^{(1)}(x_k)$$

becomes

$$p_{2n+1}^{(1)}(x_j) = \sum_{k=0}^{n} 0 \cdot f(x_k) + \sum_{k=0}^{n} \delta_{j-k} f^{(1)}(x_k) = f^{(1)}(x_j),$$

which confirms the second case in (6.54).

Now we will confirm that $h_k(x)$, and $\hat{h}_k(x)$ as defined in (6.56), and (6.57) satisfy the requirements given in (6.59). The conditions in (6.59b) imply that $\hat{h}_k(x)$ must have a double root at $x = x_j$ for $j \neq k$, and a single root at $x = x_k$. A polynomial of degree at most $2n + 1$ that satisfies these requirements, and such that $\hat{h}_k^{(1)}(x_k) = 1$ is

$$\hat{h}_k(x) = (x - x_k) \frac{(x - x_0)^2 \cdots (x - x_{k-1})^2 \cdot (x - x_{k+1})^2 \cdots (x - x_n)^2}{(x_k - x_0)^2 \cdots (x_k - x_{k-1})^2 \cdot (x_k - x_{k+1})^2 \cdots (x_k - x_n)^2}$$

$$= (x - x_k) L_k^2(x).$$

Certainly $\hat{h}_k(x_k) = 0$. Moreover, $\hat{h}_k^{(1)}(x) = L_k^2(x) + 2(x - x_k) L_k(x) L_k^{(1)}(x)$ so $\hat{h}_k^{(1)}(x_k) = L_k^2(x_k) = 1$ [via (6.8)]. These verify (6.59b).

Now we consider (6.59a). These imply x_j for $j \neq k$ is a double root of $h_k(x)$, and we may consider (for suitable a and b to be found below)

$$h_k(x) = \frac{1}{\prod_{\substack{i=0 \\ i \neq k}}^{n} (x_i - x_k)^2} (x - x_0)^2 \cdots (x - x_{k-1})^2 (x - x_{k+1})^2$$

$$\cdots (x - x_n)^2 (ax + b)$$

which has degree at most $2n + 1$. More concisely, this polynomial is

$$h_k(x) = L_k^2(x)(ax + b).$$

From (6.59a) we require

$$1 = h_k(x_k) = L_k^2(x_k)(ax_k + b) = ax_k + b, \tag{6.60}$$

Also, $h_k^{(1)}(x) = aL_k^2(x) + 2L_k(x)L_k^{(1)}(x)(ax + b)$, and we also need [via (6.59a)]

$$h_k^{(1)}(x_k) = aL_k^2(x_k) + 2L_k(x_k)L_k^{(1)}(x_k)(ax_k + b) = 0,$$

but again since $L_k(x_k) = 1$, this expression reduces to

$$a + 2L_k^{(1)}(x_k) = 0,$$

where we have used (6.60). Hence

$$a = -2L_k^{(1)}(x_k), \quad b = 1 + 2L_k^{(1)}(x_k)x_k.$$

Therefore, we finally have

$$h_k(x) = [1 - 2L_k^{(1)}(x_k)(x - x_k)]L_k^2(x)$$

which is (6.56). Since $L_k(x_j) = 0$ for $j \neq k$ it is clear that $h_k(x_j) = 0$ for $j \neq k$. It is also easy to see that $h_k^{(1)}(x_j) = 0$ for all $j \neq k$ too. Thus, (6.59a) is confirmed for $h_k(x)$ as defined in (6.56).

An error bound for Hermite interpolation is provided by the expression

$$f(x) = p_{2n+1}(x) + \frac{1}{(2n+2)!} \prod_{i=0}^{n} (x - x_i)^2 f^{(2n+2)}(\xi) \tag{6.61}$$

for some $\xi \in (a, b)$, where $f(x) \in C^{2n+2}[a, b]$. We shall not derive (6.61) except to note that the approach is similar to the derivation of (6.14). Equation (6.14) was really derived in Section 6.3.

In its present form Hermite interpolation requires working with Lagrange polynomials, and their derivatives. As noted by Burden and Faires [6], this is rather tedious (i.e., not computationally efficient). A procedure involving Newton interpolation (recall Section 6.3) may be employed to reduce the labor that would otherwise be involved in Hermite interpolation. We do not consider this approach, but instead refer the reader to Burden and Faires [6] for the details. We use Hermite interpolation in Chapter 9 to develop numerical integration methods, and efficient Hermite interpolation is not needed for this purpose.

6.5 SPLINE INTERPOLATION

Spline (spliced *line*) *interpolation* is a particular kind of *piecewise polynomial interpolation*. We may wish, for example, to approximate $f(x)$ for $x \in [a, b] \subset \mathbf{R}$ when given the sample points $\{(x_k, f(x_k)) | k \in \mathbf{Z}_{n+1}\}$ by fitting straight-line segments in between $(x_k, f(x_k))$, and $(x_{k+1}, f(x_{k+1}))$ for $k = 0, 1, \ldots, n - 1$. An

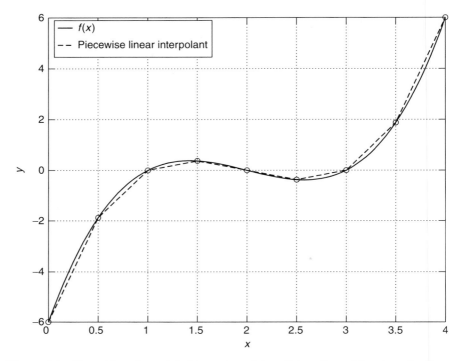

Figure 6.1   The cubic polynomial $f(x) = (x - 1)(x - 2)(x - 3)$, and its piecewise linear interpolant (dashed line) at the nodes $x_k = x_0 + hk$ for which $x_0 = 0$, and $h = \frac{1}{2}$, where $k = 0, 1, \ldots, n$ with $n = 8$.

example of this appears in Fig. 6.1. This has a number of disadvantages. Although $f(x)$ may be differentiable at $x = x_k$, the piecewise linear approximation will not be (in general). Also, the graph of the interpolant has visually displeasing "kinks" in it. If interpolation is for a computer graphics application, or to define the physical surface of an automobile body or airplane, then such kinks are seldom acceptable. Splines are a means to deal with this problem. It is also worth noting that, more recently, splines have found a role in the design of wavelet functions [7, 8], which were briefly mentioned in Chapter 1.

The following definition is taken from Epperson [9], and our exposition of spline functions in this section follows that in [9] fairly closely. As always $f^{(i)}(x) = d^i f(x)/dx^i$ (i.e., this is the notation for the ith derivative of $f(x)$).

Definition 6.1: Spline   Suppose that we are given $\{(x_k, f(x_k)) | k \in \mathbf{Z}_{n+1}\}$. The piecewise polynomial function $p_m(x)$ is called a *spline* if

(S1) $p_m(x_k) = f(x_k)$ for all $k \in \mathbf{Z}_{n+1}$ (interpolation).
(S2) $\lim_{x \to x_k^-} p_m^{(i)}(x) = \lim_{x \to x_k^+} p_m^{(i)}(x)$ for all $i \in \mathbf{Z}_{N+1}$ (smoothness).

(S3) $p_m(x)$ is a polynomial of degree no larger than m on every subinterval $[x_k, x_{k+1}]$ for $k \in \mathbf{Z}_n$ (interval of definition).

We say that m is the *degree of approximation* and N is the *degree of smoothness* of the spline $p_m(x)$.

There is a relationship between m and N. As there are n subintervals $[x_k, x_{k+1}]$, and each of these is the domain of definition of a degree m polynomial, we see that there are $D_f = n(m + 1)$ *degrees of freedom*. Each polynomial is specified by $m + 1$ coefficients, and there are n of these polynomials; hence D_f is the number of parameters to solve for in total. From Definition 6.1 there are $n + 1$ interpolation conditions [axiom (S1)]. And there are $n - 1$ junction points $x_1, \ldots x_{n-1}$ (sometimes also called *knots*), with $N + 1$ continuity conditions being imposed on each of them [axiom (S2)]. As a result, there are $D_c = (n + 1) + (n - 1)(N + 1)$ *constraints*. Consider

$$D_f - D_c = n(m + 1) - [(n + 1) + (n - 1)(N + 1)] = n(m - N - 1) + N.$$
(6.62)

It is a common practice to enforce the condition $m - N - 1 = 0$; that is, we let

$$m = N + 1.$$
(6.63)

This relates the degree of approximation to the degree of smoothness in a simple manner. Below we will focus our attention exclusively on the special case of the *cubic splines* for which $m = 3$. From (6.63) we must therefore have $N = 2$. With condition (6.63), then, from (6.62) we have

$$D_f - D_c = N.$$
(6.64)

As a result, it is necessary to impose N further constraints on the design problem. How this is done is considered in detail below. Since we will look only at $m = 3$ with $N = 2$, we must impose two additional constraints. This will be done by imposing one constraint at each endpoint of the interval $[a, b]$. There is more than one way to do this as will be seen later.

From Definition 6.1 it superficially appears that we need to compute n different polynomials. However, it is possible to recast our problem in terms of *B-splines*. A B-spline acts as a prototype in the formation of a basis set of splines.

Aside from the assumption that $m = 3$, $N = 2$, let us further assume that

$$a = x_0 < x_1 < \cdots < x_{n-1} < x_n = b$$
(6.65)

with $x_{k+1} - x_k = h$ for $k = 0, 1, \ldots, n - 1$. This is the *uniform grid* assumption. We will also need to account for *boundary conditions*, and this requires us to introduce the additional grid points

$$x_{-3} = a - 3h, \quad x_{-2} = a - 2h, \quad x_{-1} = a - h$$
(6.66)

and

$$x_{n+3} = b + 3h, \quad x_{n+2} = b + 2h, \quad x_{n+1} = b + h. \qquad (6.67)$$

Our prototype cubic B-spline will be the function

$$S(x) = \begin{cases} 0, & x \le -2 \\ (x+2)^3, & -2 \le x \le -1 \\ 1 + 3(x+1) + 3(x+1)^2 - 3(x+1)^3, & -1 \le x \le 0 \\ 1 + 3(1-x) + 3(1-x)^2 - 3(1-x)^3, & 0 \le x \le 1 \\ (2-x)^3, & 1 \le x \le 2 \\ 0, & x \ge 2 \end{cases} \qquad (6.68)$$

This function has nodes at $x \in \{-2, -1, 0, 1, 2\}$. A plot of it appears in Fig. 6.2, and we see that it has a bell shape similar to the Gaussian pulse we saw in Chapter 3. We may verify that $S(x)$ satisfies Definition 6.1 as follows. Plainly, it is piecewise cubic (i.e., $m = 3$), so axiom (S3) holds. The first and second derivatives are,

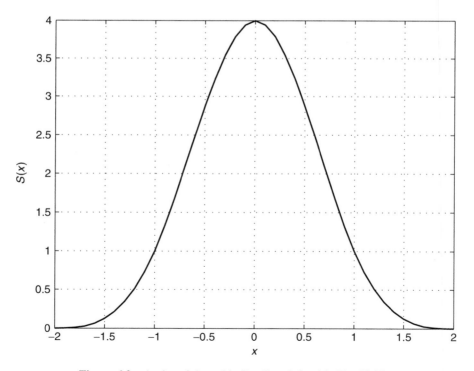

Figure 6.2 A plot of the cubic B-spline defined in Eq. (6.68).

respectively

$$
S^{(1)}(x) = \begin{cases}
0, & x \le -2 \\
3(x + 2)^2, & -2 \le x \le -1 \\
3 + 6(x + 1) - 9(x + 1)^2, & -1 \le x \le 0 \\
-3 - 6(1 - x) + 9(1 - x)^2, & 0 \le x \le 1 \\
-3(2 - x)^2, & 1 \le x \le 2 \\
0, & x \ge 2
\end{cases}
\tag{6.69}
$$

and

$$
S^{(2)}(x) = \begin{cases}
0, & x \le -2 \\
6(x + 2), & -2 \le x \le -1 \\
6 - 18(x + 1), & -1 \le x \le 0 \\
6 - 18(1 - x), & 0 \le x \le 1 \\
6(2 - x), & 1 \le x \le 2 \\
0, & x \ge 2
\end{cases}
\tag{6.70}
$$

We note that from (6.68)–(6.70) that

$$S(0) = 4, \qquad S(\pm 1) = 1, \qquad S(\pm 2) = 0, \tag{6.71a}$$

$$S^{(1)}(0) = 0, \qquad S^{(1)}(\pm 1) = \mp 3, \qquad S^{(1)}(\pm 2) = 0, \tag{6.71b}$$

$$S^{(2)}(0) = -12, \qquad S^{(2)}(\pm 1) = 6, \qquad S^{(2)}(\pm 2) = 0. \tag{6.71c}$$

So it is apparent that for $i = 0, 1, 2$ we have

$$\lim_{x \to x_k^-} S^{(i)}(x) = \lim_{x \to x_k^+} S^{(i)}(x)$$

for all $x_k \in \{-2, -1, 0, 1, 2\}$. Thus, the smoothness axiom (S2) is met for $N = 2$.

Now we need to consider how we may employ $S(x)$ to approximate any $f(x)$ for $x \in [a, b] \subset \mathbf{R}$ when working with the grid specified in (6.65)–(6.67). To this end we define

$$S_i(x) = S\left(\frac{x - x_i}{h}\right) \tag{6.72}$$

for $i = -1, 0, 1, \ldots, n, n + 1$. Since $S_i^{(1)}(x) = \frac{1}{h} S^{(1)}\left(\frac{x - x_i}{h}\right)$, $S_i^{(2)}(x) = \frac{1}{h^2} S^{(2)}\left(\frac{x - x_i}{h}\right)$ from (6.71), we have (e.g., $x_{i \pm 1} = x_i \pm h$.)

$$S_i(x_i) = S(0) = 4, \qquad S_i(x_{i \pm 1}) = S(\pm 1) = 1, \qquad S_i(x_{i \pm 2}) = S(\pm 2) = 0, \tag{6.73a}$$

$$S_i^{(1)}(x_i) = 0, \qquad S_i^{(1)}(x_{i\pm1}) = \mp\frac{3}{h}, \qquad S_i^{(1)}(x_{i\pm2}) = 0, \qquad (6.73b)$$

$$S_i^{(2)}(x_i) = -\frac{12}{h^2}, \qquad S_i^{(2)}(x_{i\pm1}) = \frac{6}{h^2}, \qquad S_i^{(2)}(x_{i\pm2}) = 0. \qquad (6.73c)$$

We construct a cubic B-spline interpolant for any $f(x)$ by defining spline $p_3(x)$ to be a linear combination of $S_i(x)$ for $i = -1, 0, 1, \ldots, n, n+1$, i.e., for suitable coefficients a_i we have

$$p_3(x) = \sum_{i=-1}^{n+1} a_i S_i(x). \qquad (6.74)$$

The series coefficients a_i are determined in order to satisfy axiom (S1) of Definition 6.1; thus, we have

$$f(x_k) = \sum_{i=-1}^{n+1} a_i S_i(x_k) \qquad (6.75)$$

for $k \in \mathbf{Z}_{n+1}$.

If we apply (6.73a) to (6.75), we observe that

$$
\begin{aligned}
f(x_0) &= \sum_{i=-1}^{n+1} a_i S_i(x_0) &&= a_{-1}S_{-1}(x_0) + a_0 S_0(x_0) + a_1 S_1(x_0) \\
f(x_1) &= \sum_{i=-1}^{n+1} a_i S_i(x_1) &&= a_0 S_0(x_1) + a_1 S_1(x_1) + a_2 S_2(x_1) \\
&\ \ \vdots \\
f(x_n) &= \sum_{i=-1}^{n+1} a_i S_i(x_n) &&= a_{n-1}S_{n-1}(x_n) + a_n S_n(x_n) + a_{n+1}S_{n+1}(x_n)
\end{aligned}
$$
$$(6.76)$$

For example, for $f(x_0)$ in (6.76), we note that $S_k(x_0) = 0$ for $k \geq 2$ since ($x_k = x_0 + kh$)

$$S_k(x_0) = S\left(\frac{x_0 - x_k}{h}\right) = S\left(\frac{x_0 - (kh + x_0)}{h}\right) = S(-k) = 0.$$

More generally

$$f(x_k) = a_{k-1}S_{k-1}(x_k) + a_k S_k(x_k) + a_{k+1}S_{k+1}(x_k) \qquad (6.77)$$

for which $k \in \mathbf{Z}_{n+1}$. Again via (6.73a) we see that

$$S_{k-1}(x_k) = S\left(\frac{(x_0 + kh) - (x_0 + (k-1)h)}{h}\right) = S(1) = 1,$$

$$S_k(x_k) = S\left(\frac{(x_0 + kh) - (x_0 + kh)}{h}\right) = S(0) = 4,$$

$$S_{k+1}(x_k) = S\left(\frac{(x_0 + kh) - (x_0 + (k+1)h)}{h}\right) = S(-1) = 1.$$

Thus, (6.77) becomes

$$a_{k-1} + 4a_k + a_{k+1} = f(x_k) \tag{6.78}$$

again for $k = 0, 1, \ldots, n$. In matrix form we have

$$\underbrace{\begin{bmatrix} 1 & 4 & 1 & 0 & \cdots & 0 & 0 & 0 \\ 0 & 1 & 4 & 1 & \cdots & 0 & 0 & 0 \\ \vdots & \vdots & \vdots & \vdots & & \vdots & \vdots & \vdots \\ 0 & 0 & 0 & 0 & \cdots & 4 & 1 & 0 \\ 0 & 0 & 0 & 0 & \cdots & 1 & 4 & 1 \end{bmatrix}}_{=A} \underbrace{\begin{bmatrix} a_{-1} \\ a_0 \\ \vdots \\ a_n \\ a_{n+1} \end{bmatrix}}_{=a} = \underbrace{\begin{bmatrix} f(x_0) \\ f(x_1) \\ \vdots \\ f(x_{n-1}) \\ f(x_n) \end{bmatrix}}_{=f}. \tag{6.79}$$

We note that $A \in \mathbf{R}^{(n+1) \times (n+3)}$, so there are $n + 1$ equations in $n + 3$ unknowns ($a \in \mathbf{R}^{n+3}$). The *tridiagonal linear system* in (6.79) cannot be solved in its present form as we need two additional constraints. This is to be expected from our earlier discussion surrounding (6.62)–(6.64). Recall that we have chosen $m = 3$, so $N = 2$, and so, via (6.64), we have $D_f - D_c = N = 2$, implying the need for two more constraints. There are two common approaches to obtaining these constraints:

1. We enforce $p_3^{(2)}(x_0) = p_3^{(2)}(x_n) = 0$ (*natural spline*).
2. We enforce $p_3^{(1)}(x_0) = f^{(1)}(x_0)$, and $p_3^{(1)}(x_n) = f^{(1)}(x_n)$ (*complete spline*, or *clamped spline*).

Of these two choices, the natural spline is a bit easier to work with but can lead to larger approximation errors near the interval endpoints x_0, and x_n compared to working with the complete spline. The complete spline can avoid the apparent need to know the derivatives $f^{(1)}(x_0)$, and $f^{(1)}(x_n)$ by using numerical approximations to the derivative (see Section 9.6).

We will first consider the case of the natural spline. From (6.74), we obtain

$$p_3(x_0) = a_{-1} S_{-1}(x_0) + a_0 S_0(x_0) + a_1 S_1(x_0),$$

so that

$$p_3^{(2)}(x_0) = a_{-1} S_{-1}^{(2)}(x_0) + a_0 S_0^{(2)}(x_0) + a_1 S_1^{(2)}(x_0). \tag{6.80}$$

Since $S_i^{(2)}(x_0) = \frac{1}{h^2} S^{(2)}\left(\frac{x_0 - x_i}{h}\right)$, we have

$$S_{-1}^{(2)}(x_0) = \frac{1}{h^2} S^{(2)}\left(\frac{x_0 - (x_0 - h)}{h}\right) = \frac{1}{h^2} S^{(2)}(1) = \frac{6}{h^2}, \tag{6.81a}$$

$$S_0^{(2)}(x_0) = \frac{1}{h^2} S^{(2)}\left(\frac{x_0 - x_0}{h}\right) = \frac{1}{h^2} S^{(2)}(0) = -\frac{12}{h^2}, \tag{6.81b}$$

$$S_1^{(2)}(x_0) = \frac{1}{h^2} S^{(2)}\left(\frac{x_0 - (x_0 + h)}{h}\right) = \frac{1}{h^2} S^{(2)}(-1) = \frac{6}{h^2}, \tag{6.81c}$$

where (6.71c) was used. Thus, (6.80) reduces to

$$h^{-2}[6a_{-1} - 12a_0 + 6a_1] = 0$$

or

$$a_{-1} = 2a_0 - a_1. \tag{6.82}$$

Similarly

$$p_3(x_n) = a_{n-1} S_{n-1}(x_n) + a_n S_n(x_n) + a_{n+1} S_{n+1}(x_n),$$

so that

$$p_3^{(2)}(x_n) = a_{n-1} S_{n-1}^{(2)}(x_n) + a_n S_n^{(2)}(x_n) + a_{n+1} S_{n+1}^{(2)}(x_n). \tag{6.83}$$

Since $S_i^{(2)}(x_n) = \frac{1}{h^2} S^{(2)}\left(\frac{x_n - x_i}{h}\right)$, we have

$$S_{n-1}^{(2)}(x_n) = \frac{1}{h^2} S^{(2)}\left(\frac{(x_0 + nh) - (x_0 + (n-1)h)}{h}\right) = \frac{1}{h^2} S^{(2)}(1) = \frac{6}{h^2}, \tag{6.84a}$$

$$S_n^{(2)}(x_n) = \frac{1}{h^2} S^{(2)}\left(\frac{(x_0 + nh) - (x_0 + nh)}{h}\right) = \frac{1}{h^2} S^{(2)}(0) = -\frac{12}{h^2}, \tag{6.84b}$$

$$S_{n+1}^{(2)}(x_n) = \frac{1}{h^2} S^{(2)}\left(\frac{(x_0 + nh) - (x_0 + (n+1)h)}{h}\right) = \frac{1}{h^2} S^{(2)}(-1) = \frac{6}{h^2}, \tag{6.84c}$$

where (6.71c) was again employed. Thus, (6.83) reduces to

$$h^{-2}[6a_{n-1} - 12a_n + 6a_{n+1}] = 0$$

or

$$a_{n+1} = 2a_n - a_{n-1}. \tag{6.85}$$

Now since [from (6.79)]

$$a_{-1} + 4a_0 + a_1 = f(x_0)$$

using (6.82) we have

$$6a_0 = f(x_0), \tag{6.86a}$$

and similarly

$$a_{n-1} + 4a_n + a_{n+1} = f(x_n),$$

so via (6.85) we have

$$6a_n = f(x_n). \tag{6.86b}$$

Using (6.86) we may rewrite (6.79) as the linear system

$$\underbrace{\begin{bmatrix} 4 & 1 & 0 & \cdots & 0 & 0 & 0 \\ 1 & 4 & 1 & \cdots & 0 & 0 & 0 \\ \vdots & \vdots & \vdots & & \vdots & \vdots & \vdots \\ 0 & 0 & 0 & \cdots & 1 & 4 & 1 \\ 0 & 0 & 0 & \cdots & 0 & 1 & 4 \end{bmatrix}}_{=A} \underbrace{\begin{bmatrix} a_1 \\ a_2 \\ \vdots \\ a_{n-2} \\ a_{n-1} \end{bmatrix}}_{=a} = \underbrace{\begin{bmatrix} f(x_1) - \frac{1}{6}f(x_0) \\ f(x_2) \\ \vdots \\ f(x_{n-2}) \\ f(x_{n-1}) - \frac{1}{6}f(x_n) \end{bmatrix}}_{=f}. \tag{6.87}$$

Here we have $A \in \mathbf{R}^{(n-1)\times(n-1)}$, so now the tridiagonal system $Aa = f$ has a unique solution, assuming that A^{-1} exists. The existence of A^{-1} will be justified below.

Now we will consider the case of the complete spline. In the case of

$$p_3^{(1)}(x_0) = a_{-1}S_{-1}^{(1)}(x_0) + a_0 S_0^{(1)}(x_0) + a_1 S_1^{(1)}(x_0) = f^{(1)}(x_0), \tag{6.88}$$

since $S_i^{(1)}(x_0) = \frac{1}{h}S^{(1)}\left(\frac{x_0 - x_i}{h}\right)$, we have, using (6.71b)

$$S_{-1}^{(1)}(x_0) = \frac{1}{h}S^{(1)}\left(\frac{x_0 - (x_0 - h)}{h}\right) = \frac{1}{h}S^{(1)}(1) = -\frac{3}{h},$$

$$S_0^{(1)}(x_0) = \frac{1}{h}S^{(1)}\left(\frac{x_0 - x_0}{h}\right) = \frac{1}{h}S^{(1)}(0) = 0,$$

$$S_1^{(1)}(x_0) = \frac{1}{h}S^{(1)}\left(\frac{x_0 - (x_0 + h)}{h}\right) = \frac{1}{h}S^{(1)}(-1) = \frac{3}{h},$$

so (6.88) becomes

$$p_3^{(1)}(x_0) = 3h^{-1}[-a_{-1} + a_1] = f^{(1)}(x_0). \tag{6.89}$$

Similarly

$$p_3^{(1)}(x_n) = a_{n-1}S_{n-1}^{(1)}(x_n) + a_n S_n^{(1)}(x_n) + a_{n+1}S_{n+1}^{(1)}(x_n) = f^{(1)}(x_n) \tag{6.90}$$

reduces to

$$p_3^{(1)}(x_n) = 3h^{-1}[-a_{n-1} + a_{n+1}] = f^{(1)}(x_n). \tag{6.91}$$

From (6.89) and (6.91), we obtain

$$a_{-1} = a_1 - \tfrac{1}{3}hf^{(1)}(x_0), \qquad a_{n+1} = a_{n-1} + \tfrac{1}{3}hf^{(1)}(x_n). \tag{6.92}$$

If we substitute (6.92) into (6.79), we obtain

$$\underbrace{\begin{bmatrix} 4 & 2 & 0 & \cdots & 0 & 0 & 0 \\ 1 & 4 & 1 & \cdots & 0 & 0 & 0 \\ \vdots & \vdots & \vdots & & \vdots & \vdots & \vdots \\ 0 & 0 & 0 & \cdots & 1 & 4 & 1 \\ 0 & 0 & 0 & \cdots & 0 & 2 & 4 \end{bmatrix}}_{=A} \underbrace{\begin{bmatrix} a_0 \\ a_1 \\ \vdots \\ a_{n-1} \\ a_n \end{bmatrix}}_{=a} = \underbrace{\begin{bmatrix} f(x_0) + \tfrac{1}{3}hf^{(1)}(x_0) \\ f(x_1) \\ \vdots \\ f(x_{n-1}) \\ f(x_n) - \tfrac{1}{3}hf^{(1)}(x_n) \end{bmatrix}}_{=f}. \tag{6.93}$$

Now we have $A \in \mathbf{R}^{(n+1)\times(n+1)}$, and we see that $Aa = f$ of (6.93) will have a unique solution provided that A^{-1} exists.

Matrix A is tridiagonal, and so is quite *sparse* (i.e., it has many zero-valued entries) because of the "locality" of the function $S(x)$. This locality makes it possible to evaluate $p_3(x)$ in (6.74) efficiently. If we know that $x \in [x_k, x_{k+1}]$, then

$$p_3(x) = a_{k-1}S_{k-1}(x) + a_k S_k(x) + a_{k+1}S_{k+1}(x) + a_{k+2}S_{k+2}(x). \tag{6.94}$$

We write supp $g(x) = [a, b]$ to represent the fact that $g(x) = 0$ for all $x < a$, and $x > b$, while $g(x)$ might not be zero-valued for $x \in [a, b]$. From (6.68) we may therefore say that supp $S(x) = [-2, 2]$, and so from (6.72), supp $S_i(x) = [x_i - 2h, x_i + 2h]$, so we see that $S_i(x)$ is not necessarily zero-valued for $x \in [x_i - 2h, x_i + 2h]$. From this, (6.94), and the fact that $x_i = x_0 + ih$, we obtain

$$\text{supp } p_3(x) = \cup_{i=k-1}^{k+2}\text{supp } S_i(x) = [x_0 + (k-3)h, x_0 + (k+1)h] \cup$$
$$\cdots \cup [x_0 + kh, x_0 + (k+4)h] \tag{6.95}$$

(if A_i are sets, then $\cup_{i=n}^m A_i = A_n \cup \cdots \cup A_m$), which "covers" the interval $[x_k, x_{k+1}] = [x_0 + kh, x_0 + (k+1)h]$. Because the sampling grid (6.65)–(6.67), is uniform it is easy to establish that (via $x \geq x_0 + kh$)

$$k = \left\lfloor \frac{x - x_0}{h} \right\rfloor, \tag{6.96}$$

where $\lfloor x \rfloor = $ the largest integer that is $\leq x \in \mathbf{R}$. Since supp $S(x)$ is an interval of finite length, we say that $S(x)$ is *compactly supported*. This locality of support is also a part of what makes splines useful in wavelet constructions.

We now consider the solution of $Aa = f$ in either of (6.87) or (6.93). Obviously, Gaussian elimination is a possible method, but this is not efficient since Gaussian elimination is a general procedure that does not take advantage of any matrix structure. The sparse tridiagonal structure of A can be exploited for a more efficient solution. We will consider a general method of tridiagonal linear system solution here, but it is one that is based on modifying the general Gaussian elimination method. We begin by defining

$$\underbrace{\begin{bmatrix} a_{00} & a_{01} & 0 & \cdots & 0 & 0 \\ a_{10} & a_{11} & a_{12} & \cdots & 0 & 0 \\ 0 & a_{21} & a_{22} & \cdots & 0 & 0 \\ \vdots & \vdots & \vdots & & \vdots & \vdots \\ 0 & 0 & 0 & \cdots & a_{n-1,n-1} & a_{n-1,n} \\ 0 & 0 & 0 & \cdots & a_{n,n-1} & a_{n,n} \end{bmatrix}}_{=A} \underbrace{\begin{bmatrix} x_0 \\ x_1 \\ x_2 \\ \vdots \\ x_{n-1} \\ x_n \end{bmatrix}}_{=x} = \underbrace{\begin{bmatrix} f_0 \\ f_1 \\ f_2 \\ \vdots \\ f_{n-1} \\ f_n \end{bmatrix}}_{=f}. \quad (6.97)$$

Clearly, $A \in \mathbf{R}^{(n+1) \times (n+1)}$. There is some terminology associated with tridiagonal matrices. The main diagonal consists of the elements $a_{i,i}$, while the diagonal above this consists of the elements $a_{i,i+1}$, and is often called the *superdiagonal*. Similarly, the diagonal below the main diagonal consists of the elements $a_{i+1,i}$, and is often called the *subdiagonal*.

Our approach to solving $Ax = f$ in (6.97) will be to apply Gaussian elimination to the augmented linear system $[A|f]$ in order to reduce A to upper triangular form, and then backward substitution will be used to solve for x. This is much the same procedure as considered in Section 4.5, except that the tridiagonal structure of A makes matters easier. To see this, consider the special case of $n = 3$ as an example, that is ($A^0 = A$, $f^0 = f$)

$$[A^0 | f^0] = \begin{bmatrix} a_{00}^0 & a_{01}^0 & 0 & 0 & | & f_0^0 \\ a_{10}^0 & a_{11}^0 & a_{12}^0 & 0 & | & f_1^0 \\ 0 & a_{21}^0 & a_{22}^0 & a_{23}^0 & | & f_2^0 \\ 0 & 0 & a_{32}^0 & a_{33}^0 & | & f_3^0 \end{bmatrix}. \quad (6.98)$$

We may apply elementary row operations to eliminate a_{10}^0 in (6.98). Thus, (6.98) becomes

$$[A^1 | f^1] = \begin{bmatrix} a_{00}^0 & a_{01}^0 & 0 & 0 & | & f_0^0 \\ 0 & a_{11}^0 - \frac{a_{10}^0}{a_{00}^0} a_{01}^0 & a_{12}^0 & 0 & | & f_1^0 - \frac{a_{10}^0}{a_{00}^0} f_0^0 \\ 0 & a_{21}^0 & a_{22}^0 & a_{23}^0 & | & f_2^0 \\ 0 & 0 & a_{32}^0 & a_{33}^0 & | & f_3^0 \end{bmatrix}$$

$$
= \begin{bmatrix}
a_{00}^1 & a_{01}^1 & 0 & 0 & | & f_0^1 \\
0 & a_{11}^1 & a_{12}^1 & 0 & | & f_1^1 \\
0 & a_{21}^1 & a_{22}^1 & a_{23}^1 & | & f_2^1 \\
0 & 0 & a_{32}^1 & a_{33}^1 & | & f_3^1
\end{bmatrix}. \tag{6.99}
$$

Now we apply elementary row operations to eliminate a_{21}^1. Thus, (6.99) becomes

$$
[A^2 | f^2] = \begin{bmatrix}
a_{00}^1 & a_{01}^1 & 0 & 0 & | & f_0^1 \\
0 & a_{11}^1 & a_{12}^1 & 0 & | & f_1^1 \\
0 & 0 & a_{22}^1 - \frac{a_{21}^1}{a_{11}^1}a_{12}^1 & a_{23}^1 & | & f_2^1 - \frac{a_{21}^1}{a_{11}^1}f_1^1 \\
0 & 0 & a_{32}^1 & a_{33}^1 & | & f_3^1
\end{bmatrix}
$$

$$
= \begin{bmatrix}
a_{00}^2 & a_{01}^2 & 0 & 0 & | & f_0^2 \\
0 & a_{11}^2 & a_{12}^2 & 0 & | & f_1^2 \\
0 & 0 & a_{22}^2 & a_{23}^2 & | & f_2^2 \\
0 & 0 & a_{32}^2 & a_{33}^2 & | & f_3^2
\end{bmatrix}. \tag{6.100}
$$

Finally, we eliminate a_{32}^2, in which case (6.100) becomes

$$
[A^3 | f^3] = \begin{bmatrix}
a_{00}^2 & a_{01}^2 & 0 & 0 & | & f_0^2 \\
0 & a_{11}^2 & a_{12}^2 & 0 & | & f_1^2 \\
0 & 0 & a_{22}^2 & a_{23}^2 & | & f_2^2 \\
0 & 0 & 0 & a_{33}^2 - \frac{a_{32}^2}{a_{22}^2}a_{23}^2 & | & f_3^2 - \frac{a_{32}^2}{a_{22}^2}f_2^2
\end{bmatrix}
$$

$$
= \begin{bmatrix}
a_{00}^3 & a_{01}^3 & 0 & 0 & | & f_0^3 \\
0 & a_{11}^3 & a_{12}^3 & 0 & | & f_1^3 \\
0 & 0 & a_{22}^3 & a_{23}^3 & | & f_2^3 \\
0 & 0 & 0 & a_{33}^3 & | & f_3^3
\end{bmatrix}. \tag{6.101}
$$

We have $A^3 = U$, an upper triangular matrix. Thus, $Ux = f^3$ can be solved by backward substitution. The reader should write a pseudocode program to implement this approach for any n. Ideally, the code ought to be written such that only the vector f, and the main, super-, and subdiagonals of A are stored.

We observe that the algorithm we have just constructed will work only if $a_{i,i}^i \neq 0$ for all $i = 0, 1, \ldots, n$. Thus, our approach may not be stable. We expect this potential problem because our algorithm does not employ any pivoting. However, our application here is to use the algorithm to solve either (6.87) or (6.93). For A in either of these cases, we never get $a_{i,i}^i = 0$. We may justify this claim as follows. In Definition 6.2 it is understood that $a_{i,j} = 0$ for $i, j \notin \mathbf{Z}_{n+1}$.

Definition 6.2: Diagonal Dominance The tridiagonal matrix $A = [a_{i,j}]_{i,j=0,\ldots,n} \in \mathbf{R}^{(n+1) \times (n+1)}$ is *diagonally dominant* if

$$a_{i,i} > |a_{i,i-1}| + |a_{i,i+1}| > 0$$

for $i = 0, 1, \ldots, n$.

It is clear that A in (6.87), or in (6.93) is diagonally dominant. For the algorithm we have developed to solve $Ax = f$, in general we can say

$$a_{k+1,k+1}^{k+1} = a_{k+1,k+1}^k - \frac{a_{k+1,k}^k}{a_{k,k}^k} a_{k,k+1}^k, \tag{6.102}$$

with $a_{k+1,k}^{k+1} = 0$, and

$$a_{k,k+1}^k = a_{k,k+1} \tag{6.103}$$

for $k = 0, 1, \ldots, n - 1$. Condition (6.103) states that the algorithm does not modify the superdiagonal elements of A.

Theorem 6.2: If tridiagonal matrix A is diagonally dominant, then the algorithm for solving $Ax = f$ will not yield $a_{i,i}^i = 0$ for any $i = 0, 1, \ldots, n$.

Proof We give only an outline of the main idea. The complete proof needs mathematical induction.

From Definition 6.2

$$a_{0,0}^0 > |a_{0,-1}^0| + |a_{0,1}^0| = |a_{0,1}^0| > 0,$$

so from (6.102)

$$a_{1,1}^1 = a_{1,1}^0 - \frac{a_{0,1}^0}{a_{0,0}^0} a_{1,0}^0,$$

with $a_{1,0}^1 = 0$. Thus, A^1 can be obtained from $A^0 = A$. Now again from Definition 6.2

$$a_{1,1}^0 > |a_{1,0}^0| + |a_{1,2}^0| > 0,$$

so that, because $0 < |a_{0,1}^0/a_{0,0}^0| < 1$, we can say that

$$a_{1,1}^1 \geq a_{1,1}^0 - \left| \frac{a_{0,1}^0}{a_{0,0}^0} \right| |a_{1,0}^0|$$

$$\geq |a_{1,0}^0| + |a_{1,2}^0| - \left| \frac{a_{0,1}^0}{a_{0,0}^0} \right| |a_{1,0}^0|$$

$$= \left(1 - \left| \frac{a_{0,1}^0}{a_{0,0}^0} \right| \right) |a_{1,0}^0| + |a_{1,2}^0|$$

$$> |a_{1,2}^0| = |a_{1,2}^1| = |a_{1,0}^1| + |a_{1,2}^1| > 0,$$

and A^1 is diagonally dominant. The diagonal dominance of A^0 implies the diagonal dominance of A^1. In general, we see that if A^k is diagonally dominant, then A^{k+1} will be diagonally dominant as well, and so A^{k+1} can be formed from A^k for all $k = 0, 1, \ldots, n - 1$. Thus, $a_{i,i}^i \neq 0$ for all $i = 0, 1, \ldots, n$.

If A is diagonally dominant, it will be well-conditioned, too (not obvious), and so our algorithm to solve $Ax = f$ is actually quite stable in this case.

Before considering an example of spline interpolation, we specify a result concerning the accuracy of approximation with cubic splines. This is a modified version of Theorem 3.13 in Burden and Faires [6], or of Theorem 4.7 in Epperson [9].

Theorem 6.3:   If $f(x) \in C^4[a, b]$ with $\max_{x \in [a,b]} |f^{(4)}(x)| \leq M$, and $p_3(x)$ is the unique complete spline interpolant for $f(x)$, then

$$\max_{x \in [a,b]} |f(x) - p_3(x)| \leq \frac{5}{384} M h^4.$$

Proof   The proof is omitted, but Epperson [9] suggests referring to the article by Hall [10].

We conclude this section with an illustration of the quality of approximation of the cubic splines.

Example 6.5   Figure 6.3 shows the natural and complete cubic spline interpolants to the function $f(x) = \exp(-x^2)$ for $x \in [-1, 1]$. We have chosen the nodes $x_k = -1 + \frac{1}{2}k$, for $k = 0, 1, 2, 3, 4$ (i.e., $n = 4$, and $h = \frac{1}{2}$). Clearly, $f^{(1)}(x) = -2x \exp(-x^2)$. Thus, at the nodes

$$f(\pm 1) = 0.36787944, \qquad f(\pm \tfrac{1}{2}) = 0.60653066, \qquad f(0) = 1.00000000,$$

and

$$f^{(1)}(\pm 1) = \mp 0.73575888.$$

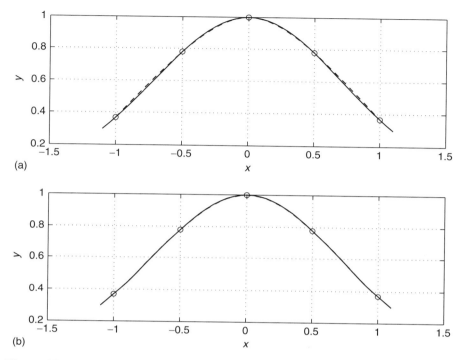

Figure 6.3   Natural (a) and complete (b) spline interpolants (dashed lines) for the function $f(x) = e^{-x^2}$ (solid lines) on the interval $[-1, 1]$. The circles correspond to node locations. Here $n = 4$, with $h = \frac{1}{2}$, and $x_0 = -1$. The nodes are at $x_k = x_0 + hk$ for $k = 0, \ldots, n$.

Of course, the spline series for our example has the form

$$p_3(x) = \sum_{k=-1}^{5} a_k S_k(x),$$

so we need to determine the coefficients a_k from both (6.87), and (6.93).

Considering the natural spline interpolant first, from (6.87), we have

$$\begin{bmatrix} 4 & 1 & 0 \\ 1 & 4 & 1 \\ 0 & 1 & 4 \end{bmatrix} \begin{bmatrix} a_1 \\ a_2 \\ a_3 \end{bmatrix} = \begin{bmatrix} f(x_1) - \frac{1}{6}f(x_0) \\ f(x_2) \\ f(x_3) - \frac{1}{6}f(x_4) \end{bmatrix}.$$

Additionally

$$a_0 = \frac{1}{6}f(x_0) = \frac{1}{6}f(-1), \quad a_4 = \frac{1}{6}f(x_4) = \frac{1}{6}f(1),$$

and

$$a_{-1} = f(x_0) - 4a_0 - a_1, \quad a_5 = f(x_4) - 4a_4 - a_3.$$

The natural spline series coefficients are therefore

k	a_k
-1	-0.01094139
0	0.06131324
1	0.13356787
2	0.18321607
3	0.13356787
4	0.06131324
5	-0.01094139

Now, on considering the case of the complete spline, from (6.93) we have

$$
\begin{bmatrix}
4 & 2 & 0 & 0 & 0 \\
1 & 4 & 1 & 0 & 0 \\
0 & 1 & 4 & 1 & 0 \\
0 & 0 & 1 & 4 & 1 \\
0 & 0 & 0 & 2 & 4
\end{bmatrix}
\begin{bmatrix}
a_0 \\
a_1 \\
a_2 \\
a_3 \\
a_4
\end{bmatrix}
=
\begin{bmatrix}
f(x_0) + \frac{1}{3}hf^{(1)}(x_0) \\
f(x_1) \\
f(x_2) \\
f(x_3) \\
f(x_4) - \frac{1}{3}hf^{(1)}(x_4)
\end{bmatrix}.
$$

Additionally

$$
a_{-1} = a_1 - \tfrac{1}{3}hf^{(1)}(x_0), \quad a_5 = a_3 + \tfrac{1}{3}hf^{(1)}(x_4).
$$

The complete spline series coefficients are therefore

k	a_k
-1	0.01276495
0	0.05493076
1	0.13539143
2	0.18230428
3	0.13539143
4	0.05493076
5	0.01276495

This example demonstrates what was suggested earlier, and that is that the complete spline interpolant tends to be more accurate than the natural spline interpolant. However, accuracy of the complete spline interpolant is contingent on accurate estimation, or knowledge of the derivatives $f^{(1)}(x_0)$ and $f^{(1)}(x_n)$.

REFERENCES

1. D. R. Hill, *Experiments in Computational Matrix Algebra* (C. B. Moler, consulting ed.), Random House, New York, 1988.
2. G. H. Golub and C. F. Van Loan, *Matrix Computations*, 2nd ed., Johns Hopkins Univ. Press, Baltimore, MD, 1989.

3. F. B. Hildebrand, *Introduction to Numerical Analysis*, 2nd ed., McGraw-Hill, New York, 1974.

4. G. E. Forsythe, M. A. Malcolm, and C. B. Moler, *Computer Methods for Mathematical Computations*, Prentice-Hall, Englewood Cliffs, NJ, 1977.

5. E. Isaacson and H. B. Keller, *Analysis of Numerical Methods*, Wiley, New York, 1966.

6. R. L. Burden and J. D. Faires, *Numerical Analysis*, 4th ed., PWS-KENT Publi., Boston, MA, 1989.

7. C. K. Chui, *An Introduction to Wavelets*, Academic Press, Boston, MA, 1992.

8. M. Unser and T. Blu, "Wavelet Theory Demystified," *IEEE Trans. Signal Process.* **51**, 470–483 (Feb. 2003).

9. J. F. Epperson, *An Introduction to Numerical Methods and Analysis*, Wiley, New York, 2002.

10. C. A. Hall, "On Error Bounds for Spline Interpolation," *J. Approx. Theory.* **1**, 209–218 (1968).

PROBLEMS

6.1. Find the Lagrange interpolant $p_2(t) = \sum_{j=0}^{2} x_j L_j(t)$ for data set $\{(-1, \frac{1}{2}), (0, 1), (1, -1)\}$. Find $p_{2,j}$ in $p_2(t) = \sum_{j=0}^{2} p_{2,j} t^j$.

6.2. Find the Lagrange interpolant $p_3(t) = \sum_{j=0}^{3} x_j L_j(t)$ for data set $\{(0, 1), (\frac{1}{2}, 2), (1, \frac{3}{2}), (2, -1)\}$. Find $p_{3,j}$ in $p_3(t) = \sum_{j=0}^{3} p_{3,j} t^j$.

6.3. We want to interpolate $x(t)$ at $t = t_1, t_2$, and $x^{(1)}(t) = dx(t)/dt$ at $t = t_0, t_3$ using $p_3(t) = \sum_{j=0}^{3} p_{3,j} t^j$. Let $x_j = x(t_j)$ and $x_j^{(1)} = x^{(1)}(t_j)$. Find the linear system of equations satisfied by $(p_{3,j})$.

6.4. Find a general expression for $L_j^{(1)}(t) = dL_j(t)/dt$.

6.5. In Section 6.2 it was mentioned that fast algorithms exist to solve Vandermonde linear systems of equations. The Vandermonde system $Ap = x$ is given in expanded form as

$$
\begin{bmatrix}
1 & t_0 & t_0^2 & \cdots & t_0^{n-1} & t_0^n \\
1 & t_1 & t_1^2 & \cdots & t_1^{n-1} & t_1^n \\
\vdots & \vdots & \vdots & & \vdots & \vdots \\
1 & t_{n-1} & t_{n-1}^2 & \cdots & t_{n-1}^{n-1} & t_{n-1}^n \\
1 & t_n & t_n^2 & \cdots & t_n^{n-1} & t_n^n
\end{bmatrix}
\begin{bmatrix}
p_0 \\
p_1 \\
\vdots \\
p_{n-1} \\
p_n
\end{bmatrix}
=
\begin{bmatrix}
x_0 \\
x_1 \\
\vdots \\
x_{n-1} \\
x_n
\end{bmatrix}, \quad (6.P.1)
$$

and $p_n(t) = \sum_{j=0}^{n} p_j t^j$ interpolates the points $\{(t_j, x_j) | k = 0, 1, 2, \ldots, n\}$. A fast algorithm to solve (6.P.1) is

```
for k := 0 to n − 1 do begin
    for i := n downto to k + 1 do begin
        x_i := (x_i − x_{i−1})/(t_i − t_{i−k−1});
        end;
    end;
for k := n − 1 downto 0 do begin
    for i := k to n − 1 do begin
        x_i := x_i − t_k x_{i+1};
        end;
    end;
```

The algorithm overwrites vector $x = [x_0 \; x_1 \; \cdots \; x_n]^T$ with the vector $p = [p_0 \; p_1 \; \cdots \; p_n]^T$.

(a) Count the number of arithmetic operations needed by the fast algorithm. What is the asymptotic time complexity of it, and how does this compare with Gaussian elimination as a method to solve (6.P.1)?

(b) Test the fast algorithm out on the system

$$\begin{bmatrix} 1 & 1 & 1 & 1 \\ 1 & 2 & 4 & 8 \\ 1 & 3 & 9 & 27 \\ 1 & 4 & 16 & 64 \end{bmatrix} \begin{bmatrix} p_0 \\ p_1 \\ p_2 \\ p_3 \end{bmatrix} = \begin{bmatrix} 10 \\ 26 \\ 58 \\ 112 \end{bmatrix}.$$

(c) The "top" k loop in the fast algorithm produces the *Newton form* (Section 6.3) of the representation for $p_n(t)$. For the system in (b), confirm that

$$p_n(t) = \sum_{k=0}^{n} x_k \prod_{i=0}^{k-1} (t - t_i),$$

where $\{x_k | k \in \mathbf{Z}_{n+1}\}$ are the outputs from the top k loop. Since in (b) $n = 3$, we must have for this particular special case

$$p_3(t) = x_0 + x_1(t - t_0) + x_2(t - t_0)(t - t_1) + x_3(t - t_0)(t - t_1)(t - t_2).$$

(*Comment:* It has been noted by Björck and Pereyra that the fast algorithm often yields accurate results even when A is ill-conditioned.)

6.6. Prove that for

$$A_n = \begin{bmatrix} 1 & t_0 & t_0^2 & \cdots & t_0^{n-1} & t_0^n \\ 1 & t_1 & t_1^2 & \cdots & t_1^{n-1} & t_1^n \\ \vdots & \vdots & \vdots & & \vdots & \vdots \\ 1 & t_{n-1} & t_{n-1}^2 & \cdots & t_{n-1}^{n-1} & t_{n-1}^n \\ 1 & t_n & t_n^2 & \cdots & t_n^{n-1} & t_n^n \end{bmatrix},$$

we have

$$\det(A_n) = \prod_{0 \le i < j \le n} (t_i - t_j).$$

(*Hint:* Use mathematical induction.)

6.7. Write a MATLAB function to interpolate the data $\{(t_j, x_j) | j \in \mathbf{Z}_{n+1}\}$ with polynomial $p_n(t)$ via Lagrange interpolation. The function must accept t as input, and return $p_n(t)$.

6.8. Runge's phenomenon was mentioned in Section 6.2 with respect to interpolating $f(t) = 1/(1 + t^2)$ on $t \in [-5, 5]$. Use polynomial $p_n(t)$ to interpolate $f(t)$ at the points $t_k = t_0 + kh$, $k = 0, 1, \ldots, n$, where $t_0 = -5$, and $t_n = 5$ (so $h = (t_n - t_0)/n$). Do this for $n = 5, 8, 10$. Use the MATLAB function from the previous problem. Plot $f(t)$ and $p_n(t)$ on the same graph for all of $n = 5, 8, 10$. Comment on the accuracy of interpolation as n increases.

6.9. Suppose that we wish to interpolate $f(t) = \sin t$ for $t \in [0, \pi/2]$ using polynomial $p_n(t) = \sum_{j=0}^{n} p_{n,j} t^j$. The approximation error is $e_n(t) = f(t) - p_n(t)$. Since $|f^{(n)}(t)| \le 1$ for all $t \in \mathbf{R}$ via (6.14), it follows that

$$|e_n(t)| \le \frac{1}{(n+1)!} \prod_{i=0}^{n} |t - t_i| = b(t). \tag{6.P.2}$$

Let the grid (sample) points be $t_0 = 0$, $t_n = \pi/2$, and $t_k = t_0 + kh$ for $k = 0, 1, \ldots, n$ so that $h = (t_n - t_0)/n = \frac{\pi}{2n}$. Using MATLAB

(a) For $n = 2$, on the same graph plot $f(t) - p_n(t)$, and $\pm b(t)$.
(b) For $n = 4$, on the same graph plot $f(t) - p_n(t)$, and $\pm b(t)$.

Cases (a) and (b) can be separate plots (e.g., using MATLAB subplot). Does the bound in (6.P.2) hold in both cases?

6.10. Consider $f(t) = \cosh t = \frac{1}{2}[e^t + e^{-t}]$ which is to be interpolated by $p_2(t) = \sum_{j=0}^{2} p_{2,j} t^j$ on $t \in [-1, 1]$ at the points $t_0 = -1, t_0 = 0, t_1 = 1$. Use (6.14) to find an upper bound on the *size of the approximation error* $e_n(t)$ for $t \in [-1, 1]$, where the size of the approximation error is given by the norm

$$\|e_n\|_\infty = \max_{a \le t \le b} |e_n(t)|,$$

where here $a = -1, b = 1$.

6.11. For

$$V = \begin{bmatrix} 1 & 1 & \cdots & 1 & 1 \\ t_0 & t_1 & \cdots & t_{n-1} & t_n \\ \vdots & \vdots & & \vdots & \vdots \\ t_0^n & t_1^n & \cdots & t_{n-1}^n & t_n^n \end{bmatrix} \in \mathbf{R}^{(n+1) \times (n+1)},$$

it is claimed by Gautschi that

$$\|V^{-1}\|_\infty \le \max_{0\le k\le n} \prod_{\substack{i=0 \\ i\ne k}}^{n} \frac{1+|t_i|}{|t_k - t_i|}. \tag{6.P.3}$$

Find an upper bound on $\kappa_\infty(V)$ that uses (6.P.3). How useful is an *upper* bound on the condition number? Would a lower bound on condition number be more useful? Explain.

6.12. For each function listed below, use divided difference tables (recall Example 6.3) to construct the degree n Newton interpolating polynomial for the specified points.

(a) $f(t) = \sqrt{t}$, $t_0 = 0, t_1 = 1, t_2 = 3$. Use $n = 2$.
(b) $f(t) = \cosh t$, $t_0 = -1, t_1 = \frac{-1}{2}, t_2 = \frac{1}{2}, t_3 = 1$. Use $n = 3$.
(c) $f(t) = \ln t$, $t_0 = 1, t_1 = 2, t_2 = 3$. Use $n = 2$.
(d) $f(t) = 1/(1+t)$, $t_0 = 0, t_1 = \frac{1}{2}, t_2 = 1, t_3 = 2$. Use $n = 3$.

6.13. Prove Eq. (6.33).

6.14. Consider Theorem 6.1. For $n = 2$, find $h_k(x)$ and $\hat{h}_k(x)$ as direct-form (i.e., a form such as (6.4)) polynomials in x. Of course, do this for all $k = 0, 1, 2$. Find Hermite interpolation polynomial $p_{2n+1}(x)$ for $n = 2$ that interpolates $f(x) = \sqrt{x}$ at the points $x_0 = 1, x_1 = 3/2, x_2 = 2$.

6.15. Find the Hermite interpolating polynomial $p_5(x)$ to interpolate $f(x) = e^x$ at the points $x_0 = 0, x_1 = 0.1, x_2 = 0.2$. Use MATLAB to compare the accuracy of approximation of $p_5(x)$ to that of $p_2^a(x)$ given in Example 6.3 (or Example 6.4).

6.16. The following matrix is important in solving spline interpolation problems [recall (6.87)]:

$$A_n = \begin{bmatrix} 4 & 1 & 0 & \cdots & 0 & 0 \\ 1 & 4 & 1 & \cdots & 0 & 0 \\ 0 & 1 & 4 & \cdots & 0 & 0 \\ \vdots & \vdots & \vdots & & \vdots & \vdots \\ 0 & 0 & 0 & \cdots & 4 & 1 \\ 0 & 0 & 0 & \cdots & 1 & 4 \end{bmatrix} \in \mathbf{R}^{n\times n}.$$

Suppose that $D_n = \det(A_n)$.

(a) Find D_1, D_2, D_3, and D_4 by direct (hand) calculation using the basic rules for computing determinants.

(b) Show that

$$D_{n+2} - 4D_{n+1} + D_n = 0$$

(i.e., the determinants we seek may be generated by a *second-order difference equation*).

(c) For suitable constants $\alpha, \beta \in \mathbf{R}$, it can be shown that

$$D_n = \alpha(2 + \sqrt{3})^n + \beta(2 - \sqrt{3})^n \qquad (6.P.4)$$

($n \in \mathbf{N}$). Find α and β.

[*Hint:* Set up two linear equations in the two unknowns α and β using (6.P.4) for $n = 1, 2$ and the results from (a).]

(d) Prove that $D_n > 0$ for all $n \in \mathbf{N}$.

(e) Is $A_n > 0$ for all $n \in \mathbf{N}$? Justify your answer.

[*Hint:* Recall part (a) in Problem 4.13 (of Chapter 4).]

6.17. Repeat Example 6.5, except use

$$f(x) = \frac{1}{1 + x^2},$$

and $x \in [-5, 5]$ with nodes $x_k = -5 + \frac{5}{2}k$ for $k = 0, 1, 2, 3, 4$ (i.e., $n = 4$, and $h = \frac{5}{2}$). How do the results compare to the results in Problem 6.8 (assuming that you have done Problem 6.8)? Of course, you should use MATLAB to "do the dirty work." You may use built-in MATLAB linear system solvers.

6.18. Repeat Example 6.5, except use

$$f(x) = \sqrt{1 + x}.$$

Use MATLAB to aid in the task.

6.19. Write a MATLAB function to solve tridiagonal linear systems of equations based on the theory for doing so given in Section 6.5. Test your algorithm out on the linear systems given in Example 6.5.

7 Nonlinear Systems of Equations

7.1 INTRODUCTION

In this chapter we consider the problem of finding x to satisfy the equation

$$f(x) = 0 \tag{7.1}$$

for arbitrary f, but such that $f(x) \in \mathbf{R}$. Usually we will restrict our discussion to the case where $x \in \mathbf{R}$, but $x \in \mathbf{C}$ can also be of significant interest in practice (especially if f is a polynomial). However, any x satisfying (7.1) is called a *root* of the equation. Such an x is also called a *zero* of the function f. More generally, we are also interested in solutions to systems of equations:

$$f_0(x_0, x_1, \ldots, x_{n-1}) = 0,$$

$$f_1(x_0, x_1, \ldots, x_{n-1}) = 0,$$

$$\vdots$$

$$f_{n-1}(x_0, x_1, \ldots, x_{n-1}) = 0. \tag{7.2}$$

Again, $f_i(x_0, x_1, \ldots, x_{n-1}) \in \mathbf{R}$, and we will assume $x_k \in \mathbf{R}$ for all $i, k \in \mathbf{Z}_n$. Various solution methods will be considered such the *bisection method*, *fixed-point method*, and the *Newton–Raphson method*. All of these methods are *iterative* in that they generate a sequence of points (or more generally vectors) that converge (we hope) to the desired solution. Consequently, ideas from Chapter 3 are relevant here. The number of iterations needed to achieve a solution with a given accuracy is considered. Iterative procedures can break down (i.e., fail) in various ways. The sequence generated by a procedure may diverge, oscillate, or display chaotic behavior (which in the past was sometimes described as "wandering behavior"). Examples of breakdown phenomena will therefore also be considered. Some attention will be given to chaotic phenomena, as these are of growing engineering interest. Applications of chaos are now being considered for such areas as cryptography and spread-spectrum digital communications. These proposed applications are still controversial, but at the very least some knowledge of chaos gives a deeper insight into the behavior of nonlinear dynamic systems.

An Introduction to Numerical Analysis for Electrical and Computer Engineers, by C.J. Zarowski
ISBN 0-471-46737-5 © 2004 John Wiley & Sons, Inc.

The equations considered in this chapter are *nonlinear*, and so are in contrast with those of Chapter 4. Chapter 4 considered linear systems of equations only. The reader is probably aware of the fact that a linear system either has a unique solution, no solution, or an infinite number of solutions. Which case applies depends on the size and rank of the matrix in the linear system. Chapter 4 emphasized the handling of square and invertible matrices for which the solution exists, and is unique. In Chapter 4 we saw that well-defined procedures exist (e.g., Gaussian elimination) that give the solution in a finite number of steps.

The solution of nonlinear equations is significantly more complicated than the solution of linear systems. Existence and uniqueness problems typically have no easy answers. For example, $e^{2x} + 1 = 0$ has no real-valued solutions, but if we allow $x \in \mathbf{C}$, then $x = \frac{1}{2}k\pi j$ for which k is an odd integer (since $e^{2x} + 1 = e^{j\pi k} + 1 = \cos(k\pi) + j\sin(k\pi) + 1 = \cos(k\pi) = -1 + 1 = 0$ as k is odd). On the other hand

$$e^{-t} - \sin(t) = 0$$

also has an infinite number of solutions, but for which $t \in \mathbf{R}$ (see Fig. 7.1). However, the solutions are not specifiable with a nice formula. In Fig. 7.1 the solutions correspond to the point where the two curves intersect each other.

Polynomial equations are of special interest.[1] For example, $x^2 + 1 = 0$ has only complex solutions $x = \pm j$. Multiple roots are also possible. For example

$$x^3 - 3x^2 + 3x - 1 = (x - 1)^3 = 0$$

has a *real root of multiplicity 3* at $x = 1$. We remark that the methods of this chapter are general and so (in principle) can be applied to find the solution of any nonlinear equation, polynomial or otherwise. But the special importance of polynomial equations has caused the development (over the centuries) of algorithms dedicated to polynomial equation solution. Thus, special algorithms exist to solve

$$p_n(x) = \sum_{k=0}^{n} p_{n,k} x^k = 0, \tag{7.3}$$

[1] Why are polynomial equations of special interest? There are numerous answers to this. But the reader already knows one reason why from basic electric circuits. For example, an unforced *RLC* (resistance × inductance × capacitance) circuit has a response due to energy initially stored in the energy storage elements (the inductors and capacitors). If $x(t)$ is the voltage drop across an element or the current through an element, then

$$a_n \frac{d^n x(t)}{dt^n} + a_{n-1} \frac{d^{n-1} x(t)}{dt^{n-1}} + \cdots + a_1 \frac{dx(t)}{dt} + a_0 = 0.$$

The coefficients a_k depend on the circuit elements. The solution to the differential equation depends on the roots of the characteristic equation:

$$a_n \lambda^n + a_{n-1} \lambda^{n-1} + \cdots + a_1 \lambda + a_0 = 0.$$

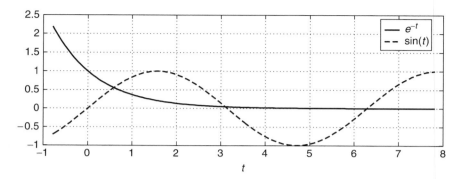

Figure 7.1   Plot of the individual terms in the equation $f(t) = e^{-t} - \sin(t) = 0$. The infinite number of solutions possible corresponds to the point where the plotted curves intersect.

and these algorithms do not apply to general nonlinear equations. Sometimes these are based on the general methods we consider in this chapter. Other times completely different methods are employed (e.g., replacing the problem of finding polynomial zeros by the equivalent problem of finding matrix eigenvalues as suggested in Jenkins and Traub [22]). However, we do not consider these special polynomial equation solvers here. We only mention that some interesting references on this matter are Wilkinson [1] and Cohen [2]. These describe the concept of ill-conditioned polynomials, and how to apply deflation procedures to produce more accurate estimates of roots. The difficulties posed by multiple roots are considered in Hull and Mathon [3]. Modern math-oriented software tools (e.g., MATLAB) often take advantage of theories such as described in Refs. 1–3, (as well as in other sources). In MATLAB polynomial zeros may be found using the roots and mroots functions. Function mroots is a modern root finder that reliably determines multiple roots.

7.2 BISECTION METHOD

The *bisection method* is a simple intuitive approach to solving

$$f(x) = 0. \tag{7.4}$$

It is assumed that $f(x) \in \mathbf{R}$, and that $x \in \mathbf{R}$. This method is based on the following theorem.

Theorem 7.1: Intermediate Value Theorem   If $f|[a, b] \to \mathbf{R}$ is continuous on the closed, bounded interval $[a, b]$, and $y_0 \in \mathbf{R}$ is such that $f(a) \leq y_0 \leq f(b)$, then there is an $x_0 \in [a, b]$ such that $f(x_0) = y_0$. In other words, a continuous function on a closed and bounded interval takes on all values between $f(a)$ and $f(b)$ at least once.

The bisection method works as follows. Suppose that we have an initial interval $[a_0, b_0]$ such that

$$f(a_0)f(b_0) < 0, \qquad (7.5)$$

which means that $f(a_0)$ and $f(b_0)$ have opposite signs (i.e., one is positive while the other is negative). By Theorem 7.1 there must be a $p \in (a_0, b_0)$ so that $f(p) = 0$. We say that $[a_0, b_0]$ *brackets the root* p. Suppose that

$$p_0 = \tfrac{1}{2}(a_0 + b_0). \qquad (7.6)$$

This is the midpoint of the interval $[a_0, b_0]$. Consider the following cases:

1. If $f(p_0) = 0$, then $p = p_0$ and we have found a root. We may stop at this point.
2. If $f(a_0)f(p_0) < 0$, then it must be the case that $p \in [a_0, p_0]$, so we define the new interval $[a_1, b_1] = [a_0, p_0]$. This new interval brackets the root.
3. If $f(p_0)f(b_0) < 0$ then it must be the case that $p \in [p_0, b_0]$ so we define the new interval $[a_1, b_1] = [p_0, b_0]$. This new interval brackets the root.

The process is repeated by considering the midpoint of the new interval, which is

$$p_1 = \tfrac{1}{2}(a_1 + b_1), \qquad (7.7)$$

and considering the three cases again. In principle, the process terminates when case 1 is encountered. In practice, case 1 is unlikely in part because of the effects of rounding errors, and so we need a more practical criterion to stop the process. This will be considered a little later on. For now, pseudocode describing the basic algorithm may be stated as follows:

```
input [a₀, b₀] which brackets the root p;
p₀ := (a₀ + b₀)/2;
k := 0;
while stopping criterion is not met do begin
      if f(aₖ)f(pₖ) < 0 then begin
            aₖ₊₁ := aₖ;
            bₖ₊₁ := pₖ;
            end;
      else begin
            aₖ₊₁ := pₖ;
            bₖ₊₁ := bₖ;
            end;
      pₖ₊₁ := (aₖ₊₁ + bₖ₊₁)/2;
      k := k + 1;
      end;
```

When the algorithm terminates, the last value of p_{k+1} computed is an estimate of p.

We see that the bisection algorithm constructs sequence $(p_n) = (p_0, p_1, p_2, \ldots)$ such that

$$\lim_{n \to \infty} p_n = p, \tag{7.8}$$

where p_n is the midpoint of $[a_n, b_n]$, and $f(p) = 0$. Formal proof that this process works (i.e., yields a unique p such that $f(p) = 0$) is due to the following theorem.

Theorem 7.2: Cantor's Intersection Theorem   Suppose that $([a_k, b_k])$ is a sequence of closed and bounded intervals such that

$$[a_0, b_0] \supset [a_1, b_1] \supset \cdots \supset [a_n, b_n] \supset \cdots,$$

with $\lim_{n \to \infty}(b_n - a_n) = 0$. There is a unique point $p \in [a_n, b_n]$ for all $n \in \mathbf{Z}^+$:

$$\bigcap_{n=0}^{\infty} [a_n, b_n] = \{p\}.$$

The bisection method produces (p_n) such that $a_n < p_n < b_n$, and $p \in [a_n, b_n]$ for all $n \in \mathbf{Z}^+$. Consequently, since $p_n = \frac{1}{2}(a_n + b_n)$

$$|p_n - p| \le |b_n - a_n| \le \frac{b - a}{2^n} \tag{7.9}$$

for $n \in \mathbf{Z}^+$, so that $\lim_{n \to \infty} p_n = p$. Recall that f is assumed continuous on $[a, b]$, so $\lim_{n \to \infty} f(p_n) = f(p)$. So now observe that

$$|p_n - a_n| \le \frac{1}{2^n}|b - a|, \, |b_n - p_n| \le \frac{1}{2^n}|b - a| \tag{7.10}$$

so, via $|x - y| = |(x - z) - (y - z)| \le |x - z| + |y - z|$ (triangle inequality), we have

$$|p - a_n| \le |p - p_n| + |p_n - a_n| \le \frac{1}{2^n}(b - a) + \frac{1}{2^n}(b - a) = \frac{1}{2^{n-1}}(b - a), \tag{7.11a}$$

and similarly

$$|p - b_n| \le |p - p_n| + |p_n - b_n| \le \frac{1}{2^{n-1}}(b - a). \tag{7.11b}$$

Thus

$$\lim_{n \to \infty} a_n = \lim_{n \to \infty} b_n = p. \tag{7.12}$$

At each step the root p is bracketed. This implies that there is a subsequence [of (p_n)] denoted (x_n) converging to p so that $f(x_n) > 0$ for all $n \in \mathbf{Z}^+$. Similarly, there is a subsequence (y_n) converging to p such that $f(y_n) < 0$ for all $n \in \mathbf{Z}^+$. Thus

$$f(p) = \lim_{n \to \infty} f(x_n) \ge 0, \, f(p) = \lim_{n \to \infty} f(y_n) \le 0,$$

which implies that $f(p) = 0$. We must conclude that the bisection method produces sequence (p_n) converging to p such that $f(p) = 0$. Hence, the bisection method always works.

Example 7.1   We want to find $5^{1/3}$ (cube root of five). This is equivalent to solving the equation $f(x) = x^3 - 5 = 0$. We note that $f(1) = -4$ and $f(2) = 3$, so we may use $[a_0, b_0] = [a, b] = [1, 2]$ to initially bracket the root. We remark that the "exact" value is $5^{1/3} = 1.709976$ (seven significant figures). Consider the following iterations of the bisection method:

$$[a_0, b_0] = [1, 2], \qquad p_0 = 1.500000, \qquad f(p_0) = -1.625000$$
$$[a_1, b_1] = [p_0, b_0], \qquad p_1 = 1.750000, \qquad f(p_1) = 0.359375$$
$$[a_2, b_2] = [a_1, p_1], \qquad p_2 = 1.625000, \qquad f(p_2) = -0.708984$$
$$[a_3, b_3] = [p_2, b_2], \qquad p_3 = 1.687500, \qquad f(p_3) = -0.194580$$
$$[a_4, b_4] = [p_3, b_3], \qquad p_4 = 1.718750, \qquad f(p_4) = 0.077362$$
$$[a_5, b_5] = [a_4, p_4], \qquad p_5 = 1.703125, \qquad f(p_5) = -0.059856$$

We see that $|p_5 - p| = |1.703125 - 1.709976| = 0.006851$. From (7.9)

$$|p_5 - p| \le \frac{b - a}{2^5} = \frac{2 - 1}{2^5} = \frac{1}{2^5} = 0.0312500.$$

The exact error certainly agrees with this bound.

When should we stop iterating? In other words, what *stopping criterion* should be chosen? Some possibilities are (for $\epsilon > 0$)

$$\left| \tfrac{1}{2} (b_n - a_n) \right| < \epsilon, \tag{7.13a}$$

$$|p_n - p| < \epsilon, \tag{7.13b}$$

$$f(p_n) < \epsilon, \tag{7.13c}$$

$$\left| \frac{p_n - p_{n-1}}{p_n} \right| < \epsilon \quad (p_n \ne 0). \tag{7.13d}$$

We would stop iterating when the inequalities are satisfied. Usually (7.13d) is recommended. Condition (7.13a) is not so good as termination depends on the size of the nth interval, while it is the accuracy of the estimate of p that is of most interest. Condition (7.13b) requires knowing p in advance, which is not reasonable since it is p that we are trying to determine. Condition (7.13c) is based on $f(p_n)$, and again we are more interested in how well p_n approximates p. Thus, we are left with (7.13d). This condition leads to termination when p_n is relatively not much different from p_{n-1}.

How may we characterize the computational efficiency of an iterative algorithm? In Chapter 4 the algorithms terminated in a finite number of steps (with the exception of the iterative procedures suggested in Section 4.7 which do not terminate unless a stopping condition is imposed), so flop counting was a reasonable measure. Where iterative procedures such as the bisection method are concerned, we prefer

Definition 7.1: Rate of Convergence   Suppose that (x_n) converges to 0: $\lim_{n \to \infty} x_n = 0$. Suppose that (p_n) converges to p, i.e., $\lim_{n \to \infty} p_n = p$. If there is a $K \in \mathbf{R}$, but $K > 0$, and $N \in \mathbf{Z}^+$ such that

$$|p_n - p| \le K|x_n|$$

for all $n \ge N$, then we say that (p_n) *converges to p with rate of convergence $O(x_n)$*.

This is an alternative use of the "big O" notation that was first seen in Chapter 4. Recalling (7.9), we obtain

$$|p_n - p| \le (b - a)\frac{1}{2^n}, \tag{7.14}$$

so that $K = b - a$, $x_n = \frac{1}{2^n}$, and $N = 0$. Thus, the bisection method generates sequence (p_n) that converges to p (with $f(p) = 0$) at the rate $O(1/2^n)$.

From (7.14), if we want $|p_n - p| < \epsilon$, then we may choose n so that

$$|p_n - p| \le \frac{b - a}{2^n} < \epsilon,$$

implying $2^n > (b - a)/\epsilon$, or we may choose n so

$$n = \left\lceil \log_2 \left(\frac{b - a}{\epsilon} \right) \right\rceil, \tag{7.15}$$

where $\lceil x \rceil$ = smallest integer greater than or equal to x. This can be used as an alternative means to terminate the bisection algorithm. But the conservative nature of (7.15) suggests that the algorithm that employs it may compute more iterations than are really necessary for the desired accuracy.

7.3 FIXED-POINT METHOD

Here we consider the *Banach fixed-point theorem* [4] as the theoretical basis for a nonlinear equation solver. Suppose that X is a set, and $T|X \to X$ is a mapping of X into itself. A *fixed point* of T is an $x \in X$ such that

$$Tx = x. \tag{7.16}$$

For example, suppose $X = [0, 1] \subset \mathbf{R}$, and

$$Tx = \tfrac{1}{2}x(1 - x). \tag{7.17}$$

We certainly have $Tx \in [0, 1]$ for any $x \in X$. The solution to

$$x = \tfrac{1}{2}x(1 - x)$$

(i.e., to $Tx = x$) is $x = 0$. So T has fixed point $x = 0$. (We reject the solution $x = -1$ since $-1 \notin [0, 1]$.)

Definition 7.2: Contraction   Let $X = (X, d)$ be a metric space. Mapping $T|X \to X$ is called a *contraction* (or a *contraction mapping*, or a *contractive mapping*) *on* X if there is an $\alpha \in \mathbf{R}$ such that $0 < \alpha < 1$, and for all $x, y \in X$

$$d(Tx, Ty) \le \alpha d(x, y). \tag{7.18}$$

Applying T to "points" x and y brings them closer together if T is a contractive mapping. If $\alpha = 1$, the mapping is sometimes called *nonexpansive*.

Theorem 7.3: Banach Fixed-Point Theorem   Consider the metric space $X = (X, d)$, where $X \not\in \emptyset$. Suppose that X is complete, and $T|X \to X$ is a contraction on X. Then T has a unique fixed point.

Proof   We must construct (x_n), and show that it is Cauchy so that (x_n) converges in X (recall Section 3.2). Then we prove that x is the only fixed point of T. Suppose that $x_0 \in X$ is any point from X. Consider the sequence produced by the repeated application of T to x_0:

$$x_0, x_1 = Tx_0, x_2 = Tx_1 = T^2 x_0, \ldots, x_n = Tx_{n-1} = T^n x_0, \ldots. \tag{7.19}$$

From (7.19) and (7.18), we obtain

$$
\begin{aligned}
d(x_{k+1}, x_k) &= d(Tx_k, Tx_{k-1}) \\
&\le \alpha d(x_k, x_{k-1}) \\
&= \alpha d(Tx_{k-1}, Tx_{k-2}) \\
&\le \alpha^2 d(x_{k-1}, x_{k-2}) \\
&\ \ \vdots \\
&\le \alpha^k d(x_1, x_0).
\end{aligned}
\tag{7.20}
$$

Using the triangle inequality with $n > k$, from (7.20)

$$d(x_k, x_n) \leq d(x_k, x_{k+1}) + d(x_{k+1}, x_{k+2}) + \cdots + d(x_{n-1}, x_n)$$
$$\leq (\alpha^k + \alpha^{k+1} \cdots + \alpha^{n-1}) d(x_0, x_1)$$
$$= \alpha^k \frac{1 - \alpha^{n-k}}{1 - \alpha} d(x_0, x_1)$$

(where the last equality follows by application of the formula for geometric series, seen very frequently in previous chapters). Since $0 < \alpha < 1$, we have $1 - \alpha^{n-k} < 1$. Thus

$$d(x_k, x_n) \leq \frac{\alpha^k}{1 - \alpha} d(x_0, x_1) \tag{7.21}$$

$(n > k)$. Since $0 < \alpha < 1$, $0 < 1 - \alpha < 1$, too. In addition, $d(x_0, x_1)$ is fixed (since we have chosen x_0). We may make the right-hand side of (7.21) arbitrarily small by making k sufficiently big (keeping $n > k$). Consequently, (x_n) is Cauchy. X is assumed to be complete, so $x_n \to x \in X$.

From the triangle inequality and (7.18)

$$d(Tx, x) \leq d(x, x_k) + d(x_k, Tx)$$
$$\leq d(x, x_k) + \alpha d(x_{k-1}, x) \qquad \text{(recall that } Tx = x)$$
$$< \epsilon \qquad \text{(any } \epsilon > 0)$$

if $k \to \infty$, since $x_n \to x$. Consequently, $d(x, Tx) = 0$ implies that $Tx = x$ (recall (M1) in Definition 1.1). Immediately, x is a fixed point of T.

Point x is unique. Let us assume that there is another fixed point $\hat{x}$, i.e., $T\hat{x} = \hat{x}$. But from (7.18)

$$d(x, \hat{x}) = d(Tx, T\hat{x}) \leq \alpha d(x, \hat{x}),$$

implying $d(x, \hat{x}) = 0$ because $0 < \alpha < 1$. Thus $x = \hat{x}$ [(M1) From Definition 1.1 again].

Theorem 7.3 is also called the *contraction theorem*, a special instance of which was seen in Section 4.7. The theorem applies to complete metric spaces, and so is applicable to Banach and Hilbert spaces. (Recall that inner products induce norms, and norms induce metrics. Hilbert and Banach spaces are complete. So they must be complete metric spaces as well.)

Note that we define $T^0 x = x$ for any $x \in X$. Thus T^0 is the identity mapping (identity operator).

Corollary 7.1 Under the conditions of Theorem 7.3 sequence (x_n) from $x_n = T^n x_0$ [i.e., the sequence in (7.19)] for any $x_0 \in X$ converges to a unique $x \in X$ such that $Tx = x$. We have the following error estimates (i.e., bounds):

1. Prior estimate:

$$d(x_k, x) \le \frac{\alpha^k}{1 - \alpha} d(x_0, x_1) \tag{7.22a}$$

2. Posterior estimate:

$$d(x_k, x) \le \frac{\alpha}{1 - \alpha} d(x_k, x_{k-1}). \tag{7.22b}$$

Proof   The prior estimate is an immediate consequence of Theorem 7.3 since it lies within the proof of this theorem [let $n \to \infty$ in (7.21)].

Now consider (7.22b). In (7.22a) let $k = 1$, $y_0 = x_0$, and $y_1 = x_1$. Thus, $d(x_1, x) = d(y_1, x), d(x_0, x_1) = d(y_0, y_1)$, and so

$$d(y_1, x) \le \frac{\alpha}{1 - \alpha} d(y_0, y_1). \tag{7.23}$$

Let $y_0 = x_{k-1}$, so $y_1 = T y_0 = T x_{k-1} = x_k$ and (7.23) becomes

$$d(x_k, x) \le \frac{\alpha}{1 - \alpha} d(x_k, x_{k-1}),$$

which is (7.22b).

The error bounds in Corollary 7.1 are useful in application of the contraction theorem to computational problems. For example, (7.22a) can estimate the number of iteration steps needed to achieve a given accuracy in the estimate of the solution to a nonlinear equation. The result would be analogous to Eq. (7.15) in the previous section.

A difficulty with the theory so far is that $T | X \to X$ is not always contractive over the entire metric space X, but only on a subset, say, $Y \subset X$. However, a basic result from functional analysis is that *any closed subset of X is complete*. Thus, for any closed $Y \subset X$, there is a fixed point $x \in Y \subset X$, and $x_n \to x$ ($x_n = T^n x_0$ for suitable $x_0 \in Y$). For this idea to work, we *must choose* $x_0 \in Y$ so that $x_n \in Y$ for all $n \in \mathbf{Z}^+$.

What does "closed subset" (formally) mean? A *neighborhood* of $x \in X$ (metric space X) is the set $N_\epsilon(x) = \{y | d(x, y) < \epsilon, \epsilon > 0\} \subset X$. Parameter ϵ is the *radius* of the neighborhood. We say that x is a *limit point* of $Y \subset X$ if *every neighborhood* of x contains $y \ne x$ such that $y \in Y$. Y is *closed* if every limit point of Y belongs to Y. These definitions are taken from Rudin [5, p. 32].

Example 7.2   Suppose that $X = \mathbf{R}$, $Y = (0, 1) \subset X$. We recall that X is a complete metric space if $d(x, y) = |x - y|$ ($x, y \in X$). Is Y closed? Limit points of Y are $x = 0$ and $x = 1$. [Any $x \in (0, 1)$ is also a limit point of Y.] But $0 \notin Y$, and $1 \notin Y$. Therefore, Y is not closed. On the other hand, $Y = [0, 1]$ *is* a closed subset of X since the limit points are now in Y.

Fixed-point theorems (and related ideas) such as we have considered so far have a large application range. They can be used to prove the existence and uniqueness of solutions to both integral and differential equations [4, pp. 314–326]. They can also provide (sometimes) computational algorithms for their solution. Fixed-point results can have applications in signal reconstruction and image processing [6], digital filter design [7], the interpolation of bandlimited sequences [8], and the solution to so-called convex feasibility problems in general [9]. However, we will consider the application of fixed-point theorems only to the problem of solving nonlinear equations.

If we wish to find $p \in \mathbf{R}$ so that

$$f(p) = 0, \tag{7.24}$$

then we may define $g(x)$ $(g(x)|\mathbf{R} \to \mathbf{R})$ with fixed point p such that

$$g(x) = x - f(x), \tag{7.25}$$

and we then see $g(p) = p - f(p) = p \Rightarrow f(p) = 0$. Conversely, if there is a function $g(x)$ such that

$$g(p) = p, \tag{7.26}$$

then

$$f(x) = x - g(x) \tag{7.27}$$

will have a zero at $x = p$. So, if we wish to solve $f(x) = 0$, one approach would be to find a suitable $g(x)$ as in (7.27) (or (7.25)), and find fixed points for it.

Theorem 7.3 informs us about the existence and uniqueness of fixed points for mappings on complete metric spaces (and ultimately on closed subsets of such spaces). Furthermore, the theorem leads us to a well-defined computational algorithm to find fixed points. At the outset, space $X = \mathbf{R}$ with metric $d(x, y) = |x - y|$ is complete, and for us g and f are mappings on X. So if g and f are related according to (7.27), then any fixed point of g will be a zero of f, and can be found by iterating as spelled out in Theorem 7.3. However, the discussion following Corollary 7.1 warned us that g may not be contractive on $X = \mathbf{R}$ but only on some subset $Y \subset X$. In fact, g is only rarely contractive on all of $\mathbf{R}$. We therefore usually need to find $Y = [a, b] \subset X = \mathbf{R}$ such that g is contractive on Y, and *then* we compute $x_n = g^n x_0 \in Y$ for $n \in \mathbf{Z}^+$. Then $x_n \to x$, and $f(x) = 0$. We again emphasize that $x_n \in Y$ is necessary for all n. If g is contractive on $[a, b] \subset X$, then, from Definition 7.2, this means that for any $x, y \in [a, b]$

$$|g(x) - g(y)| \leq \alpha |x - y| \tag{7.28}$$

for some real-valued α such that $0 < \alpha < 1$.

Example 7.3   Suppose that we want roots of $f(x) = \lambda x^2 + (1 - \lambda)x = 0$ (assume $\lambda > 0$). Of course, this is a quadratic equation and so may be easily solved by the usual formula for the roots of such an equation (recall Section 4.2). However, this example is an excellent illustration of the behavior of fixed-point schemes. It is quite easy to verify that

$$f(x) = x - \underbrace{\lambda x(1 - x)}_{=g(x)}.$$ (7.29)

We observe that $g(x)$ is a quadratic in x, and also

$$g^{(1)}(x) = (2x - 1)\lambda,$$

so that $g(x)$ has a maximum at $x = \frac{1}{2}$ for which $g(\frac{1}{2}) = \lambda/4$. Therefore, if we allow only $0 < \lambda \le 4$, then $g(x) \in [0, 1]$ for all $x \in [0, 1]$. Certainly $[0, 1] \subset \mathbf{R}$. A sketch of $g(x)$ and of $y = x$ appears in Fig. 7.2 for various λ. The intersection of these two curves locates the fixed points of g on $[0, 1]$.

We will suppose $Y = [0, 1]$. Although $g(x) \in [0, 1]$ for all $x \in [0, 1]$ under the stated conditions, g is not necessarily always contractive on the closed interval $[0, 1]$. Suppose $\lambda = \frac{1}{2}$; then, for all $x \in [0, 1]$, $g(x) \in [0, \frac{1}{8}]$ (see the dotted line in Fig. 7.2, and we can calculate this from knowledge of g). The mapping is contractive for this case on all of $[0, 1]$. [This is justified below with respect to Eq. (7.30).] Also, $g(x) = x$ only for $x = 0$. If we select any $x_0 \in [0, 1]$, then, for

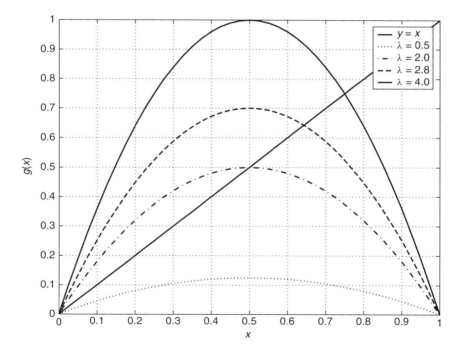

Figure 7.2 Plot of $g(x) = \lambda x(1 - x)$ for various λ, and a plot of $y = x$. The places where $y = x$ and $g(x)$ intersect define the fixed points of $g(x)$.

$x_n = g^n x_0$, we can expect $x_n \to 0$. For example, suppose $x_0 = 0.7500$; then the first few iterates are

$$x_0 = 0.7500$$

$$x_1 = 0.5x_0(1 - x_0) = 0.0938$$

$$x_2 = 0.5x_1(1 - x_1) = 0.0425$$

$$x_3 = 0.5x_2(1 - x_2) = 0.0203$$

$$x_4 = 0.5x_3(1 - x_3) = 0.0100$$

$$x_5 = 0.5x_4(1 - x_4) = 0.0049$$

$$\vdots$$

The process is converging to the unique fixed point at $x = 0$.

Suppose now that $\lambda = 2$; then $g(x) = 2x(1 - x)$, and now $g(x)$ has two fixed points on $[0, 1]$, which are $x = 0$, and $x = \frac{1}{2}$. For all $x \in [0, 1]$ we have $g(x) \in [0, \frac{1}{2}]$, but g is not contractive on $[0, 1]$. For example, suppose $x = 0.8$, $y = 0.9$, then $g(.8) = 0.3200$, and $g(.9) = 0.1800$. Thus, $|x - y| = 0.1$, but $|g(x) - g(y)| = 0.14 > 0.1$. On the other hand, suppose that $x_0 = 0.7500$; then the first few iterates are

$$x_0 = 0.7500$$

$$x_1 = 2x_0(1 - x_0) = 0.3750$$

$$x_2 = 2x_1(1 - x_1) = 0.4688$$

$$x_3 = 2x_2(1 - x_2) = 0.4980$$

$$x_4 = 2x_3(1 - x_3) = 0.5000$$

$$x_5 = 2x_4(1 - x_4) = 0.5000$$

$$\vdots$$

This process converges to one of the fixed points of g even though g is not contractive on $[0, 1]$.

From (7.28)

$$\alpha \geq \left| \frac{g(x) - g(y)}{x - y} \right| \qquad (x \neq y),$$

from which, if we substitute $g(x) = \lambda x(1 - x)$, and $g(y) = \lambda y(1 - y)$, then

$$\alpha = \sup_{x \neq y} \lambda |1 - x - y|. \qquad (7.30)$$

If $\lambda = \frac{1}{2}$, then $\alpha = \frac{1}{2}$. If $\lambda = 2$, we cannot have α, so that $0 < \alpha < 1$. If $g(x) = 2x(1 - x)$ (i.e., $\lambda = 2$ again), but now instead $Y = [0.4, 0.6]$, then, for all

$x \in Y = [0.4, 0.6]$, we have $g(x) \in [0.48, 0.50] \subset Y$, implying $g|Y \rightarrow Y$. From (7.30) we have for this situation $\alpha = 0.4$. The mapping g is contractive on Y. Suppose $x_0 = 0.450000$; then

$$x_0 = 0.450000$$

$$x_1 = 2x_0(1 - x_0) = 0.495000$$

$$x_2 = 2x_1(1 - x_1) = 0.499950$$

$$x_3 = 2x_2(1 - x_2) = 0.500000$$

$$x_4 = 2x_3(1 - x_3) = 0.500000$$

$$\vdots$$

The reader ought to compare the results of this example to the error bounds from Corollary 7.1 as an exercise.

More generally, and again from (7.28), we have (for $x, y \in Y = [a, b] \subset \mathbf{R}$)

$$\alpha = \sup_{x \neq y} \left| \frac{g(x) - g(y)}{x - y} \right|. \tag{7.31}$$

Now recall the *mean-value theorem* (i.e., Theorem 3.3). If $g(x)$ is continuous on $[a, b]$, and $g^{(1)}(x)$ is continuous on (a, b), then there is a $\xi \in (a, b)$ such that

$$g^{(1)}(\xi) = \frac{g(b) - g(a)}{b - a}.$$

Consequently, instead of (7.31) we may use

$$\alpha = \sup_{x \in (a,b)} |g^{(1)}(x)|. \tag{7.32}$$

Example 7.4   We may use (7.32) to rework some of the results in Example 7.3. Since $g^{(1)}(x) = \lambda(2x - 1)$, if $Y = [0, 1]$, and $\lambda = \frac{1}{2}$, then

$$\alpha = \sup_{x \in (0,1)} \left| x - \frac{1}{2} \right| = \frac{1}{2}.$$

If $\lambda = 2$, then $\alpha = 2$. If now $Y = [0.4, 0.6]$ with $\lambda = 2$, then

$$\alpha = \sup_{x \in (0.4, 0.6)} |4x - 2| = 0.4.$$

Now suppose that $\lambda = 2.8$, and consider $Y = [0.61, 0.67]$, which contains a fixed-point of g (see Fig. 7.2, which contains a curve for this case). We have

$$\alpha = \sup_{x \in (0.61, 0.67)} |5.6x - 2.8| = 0.9520,$$

and if $x \in Y$, then $g(x) \in [0.619080, 0.666120] \subset Y$ so that $g|Y \to Y$, and so g is contractive on Y. Thus, we consider the iterates

$$x_0 = 0.650000$$

$$x_1 = 2.8x_0(1 - x_0) = 0.637000$$

$$x_2 = 2.8x_1(1 - x_1) = 0.647447$$

$$x_3 = 2.8x_2(1 - x_2) = 0.639126$$

$$x_4 = 2.8x_3(1 - x_3) = 0.645803$$

$$x_5 = 2.8x_4(1 - x_4) = 0.640476$$

$$\vdots$$

The true fixed point (to 6 significant figures) is $x = 0.642857$ (i.e., $x_n \to x$). We may check these numbers against the bounds of Corollary 7.1. Therefore, we consider the distances

$$d(x_5, x) = 0.002381, \qquad d(x_0, x_1) = 0.013000, \qquad d(x_5, x_4) = 0.005327,$$

and from (7.22a)

$$d(x_5, x) \leq \frac{\alpha^k}{1 - \alpha} d(x_0, x_1) = 0.211780,$$

and from (7.22b)

$$d(x_5, x) \leq \frac{\alpha}{1 - \alpha} d(x_5, x_4) = 0.105652.$$

These error bounds are very loose, but they are nevertheless consistent with the true error $d(x_5, x) = 0.002381$.

We have worked with $g(x) = \lambda x(1 - x)$ in the previous two examples (Example 7.3 and Example 7.4). But this is not the only possible choice. It may be better to make other choices.

Example 7.5   Again, assume $f(x) = \lambda x^2 + (1 - \lambda)x = 0$ as in the previous two examples. Observe that

$$\lambda x^2 = (\lambda - 1)x$$

implies that

$$x = \sqrt{\frac{\lambda - 1}{\lambda}} x^{1/2} = g(x). \tag{7.33}$$

If $\lambda = 4$, then $f(x) = 0$ for $x = 0$ and for $x = \frac{3}{4}$. For $g(x)$ in (7.29), $g^{(1)}(x) = 8x - 4$, and $g^{(1)}(\frac{3}{4}) = 2$. We cannot find a closed interval Y containing $x = \frac{3}{4}$ on which g is contractive with $g|Y \to Y$ (the slope of the curve $g(x)$ is too steep in the vicinity of the fixed point). But if we choose (7.33) instead, then

$$g(x) = \frac{\sqrt{3}}{2}x^{1/2}, \qquad g^{(1)}(x) = \frac{\sqrt{3}}{4}\frac{1}{x^{1/2}},$$

and if $Y = [0.7, 0.8]$, then for $x \in Y$, we have $g(x) \in [0.7246, 0.7746] \subset Y$, and

$$\alpha = \sup_{x \in (0.7, 0.8)} \frac{\sqrt{3}}{4}\frac{1}{\sqrt{x}} = 0.5175.$$

So, g in (7.33) is contractive on Y. We observe the iterates

$$x_0 = 0.7800$$

$$x_1 = \frac{\sqrt{3}}{2}x_0^{1/2} = 0.7649$$

$$x_2 = \frac{\sqrt{3}}{2}x_1^{1/2} = 0.7574$$

$$x_3 = \frac{\sqrt{3}}{2}x_2^{1/2} = 0.7537$$

$$x_4 = \frac{\sqrt{3}}{2}x_3^{1/2} = 0.7518$$

$$x_5 = \frac{\sqrt{3}}{2}x_4^{1/2} = 0.7509,$$

$$\vdots$$

which converge to $x = \frac{3}{4}$ (i.e., $x_n \to x$).

7.4   NEWTON–RAPHSON METHOD

This method is yet another iterative approach to finding roots of nonlinear equations. In fact, it is a version of the fixed-point method that was considered in the previous section. However, it is of sufficient importance to warrant separate consideration within its own section.

7.4.1   The Method

One way to derive the Newton–Raphson method is by a geometric approach. The method attempts to solve

$$f(x) = 0 \tag{7.34}$$

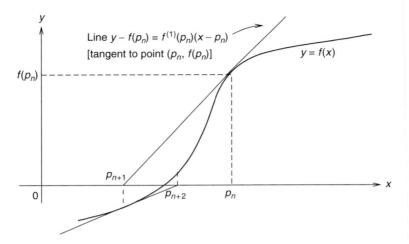

Figure 7.3 Geometric interpretation of the Newton–Raphson method.

($x \in \mathbf{R}$, and $f(x) \in \mathbf{R}$) by approximating the root p (i.e., $f(p) = 0$) by a succession of x intercepts of tangent lines to the curve $y = f(x)$ at x the current approximation to p. Specifically, if p_n is the current estimate of p, then the tangent line to the point $(p_n, f(p_n))$ on the curve $y = f(x)$ is

$$y = f(p_n) + f^{(1)}(p_n)(x - p_n) \tag{7.35}$$

(see Fig. 7.3). The next approximation to p is $x = p_{n+1}$ such that $y = 0$, that is, from (7.35)

$$0 = f(p_n) + f^{(1)}(p_n)(p_{n+1} - p_n),$$

so for $n \in \mathbf{Z}^+$

$$p_{n+1} = p_n - \frac{f(p_n)}{f^{(1)}(p_n)}. \tag{7.36}$$

To start this process off, we need an initial guess at p, that is, we need to select p_0. Clearly this approach requires $f^{(1)}(p_n) \neq 0$ for all n; otherwise the process will terminate. Continuation will not be possible, except perhaps by choosing a new starting point p_0. This is one of the ways in which the method can break down. It might be called *premature termination*. (Breakdown phenomena are discussed further later in the text.)

Another derivation of (7.36) is as follows. Recall the theory of Taylor series from Section 3.5. Suppose that $f(x)$, $f^{(1)}(x)$, and $f^{(2)}(x)$ are all continuous on $[a, b]$. Suppose that p is a root of $f(x) = 0$, and that p_n approximates p. From Taylor's theorem [i.e., (3.71)]

$$f(x) = f(p_n) + f^{(1)}(p_n)(x - p_n) + \tfrac{1}{2}f^{(2)}(\xi)(x - p_n)^2, \tag{7.37}$$

where $\xi = \xi(x) \in (p_n, x)$. Since $f(p) = 0$ from (7.37), we have

$$0 = f(p) = f(p_n) + f^{(1)}(p_n)(p - p_n) + \tfrac{1}{2} f^{(2)}(\xi)(p - p_n)^2. \qquad (7.38)$$

If $|p - p_n|$ is small, we may neglect that last term of (7.38), and hence

$$0 \approx f(p_n) + f^{(1)}(p_n)(p - p_n),$$

implying

$$p \approx p_n - \frac{f(p_n)}{f^{(1)}(p_n)}. \qquad (7.39)$$

We treat the right-hand side of (7.39) as p_{n+1}, the next approximation to p. Thus, we again arrive at (7.36). The assumption that $(p - p_n)^2$ is negligible is important. If p_n is not close enough to p, then the method may not converge. In particular, the choice of starting point p_0 is important.

As already mentioned, the Newton–Raphson method is a special instance of the fixed-point iteration method, where

$$g(x) = x - \frac{f(x)}{f^{(1)}(x)}. \qquad (7.40)$$

So

$$p_{n+1} = g(p_n) \qquad (7.41)$$

for $n \in \mathbf{Z}^+$. Stopping criteria are (for $\epsilon > 0$)

$$|p_n - p_{n-1}| < \epsilon, \qquad (7.42a)$$

$$|f(p_n)| < \epsilon, \qquad (7.42b)$$

$$\left| \frac{p_n - p_{n-1}}{p_n} \right| < \epsilon, \quad p_n \neq 0. \qquad (7.42c)$$

As with the bisection method, we prefer (7.42c).

Theorem 7.4: Convergence Theorem for the Newton–Raphson Method
Let f be continuous on $[a, b]$. Let $f^{(1)}(x)$, and $f^{(2)}(x)$ exist and be continuous for all $x \in (a, b)$. If $p \in [a, b]$ with $f(p) = 0$, and $f^{(1)}(p) \neq 0$, then there is a $\delta > 0$ such that (7.36) generates sequence (p_n) with $p_n \to p$ for any $p_0 \in [p - \delta, p + \delta]$.

Proof We have

$$p_{n+1} = g(p_n), \quad n \in \mathbf{Z}^+$$

with $g(x) = x - \frac{f(x)}{f^{(1)}(x)}$. We need $Y = [p - \delta, p + \delta] \subset \mathbf{R}$ with $g|Y \to Y$, and g is contractive on Y. (Then we may immediately apply the convergence results from the previous section.)

Since $f^{(1)}(p) \neq 0$, and since $f^{(1)}(x)$ is continuous at p, there will be a $\delta_1 > 0$ such that $f^{(1)}(x) \neq 0$ for all $x \in [p - \delta_1, p + \delta_1] \subset [a, b]$, so g is defined and continuous for $x \in [p - \delta_1, p + \delta_1]$. Also

$$g^{(1)}(x) = 1 - \frac{[f^{(1)}(x)]^2 - f(x)f^{(2)}(x)}{[f^{(1)}(x)]^2} = \frac{f(x)f^{(2)}(x)}{[f^{(1)}(x)]^2}$$

for $x \in [p - \delta_1, p + \delta_1]$. In addition, $f^{(2)}(x)$ is continuous on (a, b), so $g^{(1)}(x)$ is continuous on $[p - \delta_1, p + \delta_1]$. We assume $f(p) = 0$, so

$$g^{(1)}(p) = \frac{f(p)f^{(2)}(p)}{[f^{(1)}(p)]^2} = 0.$$

We have $g^{(1)}(x)$ continuous at $x = p$, implying that $\lim_{x \to p} g^{(1)}(x) = g^{(1)}(p) = 0$, so there is a $\delta > 0$ with $0 < \delta < \delta_1$ such that

$$|g^{(1)}(x)| \leq \alpha < 1 \tag{7.43}$$

for $x \in [p - \delta, p + \delta] = Y$ for some α such that $0 < \alpha < 1$. If $x \in Y$, then by the mean-value theorem, there is a $\xi \in (x, p)$ such that

$$|g(x) - p| = |g(x) - g(p)| = |g^{(1)}(\xi)||x - p|$$
$$\leq \alpha|x - p| < |x - p| < \delta,$$

so that we have $g(x) \in Y$ for all $x \in Y$. That is, $g|Y \to Y$. Because of (7.43)

$$\alpha = \sup_{x \in Y} |g^{(1)}(x)|,$$

and $0 < \alpha < 1$, so that g is contractive on Y. Immediately, sequence (p_n) from $p_{n+1} = g(p_n)$ for all $p_0 \in Y$ converges to p (i.e., $p_n \to p$).

Essentially, the Newton–Raphson method is guaranteed to converge to a root if p_0 is close enough to it and f is sufficiently smooth. Theorem 7.4 is weak because it does not specify how to select p_0.

The Newton–Raphson method, if it converges, tends to do so quite quickly. However, the method needs $f^{(1)}(p_n)$ as well as $f(p_n)$. It might be the case that $f^{(1)}(p_n)$ requires much effort to evaluate. The *secant method* is a variation on the Newton–Raphson method that replaces $f^{(1)}(p_n)$ with an approximation. Specifically, since

$$f^{(1)}(p_n) = \lim_{x \to p_n} \frac{f(x) - f(p_n)}{x - p_n},$$

if $p_{n-1} \approx p_n$, then

$$f^{(1)}(p_n) \approx \frac{f(p_{n-1}) - f(p_n)}{p_{n-1} - p_n}. \tag{7.44}$$

This is the slope of the chord that connects the points $(p_{n-1}, f(p_{n-1}))$, and $(p_n, f(p_n))$ on the graph of $f(x)$. We may substitute (7.44) into (7.36) obtaining (for $n \in \mathbf{N}$)

$$p_{n+1} = p_n - \frac{(p_n - p_{n-1})f(p_n)}{f(p_n) - f(p_{n-1})}. \tag{7.45}$$

We need p_0 and p_1 to initialize the iteration process in (7.45). Usually these are chosen to bracket the root, but this does not guarantee convergence. If the method does converge, it tends to do so more slowly than the Newton–Raphson method, but this is the penalty to be paid for avoiding the computation of derivatives $f^{(1)}(x)$. We remark that the *method of false position* is a modification of the secant method that is guaranteed to converge because successive approximations to p (i.e., p_{n-1} and p_n) are chosen to always bracket the root. But it is possible that convergence may be slow. We do not cover this method in this book. We merely mention that it appears elsewhere in the literature [10–12].

7.4.2   Rate of Convergence Analysis

What can be said about the speed of convergence of fixed-point schemes in general, and of the Newton–Raphson method in particular? Consider the following definition.

Definition 7.3: Suppose that (p_n) is such that $p_n \to p$ with $p_n \neq p$ ($n \in \mathbf{Z}^+$). If there are $\lambda > 0$, and $\delta > 0$ such that

$$\lim_{n \to \infty} \frac{|p_{n+1} - p|}{|p_n - p|^\delta} = \lambda,$$

then we say that (p_n) converges to p of *order δ with asymptotic error constant λ*. Additionally

- If $\delta = 1$, we say that (p_n) is *linearly convergent.*
- If $\delta > 1$, we have *superlinear convergence.*
- If $\delta = 2$, we have *quadratic convergence.*

From this definition, if n is big enough, then

$$|p_{n+1} - p| \approx \lambda |p_n - p|^\delta. \tag{7.46}$$

Thus, we would like δ to be large and λ to be small for fast convergence. Since δ is in the exponent, this parameter is more important than λ for determining the rate of convergence.

Consider the fixed-point iteration

$$p_{n+1} = g(p_n),$$

where g satisfies the requirements of a contraction mapping (so the Banach fixed-point theorem applies). We can therefore say that

$$\lim_{n \to \infty} \frac{|p_{n+1} - p|}{|p_n - p|} = \lim_{n \to \infty} \frac{|g(p_n) - g(p)|}{|p_n - p|}, \qquad (7.47)$$

so from the mean-value theorem we have ξ_n between p_n and p for which

$$g(p_n) - g(p) = g^{(1)}(\xi_n)(p_n - p),$$

which, if used in (7.47), implies that

$$\lim_{n \to \infty} \frac{|p_{n+1} - p|}{|p_n - p|} = \lim_{n \to \infty} |g^{(1)}(\xi_n)|. \qquad (7.48)$$

Because ξ_n is between p_n and p and $p_n \to p$, we must have $\xi_n \to p$ as well. Also, we will assume $g^{(1)}(x)$ is continuous at p, so (7.48) now becomes

$$\lim_{n \to \infty} \frac{|p_{n+1} - p|}{|p_n - p|} = |g^{(1)}(p)|, \qquad (7.49)$$

which is a constant. Applying Definition 7.3, we conclude that the fixed-point method is typically only linearly convergent because $\delta = 1$, and the asymptotic error constant is $\lambda = |g^{(1)}(p)|$, provided $g^{(1)}(p) \neq 0$. However, if $g^{(1)}(p) = 0$, we expect faster convergence. It turns out that this is often the case for the Newton–Raphson method, as will now be demonstrated.

The iterative scheme for the Newton–Raphson method is (7.36), which is a particular case of fixed-point iteration where now

$$g(x) = x - \frac{f(x)}{f^{(1)}(x)}, \qquad (7.50)$$

and for which

$$g^{(1)}(x) = \frac{f(x) f^{(2)}(x)}{[f^{(1)}(x)]^2}, \qquad (7.51)$$

so that $g^{(1)}(p) = 0$ because $f(p) = 0$. Thus, superlinear convergence is anticipated for this particular fixed-point scheme. Suppose that we have the Taylor expansion

$$g(x) = g(p) + g^{(1)}(p)(x - p) + \tfrac{1}{2} g^{(2)}(\xi)(x - p)^2$$

for which ξ is between p and x. Since $g(p) = p$ and $g^{(1)}(p) = 0$, this becomes

$$g(x) = p + \tfrac{1}{2}g^{(2)}(\xi)(x - p)^2.$$

For $x = p_n$, this in turn becomes

$$p_{n+1} = g(p_n) = p + \tfrac{1}{2}g^{(2)}(\xi_n)(p - p_n)^2 \qquad (7.52)$$

for which ξ_n lies between p and p_n. Equation (7.52) can be rearranged as

$$p_{n+1} - p = \tfrac{1}{2}g^{(2)}(\xi_n)(p - p_n)^2,$$

and so

$$\lim_{n\to\infty} \frac{|p_{n+1} - p|}{|p_n - p|^2} = \frac{1}{2}|g^{(2)}(p)|, \qquad (7.53)$$

since $\xi_n \to p$, and we are also assuming that $g^{(2)}(x)$ is continuous at p. Immediately, $\delta = 2$, and the Newton–Raphson method is quadratically convergent. The asymptotic error constant is plainly equal to $\frac{1}{2}|g^{(2)}(p)|$, provided $g^{(2)}(p) \neq 0$. If $g^{(2)}(p) = 0$, then an even higher order of convergence may be expected. It is emphasized that these convergence results depend on $g(x)$ being smooth enough. Given this, since the convergence is at least quadratic, the number of accurate decimal digits in the approximation to a root approximately doubles at every iteration.

7.4.3 Breakdown Phenomena

The Newton–Raphson method can fail to converge (i.e., break down) in various ways. Premature termination was mentioned earlier. An example appears in Fig. 7.4. This figure also illustrates *divergence*, which means that [if $f(p) = 0$]

$$\lim_{n\to\infty} |p_n - p| = \infty.$$

In other words, the sequence of iterates (p_n) generated by the Newton–Raphson method moves progressively farther and farther away from the desired root p.

Another failure mechanism called *oscillation* may be demonstrated as follows. Suppose that

$$f(x) = x^3 - 2x + 2, \ f^{(1)}(x) = 3x^2 - 2. \qquad (7.54)$$

Newton's method for this specific case is

$$p_{n+1} = p_n - \frac{p_n^3 - 2p_n + 2}{3p_n^2 - 2} = \frac{2p_n^3 - 2}{3p_n^2 - 2}. \qquad (7.55)$$

If we were to select $p_0 = 0$ as a starting point, then

$$p_0 = 0, \quad p_1 = 1, \quad p_2 = 0, \quad p_3 = 1, \quad p_4 = 0, \quad p_5 = 1, \dots .$$

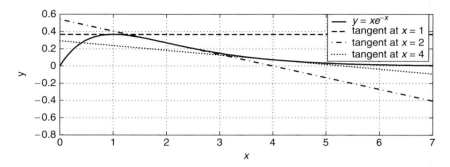

Figure 7.4   Illustration of breakdown phenomena in the Newton–Raphson method; here, $f(x) = xe^{-x}$. The figure shows premature termination if $p_n = x = 1$ since $f^{(1)}(1) = 0$. Also shown is divergence for the case where $p_0 > 1$. (See how the x intercepts of the tangent lines for $x > 1$ become bigger as x increases.)

More succinctly, $p_n = \frac{1}{2}(1 - (-1)^n)$. The sequence of iterates is oscillating; it does not diverge, and it does not converge, either. The sequence is periodic (it may be said to be period 2), and in this case quite simple, but far more complicated oscillations are possible with longer periods.

Premature termination, divergence, and oscillation are not the only possible breakdown mechanisms. Another possibility is that the sequence (p_n) can be *chaotic*. Loosely speaking, *chaos* is a nonperiodic oscillation with a complicated structure. Chaotic oscillations look a lot like random noise. This will be considered in more detail in Section 7.6.

7.5   SYSTEMS OF NONLINEAR EQUATIONS

We now extend the fixed-point and Newton–Raphson methods to solving nonlinear *systems* of equations. We emphasize two equations in two unknowns (i.e., two-dimensional problems). But much of what is said here applies to higher dimensions.

7.5.1   Fixed-Point Method

As remarked, it is easier to begin by first considering the two-dimensional problem. More specifically, we wish to solve

$$\hat{f}_0(x_0, x_1) = 0, \quad \hat{f}_1(x_0, x_1) = 0, \tag{7.56}$$

which we will assume may be rewritten in the form

$$x_0 - f_0(x_0, x_1) = 0, \quad x_1 - f_1(x_0, x_1) = 0, \tag{7.57}$$

and we see that solving these is equivalent to finding where the curves in (7.57) intersect in $\mathbf{R}^2$ (i.e., $[x_0 \quad x_1]^T \in \mathbf{R}^2$). A general picture is in Fig. 7.5. As in the

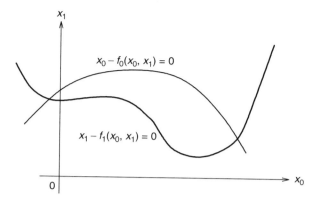

Figure 7.5 Typical curves in $\mathbf{R}^2$ corresponding to the system of equations in (7.57).

case of one-dimensional problems, there will usually be more than one way to rewrite (7.56) in the form of (7.57).

Example 7.6 Suppose

$$\hat{f}_0(x_0, x_1) = x_0 - x_0^2 - \tfrac{1}{4}x_1^2 = 0 \Rightarrow f_0(x_0, x_1) = x_0^2 + \tfrac{1}{4}x_1^2,$$
$$\hat{f}_1(x_0, x_1) = x_1 - x_0^2 + x_1^2 = 0 \Rightarrow f_1(x_0, x_1) = x_0^2 - x_1^2.$$

We see that

$$x_0^2 + \tfrac{1}{4}x_1^2 - x_0 = 0 \Rightarrow (x_0 - \tfrac{1}{2})^2 + \tfrac{1}{4}x_1^2 = \tfrac{1}{4}$$

(an ellipse in $\mathbf{R}^2$), and

$$x_0^2 - x_1^2 - x_1 = 0 \Rightarrow x_0^2 - (x_1 + \tfrac{1}{2})^2 = -\tfrac{1}{4}$$

(a hyperbola in $\mathbf{R}^2$). These are plotted in Fig. 7.6. The solutions to the system are the points where the ellipse and hyperbola intersect each other. Clearly, the solution is not unique.

It is frequently the case that nonlinear systems of equations will have more than one solution just as this was very possible and very common in the case of a single equation.

Vector notation leads to compact descriptions. Specifically, we define $\bar{x} = [x_0 x_1]^T \in \mathbf{R}^2$, $f_0(\bar{x}) = f_0(x_0, x_1)$, $f_1(\bar{x}) = f_1(x_0, x_1)$, and

$$F(\bar{x}) = \begin{bmatrix} f_0(x_0, x_1) \\ f_1(x_0, x_1) \end{bmatrix} = \begin{bmatrix} f_0(\bar{x}) \\ f_1(\bar{x}) \end{bmatrix}. \tag{7.58}$$

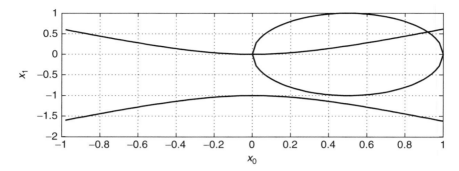

Figure 7.6   The curves of Example 7.6 (an ellipse and a hyperbola). The solutions to the equations in Example 7.6 are the points where these curves intersect.

Then the nonlinear system in (7.57) becomes

$$\overline{x} = F(\overline{x}), \tag{7.59}$$

and the fixed point $\overline{p} \in \mathbf{R}^2$ of F satisfies

$$\overline{p} = F(\overline{p}). \tag{7.60}$$

We recall that $\mathbf{R}^2$ is a normed space if

$$||\overline{x}||^2 = \sum_{i=0}^{1} x_i^2 = x_0^2 + x_1^2 \tag{7.61}$$

($\overline{x} \in \mathbf{R}^2$). We may consider a sequence of vectors $(\overline{x}_n)$ (i.e., $\overline{x}_n \in \mathbf{R}^2$ for $n \in \mathbf{Z}^+$), and it converges to $\overline{x}$ iff

$$\lim_{n \to \infty} ||\overline{x}_n - \overline{x}|| = 0.$$

We recall from Chapter 3 that $\mathbf{R}^2$ with the norm in (7.61) is a complete space, so every Cauchy sequence $(\overline{x}_n)$ in it will converge.

As with the scalar case considered in Section 7.3, we consider the sequence of iterates $(\overline{p}_n)$ such that

$$\overline{p}_{n+1} = F(\overline{p}_n), \quad n \in \mathbf{Z}^+.$$

The previous statements (and the following theorem) apply if $\mathbf{R}^2$ is replaced by $\mathbf{R}^m$ ($m > 2$); that is, the space can be of higher dimension to accommodate m equations in m unknowns. Naturally, for $\mathbf{R}^m$ the norm in (7.61) must change according to $||\overline{x}||^2 = \sum_{k=0}^{m-1} x_k^2$. The following theorem is really a special instance of the Banach fixed-point theorem seen earlier.

Theorem 7.5: Suppose that $\mathcal{R}$ is a closed subset of $\mathbf{R}^2$, $F|\mathcal{R} \to \mathcal{R}$, and F is contractive on $\mathcal{R}$; then

$$\bar{x} = F(\bar{x})$$

has a unique solution $\bar{p} \in \mathbf{R}$. The sequence $(\bar{p}_n)$, where

$$\bar{p}_{n+1} = F(\bar{p}_n), \quad \bar{p}_0 \in \mathcal{R}, \quad n \in \mathbf{Z}^+ \tag{7.62}$$

is such that

$$\lim_{n \to \infty} ||\bar{p}_n - \bar{p}|| = 0$$

and

$$||\bar{p}_n - \bar{p}|| \leq \frac{\alpha^n}{1-\alpha} ||\bar{p}_1 - \bar{p}_0||, \tag{7.63}$$

where $||F(\bar{p}_1) - F(\bar{p}_2)|| \leq \alpha ||\bar{p}_1 - \bar{p}_2||$ for any $\bar{p}_1, \bar{p}_2 \in \mathcal{R}$, and $0 < \alpha < 1$.

Proof As noted, this theorem is really a special instance of the Banach fixed-point theorem (Theorem 7.3), so we only outline the proof.

It was mentioned in Section 7.3 that any closed subset of a complete metric space is also complete. $\mathbf{R}^2$ with norm (7.61) is a complete metric space, and since $\mathcal{R} \subset \mathbf{R}^2$ is closed, $\mathcal{R}$ must be complete. According to Theorem 7.3, F has a unique fixed point $\bar{p} \in \mathcal{R}$ [i.e., $F(\bar{p}) = \bar{p}$], and sequence $(\bar{p}_n)$ from $\bar{p}_{n+1} = F(\bar{p}_n)$ (with $\bar{p}_0 \in \mathcal{R}$, and $n \in \mathbf{Z}^+$) converges to $\bar{p}$. The error bound

$$||\bar{p}_n - \bar{p}|| \leq \frac{\alpha^n}{1-\alpha} ||\bar{p}_1 - \bar{p}_0||$$

is an immediate consequence of Corollary 7.1.

A typical choice for $\mathcal{R}$ would be the *bounded and closed rectangular region*

$$\mathcal{R} = \{[x_0 x_1]^T | a_0 \leq x_0 \leq b_0, a_1 \leq x_1 \leq b_1\}. \tag{7.64}$$

The next theorem applies the Schwarz inequality of Chapter 1 to the estimation of α. It must be admitted that applying the following theorem is often quite difficult in practice, and in the end it is often better to simply implement the iterative method in (7.62) and experiment with it rather than go through the laborious task of computing α according to the theorem's dictates. However, exceptions cannot be ruled out, and so knowledge of the theorem might be helpful.

Theorem 7.6: Suppose that $\mathcal{R} \subset \mathbf{R}^2$ is as defined in (7.64); then, if

$$\alpha = \max_{\bar{x} \in \mathcal{R}} \left[\left(\frac{\partial f_0}{\partial x_0}\right)^2 + \left(\frac{\partial f_0}{\partial x_1}\right)^2 + \left(\frac{\partial f_1}{\partial x_0}\right)^2 + \left(\frac{\partial f_1}{\partial x_1}\right)^2 \right]^{1/2}, \tag{7.65}$$

we have

$$||F(\overline{x}_1) - F(\overline{x}_2)|| \le \alpha ||\overline{x}_1 - \overline{x}_2||$$

for all $\overline{x}_1, \overline{x}_2 \in \mathcal{R}$.

Proof   We use a two-dimensional version of the Taylor expansion theorem, which we will not attempt to justify in this book.

Given $\overline{x}_1 = [x_{1,0} \ x_{1,1}]^T$, $\overline{x}_2 = [x_{2,0} \ x_{2,1}]^T$, there is a point $\xi \in \mathcal{R}$ on the line segment that joins $\overline{x}_1$ to $\overline{x}_2$ such that

$$F(\overline{x}_1) = F(\overline{x}_2) + F^{(1)}(\xi)(\overline{x}_1 - \overline{x}_2)$$

$$= F(\overline{x}_2) + \begin{bmatrix} \dfrac{\partial f_0(\xi)}{\partial x_0} & \dfrac{\partial f_0(\xi)}{\partial x_1} \\[3mm] \dfrac{\partial f_1(\xi)}{\partial x_0} & \dfrac{\partial f_1(\xi)}{\partial x_1} \end{bmatrix} \begin{bmatrix} x_{1,0} - x_{2,0} \\[2mm] x_{1,1} - x_{2,1} \end{bmatrix}$$

so

$$||F(\overline{x}_1) - F(\overline{x}_2)||^2 = \left[\frac{\partial f_0(\xi)}{\partial x_0}(x_{1,0} - x_{2,0}) + \frac{\partial f_0(\xi)}{\partial x_1}(x_{1,1} - x_{2,1}) \right]^2$$

$$+ \left[\frac{\partial f_1(\xi)}{\partial x_0}(x_{1,0} - x_{2,0}) + \frac{\partial f_1(\xi)}{\partial x_1}(x_{1,1} - x_{2,1}) \right]^2$$

$$\le \left[\left(\frac{\partial f_0(\xi)}{\partial x_0} \right)^2 + \left(\frac{\partial f_0(\xi)}{\partial x_1} \right)^2 \right] ||\overline{x}_1 - \overline{x}_2||^2$$

$$+ \left[\left(\frac{\partial f_1(\xi)}{\partial x_0} \right)^2 + \left(\frac{\partial f_1(\xi)}{\partial x_1} \right)^2 \right] ||\overline{x}_1 - \overline{x}_2||^2$$

via the Schwarz inequality (Theorem 1.1). Consequently

$$||F(\overline{x}_1) - F(\overline{x}_2)||^2 \le \left[\left(\frac{\partial f_0(\xi)}{\partial x_0} \right)^2 + \left(\frac{\partial f_0(\xi)}{\partial x_1} \right)^2 \right.$$

$$\left. + \left(\frac{\partial f_1(\xi)}{\partial x_0} \right)^2 + \left(\frac{\partial f_1(\xi)}{\partial x_1} \right)^2 \right] ||\overline{x}_1 - \overline{x}_2||^2$$

$$\le \left\{ \max_{\overline{x} \in \mathcal{R}} \left[\left(\frac{\partial f_0(\overline{x})}{\partial x_0} \right)^2 + \left(\frac{\partial f_0(\overline{x})}{\partial x_1} \right)^2 \right. \right.$$

$$\left. \left. + \left(\frac{\partial f_1(\overline{x})}{\partial x_0} \right)^2 + \left(\frac{\partial f_1(\overline{x})}{\partial x_1} \right)^2 \right] \right\} ||\overline{x}_1 - \overline{x}_2||^2$$

$$= \alpha^2 ||\overline{x}_1 - \overline{x}_2||^2.$$

As with the one-dimensional Taylor theorem from Chapter 3, our application of it here assumes that F is sufficiently smooth.

To apply Theorem 7.5, we try to select F and $\mathcal{R}$ so that $0 < \alpha < 1$. As already noted, this is often done experimentally. The iterative process (7.62) can give us an algorithm only if we have a stopping criterion. A good choice is (for suitable $\epsilon > 0$) to stop iterating when

$$\frac{||\overline{p}_n - \overline{p}_{n-1}||}{||\overline{p}_n||} < \epsilon, \quad \overline{p}_n \neq 0. \tag{7.66}$$

It is possible to use norms in (7.66) other than the one defined by (7.61). (A Chebyshev norm might be a good alternative.) Since our methodology is often to make guesses about F and $\mathcal{R}$, it is very possible that a given guess will be wrong. In other words, convergence may never occur. Thus, a loop that implements the algorithm should also be programmed to terminate when the number of iterations exceeds some reasonable threshold. In this event, the program must also be written to print a message saying that convergence has not occurred because the upper limit on the allowed number of iterations was exceeded. This is an important example of *exception handling* in numerical computing.

Example 7.7 If we consider $f_0(x_0, x_1)$ and $f_1(x_0, x_1)$ as in Example 7.6, then the sequence of iterates for the starting vector $\overline{p}_0 = [1 \quad 1]^T$ is

$$\begin{aligned} p_{1,0} &= 1.2500, & p_{1,1} &= 0.0000 \\ p_{2,0} &= 1.5625, & p_{2,1} &= 1.5625 \\ p_{3,0} &= 3.0518, & p_{3,1} &= 0.0000 \\ p_{4,0} &= 9.3132, & p_{4,1} &= 9.3132 \end{aligned}$$

$$\vdots$$

This vector sequence is not converging to the root near $[0.9 \quad 0.5]^T$ (see Fig. 7.6). Many experiments with different starting values do not lead to a solution. So we need to change F.

We may rewrite $x_0 - x_0^2 - \frac{1}{4}x_1^2 = 0$ as

$$x_0 = \frac{x_0^2}{x_0^2 + \frac{1}{4}x_1^2} = f_0(x_0, x_1),$$

and we may rewrite $x_1 - x_0^2 + x_1^2 = 0$ as

$$x_1 = \frac{x_0^2}{1 + x_1} = f_1(x_0, x_1).$$

These redefine F [recall (7.58) for the general form of F]. For this choice of F and again with $\overline{p}_0 = [1 \quad 1]^T$, the sequence of vectors is

$$p_{1,0} = 0.8000, \qquad p_{1,1} = 0.5000$$

$$p_{2,0} = 0.9110, \qquad p_{2,1} = 0.4627$$

$$p_{3,0} = 0.9480, \qquad p_{3,1} = 0.5818$$

$$p_{4,0} = 0.9140, \qquad p_{4,1} = 0.5682$$

$$\vdots$$

$$p_{14,0} = 0.9189, \qquad p_{14,1} = 0.5461$$

In this case the vector sequence now converges to the root with four-decimal-place accuracy by the 14th iteration.

7.5.2   Newton–Raphson Method

Consider yet again two equations in two unknowns

$$f_0(x_0, x_1) = 0,$$

$$f_1(x_0, x_1) = 0, \qquad\qquad (7.67)$$

each of which defines a curve in the plane (again $\overline{x} = [x_0 \quad x_1]^T \in \mathbf{R}^2$). Solutions to (7.67) are points of intersection of the two curves in $\mathbf{R}^2$. We will denote a point of intersection by $\overline{p} = [p_0 \quad p_1]^T \in \mathbf{R}^2$, which is a root of the system (7.67).

Suppose that $\overline{x}_0 = [x_{0,0} \quad x_{0,1}]^T$ is an initial approximation to the root $\overline{p}$. Assume that f_0 and f_1 are smooth enough to possess a two-dimensional Taylor series expansion,

$$f_0(x_0, x_1) = f_0(x_{0,0}, x_{0,1}) + \frac{\partial f_0}{\partial x_0}(x_{0,0}, x_{0,1})(x_0 - x_{0,0})$$

$$+ \frac{\partial f_0}{\partial x_1}(x_{0,0}, x_{0,1})(x_1 - x_{0,1}) + \frac{1}{2!} \left\{ \frac{\partial^2 f_0(x_{0,0}, x_{0,1})}{\partial x_0^2}(x_0 - x_{0,0})^2 \right.$$

$$+ 2\frac{\partial^2 f_0(x_{0,0}, x_{0,1})}{\partial x_0 \partial x_1}(x_0 - x_{0,0})(x_1 - x_{0,1})$$

$$+ \left. \frac{\partial^2 f_0(x_{0,0}, x_{0,1})}{\partial x_1^2}(x_1 - x_{0,1})^2 \right\} + \cdots$$

$$f_1(x_0, x_1) = f_1(x_{0,0}, x_{0,1}) + \frac{\partial f_1}{\partial x_0}(x_{0,0}, x_{0,1})(x_0 - x_{0,0})$$

$$+ \frac{\partial f_1}{\partial x_1}(x_{0,0}, x_{0,1})(x_1 - x_{0,1}) + \frac{1}{2!} \left\{ \frac{\partial^2 f_1(x_{0,0}, x_{0,1})}{\partial x_0^2}(x_0 - x_{0,0})^2 \right.$$

$$+ 2\frac{\partial^2 f_1(x_{0,0}, x_{0,1})}{\partial x_0 \partial x_1}(x_0 - x_{0,0})(x_1 - x_{0,1})$$

$$+ \frac{\partial^2 f_1(x_{0,0}, x_{0,1})}{\partial x_1^2}(x_1 - x_{0,1})^2 \Bigg\} + \cdots$$

which have the somewhat more compact form using vector notation as

$$f_0(\overline{x}) = f_0(\overline{x}_0) + \frac{\partial f_0(\overline{x}_0)}{\partial x_0}(x_0 - x_{0,0}) + \frac{\partial f_0(\overline{x}_0)}{\partial x_1}(x_1 - x_{0,1})$$

$$+ \frac{1}{2!}\Bigg\{\frac{\partial^2 f_0(\overline{x}_0)}{\partial x_0^2}(x_0 - x_{0,0})^2 + 2\frac{\partial^2 f_0(\overline{x}_0)}{\partial x_0 \partial x_1}(x_0 - x_{0,0})(x_1 - x_{0,1})$$

$$+ \frac{\partial^2 f_0(\overline{x}_0)}{\partial x_1^2}(x_1 - x_{0,1})^2 \Bigg\} + \cdots \tag{7.68a}$$

$$f_1(\overline{x}) = f_1(\overline{x}_0) + \frac{\partial f_1(\overline{x}_0)}{\partial x_0}(x_0 - x_{0,0}) + \frac{\partial f_1(\overline{x}_0)}{\partial x_1}(x_1 - x_{0,1})$$

$$+ \frac{1}{2!}\Bigg\{\frac{\partial^2 f_1(\overline{x}_0)}{\partial x_0^2}(x_0 - x_{0,0})^2 + 2\frac{\partial^2 f_1(\overline{x}_0)}{\partial x_0 \partial x_1}(x_0 - x_{0,0})(x_1 - x_{0,1})$$

$$+ \frac{\partial^2 f_1(\overline{x}_0)}{\partial x_1^2}(x_1 - x_{0,1})^2 \Bigg\} + \cdots . \tag{7.68b}$$

If $\overline{x}_0$ is close to $\overline{p}$, then from (7.68), we have

$$0 = f_0(\overline{p}) \approx f_0(\overline{x}_0) + \frac{\partial f_0(\overline{x}_0)}{\partial x_0}(p_0 - x_{0,0}) + \frac{\partial f_0(\overline{x}_0)}{\partial x_1}(p_1 - x_{0,1})$$

$$+ \frac{1}{2!}\Bigg\{\frac{\partial^2 f_0(\overline{x}_0)}{\partial x_0^2}(p_0 - x_{0,0})^2 + 2\frac{\partial^2 f_0(\overline{x}_0)}{\partial x_0 \partial x_1}(p_0 - x_{0,0})(p_1 - x_{0,1})$$

$$+ \frac{\partial^2 f_0(\overline{x}_0)}{\partial x_1^2}(p_1 - x_{0,1})^2 \Bigg\} + \cdots \tag{7.69a}$$

$$0 = f_1(\overline{p}) \approx f_1(\overline{x}_0) + \frac{\partial f_1(\overline{x}_0)}{\partial x_0}(p_0 - x_{0,0}) + \frac{\partial f_1(\overline{x}_0)}{\partial x_1}(p_1 - x_{0,1})$$

$$+ \frac{1}{2!}\Bigg\{\frac{\partial^2 f_1(\overline{x}_0)}{\partial x_0^2}(p_0 - x_{0,0})^2 + 2\frac{\partial^2 f_1(\overline{x}_0)}{\partial x_0 \partial x_1}(p_0 - x_{0,0})(p_1 - x_{0,1})$$

$$+ \frac{\partial^2 f_1(\overline{x}_0)}{\partial x_1^2}(p_1 - x_{0,1})^2 \Bigg\} + \cdots \tag{7.69b}$$

If we neglect the higher-order terms (second derivatives and higher-order derivatives), then we obtain

$$\frac{\partial f_0(\overline{x}_0)}{\partial x_0}(p_0 - x_{0,0}) + \frac{\partial f_0(\overline{x}_0)}{\partial x_1}(p_1 - x_{0,1}) \approx -f_0(\overline{x}_0), \tag{7.70a}$$

$$\frac{\partial f_1(\overline{x}_0)}{\partial x_0}(p_0 - x_{0,0}) + \frac{\partial f_1(\overline{x}_0)}{\partial x_1}(p_1 - x_{0,1}) \approx -f_1(\overline{x}_0). \tag{7.70b}$$

As a shorthand notation, define

$$f_{i,j}(\overline{x}_0) = \frac{\partial f_i(\overline{x}_0)}{\partial x_j} \tag{7.71}$$

so (7.70) becomes

$$(p_0 - x_{0,0})f_{0,0}(\overline{x}_0) + (p_1 - x_{0,1})f_{0,1}(\overline{x}_0) \approx -f_0(\overline{x}_0), \tag{7.72a}$$

$$(p_0 - x_{0,0})f_{1,0}(\overline{x}_0) + (p_1 - x_{0,1})f_{1,1}(\overline{x}_0) \approx -f_1(\overline{x}_0). \tag{7.72b}$$

Multiply (7.72a) by $f_{1,1}(\overline{x}_0)$, and multiply (7.72b) by $f_{0,1}(\overline{x}_0)$. Subtracting the second equation from the first results in

$$(p_0 - x_{0,0})[f_{0,0}(\overline{x}_0)f_{1,1}(\overline{x}_0) - f_{1,0}(\overline{x}_0)f_{0,1}(\overline{x}_0)]$$
$$\approx -f_0(\overline{x}_0)f_{1,1}(\overline{x}_0) + f_1(\overline{x}_0)f_{0,1}(\overline{x}_0). \tag{7.73a}$$

Now multiply (7.72a) by $f_{1,0}(\overline{x}_0)$, and multiply (7.72b) by $f_{0,0}(\overline{x}_0)$. Subtracting the second equation from the first results in

$$(p_1 - x_{0,1})[f_{0,1}(\overline{x}_0)f_{1,0}(\overline{x}_0) - f_{0,0}(\overline{x}_0)f_{1,1}(\overline{x}_0)]$$
$$\approx -f_0(\overline{x}_0)f_{1,0}(\overline{x}_0) + f_1(\overline{x}_0)f_{0,0}(\overline{x}_0). \tag{7.73b}$$

From (7.73b), we obtain

$$p_0 \approx x_{0,0} + \frac{-f_0(\overline{x}_0)f_{1,1}(\overline{x}_0) + f_1(\overline{x}_0)f_{0,1}(\overline{x}_0)}{f_{0,0}(\overline{x}_0)f_{1,1}(\overline{x}_0) - f_{0,1}(\overline{x}_0)f_{1,0}(\overline{x}_0)}, \tag{7.74a}$$

$$p_1 \approx x_{0,1} + \frac{-f_1(\overline{x}_0)f_{0,0}(\overline{x}_0) + f_0(\overline{x}_0)f_{1,0}(\overline{x}_0)}{f_{0,0}(\overline{x}_0)f_{1,1}(\overline{x}_0) - f_{0,1}(\overline{x}_0)f_{1,0}(\overline{x}_0)}. \tag{7.74b}$$

We may assume that the right-hand side of (7.74) is the next approximation to $\overline{p}$

$$x_{1,0} \approx x_{0,0} + \left.\frac{-f_0 f_{1,1} + f_1 f_{0,1}}{f_{0,0}f_{1,1} - f_{0,1}f_{1,0}}\right|_{\overline{x}_0}, \tag{7.75a}$$

$$x_{1,1} \approx x_{0,1} + \left.\frac{-f_1 f_{0,0} + f_0 f_{1,0}}{f_{0,0}f_{1,1} - f_{0,1}f_{1,0}}\right|_{\overline{x}_0} \tag{7.75b}$$

$(\overline{x}_1 = [x_{1,0}x_{1,1}]^T)$, where the functions and derivatives are to be evaluated at $\overline{x}_0$. We may continue this process to generate $(\overline{x}_n)$ for $n \in \mathbf{Z}^+$ (so in general $\overline{x}_n = [x_{n,0} \quad x_{n,1}]^T$) according to

$$x_{n+1,0} = x_{n,0} + \frac{-f_0(\overline{x}_n)f_{1,1}(\overline{x}_n) + f_1(\overline{x}_n)f_{0,1}(\overline{x}_n)}{f_{0,0}(\overline{x}_n)f_{1,1}(\overline{x}_n) - f_{0,1}(\overline{x}_n)f_{1,0}(\overline{x}_n)}, \tag{7.76a}$$

$$x_{n+1,1} = x_{n,1} + \frac{-f_1(\overline{x}_n)f_{0,0}(\overline{x}_n) + f_0(\overline{x}_n)f_{1,0}(\overline{x}_n)}{f_{0,0}(\overline{x}_n)f_{1,1}(\overline{x}_n) - f_{0,1}(\overline{x}_n)f_{1,0}(\overline{x}_n)}. \tag{7.76b}$$

As in the previous subsection, we define

$$F(\overline{x}_n) = \begin{bmatrix} f_0(x_{n,0}, x_{n,1}) \\ f_1(x_{n,0}, x_{n,1}) \end{bmatrix} = \begin{bmatrix} f_0(\overline{x}_n) \\ f_1(\overline{x}_n) \end{bmatrix}. \tag{7.77}$$

Also

$$F^{(1)}(\overline{x}_n) = \begin{bmatrix} f_{0,0}(\overline{x}_n) & f_{0,1}(\overline{x}_n) \\ f_{1,0}(\overline{x}_n) & f_{1,1}(\overline{x}_n) \end{bmatrix} = J_F(\overline{x}_n), \tag{7.78}$$

which is the *Jacobian matrix* J_F evaluated at $\overline{x} = \overline{x}_n$. We see that

$$[J_F(\overline{x}_n)]^{-1} = \frac{1}{f_{0,0}(\overline{x}_n)f_{1,1}(\overline{x}_n) - f_{0,1}(\overline{x}_n)f_{1,0}(\overline{x}_n)} \begin{bmatrix} f_{1,1}(\overline{x}_n) & -f_{0,1}(\overline{x}_n) \\ -f_{1,0}(\overline{x}_n) & f_{0,0}(\overline{x}_n) \end{bmatrix}, \tag{7.79}$$

so in vector notation (7.76) becomes

$$\overline{x}_{n+1} = \overline{x}_n - [J_F(\overline{x}_n)]^{-1} F(\overline{x}_n) \tag{7.80}$$

for $n \in \mathbf{Z}^+$. If $\overline{x}_n \in \mathbf{R}^m$ (i.e., if we consider m equations in m unknowns), then

$$J_F(\overline{x}_n) = \begin{bmatrix} f_{0,0}(\overline{x}_n) & f_{0,1}(\overline{x}_n) & \cdots & f_{0,m-1}(\overline{x}_n) \\ f_{1,0}(\overline{x}_n) & f_{1,1}(\overline{x}_n) & \cdots & f_{1,m-1}(\overline{x}_n) \\ \vdots & \vdots & & \vdots \\ f_{m-1,0}(\overline{x}_n) & f_{m-1,1}(\overline{x}_n) & \cdots & f_{m-1,m-1}(\overline{x}_n) \end{bmatrix}, \tag{7.81a}$$

and

$$F(\overline{x}_n) = \begin{bmatrix} f_0(\overline{x}_n) \\ f_1(\overline{x}_n) \\ \vdots \\ f_{m-1}(\overline{x}_n) \end{bmatrix}. \tag{7.81b}$$

Of course, $\overline{x}_n = [x_{n,0} \ x_{n,1} \ \cdots \ x_{n,m-1}]^T \in \mathbf{R}^m$.

Equation (7.80) reduces to (7.46) when we have only one equation in one unknown. We see that the method will fail if $J_F(\overline{x}_n)$ is singular at $\overline{x}_n$. As in the one-dimensional problem of Section 7.4, the success of the method depends on picking a good starting point $\overline{x}_0$. If convergence occurs, then it is quadratic as in the one-dimensional (i.e., scalar) case. It is sometimes possible to force the method to converge even if the starting point is poorly selected, but this will not be considered here. The computational complexity of the method is quite high. If $\overline{x}_n \in \mathbf{R}^m$, then from (7.80) and (7.81), we require $m^2 + m$ function evaluations, and we need to invert an $m \times m$ Jacobian matrix at every iteration. We know from Chapter 4 that matrix inversion needs $O(m^3)$ operations. Ill conditioning of the Jacobian is very much a potential problem as well.

Example 7.8   Refer to the examples in Section 7.5.1. In Example 7.6 there we wanted to solve

$$f_0(x_0, x_1) = x_0 - x_0^2 - \tfrac{1}{4}x_1^2 = 0, \ f_1(x_0, x_1) = x_1 - x_0^2 + x_1^2 = 0.$$

Consequently

$$f_0(\overline{x}_n) = x_{n,0} - x_{n,0}^2 - \tfrac{1}{4}x_{n,1}^2, \ f_1(\overline{x}_n) = x_{n,1} - x_{n,0}^2 + x_{n,1}^2,$$

and the derivatives are

$$f_{0,0}(\overline{x}_n) = 1 - 2x_{n,0}, \ f_{1,0}(\overline{x}_n) = -2x_{n,0}$$

$$f_{0,1}(\overline{x}_n) = -\tfrac{1}{2}x_{n,1}, \ f_{1,1}(\overline{x}_n) = 1 + 2x_{n,1}.$$

Via (7.76), the desired equations are

$$x_{n+1,0} = x_{n,0}$$
$$+ \frac{-(x_{n,0} - x_{n,0}^2 - \tfrac{1}{4}x_{n,1}^2)(1 + 2x_{n,1}) + (x_{n,1} - x_{n,0}^2 + x_{n,1}^2)(-\tfrac{1}{2}x_{n,1})}{(1 - 2x_{n,0})(1 + 2x_{n,1}) - x_{n,0}x_{n,1}},$$

$$\text{(7.82a)}$$

$$x_{n+1,1} = x_{n,1}$$
$$+ \frac{-(x_{n,1} - x_{n,0}^2 + x_{n,1}^2)(1 - 2x_{n,0}) + (x_{n,0} - x_{n,0}^2 - \tfrac{1}{4}x_{n,1}^2)(-2x_{n,0})}{(1 - 2x_{n,0})(1 + 2x_{n,1}) - x_{n,0}x_{n,1}}. \quad \text{(7.82b)}$$

If we execute the iterative procedure in (7.82), we obtain

$$x_{0,0} = 0.8000, \quad x_{0,1} = 0.5000$$

$$x_{1,0} = 0.9391, \quad x_{1,1} = 0.5562$$

$$x_{2,0} = 0.9193, \quad x_{2,1} = 0.5463$$

$$x_{3,0} = 0.9189, \quad x_{3,1} = 0.5461$$

We see that the answer is correct to four decimal places in only three iterations. This is much faster than the fixed-point method seen in Example 7.7.

7.6   CHAOTIC PHENOMENA AND A CRYPTOGRAPHY APPLICATION

Iterative processes such as $x_{n+1} = g(x_n)$ (which includes the Newton–Raphson method) can converge to a fixed point [i.e., to x such that $g(x) = x$], or they can fail to do so in various ways. This was considered in Section 7.4.3. We are interested here in the case where (x_n) is a *chaotic sequence*, in which case g is often said to be a *chaotic map*. Formal definitions exist for chaotic maps [13, p. 50]. However, these are rather technical. They are also difficult to apply except in relatively simple cases. We shall therefore treat chaos in an intuitive/empirical (i.e., experimental) manner for simplicity.

In Section 7.3 we considered examples based on the *logistic map*

$$g(x) = \lambda x(1 - x) \tag{7.83}$$

(recall Examples 7.3–7.5). Suppose that $\lambda = 4$. Figure 7.7 shows two output sequences from this map for two slightly different initial conditions. Plot (a) shows for $x_0 = x_0^a = 0.745$; while plot (b), $x_0 = x_0^b = 0.755$. We see that $|x_0^a - x_0^b| = 0.01$, yet after only a few iterations, the two sequences are very different from each other. This is one of the distinguishing features of chaos: sensitive dependence of the resulting sequence to minor changes in the initial conditions. For $\lambda = 4$, we have $g|[0, 1] \to [0, 1]$, so divergence is impossible. Chaotic sequences do not diverge. They remain bounded; that is, there is a $M \in \mathbf{R}$ such that $0 < M < \infty$ with $|x_n| \le M$ for all $n \in \mathbf{Z}^+$. But the sequence does not converge, and it is not periodic, either. In fact, the plots in Fig. 7.7 show that the elements of sequence (x_n) seem to wander around rather aimlessly (i.e., apparently randomly). This wandering behavior has been observed in the past [14, p. 167], but was not generally recognized as being a chaotic phenomenon until more recently.

It has been known for a very long time that effective cryptographic systems should exploit randomness [15]. Since chaotic sequences have apparently random qualities, it is not surprising that they have been proposed as random-number generators (or as pseudo-random-number generators) for applications in cryptography [16, 17]. However, it is presently a matter of legitimate controversy regarding just how secure a cryptosystem based on chaos can be. One difficulty is as follows. Nominally, a chaotic map g takes on values from the set of real numbers. But if such a map is implemented on a digital computer, then, since all computers are finite-state machines, any chaotic sequence will not be truly chaotic as it will eventually repeat. Short period sequences are cryptographically weak (i.e., not secure). There is presently no effective procedure (beyond exhaustive searching) for determining when this difficulty will arise in a chaos-based system. This is not the only problem (see p. 1507 of Ref. 17 for others).

Two specific chaos-based cryptosystems are presented by De Angeli et al. [18] and Papadimitriou et al. [19]. (There have been many others proposed in recent

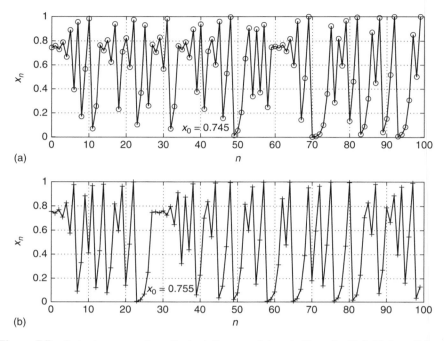

Figure 7.7 Output sequence from the logistic map $g(x) = 4x(1 - x)$ with initial conditions $x_0 = 0.745$ (a) and $x_0 = 0.755$ (b). Observe that although these initial conditions are close together, the resulting sequences become very different from each other after only a few iterations.

years.) *Key size* is the number of different possible encryption keys available in the system. Naturally, this number should be large enough to prevent a codebreaker (eavesdropper, cryptanalyst) from guessing the correct key. However, it is also well known that a large key size is most definitely not a guarantee that the cryptosystem will be secure. Papadimitriou et al. [19] demonstrate that their system has a large key size, but their security analysis [19] did not go beyond this. Their system [19] seems difficult to analyze, however. In what follows we shall present some analysis of the system in De Angeli et al. [18], and see that if implemented on a digital computer, it is not really very secure. We begin with a description of their system [18]. Their method in [18] is based on the *Hénon map* (see Fig. 7.8), which is a mapping defined on $\mathbf{R}^2$ according to

$$x_{0,n+1} = 1 - \alpha x_{0,n}^2 + x_{1,n}$$
$$x_{1,n+1} = \beta x_{0,n} \tag{7.84}$$

for $n \in \mathbf{Z}^+$ (so $[x_{0,n} x_{1,n}]^T \in \mathbf{R}^2$), and which is known to be chaotic in some neighborhood of

$$\alpha = 1.4, \quad \beta = 0.3 \tag{7.85}$$

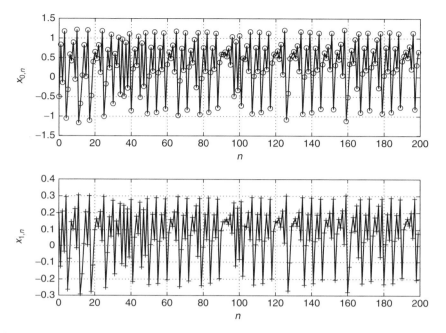

Figure 7.8 Typical state sequences from the Hénon map for $\alpha = 1.45$ and $\beta = 0.25$, with initial conditions $x_{0,0} = -0.5$ and $x_{1,0} = 0.2$.

so that constants α and β form the *encryption key* for the system. No choice in the allowed neighborhood is a valid key. An immediate problem is that there seems to be no detailed description of which points are allowed. A particular choice of key should therefore be tested to see if the resulting sequence is chaotic. In what follows the *output sequence* from the map is defined to be

$$y_n = x_{0,n}. \tag{7.86}$$

The vector $x_n = [x_{0,n} \quad x_{1,n}]^T$ is often called a *state vector*. The elements are *state variables*.

The encryption algorithm works by mixing the chaotic sequence of (7.86) with the *message sequence*, which we shall denote by (s_n). The mixing (described below) yields the *cyphertext sequence*, denoted (c_n). A problem is that the receiver must somehow recover the original message (s_n) from the cyphertext (c_n) using knowledge of the encryption algorithm, and the key $\{\alpha, \beta\}$.

Consider the following mapping:

$$\begin{aligned}
\hat{x}_{0,n+1} &= 1 - \alpha y_n^2 + \hat{x}_{1,n}, \\
\hat{x}_{1,n+1} &= \beta y_n.
\end{aligned} \tag{7.87}$$

Note that this mapping is of the same form as (7.84) except that $x_{0,n}$ is replaced by y_n, but from (7.86) these are the same (nominally). The mapping in (7.84) represents a physical system (or piece of software) at the transmitter, while (7.87) is part of the receiver (hardware or software). Now define the *error sequences*

$$\delta x_{i,n} = \hat{x}_{i,n} - x_{i,n} \qquad (i \in \{0, 1\}) \tag{7.88}$$

for which it is possible to show [using (7.84), (7.86), and (7.87)] that

$$\begin{bmatrix} \delta x_{0,n+1} \\ \delta x_{1,n+1} \end{bmatrix} = \underbrace{\begin{bmatrix} 0 & 1 \\ 0 & 0 \end{bmatrix}}_{=A} \begin{bmatrix} \delta x_{0,n} \\ \delta x_{1,n} \end{bmatrix}. \tag{7.89}$$

We observe that $A^2 = 0$. Matrix A is an example of a *nilpotent matrix* for this reason. This immediately implies that the error sequences go to zero in at most two iterations (i.e., two steps):

$$\begin{bmatrix} \delta x_{0,n+2} \\ \delta x_{1,n+2} \end{bmatrix} = \begin{bmatrix} 0 & 1 \\ 0 & 0 \end{bmatrix} \begin{bmatrix} 0 & 1 \\ 0 & 0 \end{bmatrix} \begin{bmatrix} \delta x_{0,n} \\ \delta x_{1,n} \end{bmatrix} = \begin{bmatrix} 0 \\ 0 \end{bmatrix}.$$

This fact tells us that if (y_n) is generated at the transmitter, and sent over the communications channel to the receiver, then the receiver may perfectly recover[2] the state sequence of the transmitter in not more than two steps. This is called *dead-beat synchronization*. The system in (7.87) is a specific example of a *nonlinear observer* for a nonlinear dynamic system. There is a well-developed theory of observers for linear dynamic systems [20]. The notion of an observer is a control systems concept, so we infer that control theory is central to the problem of applying chaotic systems to cryptographic applications.

All of this suggests the following algorithm for encrypting a message. Assume that we wish to encrypt a length N message sequence (s_n); that is, we only have the elements $s_0, s_1, \ldots, s_{N-2}, s_{N-1}$.

1. The transmitter generates and transmits y_0, y_1 according to (7.84) and (7.86). The transmitter also generates $y_2, y_3, \ldots, y_N, y_{N+1}$, but these are not transmitted.

2. The transmitter sends the cyphertext sequence

$$c_n = \frac{y_{n+2}}{s_n} \tag{7.90}$$

for $n = 0, 1, \ldots, N - 1$. Of course, this assumes we never have $s_n = 0$.

Equation (7.90) is not the only possible means of mixing the message and the chaotic sequence together. The decryption algorithm at the receiver is as follows:

---

[2]This assumes a perfect communications channel (which is not realistic), and that rounding errors in the computation are irrelevant (which will be true if the receiver implements arithmetic in the same manner as the transmitter).

1. The receiver regenerates the transmitter's state sequence using (7.87), its knowledge of the key, and the sequence elements y_0, y_1. Specifically, for $n = 0, 1$ compute

$$\hat{x}_{0,n+1} = 1 - \alpha y_n^2 + \hat{x}_{1,n},$$
$$\hat{x}_{1,n+1} = \beta y_n \tag{7.91a}$$

while for $n = 2, 3, \ldots, N - 1, N$ compute

$$\hat{x}_{0,n+1} = 1 - \alpha \hat{x}_{0,n}^2 + \hat{x}_{1,n},$$
$$\hat{x}_{1,n+1} = \beta \hat{x}_{0,n}. \tag{7.91b}$$

Recover y_n for $n = 2, 3, \ldots, N, N + 1$ according to

$$y_n = \hat{x}_{0,n}. \tag{7.92}$$

2. Recover the original message via

$$s_n = \frac{y_{n+2}}{c_n}, \tag{7.93}$$

where $n = 0, 1, \ldots, N - 2, N - 1$.

The initial states at the transmitter and receiver are arbitrary; that is, any $x_{0,0}$, $x_{1,0}$ and $\hat{x}_{1,0}$ may be selected in (7.84) and (7.91a). The elements y_0, y_1, and the cyphertext are sent over the channel. If these are lost because of a corrupt channel, then the receiver should request retransmission of this information. It is extremely important that the transmitter resend the same synchronizing elements y_0, y_1 and not a different pair. The reason will become clear below.

We remark that since we are assuming that the algorithms are implemented on a digital computer, so each sequence element is a binary word of some form. The synchronizing elements and cyphertext are thus a bitstream. Methods exist to encode such data for transmission over imperfect channels such that the probability of successful transmission can be made quite high. These are called *error control coding schemes*.

In general, the receiver will fail to recover the message if (1) the way in which arithmetic is performed by the receiver is not the same as at the transmitter, (2) the channel corrupts the transmission, or (3) the receiver does not know the key $\{\alpha, \beta\}$. Item 1 is important since the failure to properly duplicate arithmetic operations at the receiver will cause machine rounding errors to accumulate and prevent data recovery. This is really a case of improper synchronization. The plots of Figs. 7.9–7.11 illustrate some of these effects.

It is noteworthy that even though the receiver may not be a perfect match to the transmitter, some of the samples (at the beginning of the message) are recovered.

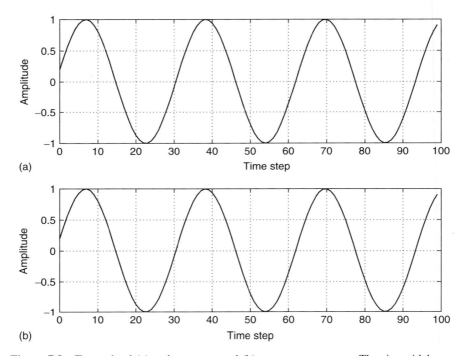

(a)

(b)

Figure 7.9 Transmitted (a) and reconstructed (b) message sequences. The sinusoidal message sequence (s_n) is perfectly reconstructed at the receiver; because the channel is perfect, the receiver knows the key (here $\alpha = 1.4$, $\beta = 0.3$), and arithmetic is performed in identical fashion at both the transmitter and receiver.

This is actually an indication that the system is not really very secure. We now consider security of the method in greater detail.

What is the key size? This question seems hard to answer accurately, but a simple analysis is as follows. Suppose that the Hénon map is chaotic in the rectangular region of the $\alpha\beta$ plane defined by $\alpha \in [1.4 - \Delta\alpha, 1.4 + \Delta\alpha]$, and $\beta \in [0.3 - \Delta\beta, 0.3 + \Delta\beta]$, with $\Delta\alpha, \Delta\beta > 0$. We do not specifically know the interval limits, and it is an ad hoc assumption that the chaotic neighborhood is rectangular (this is a false assumption). Suppose that $\underline{\beta}$ is the smallest M-bit binary fraction such that $\underline{\beta} \geq 0.3 - \beta$, and that $\overline{\beta}$ is the largest M-bit binary fraction such that $\overline{\beta} \leq 0.3 + \Delta\beta$. In this case the number of M-bit fractions from $0.3 - \Delta\beta$ to $0.3 + \Delta\beta$ is about $2^M(\overline{\beta} - \underline{\beta}) + 1$. Thus

$$2^M(\overline{\beta} - \underline{\beta} + 1) \approx 2^M[(0.3 + \Delta\beta) - (0.3 - \Delta\beta)] + 1 = 2^{M+1}\Delta\beta + 1 = K_\beta,$$
$$(7.94a)$$

and by similar reasoning for α

$$K_\alpha = 2^{M+1}\Delta\alpha + 1, \qquad (7.94b)$$

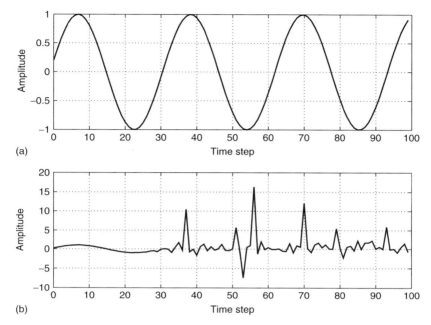

Figure 7.10   Transmitted (a) and reconstructed (b) message sequences. Here, the conditions of Fig. 7.9 hold except that the receiver uses a mismatched key $\alpha = 1.400001$, $\beta = 0.300001$. Thus, the message is eventually lost. (Note that the first few message samples seem to be recovered accurately.).

which implies that the key size is (very approximately)

$$K = K_\alpha K_\beta. \tag{7.95}$$

Even if $\Delta\alpha$ and $\Delta\beta$ are small, and the structure of the chaotic neighborhood is not rectangular, we can make M big enough (in principle) to generate a big key space. Apparently, Ruelle [21, p. 19] has shown that the Hénon map is periodic (i.e., *not chaotic*) for $\alpha = 1.3$, $\beta = 0.3$, so the size of the chaotic region is not very big. It is irregular and "full of holes" (the "holes" are key parameters that don't give chaotic outputs). In any case it seems that a large key size is possible. But as cautioned earlier, this is no guarantee of security.

What does the transmitter send over the channel? This is the same as asking what the eavesdropper knows. From the encryption algorithm the eavesdropper knows the synchronizing elements y_0, y_1, and the cyphertext. The eavesdropper also knows the algorithm, but not the key. Is this enough to find the key? It is now obvious to ask if knowledge of y_0, y_1 gives the key away. This question may be easily (?) resolved as follows.

From (7.84) and (7.86) for $n \in \mathbf{N}$, we have

$$y_{n+1} = 1 - \alpha y_n^2 + \beta y_{n-1}. \tag{7.96}$$

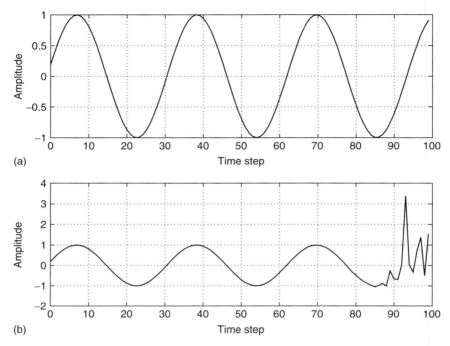

(a)

(b)

Figure 7.11 Transmitted (a) and reconstructed (b) message sequences. Here, the conditions of Fig. 7.9 hold except the receiver and the transmitter do not perform arithmetic in identical fashion. In this case the order of two operations was reversed at the receiver. Thus, the message is eventually lost. (Note again that the first few message samples seem to be recovered accurately.).

The encryption algorithm generates $y_0, y_1, \ldots, y_{N+1}$, which, from (7.96), must lead to the key satisfying the linear system of equations

$$
\underbrace{\begin{bmatrix} y_1^2 & -y_0 \\ y_2^2 & -y_1 \\ \vdots & \vdots \\ y_{N-1}^2 & -y_{N-2} \\ y_N^2 & -y_{N-1} \end{bmatrix}}_{=Y} \underbrace{\begin{bmatrix} \alpha \\ \beta \end{bmatrix}}_{=a} = \underbrace{\begin{bmatrix} 1 - y_2 \\ 1 - y_3 \\ \vdots \\ 1 - y_N \\ 1 - y_{N+1} \end{bmatrix}}_{=y}. \tag{7.97}
$$

Compactly, we have $Ya = y$. This is an overdetermined system. The eavesdropper has only y_0, y_1, so from (7.97) the eavesdropper needs to solve

$$
\begin{bmatrix} y_1^2 & -y_0 \\ y_2^2 & -y_1 \end{bmatrix} \begin{bmatrix} \alpha \\ \beta \end{bmatrix} = \begin{bmatrix} 1 - y_2 \\ 1 - y_3 \end{bmatrix}. \tag{7.98}
$$

But the eavesdropper does not know y_2, y_3 as these were not transmitted. These elements become mixed with the message, and so are not available to the eavesdropper. We immediately conclude that deadbeat synchronization is secure.[3] There is an alternative synchronization scheme [18] that may be proved to be insecure by this method of analysis.

Does the cyphertext give key information away? This seems not to have a complete answer either. However, we may demonstrate that the system of D'Angeli et al. [18] (using deadbeat synchronization) is vulnerable to a *known plaintext*[4] *attack*. Such a vulnerability is ordinarily sufficient to preclude using a method in a high-security application. The analysis assumes partial prior knowledge of the message (s_n). Let us specifically assume that

$$s_n = \sum_{k=0}^{p} a_k n^k, \tag{7.99}$$

but we do not know a_k or p. That is, the structure of our message is a polynomial sequence, but we do not know more than this. Combining (7.90) with (7.96) gives us

$$c_n s_n = 1 - \alpha c_{n-1}^2 s_{n-1}^2 + \beta c_{n-2} s_{n-2}. \tag{7.100}$$

It is not difficult to confirm that

$$s_{n-1}^2 = \sum_{k=0}^{2p} \underbrace{\left\{ \sum_{i+j=k} a_i a_j \right\}}_{=b_k} (n-1)^k \tag{7.101}$$

so that (7.100) becomes

$$\sum_{k=0}^{p} a_k n^k c_n + \sum_{k=0}^{2p} \alpha b_k (n-1)^k c_{n-1}^2 - \sum_{k=0}^{p} \beta a_k (n-2)^k c_{n-2} = 1. \tag{7.102}$$

Remembering that the eavesdropper has cyphertext (c_n) (obtained by eavesdropping), the eavesdropper can use (7.102) to set up a linear system of equations in the key and message parameters a_k. The code is therefore easily broken in this case.

If the message has a more complex structure, (7.102) will generally be replaced by some hard-to-solve nonlinear problem wherein the methods of previous sections (e.g., Newton–Raphson) can be used to break the code. We conclude that, in spite of technical problems from the eavesdropper's point of view (ill conditioning, incomplete cyphertext sequences, etc.), the scheme in Ref. 18 is not secure.

---

[3]This conclusion assumes there are no other distinct ways to work with the encryption algorithm equations in such a manner as to give an equation that an eavesdropper can solve for the key knowing only y_0, y_1, and the cyphertext.

[4]The message is also called *plaintext*.

REFERENCES

1. J. H. Wilkinson, "The Perfidious Polynomial," in *Studies in Mathematics*, G. H. Golub, ed., Vol. 24, Mathematical Association of America, 1984.

2. A. M. Cohen, "Is the Polynomial so Perfidious?" *Numerische Mathematik* **68**, 225–238 (1994).

3. T. E. Hull and R. Mathon, "The Mathematical Basis and Prototype Implementation of a New Polynomial Root Finder with Quadratic Convergence," *ACM Trans. Math. Software* **22**, 261–280 (Sept. 1996).

4. E. Kreyszig, *Introductory Functional Analysis with Applications*, Wiley, New York, 1978.

5. W. Rudin, *Principles of Mathematical Analysis*, 3rd ed., McGraw-Hill, New York, 1976.

6. D. C. Youla and H. Webb, "Image Restoration by the Method of Convex Projections: Part I—Theory," *IEEE Trans. Med. Imag.* **MI-1**, 81–94 (Oct. 1982).

7. A. E. Cetin, O. N. Gerek and Y. Yardimci, "Equiripple FIR Filter Design by the FFT Algorithm," *IEEE Signal Process. Mag.* **14**, 60–64 (March 1997).

8. K. Gröchenig, "A Discrete Theory of Irregular Sampling," *Linear Algebra Appl.* **193**, 129–150 (1993).

9. H. H. Bauschke and J. M. Borwein, "On Projection Algorithms for Solving Convex Feasibility Problems," *SIAM Rev.* **38**, 367–426 (Sept. 1996).

10. E. Kreyszig, *Advanced Engineering Mathematics*, 4th ed., Wiley, New York, 1979.

11. E. Isaacson and H. B. Keller, *Analysis of Numerical Methods*, Wiley, New York, 1966.

12. F. B. Hildebrand, *Introduction to Numerical Analysis*, 2nd ed., McGraw-Hill, New York, 1974.

13. R. L. Devaney, *An Introduction to Chaotic Dynamical Systems*, 2nd ed., Addison-Wesley, Redwood City, CA, 1989.

14. G. E. Forsythe, M. A. Malcolm and C. B. Moler, *Computer Methods for Mathematical Computations*, Prentice-Hall, Englewood Cliffs, NJ, 1977.

15. G. Brassard, *Modern Cryptology: A Tutorial*, Lecture Notes in Computer Science (series), Vol. 325, G. Goos and J. Hartmanis, eds., Springer-Verlag, New York, 1988.

16. L. Kocarev, "Chaos-Based Cryptography: A Brief Overview," *IEEE Circuits Syst. Mag.* **1**(3), 6–21 (2001).

17. F. Dachselt and W. Schwarz, "Chaos and Cryptography," *IEEE Trans. Circuits Syst.* (Part I: Fundamental Theory and Applications) **48**, 1498–1509 (Dec. 2001).

18. A. De Angeli, R. Genesio and A. Tesi, "Dead-Beat Chaos Synchronization in Discrete-Time Systems," *IEEE Trans. Circuits Syst.* (Part I: Fundamental Theory and Applications) **42**, 54–56 (Jan. 1995).

19. S. Papadimitriou, A. Bezerianos and T. Bountis, "Secure Communications with Chaotic Systems of Difference Equations," *IEEE Trans. Comput.* **46**, 27–38 (Jan. 1997).

20. M. S. Santina, A. R. Stubberud and G. H. Hostetter, *Digital Control System Design*, 2nd ed., Saunders College Publ., Fort Worth, TX, 1994.

21. D. Ruelle, *Chaotic Evolution and Strange Attractors: The Statistical Analysis of Time Series for Deterministic Nonlinear Systems*, Cambridge Univ. Press, New York, 1989.

22. M. Jenkins and J. Traub, "A Three-Stage Variable Shift Algorithm for Polynomial Zeros and Its Relation to Generalized Rayleigh Iteration," *Numer. Math.* **14**, 252–263 (1970).

PROBLEMS

7.1. For the functions and starting intervals below solve $f(x) = 0$ using the bisection method. Use stopping criterion (7.13d) with $\epsilon = 0.005$. Do the calculations with a pocket calculator.

(a) $f(x) = \log_e x + 2x + 1$, $[a_0, b_0] = [0.2, 0.3]$.

(b) $f(x) = x^3 - \cos x$, $[a_0, b_0] = [0.8, 1.0]$.

(c) $f(x) = x - e^{-x/5}$, $[a_0, b_0] = [\frac{3}{4}, 1]$.

(d) $f(x) = x^6 - x - 1$, $[a_0, b_0] = [1, \frac{3}{2}]$.

(e) $f(x) = \frac{\sin x}{x} + \exp(-x)$, $[a_0, b_0] = [3, 4]$.

(f) $f(x) = \frac{\sin x}{x} - x + 1$, $[a_0, b_0] = [1, 2]$.

7.2. Consider $f(x) = \sin(x)/x$. This function has a minimum value for some $x \in [\pi, 2\pi]$. Use the bisection method to find this x. Use the stopping criterion in (7.13d) with $\epsilon = 0.005$. Use a pocket calculator to do the computations.

7.3. This problem introduces the variation on the bisection method called *regula falsi* (or the *method of false position*). Suppose that $[a_0, b_0]$ brackets the root p [i.e., $f(p) = 0$]. Thus, $f(a_0) f(b_0) < 0$. The first estimate of the root p, denoted by p_0, is where the line joining the points $(a_0, f(a_0))$ and $(b_0, f(b_0))$ crosses the x axis.

(a) Show that

$$p_0 = a_0 - \frac{b_0 - a_0}{f(b_0) - f(a_0)} f(a_0).$$

(b) Using stopping criterion (7.13d), write pseudocode for the method of false position.

7.4. In certain signal detection problems (e.g., radar or sonar) the probability of false alarm (FA) (i.e., of saying that a certain signal is present in the data when it actually is not) is given by

$$P_{FA} = \int_{\eta}^{\infty} \frac{1}{\Gamma(p/2)2^{p/2}} x^{\frac{p}{2}-1} e^{-x/2} \, dx, \tag{7.P.1}$$

where η is called the *detection threshold*. If p is an even number, it can be shown that (7.P.1) reduces to the finite series

$$P_{FA} = e^{-\frac{1}{2}\eta} \sum_{k=0}^{(p/2)-1} \frac{1}{k!} \left(\frac{\eta}{2}\right)^k. \tag{7.P.2}$$

The detection threshold η is a very important design parameter in signal detectors. Often it is desired to specify an acceptable value for P_{FA} (where

$0 < P_{FA} < 1$), and then it is necessary to solve nonlinear equation (7.P.2) for η. Let $p = 6$. Use the bisection method to find η for

(a) $P_{FA} = 0.001$
(b) $P_{FA} = 0.01$
(c) $P_{FA} = 0.1$

7.5. We wish to solve

$$f(x) = x^4 - \frac{5}{2}x^3 + \frac{5}{2}x - 1 = 0$$

using a fixed-point method. This requires finding $g(x)$ such that

$$g(x) - x = f(x).$$

Find four different functions $g(x)$.

7.6. Can the fixed-point method be used to find the solution to

$$f(x) = x^6 - x - 1 = 0$$

for the root located in the interval $[1, \frac{3}{2}]$? Explain.

7.7. Consider the nonlinear equation

$$f(x) = x - e^{-x/5} = 0$$

(which has a solution on interval $[\frac{3}{4}, 1]$). Use (7.32) to estimate α. Recalling that $x_n = g^n x_0$ (in the fixed-point method), if $x_0 = 1$, then use (7.22a) to estimate n so that $d(x_n, x) < 0.005$ [x is the root of $f(x) = 0$]. Use a pocket calculator to compute $x_1, \ldots, x_n$.

7.8. Consider the nonlinear equation

$$f(x) = x - 1 - \frac{1}{2}e^{-x} = 0$$

(which has a solution on interval $[1, 1.2]$). Use (7.32) to estimate α. Recalling that $x_n = g^n x_0$ (in the fixed-point method), if $x_0 = 1$ then use (7.22a) to estimate n so that $d(x_n, x) < 0.001$ [x is the root of $f(x) = 0$]. Use a pocket calculator to compute $x_1, \ldots, x_n$.

7.9. Problem 5.14 (in Chapter 5) mentioned the fact that orthogonal polynomials possess the "interleaving of zeros" property. Use this property and the bisection method to derive an algorithm to find the zeros of all Legendre polynomials $P_n(x)$ for $n = 1, 2, \ldots, N$. Express the algorithm in pseudocode. Be fairly detailed about this.

7.10. We wish to find all of the roots of

$$f(x) = x^3 - 3x^2 + 4x - 2 = 0.$$

There is one real-valued root, and two complex-valued roots. It is easy to confirm that $f(1) = 0$, but use

$$g(x) = \frac{x^3 + 3x^2 + x + 2}{2x^2 + 5}$$

to estimate the real root p using fixed-point iteration [i.e, $p_{n+1} = g(p_n)$]. Using a pocket calculator, compute only p_1, p_2, p_3, and p_4, and use the starting point $p_0 = 2$. Also, use the Newton–Raphson method to estimate the real root. Again choose $p_0 = 2$, and compute only p_1, p_2, p_3, and p_4. Once the real root is found, finding the complex-valued roots is easy. Find the complex-valued roots by making use of the formula for the roots of a quadratic equation.

7.11. Consider Eq. (7.36). Via Theorem 3.3, there is an α_n between root p ($f(p) = 0$) and the iterate p_n such that

$$f(p) - f(p_n) = f^{(1)}(\alpha_n)(p - p_n).$$

(a) Show that

$$p - p_n = (p_n - p_{n-1}) \underbrace{\left(\frac{f(p_n)}{f(p_{n-1})} \frac{f^{(1)}(p_{n-1})}{f^{(1)}(\alpha_n)} \right)}_{=A_n}.$$

[*Hint:* Use the identity $1 = \frac{f(p_{n-1})}{f(p_{n-1})} \frac{f^{(1)}(p_{n-1})}{f^{(1)}(p_{n-1})}$.]

(b) Argue that if convergence is occurring, we have

$$\lim_{n \to \infty} |A_n| = 1.$$

$\left(\text{Hence } \lim_{n \to \infty} \left| \frac{p - p_n}{p_n - p_{n-1}} \right| = 1. \right)$

(c) An alternative stopping criterion for the Newton–Raphson method is to stop iterating when

$$|f(p_n)| + |p_n - p_{n-1}| < \epsilon$$

for some suitably small $\epsilon > 0$. Is this criterion preferable to (7.42)? Explain.

7.12. For the functions listed below, and for the stated starting value p_0, use the Newton–Raphson method to solve $f(p) = 0$. Use the stopping criterion (7.42a) with $\epsilon = 0.001$. Perform all calculations using only a pocket calculator.

(a) $f(x) = x + \tan x$, $p_0 = 2$.
(b) $f(x) = x^6 - x - 1$, $p_0 = 1.5$.
(c) $f(x) = x^3 - \cos x$, $p_0 = 1$.
(d) $f(x) = x - e^{-x/5}$, $p_0 = 1$.

7.13. Use the Newton–Raphson method to find the real-valued root of the polynomial equation

$$f(x) = 1 + \frac{1}{2}x + \frac{1}{6}x^2 + \frac{1}{24}x^3 = 0.$$

Choose starting point $p_0 = -2$. Iterate 4 times. [*Comment:* Polynomial $f(x)$ arises in the stability analysis of a numerical method for solving ordinary differential equations. This will be seen in Chapter 10, Eq. (10.83).]

7.14. Write a MATLAB function to solve Problem 7.4 using the Newton–Raphson method. Use the stopping criterion (7.42a).

7.15. Prove the following theorem (Newton–Raphson error formula). Let $f(x) \in C^2[a, b]$, and $f(p) = 0$ for some $p \in [a, b]$. For $p_n \in [a, b]$ with

$$p_{n+1} = p_n - \frac{f(p_n)}{f^{(1)}(p_n)},$$

there is a ξ_n between p and p_n such that

$$p - p_{n+1} = -\frac{1}{2}(p - p_n)^2 \frac{f^{(2)}(\xi_n)}{f^{(1)}(p_n)}.$$

[*Hint:* Consider the Taylor series expansion of $f(x)$ about the point $x = p_n$

$$f(x) = f(p_n) + (x - p_n)f^{(1)}(p_n) + \tfrac{1}{2}(x - p_n)^2 f^{(2)}(\xi_n),$$

and then set $x = p$.]

7.16. This problem is about two different methods to compute $\sqrt{x}$. To begin, recall that if x is a binary floating-point number (Chapter 2), then it has the form

$$x = x_0.x_1 \cdots x_t \times 2^e,$$

where, since $x \geq 0$, we have $x_0 = 0$, and because of normalization $x_1 = 1$. Generally, $x_k \in \{0, 1\}$, and e is the exponent. If $e = 2k$ (i.e., the exponent is even), we do not adjust x. But if $e = 2k + 1$ (i.e., the exponent is odd), we shift the mantissa to the right by one bit position so that e becomes $e = 2k + 2$. Thus, x now has the form

$$x = a \times 2^e,$$

where $a \in [\frac{1}{4}, 1)$ in general, and e is an even number. Immediately, $\sqrt{x} = \sqrt{a} \times 2^{e/2}$. From this description we see that any square root algorithm need work with arguments only on the interval $[\frac{1}{4}, 1]$ without loss of generality (w.l.o.g.).

(a) Finding the square root of $x = a$ is the same problem as solving $f(x) = x^2 - a = 0$. Show that the Newton–Raphson method for doing so yields the iterative algorithm

$$p_{n+1} = \frac{1}{2}\left(p_n + \frac{a}{p_n}\right), \tag{7.P.3}$$

where $p_n \to \sqrt{a}$. [*Comment:* It can be shown via an argument based on the theorem in Problem 7.15 that to ensure convergence, we should set $p_0 = \frac{1}{3}(2a + 1)$. A simpler choice for the starting value is $p_0 = \frac{1}{2}(\frac{1}{4} + 1) = \frac{5}{8}$ (since we know $a \in [\frac{1}{4}, 1]$).]

(b) Mikami et al. (1992) suggest an alternative algorithm to find $\sqrt{x}$. They recommend that for $a \in [\frac{1}{4}, 1]$ the square root of a be obtained by the algorithm

$$p_{n+1} = \beta(a - p_n^2) + p_n. \tag{7.P.4}$$

Define error sequence (e_n) as $e_n = \sqrt{a} - p_n$. Show that

$$e_{n+1} = \beta e_n^2 + (1 - 2\beta\sqrt{a})e_n.$$

[*Comment:* Mikami et al. recommend that $p_0 = 0.666667a + 0.354167$.]

(c) What condition on β gives quadratic convergence for the algorithm in (7.P.4)?

(d) Some microprocessors are intended for applications in high-speed digital signal processing. As such, they tend to be fixed-point machines with a high-speed hardware multiplier, but no divider unit. Floating-point arithmetic tends to be avoided in this application context. In view of this, what advantage might (7.P.4) have over (7.P.3) as a means to compute square roots?

7.17. Review Problem 7.10. Use $x = x_0 + jx_1$ ($x_0, x_1 \in \mathbf{R}$) to rewrite the equation $f(x) = x^3 - 3x^2 + 4x - 2 = 0$ in the form

$$f_0(x_0, x_1) = 0, \quad f_1(x_0, x_1) = 0.$$

Use the Newton–Raphson method (as implemented in MATLAB) to solve this nonlinear system of equations for the complex roots of $f(x) = 0$. Use the starting points $\overline{x}_0 = [\ x_{0,0}\quad x_{0,1}\ ]^T = [\ 2\quad 2\ ]^T$ and $\overline{x}_0 = [\ 2\quad -2\ ]^T$. Output six iterations in both cases.

7.18. Consider the nonlinear system of equations

$$f(x, y) = x^2 + y^2 - 1 = 0,$$

$$g(x, y) = \tfrac{1}{4}x^2 + 4y^2 - 1 = 0.$$

(a) Sketch f and g on the (x, y) plane.

(b) Solve for the points of intersection of the two curves in (a) by hand calculation.

(c) Write a MATLAB function that uses the Newton–Raphson method to solve for the points of intersection. Use the starting vectors $[x_0 y_0]^T = [\pm 1 \pm 1]^T$. Output six iterations in all four cases.

7.19. If $y_{n+1} = 1 - by_n^2$ and $x_n = \left(\tfrac{1}{4}\lambda - \tfrac{1}{2}\right) y_n + \tfrac{1}{2}$ for $n \in \mathbf{Z}^+$, then, if $b = \tfrac{1}{4}\lambda^2 - \tfrac{1}{2}\lambda$, show that

$$x_{n+1} = \lambda x_n (1 - x_n).$$

7.20. De Angeli, et al. [18] suggest an alternative synchronization scheme (i.e., alternative to deadbeat synchronization). This works as follows. Suppose that at the transmitter

$$x_{0,n+1} = 1 - \alpha x_{0,n}^2 + x_{1,n},$$

$$x_{1,n+1} = \beta x_{0,n},$$

$$y_n = 1 - \alpha x_{0,n}^2.$$

The expression for y_n here replaces that in (7.86). At the receiver

$$\hat{x}_{0,n+1} = \hat{x}_{1,n} + y_n$$

$$\hat{x}_{1,n+1} = \beta \hat{x}_{0,n}.$$

(a) Error sequences are defined as

$$\delta x_{i,n} = \hat{x}_{i,n} - x_{i,n}$$

for $i \in \{0, 1\}$. Find conditions on α and β, giving $\lim_{n \to \infty} ||\delta x_n|| = 0$ ($\delta x_n = [\delta x_{0,n}\quad \delta x_{1,n}]^T$).

(b) Prove that using $y_n = 1 - \alpha x_{0,n}^2$ to synchronize the receiver and transmitter is not a secure synchronization method [i.e., an eavesdropper may collect enough elements from (y_n) to solve for the key $\{\alpha, \beta\}$].

7.21. The chaotic encryption scheme of De Angeli et al. [18], which employs deadbeat synchronization, was shown to be vulnerable to a known plaintext attack, assuming a polynomial message sequence (recall Section 7.6). Show that it is vulnerable to a known plaintext attack when the message sequence is given by

$$s_n = a \sin(\omega n + \phi).$$

Assuming that the eavesdropper already knows ω and ϕ, show that a, α and β can be obtained by solving a third-order linear system of equations. How many cyphertext elements c_n are needed to solve the system?

7.22. Write, compile, and run a C program to implement the De Angeli et al. chaotic encryption/decryption scheme in Section 7.6. (C is suggested here, as I am not certain that MATLAB can do the job so easily.) Implement both the encryption and decryption algorithms in the same program. Keep the program structure simple and direct (i.e., avoid complicated data structures, and difficult pointer operations). The program input is plaintext from a file, and the output is decrypted cyphertext (also known as "recovered plaintext") that is written to another file. The user is to input the encryption key $\{\alpha, \beta\}$, and the decryption key $\{\alpha_1, \beta_1\}$ at the terminal. Test your program out on some plaintext file of your own making. It should include keyboard characters other than letters of the alphabet and numbers. Of course, your program must convert character data into a floating-point format. The floating-point numbers are input to the encryption algorithm. Algorithm output is also a sequence of floating-point numbers. These are decrypted using the decryption algorithm, and the resulting floating-point numbers are converted back into characters. There is a complication involved in decryption. Recall that nominally the plaintext is recovered according to

$$s_n = \frac{y_{n+2}}{c_n}$$

for $n = 0, 1, \ldots, N - 1$. However, this is a floating-point operation that incurs rounding error. It is therefore necessary to implement

$$s_n = \frac{y_{n+2}}{c_n} + \text{offset}$$

for which offset is a small positive value (e.g., 0.0001). The rationale is as follows. Suppose that nominally (i.e., in the absence of roundoff) $s_n = y_{n+2}/c_n = 113.000000$. The number 113 is the ASCII (American Standard Code for Information Interchange) code for some text character. Rounding in division may instead yield the value 112.999999. The operation of converting this number to an integer type will give 112 instead of 113. Clearly, the offset cures this problem. A rounding error that gives, for instance, 113.001000 is harmless. We mention that if plaintext (s_n) is from sampled voice or video, then these rounding issues are normally irrelevant. Try using the following key sets:

$$\alpha = 1.399, \beta = 0.305,$$
$$\alpha_1 = 1.399, \beta_1 = 0.305,$$

and

$$\alpha = 1.399, \beta = 0.305,$$
$$\alpha_1 = 1.389, \beta_1 = 0.304.$$

Of course, perfect recovery is expected for the first set, but not the second set. [*Comment:* It is to be emphasized that in practice keys must be chosen that cause the Hénon map to be chaotic to good approximation on the computer. Keys must not lead to an unstable system, a system that converges to a fixed point, or to one with a short-period oscillation. Key choices leading to instability will cause floating-point overflow (leading to a crash), while the other undesirable choices likely generate security hazards. Ideally (and more practically) the cyphertext array (floating-point numbers) should be written to a file by an encryption program. A *separate* decryption program would then read in the cyphertext from the file and decrypt it, producing the recovered plaintext, which would be written to another file (i.e., three separate files: original file of input plaintext, cyphertext file, and recovered plaintext file).]

8 Unconstrained Optimization

8.1 INTRODUCTION

In engineering design it is frequently the case that an optimal design is sought with respect to some performance criterion or criteria. Problems of this class are generally referred to as *optimal design problems*, or *mathematical programming problems*. Frequently nonlinearity is involved, and so the term *nonlinear programming* also appears. There are many subcategories of problem types within this broad category, and so space limitations will restrict us to an introductory presentation mainly of so-called *unconstrained problems*. Even so, only a very few ideas from within this category will be considered. However, some of the methods treated in earlier chapters were actually examples of optimal design methods within this category. This would include least-squares ideas, for example. In fact, an understanding of least-squares ideas from the previous chapters helps a lot in understanding the present one. Although the emphasis here is on unconstrained optimization problems, some consideration of constrained problems appears in Section 8.5.

8.2 PROBLEM STATEMENT AND PRELIMINARIES

In this chapter we consider the problem

$$\min_{x \in \mathbf{R}^n} f(x) \tag{8.1}$$

for which $f(x) \in \mathbf{R}$. This notation means that we wish to find a vector x that minimizes the function $f(x)$. We shall follow previous notational practices, so here $x = [x_0 \ x_1 \ \ldots \ x_{n-2} \ x_{n-1}]^T$. In what follows we shall usually assume that $f(x)$ possesses all first- and second-order partial derivatives (with respect to the elements of x), and that these are continuous functions, too. In the present context the function $f(x)$ is often called the *objective function*. For example, in least-squares problems (recall Chapter 4) we sought to minimize $V(a) = ||f(x) - \sum_{k=0}^{N-1} a_k \phi_k(x)||^2$ with respect to $a \in \mathbf{R}^N$. Thus, $V(a)$ is an objective function. The least-squares problem had sufficient structure so that a simple solution was arrived at; specifically, to find the optimal vector a (denoted $\hat{a}$), all we had to do was solve a linear system of equations.

An Introduction to Numerical Analysis for Electrical and Computer Engineers, by C.J. Zarowski
ISBN 0-471-46737-5 © 2004 John Wiley & Sons, Inc.

Now we are interested in solving more general problems. For example, we might wish to find $x = [x_0 x_1]^T$ to minimize *Rosenbrock's function*

$$f(x) = 100(x_1 - x_0^2)^2 + (1 - x_0)^2. \tag{8.2}$$

This is a famous standard test function (taken here from Fletcher [1]) often used by those who design optimization algorithms to test out their theories. A contour plot of this function appears in Fig. 8.1. It turns out that this function has a unique minimum at $x = \hat{x} = [1 \quad 1]^T$. As before, we have used a "hat" symbol to indicate the optimal solution.

Some ideas from vector calculus are essential to understanding nonlinear optimization problems. Of particular importance is the *gradient operator*:

$$\nabla = \left[\frac{\partial}{\partial x_0} \frac{\partial}{\partial x_1} \cdots \frac{\partial}{\partial x_{n-1}} \right]^T. \tag{8.3}$$

For example, if we apply this operator to Rosenbrock's function, then

$$\nabla f(x) = \begin{bmatrix} \dfrac{\partial f(x)}{\partial x_0} \\[2ex] \dfrac{\partial f(x)}{\partial x_1} \end{bmatrix} = \begin{bmatrix} -400x_0(x_1 - x_0^2) - 2(1 - x_0) \\ 200(x_1 - x_0^2) \end{bmatrix}. \tag{8.4}$$

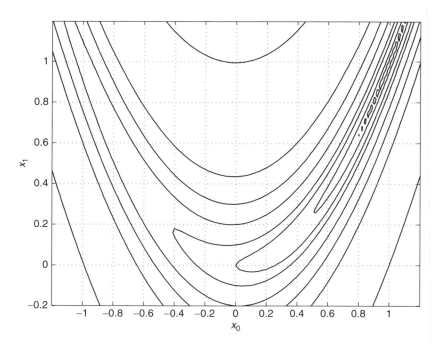

Figure 8.1   Contour plot of Rosenbrock's function.

We observe that $\nabla f(\hat{x}) = [0 \quad 0]^T$ (recall that $\hat{x} = [1 \quad 1]^T$). In other words, the gradient of the objective function at $\hat{x}$ is zero. Intuitively we expect that if $\hat{x}$ is to minimize the objective function, then we should always have $\nabla f(\hat{x}) = 0$. This turns out to be a *necessary but not sufficient condition* for a minimum. To see this, consider $g(x) = -x^2 (x \in \mathbf{R})$ for which

$$\nabla g(x) = -2x.$$

Clearly, $\nabla g(0) = 0$ yet $x = 0$ *maximizes* $g(x)$ instead of minimizing it. Consider $h(x) = x^3$, so that

$$\nabla h(x) = 3x^2$$

for which $\nabla h(0) = 0$, but $x = 0$ neither minimizes nor maximizes $h(x)$. In this case $x = 0$ corresponds to a one-dimensional version of a *saddle point*.

Thus, finding x such that $\nabla f(x) = 0$ generally gives us minima, maxima, or saddle points. These are collectively referred to as *stationary points*. We need (if possible) a condition that tells us whether a given stationary point corresponds to a minimum. Before considering this matter, we note that another problem is that a given objective function will often have several *local minima*. This is illustrated in Fig. 8.2. The function is a quartic polynomial for which there are two values of

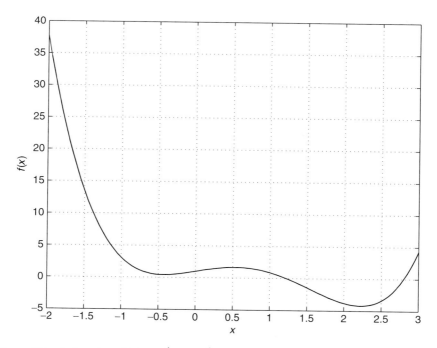

Figure 8.2   A function $f(x) = x^4 - 3.1x^3 + 2x + 1$ with two local minima. The local minimizers are near $x = -0.5$ and $x = 2.25$. The minimum near $x = 2.25$ is the unique global minimum of $f(x)$.

x such that $\nabla f(x) = 0$. We might denote these by $\hat{x}_a$ and $\hat{x}_b$. Suppose that $\hat{x}_a$ is the local minimizer near $x = -0.5$ and $\hat{x}_b$ is the local minimizer near $x = 2.25$. We see that $f(\hat{x}_a + \delta) > f(\hat{x}_a)$ for all sufficiently small (but nonzero) $\delta \in \mathbf{R}$. Similarly, $f(\hat{x}_b + \delta) > f(\hat{x}_b)$ for all sufficiently small (but nonzero) $\delta \in \mathbf{R}$. But we see that $f(\hat{x}_b) < f(\hat{x}_a)$, so $\hat{x}_b$ is the *unique global minimizer* for $f(x)$. An algorithm for minimizing $f(x)$ should seek $\hat{x}_b$; that is, in general, we seek the global minimizer (assuming that it is unique). Except in special situations, an optimization algorithm is usually not guaranteed to do better than find a local minimizer, and is not guaranteed to find a global minimizer. Note that it is entirely possible for the global minimizer not to be unique. A simple example is $f(x) = \sin(x)$ for which the stationary points are $x = (2k + 1)\pi/2$ ($k \in \mathbf{Z}$). The minimizers that are a subset of these points all give -1 for the value of $\sin(x)$.

To determine whether a stationary point is a local minimizer, it is useful to have the *Hessian matrix*

$$\nabla^2 f(x) = \left[\frac{\partial^2 f(x)}{\partial x_i \partial x_j} \right]_{i,j=0,1,\ldots,n-1} \in \mathbf{R}^{n \times n}. \tag{8.5}$$

This matrix should not be confused with the Jacobian matrix (seen in Section 7.5.2). The Jacobian and Hessian matrices are not the same. The veracity of this may be seen by comparing their definitions. For the special case of $n = 2$, we have the 2×2 Hessian matrix

$$\nabla^2 f(x) = \begin{bmatrix} \dfrac{\partial^2 f(x)}{\partial x_0^2} & \dfrac{\partial^2 f(x)}{\partial x_0 \partial x_1} \\ \dfrac{\partial^2 f(x)}{\partial x_1 \partial x_0} & \dfrac{\partial^2 f(x)}{\partial x_1^2} \end{bmatrix}. \tag{8.6}$$

If $f(x)$ in (8.6) is Rosenbrock's function, then (8.6) becomes

$$\nabla^2 f(x) = \begin{bmatrix} 1200x_0^2 - 400x_1 + 2 & -400x_0 \\ -400x_0 & 200 \end{bmatrix}. \tag{8.7}$$

The Hessian matrix for $f(x)$ helps in the following manner. Suppose that $\hat{x}$ is a local minimizer for objective function $f(x)$. We may Taylor-series-expand $f(x)$ around $\hat{x}$ according to

$$f(\hat{x} + h) = f(\hat{x}) + h^T \nabla f(\hat{x}) + \tfrac{1}{2} h^T [\nabla^2 f(\hat{x})] h + \cdots, \tag{8.8}$$

where $h \in \mathbf{R}^n$. (This will not be formally justified.) If $f(x)$ is sufficiently smooth and $\|h\|$ is sufficiently small, then the terms in (8.8) that are not shown may be entirely ignored (i.e., we neglect higher-order terms). In other words, in the neighborhood of $\hat{x}$

$$f(\hat{x} + h) \approx f(\hat{x}) + h^T \nabla f(\hat{x}) + \tfrac{1}{2} h^T [\nabla^2 f(\hat{x})] h \tag{8.9}$$

is assumed. But this is the now familiar quadratic form. For convenience we will (as did Fletcher [1]) define

$$G(x) = \nabla^2 f(x), \tag{8.10}$$

so (8.9) may be rewritten as

$$f(\hat{x} + h) \approx f(\hat{x}) + h^T \nabla f(\hat{x}) + \tfrac{1}{2}h^T G(\hat{x})h. \tag{8.11}$$

Sometimes we will write G instead of $G(x)$ if there is no danger of confusion. In Chapter 4 we proved that (8.11) has a unique minimum iff $G > 0$. In words, $f(x)$ looks like a *positive definite* quadratic form in the neighborhood of a local minimizer. Therefore, the Hessian of $f(x)$ at a local minimizer will be positive definite and thus represents a way of testing a stationary point to see if it is a minimizer. (Recall the second-derivative test for the minimum of a single-variable function from elementary calculus, which is really just a special case of this more general test.) More formally, we therefore have the following theorem.

Theorem 8.1: A sufficient condition for a local minimizer $\hat{x}$ is that $\nabla f(\hat{x}) = 0$, and $G(\hat{x}) > 0$.

This is a simplified statement of Fletcher's Theorem 2.1.1 [1, p. 14]. The proof is really just a more rigorous version of the Taylor series argument just given, and will therefore be omitted. For convenience we will also define the gradient vector

$$g(x) = \nabla f(x). \tag{8.12}$$

Sometimes we will write g for $g(x)$ if there is no danger of confusion.

If we recall Rosenbrock's function again, we may now test the claim made earlier that $\hat{x} = [1 \quad 1]^T$ is a minimizer for $f(x)$ in (8.2). For Rosenbrock's function the Hessian is in (8.7), and thus we have

$$G(\hat{x}) = \begin{bmatrix} 802 & -400 \\ -400 & 200 \end{bmatrix}. \tag{8.13}$$

The eigenvalues of this matrix are $\lambda = 0.3994, 1002$. These are both bigger than zero so $G(\hat{x}) > 0$. We have already remarked that $g(\hat{x}) = 0$, so immediately from Theorem 8.1 we conclude that $\hat{x} = [1 \quad 1]^T$ is a local minimizer for Rosenbrock's function.

8.3 LINE SEARCHES

In general, for objective function $f(x)$ we wish to allow $x \in \mathbf{R}^n$; that is, we seek the minimum of $f(x)$ by performing a search in an n-dimensional vector space. However, the one-dimensional problem is an important special case. In this section we begin by considering $n = 1$, so x is a scalar. The problem of finding the minimum of $f(x)$ [where $f(x) \in \mathbf{R}$ for all $x \in \mathbf{R}$] is sometimes called the *univariate search*. Various approaches exist for the solution of this problem, but we will consider only the *golden ratio search method* (sometimes also called the *golden section search*). We will then consider the *backtracking line search* [3] for the case where $x \in \mathbf{R}^n$ for any $n \geq 1$.

Suppose that we know $f(x)$ has a minimum over the interval $[x_l^j, x_u^j]$. Define $I^j = x_u^j - x_l^j$, which is the length of this interval. The index j represents the jth iteration of the search algorithm, so it represents the current estimate of the interval that contains the minimum. Select two points x_a^j and x_b^j ($x_a^j < x_b^j$) such that they are symmetrically placed in the interval $[x_l^j, x_u^j]$. Specifically

$$x_a^j - x_l^j = x_u^j - x_b^j. \tag{8.14}$$

A new interval $[x_l^{j+1}, x_u^{j+1}]$ is created according to the following procedure, such that for all j

$$\frac{I^j}{I^{j+1}} = \tau > 1. \tag{8.15}$$

If $f(x_a^j) \geq f(x_b^j)$ then the minimum lies in $[x_a^j, x_u^j]$, and so $I^{j+1} = x_u^j - x_a^j$. Our new points are given according to

$$x_l^{j+1} = x_a^j, \quad x_u^{j+1} = x_u^j, \quad x_a^{j+1} = x_b^j \tag{8.16a}$$

and

$$x_b^{j+1} = x_a^j + \frac{1}{\tau} I^{j+1}. \tag{8.16b}$$

If $f(x_a^j) < f(x_b^j)$, then the minimum lies in $[x_l^j, x_b^j]$, and so $I^{j+1} = x_b^j - x_l^j$. Our new points are given according to

$$x_l^{j+1} = x_l^j, \quad x_u^{j+1} = x_b^j, \quad x_b^{j+1} = x_a^j \tag{8.17a}$$

and

$$x_a^{j+1} = x_b^j - \frac{1}{\tau} I^{j+1}. \tag{8.17b}$$

Since $I^0 = x_u^0 - x_l^0$ ($j = 0$ indicates the initial case), we must have $x_a^0 = x_u^0 - \frac{1}{\tau} I^0$ and $x_b^0 = x_l^0 + \frac{1}{\tau} I^0$. Figure 8.3 illustrates the search procedure. This is for the particular case (8.17).

Because of (8.15), the rate of convergence to the minimum can be estimated. But we need to know τ; see the following theorem.

Theorem 8.2: The search interval lengths of the golden ratio search algorithm are related according to

$$I^j = I^{j+1} + I^{j+2} \tag{8.18a}$$

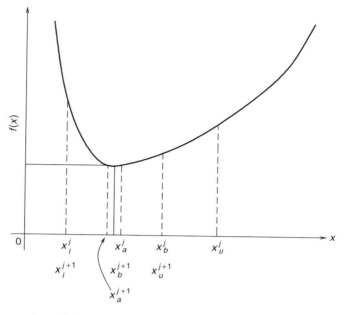

Figure 8.3 Illustrating the golden ratio search procedure.

for which the *golden ratio* τ is given by

$$\tau = \tfrac{1}{2}(1 + \sqrt{5}) \approx 1.62. \tag{8.18b}$$

Proof There are four cases to consider in establishing (8.18a). First, suppose that $f(x_a^j) \geq f(x_b^j)$, so $I^{j+1} = x_u^j - x_a^j$ with

$$x_l^{j+1} = x_a^j, \quad x_a^{j+1} = x_b^j, \quad x_b^{j+1} = x_a^j + \frac{1}{\tau}I^{j+1}, \quad x_u^{j+1} = x_u^j.$$

If it happens that $f(x_a^{j+1}) \geq f(x_b^{j+1})$, then $I^{j+2} = x_u^{j+1} - x_a^{j+1} = x_u^j - x_b^j = x_a^j - x_l^j$ (via $x_a^j - x_l^j = x_u^j - x_b^j$). This implies that $I^{j+1} + I^{j+2} = x_u^j - x_l^j = I^j$. On the other hand, if it happens that $f(x_a^{j+1}) < f(x_b^{j+1})$, then $I^{j+2} = x_b^{j+1} - x_l^{j+1} = x_u^{j+1} - x_a^{j+1} = x_a^j - x_l^j$ (via $x_a^{j+1} - x_l^{j+1} = x_u^{j+1} - x_b^{j+1}$ and $x_a^j - x_l^j = x_u^j - x_b^j$). Again we have $I^{j+1} + I^{j+2} = x_u^j - x_l^j = I^j$.
Now suppose that $f(x_a^j) < f(x_b^j)$. Therefore, $I^{j+1} = x_b^j - x_l^j$ with

$$x_l^{j+1} = x_l^j, \quad x_a^{j+1} = x_b^j - \frac{1}{\tau}I^{j+1}, \quad x_b^{j+1} = x_a^j, \quad x_u^{j+1} = x_b^j.$$

If it happens that $f(x_a^{j+1}) \geq f(x_b^{j+1})$, then $I^{j+2} = x_u^{j+1} - x_a^{j+1} = x_b^{j+1} - x_l^{j+1} = x_a^j - x_l^j = x_u^j - x_b^j$ so that $I^{j+1} + I^{j+2} = x_u^j - x_l^j = I^j$. Finally, suppose that $f(x_a^{j+1}) < f(x_b^{j+1})$, so therefore $I^{j+2} = x_b^{j+1} - x_l^{j+1} = x_a^j - x_l^j = x_u^j - x_b^j$, so yet again we conclude that $I^{j+1} + I^{j+2} = x_u^j - x_l^j = I^j$. Thus, (8.18a) is now verified.

Since

$$\frac{I^j}{I^{j+1}} = \frac{I^{j+1}}{I^{j+2}} = \tau \quad \text{and} \quad I^j = I^{j+1} + I^{j+2},$$

we have $I^{j+1} = \frac{1}{\tau} I^j$ and $I^{j+2} = \frac{1}{\tau} I^{j+1} = \frac{1}{\tau^2} I^j$, so

$$I^j = I^{j+1} + I^{j+2} = \frac{1}{\tau} I^j + \frac{1}{\tau^2} I^j,$$

immediately implying that

$$\tau^2 = \tau + 1,$$

which yields (8.18b).

The golden ratio has a long and interesting history in art as well as science and engineering, and this is considered in Schroeder [2]. For example, a famous painting by Seurat contains figures that are proportioned according to this ratio (Fig. 5.2 in Schroeder [2] includes a sketch).

The golden ratio search algorithm as presented so far assumes a user-provided starting interval. The golden ratio search algorithm has the same drawback as the bisection method for root finding (Chapter 7) in that the optimum solution must be bracketed before the algorithm can be successfully run (in general). On the other hand, an advantage of the golden ratio search method is that it does not need $f(x)$ to be differentiable. But the method can be slow to converge, in which case an improved minimizer that also does not need derivatives can be found in Brent [4].

When setting up an optimization problem it is often advisable to look for ways to reduce the dimensionality of the problem, if at all possible. We illustrate this idea with an example that is similar in some ways to the least-squares problem considered in the beginning of Section 4.6. Suppose that signal $f(t) \in \mathbf{R}$ $(t \in \mathbf{R})$ is modeled as

$$f(t) = a \sin\left(\frac{2\pi}{T} t + \phi\right) + \eta(t), \tag{8.19}$$

where $\eta(t)$ is the term that accounts for noise, interference, or measurement errors in the data $f(t)$. In other words, our data are modeled as a sinusoid plus noise. The amplitude a, phase ϕ, and period T are unknown parameters. We are assumed only to possess samples of the signal; thus, we have only the sequence elements

$$f_n = f(nT_s) = a \sin\left(\frac{2\pi}{T} nT_s + \phi\right) + \eta(nT_s) \tag{8.20}$$

for $n = 0, 1, \ldots, N - 1$. The sampling period parameter T_s is assumed to be known. As before, we may define an error sequence

$$e_n = f_n - \underbrace{a \sin\left(\frac{2\pi}{T} nT_s + \phi\right)}_{=\hat{f}_n}, \tag{8.21}$$

where again $n = 0, 1, \ldots, N - 1$. The *objective function* for our problem (using a least-squares criterion) would therefore be

$$V(a, T, \phi) = \sum_{n=0}^{N-1} \left[f_n - a \sin\left(\frac{2\pi}{T} n T_s + \phi\right) \right]^2. \tag{8.22}$$

This function depends on the unknown model parameters a, T, and ϕ in a very nonlinear manner. We might formally define our parameter vector to be $x = [a \ T \ \phi]^T \in \mathbf{R}^3$. This would lead us to conclude that we must search (by some means) a three-dimensional space to find $\hat{x}$ to minimize (8.22). But in this special problem it is possible to reduce the dimensionality of the search space from three dimensions down to only one dimension. Reducing the problem in this way makes it solvable using the golden section search method that we have just considered.

Recall the trigonometric identity

$$\sin(A + B) = \sin A \cos B + \cos A \sin B. \tag{8.23}$$

Using this, we may write

$$a \sin\left(\frac{2\pi}{T} n T_s + \phi\right) = \underbrace{a \cos \phi}_{=x_0} \sin\left(\frac{2\pi}{T} n T_s\right) + \underbrace{a \sin \phi}_{=x_1} \cos\left(\frac{2\pi}{T} n T_s\right). \tag{8.24}$$

Define $x = [x_0 \ \ x_1]^T$ and

$$v_n = \left[\sin\left(\frac{2\pi}{T} n T_s\right) \ \cos\left(\frac{2\pi}{T} n T_s\right) \right]^T. \tag{8.25}$$

We may rewrite e_n in (8.21) using these vectors:

$$e_n = f_n - v_n^T x. \tag{8.26}$$

Note that the approach used here is the same as that used to obtain e_n in Eqn. (4.99). Therefore, the reader will probably find it useful to review this material now. Continuing in this fashion, the error vector $e = [e_0 e_1 \cdots e_{N-1}]^T$, data vector $f = [f_0 f_1 \cdots f_{N-1}]^T$, and matrix of basis vectors

$$A = \begin{bmatrix} v_0^T \\ v_1^T \\ \vdots \\ v_{N-1}^T \end{bmatrix} \in \mathbf{R}^{N \times 2}$$

all may be used to write

$$e = f - Ax, \tag{8.27}$$

for which our objective function may be rewritten as

$$V(x) = e^T e = f^T f - 2x^T A^T f + x^T A^T Ax, \tag{8.28}$$

which implicitly assumes that we already know the period T. If T were known in advance, we could use the method of Chapter 4 to minimize (8.28); that is, the optimum choice (least-squares sense) for x (denoted by $\hat{x}$) satisfies the linear system

$$A^T A\hat{x} = A^T f. \tag{8.29}$$

Because of (8.24) (with $\hat{x} = [\hat{x}_0 \quad \hat{x}_1]^T$), we obtain

$$\hat{x}_0 = \hat{a} \cos \hat{\phi}, \quad \hat{x}_1 = \hat{a} \sin \hat{\phi}. \tag{8.30}$$

Since we have $\hat{x}$ from the solution to (8.29), we may use (8.30) to solve for $\hat{a}$ and $\hat{\phi}$.

However, our original problem specified that we do not know a, ϕ, or T in advance. So, how do we exploit these results to determine T as well as a and ϕ? The approach is to change how we think about $V(x)$ in (8.28). Instead of thinking of $V(x)$ as a function of x, consider it to be a function of T, but with $x = \hat{x}$ as given by (8.29). From (8.25) we see that v_n depends on T, so A is also a function of T. Thus, $\hat{x}$ is a function of T, too, because of (8.29). In fact, we may emphasize this by rewriting (8.29) as

$$[A(T)]^T A(T)\hat{x}(T) = [A(T)]^T f. \tag{8.31}$$

In other words, $A = A(T)$, and $\hat{x} = \hat{x}(T)$. The objective function $V(x)$ then becomes

$$V_1(T) = V(\hat{x}(T)) = f^T f - 2[\hat{x}(T)]^T A^T (T) f + [\hat{x}(T)]^T A^T (T) A(T)\hat{x}(T). \tag{8.32}$$

However, we may substitute (8.31) into (8.32) and simplify with the result that

$$V_1(T) = f^T f - f^T A(T)[A^T (T)A(T)]^{-1} A^T (T) f. \tag{8.33}$$

We have reduced the search space from three dimensions down to one, so we may apply the golden section search algorithm (or some other univariate search procedure) to objective function $V_1(T)$. This would result in determining the optimum period $\hat{T}$ [which minimizes $V_1(T)$]. We then compute $A(\hat{T})$, and solve for $\hat{x}$ using (8.29) as before. Knowledge of $\hat{x}$ allows us to determine $\hat{a}$ and $\hat{\phi}$ via (8.30).

Figure 8.4 illustrates a typical noisy sinusoid and the corresponding objective function $V_1(T)$. In this case the parameters chosen were $T = 24$ h, $T_s = 5$ min, $a = 1$, $\phi = -\pi/10$ radians, and $N = 500$. The noise component in the data was created using MATLAB's Gaussian random-number generator. The noise variance is $\sigma^2 = 0.5$, and the mean value is zero. We observe that the minimum value of $V_1(T)$ certainly corresponds to a value of T that is at or close to 24 h.

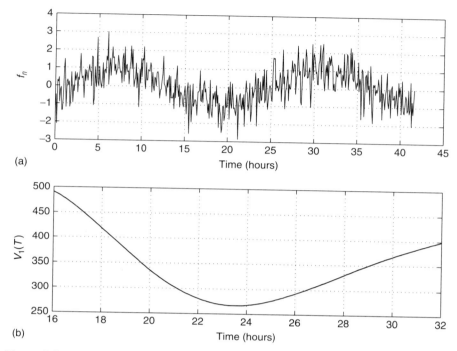

Figure 8.4 An example of a noisy sinusoid (a) and the corresponding objective function $V_1(T)$ (b).

Appendix 8.A contains a sample MATLAB program implementation of the golden section search algorithm applied to the noisy sinusoid problem depicted in Fig. 8.4. In golden.m Topt is $\hat{T}$, and eta η is the random noise sequence $\eta(nT_s)$.

We will now consider the backtracking line search algorithm. The exposition will be similar to that in Boyd and Vandenberghe [5].

Now we assume that $f(x)$ is our objective function with $x \in \mathbf{R}^n$. In a general line search we seek the minimizing vector sequence $(x^{(k)})$, $k \in \mathbf{Z}^+$, and $x^{(k)} \in \mathbf{R}^n$ (i.e., $x^{(k)} = [x_0^{(k)} \ x_1^{(k)} \ \cdots \ x_{n-1}^{(k)}]^T$) constructed according to the iterative process

$$x^{(k+1)} = x^{(k)} + t^{(k)} s^{(k)}, \tag{8.34}$$

where $t^{(k)} \in \mathbf{R}^+$, and $t^{(k)} > 0$ except when $x^{(k)}$ is optimal [i.e., minimizes $f(x)$]. The vector $s^{(k)}$ is called the *search direction*, and scalar $t^{(k)}$ is called the *step size*. Because line searches (8.34) are *descent methods*, we have

$$f(x^{(k+1)}) < f(x^{(k)}) \tag{8.35}$$

except when $x^{(k)}$ is optimal. The "points" $x^{(k+1)}$ lie along a line in the direction $s^{(k)}$ in n-dimensional space $\mathbf{R}^n$, and since the minimum of $f(x)$ must lie in the direction that satisfies (8.35), we must ensure that $s^{(k)}$ satisfies [recall (8.12)]

$$g(x^{(k)})^T s^{(k)} = [\nabla f(x^{(k)})]^T s^{(k)} < 0. \tag{8.36}$$

Geometrically, the negative-gradient vector $-g(x^{(k)})$ (which "points down") makes an acute angle (i.e., one of magnitude $<90°$) with the vector $s^{(k)}$. [Recall (4.130).] If $s^{(k)}$ satisfies (8.36) for $f(x^{(k)})$ (i.e., $f(x)$ at $x = x^{(k)}$), it is called a *descent direction for $f(x)$ at $x^{(k)}$*. A general descent algorithm has the following pseudocode description:

```
Specify starting point x⁽⁰⁾ ∈ Rⁿ;
k := 0;
while stopping criterion is not met do begin
        Find s⁽ᵏ⁾; { determine descent direction }
        Find t⁽ᵏ⁾; { line search step }
        Compute x⁽ᵏ⁺¹⁾ := x⁽ᵏ⁾ + t⁽ᵏ⁾s⁽ᵏ⁾;
          { update step }
        k := k + 1;
        end ;
```

Newton's method with the backtracking line search (Section 8.4) is a specific example of a descent method. There are others, but these will not be considered in this book.

Now we need to say more about how to choose the step size $t^{(k)}$ on the assumption that $s^{(k)}$ is known. How to determine the direction $s^{(k)}$ is the subject of Section 8.4.

So far $f|\mathbf{R}^n \to \mathbf{R}$. Subsequent considerations are simplified if we assume that $f(x)$ satisfies the following definition.

Definition 8.1: Convex Function   Function $f|\mathbf{R}^n \to \mathbf{R}$ is called a *convex function* if for all $\theta \in [0, 1]$, and for any $x, y \in \mathbf{R}^n$

$$f(\theta x + (1 - \theta)y) \leq \theta f(x) + (1 - \theta)f(y). \tag{8.37}$$

We emphasize that the domain of definition of $f(x)$ is assumed to be $\mathbf{R}^n$. It is possible to modify the definition to accommodate $f(x)$ where the domain of $f(x)$ is a proper subset of $\mathbf{R}^n$. The geometric meaning of (8.37) is shown in Fig. 8.5 for the case where $x \in \mathbf{R}$ (i.e., one-dimensional case). We see that when $f(x)$ is convex, the *chord*, which is the line segment joining $(x, f(x))$ to $(y, f(y))$, always lies above the graph of $f(x)$. Now further assume that $f(x)$ is at least twice continuously differentiable in all elements of the vector x. We observe that for any $x, y \in \mathbf{R}^n$, if $f(x)$ is convex, then

$$f(y) \geq f(x) + [\nabla f(x)]^T (y - x). \tag{8.38}$$

From the considerations of Section 8.2 it is easy to believe that $f(x)$ is convex iff

$$\nabla^2 f(x) \geq 0 \tag{8.39}$$

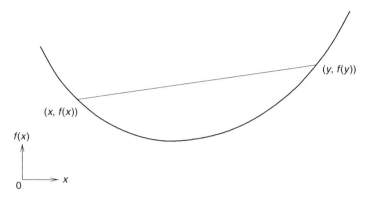

Figure 8.5 Graph of a convex function $f(x)$ ($x \in \mathbf{R}$) and the chord that connects the points $(x, f(x))$ and $(y, f(y))$.

for all $x \in \mathbf{R}^n$; that is, $f(x)$ is convex iff its Hessian matrix is at least positive semidefinite (recall Definition 4.1). The function $f(x)$ is said to be *strongly convex* iff $\nabla^2 f(x) > 0$ for all $x \in \mathbf{R}^n$.

The *backtracking line search* attempts to approximately minimize $f(x)$ along the line $\{x + ts \,|\, t \geq 0\}$ for some given s (search direction at x). The pseudocode for this algorithm is as follows:

```
Specify the search direction s;
t := 1;
while (f(x + ts) > f(x) + δt[∇f(x)]^T s)
    t := αt;
    end ;
```

In this algorithm $0 < \delta < \frac{1}{2}$, and $0 < \alpha < 1$. Commonly, $\delta \in [0.01, 0.30]$, and $\alpha \in [0.1, 0.5]$. These parameter ranges will not be justified here. As suggested earlier, how to choose s will be the subject of the next section. The method is called "backtracking" as it begins with $t = 1$, and then reduces t by factor α until $f(x + ts) \leq f(x) + \delta t \nabla^T f(x)s$. [We have $\nabla^T f(x) = [\nabla f(x)]^T$.] Recall that s is a descent direction so that (8.36) holds, specifically, $\nabla^T f(x)s < 0$, and so if t is small enough [recall (8.9)], then

$$f(x + ts) \approx f(x) + t\nabla^T f(x)s < f(x) + \delta t \nabla^T f(x)s,$$

which shows that the search must terminate eventually. We mention that the backtracking line search will terminate even if $f(x)$ is only "locally convex"—convex on some proper subset S of $\mathbf{R}^n$. This will happen provided $x \in S$ in the algorithm.

8.4 NEWTON'S METHOD

Section 8.3 suggests attempting to reduce an n-dimensional search space to a one-dimensional search space. Of course, this approach seldom works, which is why there is an elaborate body of methods on searching for minima in higher-dimensional spaces. However, these methods are too involved to consider in detail in this book, and so we will only partially elaborate on an idea from Section 8.2.

The quadratic model from Section 8.2 suggests an approach often called *Newton's method*. Suppose that $x^{(k)}$ is the current estimate of the sought-after optimum $\hat{x}$. Following (8.11), we have the Taylor approximation

$$f(x^{(k)} + \delta) \approx V(\delta) = f(x^{(k)}) + \delta^T g(x^{(k)}) + \tfrac{1}{2}\delta^T G(x^{(k)})\delta \qquad (8.40)$$

for which $\delta \in \mathbf{R}^n$ since $x^{(k)} \in \mathbf{R}^n$. Since $x^{(k)}$ is not necessarily the minimum $\hat{x}$, usually $g(x^{(k)}) \neq 0$. Vector δ is selected to minimize $V(\delta)$, and since this is a quadratic form, if $G(x^{(k)}) > 0$, then

$$G(x^{(k)})\delta = -g(x^{(k)}). \qquad (8.41)$$

The next estimate of $\hat{x}$ is given by

$$x^{(k+1)} = x^{(k)} + \delta. \qquad (8.42)$$

Pseudocode for Newton's method (in its most basic form) is

```
Input starting point x^(0);
k := 0;
While stopping criterion is not met do begin
    G(x^(k))δ := -g(x^(k));
    x^(k+1) := x^(k) + δ;
    k := k + 1;
    end;
```

The algorithm will terminate (if all goes well) with the last vector $x^{(k+1)}$ as a good approximation to $\hat{x}$. However, the Hessian $G^{(k)} = G(x^{(k)})$ may not always be positive definite, in which case this method can be expected to fail. Modification of the method is required to guarantee that at least it will converge to a local minimum. Said modifications often involve changing the method to work with line searches (i.e., Section 8.3 ideas). We will now say more about this.

As suggested in Ref. 5, we may combine the backtracking line search with the basic form of Newton's algorithm described above. A pseudocode description of the result is

```
Input starting point x^(0), and a tolerance ε > 0;
k := 0;
s^(0) := -[G(x^(0))]^(-1) g(x^(0)); { search direction at x^(0)}
λ^2 := -g^T(x^(0))s^(0);
while λ^2 > ε do begin

Use backtracking line search to find t^(k) for x^(k) and s^(k);
    x^(k+1) := x^(k) + t^(k)s^(k);
    s^(k+1) := -[G(x^(k+1))]^(-1) g(x^(k+1));
    λ^2 := -g^T(x^(k+1))s^(k+1);
    k := k + 1;
    end;
```

The algorithm assumes that $G(x^{(k)}) > 0$ for all k. In this case $[G(x^{(k)})]^{-1} > 0$ as well. If we define (for all $x \in \mathbf{R}^n$)

$$||x||^2_{G(y)} = x^T [G(y)]^{-1} x, \qquad (8.43)$$

then $||x||_{G(y)}$ satisfies the norm axioms (recall Definition 1.3), and is in fact an example of a *weighted norm*. But why do we consider λ^2 as a stopping criterion in Newton's algorithm? If we recall the term $\frac{1}{2}\delta^T G(x^{(k)})\delta$ in (8.40), since δ satisfies (8.41), at step k we must have

$$\frac{1}{2}\delta^T G(x^{(k)})\delta = \frac{1}{2}g^T(x^{(k)})[G(x^{(k)})]^{-1} g(x^{(k)}) = \frac{1}{2}||g(x^{(k)})||^2_{G(x^{(k)})} = \frac{1}{2}\lambda^2. \quad (8.44)$$

Estimate $x^{(k)}$ of $\hat{x}$ is likely to be good if (8.44) in particular is small [as opposed to merely considering squared *unweighted* norm $g(x^{(k)})^T g(x^{(k)})$]. It is known that Newton's algorithm can converge rapidly (quadratic convergence). An analysis showing this appears in Boyd and Vandenberghe [5] but is omitted here.

As it stands, Newton's method is computationally expensive since we must solve the linear system in (8.41) at every step k. This would normally involve applying the Cholesky factorization algorithm that was first mentioned (but not considered in detail) in Section 4.6. We remark in passing that the Cholesky algorithm will factorize $G^{(k)}$ according to

$$G^{(k)} = LDL^T, \qquad (8.45)$$

where L is unit lower triangular and D is a diagonal matrix. We also mention that $G^{(k)} > 0$ iff the elements of D are all positive, so the Cholesky algorithm provides a built-in positive definiteness test. The decomposition in (8.45) is a variation on the LU decomposition of Chapter 4. Recall that we proved in Section 4.5 that positive definite matrices always possess such a factorization (see Theorems 4.1 and 4.2). So Eq. (8.45) is consistent with this result. The necessity to solve a linear system of equations at every step makes us wonder if sensitivity to ill-conditioned matrices is a problem in Newton's method. It turns out that the method is often surprisingly resistant to ill conditioning (at least as reported in Ref. 5).

Example 8.1   A typical run of Newton's algorithm with the backtracking line search as applied to the problem of minimizing Rosenbrock's function [Eq. (8.2)] yields the following output:

k	$t^{(k)}$	λ^2	$x_0^{(k)}$	$x_1^{(k)}$
0	1.0000	800.00499	2.0000	2.0000
1	0.1250	1.98757	1.9975	3.9900
2	1.0000	0.41963	1.8730	3.4925
3	1.0000	0.49663	1.6602	2.7110
4	0.5000	0.38333	1.5945	2.5382
5	1.0000	0.21071	1.4349	2.0313

k	$t^{(k)}$	λ^2	$x_0^{(k)}$	$x_1^{(k)}$
6	0.5000	0.14763	1.3683	1.8678
7	1.0000	0.07134	1.2707	1.6031
8	1.0000	0.03978	1.1898	1.4092
9	1.0000	0.01899	1.1076	1.2201
10	1.0000	0.00627	1.0619	1.1255
11	1.0000	0.00121	1.0183	1.0350
12	1.0000	0.00006	1.0050	1.0099

The search parameters selected in this example are $\alpha = 0.5$, $\delta = 0.3$, and $\epsilon = .00001$, and the final estimate is $x^{(13)} = [1.0002 \quad 1.0003]^T$. For these same parameters if instead $x^{(0)} = [-1 \quad 1]^T$, then 18 iterations are needed, yielding $x^{(18)} = [0.9999 \quad 0.9998]^T$. Figure 8.6 shows the sequence of points $x^{(k)}$ for the case $x^{(0)} = [-1 \quad 1]^T$. The dashed line shows the path from starting point $[-1 \quad 1]^T$ to the minimum at $[1 \quad 1]^T$, and we see that the algorithm follows the "valley" to the optimum solution quite well.

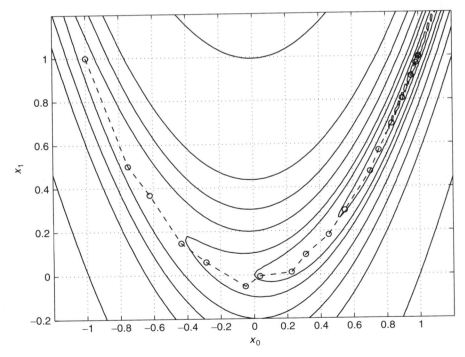

Figure 8.6 The sequence of points $(x^{(k)})$ generated by Newton's method with the back-tracking line search as applied to Rosenbrock's function using the parameters $\alpha = 0.5$, $\delta = 0.3$, and $\epsilon = 0.00001$ with $x^{(0)} = [-1 \quad 1]^T$ (see Example 8.1). The path followed is shown by the dashed line.

8.5 EQUALITY CONSTRAINTS AND LAGRANGE MULTIPLIERS

In this section we modify the original optimization problem in (8.1) according to

$$\min_{x \in \mathbf{R}^n} f(x)$$
$$\text{subject to} f_i(x) = 0 \quad \text{for all} \quad i = 0, 1, \ldots, m-1 \quad , \qquad (8.46)$$

where $f(x)$ is the *objective function* as before, and $f_i(x) = 0$ for $i = 0, 1, \ldots,$ $m - 1$ are the *equality constraints*. The functions $f_i | \mathbf{R}^n \to \mathbf{R}$ are *equality constraint functions*. The set $F = \{x | f_i(x) = 0, i = 0, \ldots, m-1\}$ is called the *feasible set*. We are interested in

$$\hat{f} = f(\hat{x}) = \min_{x \in F} f(x). \qquad (8.47)$$

There may be more than one $x = \hat{x} \in \mathbf{R}^n$ satisfying (8.47); that is, the set

$$\hat{X} = \{x | x \in F, \quad f(\hat{x}) = \hat{f}\} \qquad (8.48)$$

may have more than one element in it. We assume that our problem yields $\hat{X}$ with at least one element in it (i.e., $\hat{X} \neq \emptyset$).

Equation (8.47) is really a more compact statement of (8.46), and in words states that any minimizer $\hat{x}$ of $f(x)$ must also satisfy the equality constraints. We recall examples of this type of problem from Chapter 4 [e.g., the problem of deriving a computable expression for $\kappa_2(A)$ and in the proof of Theorem 4.5]. More examples will be seen later. Generally, in engineering, constrained optimization problems are more common than unconstrained problems. However, it is important to understand that algorithms for unconstrained problems form the core of algorithms for solving constrained problems.

We now wish to make some general statements about how to solve (8.46), and in so doing we introduce the concept of *Lagrange multipliers*. The arguments to follow are somewhat heuristic, and they follow those of Section 9.1 in Fletcher [1].

Suppose that $\hat{x} \in \mathbf{R}^n$ is at least a local minimizer for objective function $f(x) \in \mathbf{R}$. Analogously to (8.9), we have

$$f_i(\hat{x} + \delta) \approx f_i(\hat{x}) + g_i^T(\hat{x})\delta + \tfrac{1}{2}\delta^T[\nabla^2 f_i(\hat{x})]\delta, \qquad (8.49)$$

where $\delta \in \mathbf{R}^n$ is some *incremental step* away from $\hat{x}$, and $g_i(x) = \nabla f_i(x) \in \mathbf{R}^n$ $(i = 0, 1, \ldots, m-1)$ is the gradient of the ith constraint function at x. A *feasible incremental step* δ must yield $\hat{x} + \delta \in F$, and so must satisfy

$$f_i(\hat{x} + \delta) = f_i(\hat{x}) = 0 \qquad (8.50)$$

for all i. From (8.49) this implies the condition that δ must lie along *feasible direction* $s \in \mathbf{R}^n$ (at $x = \hat{x}$) such that

$$g_i^T(\hat{x})s = 0 \qquad (8.51)$$

again for all i. [We shall suppose that the vectors $g_i(\hat{x}) = \nabla f(\hat{x})$ are linearly independent for all i.] Recalling (8.36), if s were also a *descent direction at* $\hat{x}$, then

$$g^T(\hat{x})s < 0 \tag{8.52}$$

would hold ($g(x) = \nabla f(x) \in \mathbf{R}^n$). In this situation δ would reduce $f(x)$, as δ is along direction s. But this is impossible since we have assumed that $\hat{x}$ is a local minimizer for $f(x)$. [For any s at $\hat{x}$, we expect to have $g^T(\hat{x})s = 0$.] Consequently, no direction s can satisfy (8.51) and (8.52) simultaneously. This statement remains true *if* $g(\hat{x})$ is a linear combination of $g_i(\hat{x})$, that is, if, for suitable $\hat{\lambda}_i \in \mathbf{R}$ we have

$$g(\hat{x}) = \sum_{i=0}^{m-1} \hat{\lambda}_i g_i(\hat{x}).^1 \tag{8.53}$$

Thus, a *necessary condition* for $\hat{x}$ to be a local minimizer (or, more generally, a stationary point) of $f(x)$ is that [rewriting (8.53)]

$$g(\hat{x}) - \sum_{i=0}^{m-1} \hat{\lambda}_i g_i(\hat{x}) = 0 \tag{8.54}$$

for suitable $\hat{\lambda}_i \in \mathbf{R}$ which are called the *Lagrange multipliers*. We see that (8.54) can be expressed as [with ∇ as in (8.3)]

$$\nabla \left[f(\hat{x}) - \sum_{i=0}^{m-1} \hat{\lambda}_i f_i(\hat{x}) \right] = 0. \tag{8.55}$$

In other words, we replace the original problem (8.46) with the mathematically equivalent problem of minimizing

$$L(x, \lambda) = f(x) - \sum_{i=0}^{m-1} \lambda_i f_i(x), \tag{8.56}$$

---

[1] Equation (8.53) may be more formally justified as follows. Note that the same argument will also extend to make (8.53) a necessary condition for $\hat{x}$ to be a local maximizer, or saddle point for $f(x)$. Thus, (8.53) is really a necessary condition for $\hat{x}$ to be a stationary point of $f(x)$. We employ proof by contradiction. Suppose that

$$g(\hat{x}) = \hat{G}\hat{\lambda} + h,$$

where $\hat{G} = [g_0(\hat{x}) \cdots g_{m-1}(\hat{x})] \in \mathbf{R}^{n \times m}$, $\hat{\lambda} = [\hat{\lambda}_0 \cdots \hat{\lambda}_{m-1}]^T \in \mathbf{R}^m$ and $h \neq 0$. Further assume that $h \in \mathbf{R}^n$ is the component of $g(\hat{x})$ that is orthogonal to all $g_i(\hat{x})$. Thus, $\hat{G}^T h = 0$. In this instance $s = -h$ will satisfy both (8.51) and (8.52) [i.e., $g^T(\hat{x})s = -[\hat{\lambda}^T \hat{G}^T + h^T]h = -h^T h < 0$]. Satisfaction of (8.52) implies that a step δ in the direction s will reduce $f(x)$ [i.e., $f(\hat{x} + \delta) < f(\hat{x})$]. But this cannot be the case since $\hat{x}$ is a local minimizer of $f(x)$. Consequently, $h \neq 0$ is impossible, which establishes (8.53).

called the *Lagrangian function* (or *Lagrangian*), where, of course, $x \in \mathbf{R}^n$ and $\lambda \in \mathbf{R}^m$. Since $L|\mathbf{R}^n \times \mathbf{R}^m \to \mathbf{R}$, in order to satisfy (8.55), we must determine $\hat{x}, \hat{\lambda}$ such that

$$\nabla L(\hat{x}, \hat{\lambda}) = 0, \tag{8.57}$$

where now instead [of (8.3)] we have $\nabla = \begin{bmatrix} \nabla_x \\ \nabla_\lambda \end{bmatrix}$ such that

$$\nabla_x = \begin{bmatrix} \dfrac{\partial}{\partial x_0} \cdots \dfrac{\partial}{\partial x_{n-1}} \end{bmatrix}^T, \quad \nabla_\lambda = \begin{bmatrix} \dfrac{\partial}{\partial \lambda_0} \cdots \dfrac{\partial}{\partial \lambda_{m-1}} \end{bmatrix}^T. \tag{8.58}$$

Now we see that a necessary condition for a stationary point of $f(x)$ subject to our constraints is that $\hat{x}$ and $\hat{\lambda}$ form a stationary point of the Lagrangian function. Of course, to resolve whether stationary point $\hat{x}$ is a minimizer requires additional information (e.g., the Hessian). Observe that $\nabla_\lambda L(x, \lambda) = [-f_0(x) - f_1(x) \cdots - f_{m-1}(x)]^T$, so $\nabla_\lambda L(x, \lambda) = 0$ implies that $f_i(x) = 0$ for all i. This is why we take derivatives of the Lagrangian with respect to all elements of λ; it is equivalent to imposing the equality constraints as in the original problem (8.46).

We now consider a few examples of the application of the method of Lagrange multipliers.

Example 8.2   This example is from Fletcher [1, pp. 196–198]. Suppose that

$$f(x) = x_0 + x_1, \quad f_0(x) = x_0^2 - x_1 = 0,$$

so $x = [x_0 \ \ x_1]^T \in \mathbf{R}^2$, and the Lagrangian is

$$L(x, \lambda) = x_0 + x_1 - \lambda(x_0^2 - x_1).$$

Clearly, to obtain stationary points, we must solve

$$\frac{\partial L}{\partial x_0} = 1 - 2\lambda x_0 = 0,$$

$$\frac{\partial L}{\partial x_1} = 1 + \lambda = 0,$$

$$\frac{\partial L}{\partial \lambda} = x_0^2 - x_1 = 0.$$

Immediately, $\hat{\lambda} = -1$, so that $1 - 2\hat{\lambda}\hat{x}_0 = 0$ yields $\hat{x}_0 = -\frac{1}{2}$, and $\hat{x}_0^2 - \hat{x}_1 = 0$ yields $\hat{x}_1 = \frac{1}{4}$. Thus

$$\hat{x} = \begin{bmatrix} -\dfrac{1}{2} & \dfrac{1}{4} \end{bmatrix}^T.$$

Is $\hat{x}$ really a minimizer, or is it a maximizer, or saddle point?

An alternative means to solve our problem is to recognize that since $x_0^2 - x_1 = 0$, we can actually minimize the new objective function

$$f'(x_0) = x_0 + x_1|_{x_1=x_0^2} = x_0 + x_0^2$$

with respect to x_0 instead. Clearly, this is a positive definite quadratic with a well-defined and unique minimum at $x_0 = -\frac{1}{2}$. Again we conclude $\hat{x} = [-\frac{1}{2} \quad \frac{1}{4}]^T$, and it must specifically be a minimizer.

In Example 8.2, $\hat{f} = f(\hat{x}) = \hat{x}_0 + \hat{x}_1 = -\frac{1}{4} > -\infty$ only because of the constraint, $f_0(x) = x_0^2 - x_1 = 0$. Without such a constraint, we would have $\hat{f} = -\infty$.

We have described the method of Lagrange multipliers as being applied largely to minimization problems. But we have noted that this method applies to maximization problems as well because (8.57) is the necessary condition for a stationary point, and not just a minimizer. The next example is a simple maximization problem from geometry.

Example 8.3   We wish to maximize

$$F(x, y) = 4xy$$

subject to the constraint

$$C(x, y) = \tfrac{1}{4}x^2 + y^2 - 1 = 0.$$

This problem may be interpreted as the problem of maximizing the area of a rectangle of area $4xy$ such that the corners of the rectangle are on an ellipse centered at the origin of the two-dimensional plane. The ellipse is the curve $C(x, y) = 0$. (Drawing a sketch is a useful exercise.)

Following the Lagrange multiplier procedure, we construct the Lagrangian function

$$G(x, y, \lambda) = 4xy + \lambda(\tfrac{1}{4}x^2 + y^2 - 1)$$

Taking the derivatives of G and setting them to zero yields

$$\frac{\partial G}{\partial x} = 4y + \frac{1}{2}\lambda x = 0 \tag{8.59a}$$

$$\frac{\partial G}{\partial y} = 4x + 2\lambda y = 0 \tag{8.59b}$$

$$\frac{\partial G}{\partial \lambda} = \frac{1}{4}x^2 + y^2 - 1 = 0 \tag{8.59c}$$

From (8.59a,b) we have

$$\lambda = -8\frac{y}{x}, \quad \text{and} \quad \lambda = -2\frac{x}{y},$$

which means that $-2x/y = -8y/x$, or $x^2 = 4y^2$. Thus, we may replace x^2 by $4y^2$ in (8.59c), giving

$$2y^2 - 1 = 0$$

for which $y^2 = \frac{1}{2}$, and so $x^2 = 2$. From these equations we easily obtain the locations of the corners of the rectangle on the ellipse. The area of the rectangle is also seen to be four units.

Example 8.4   Recall from Chapter 4 that we sought a method to determine (compute) the matrix 2-norm

$$||A||_2 = \max_{||x||_2=1} ||Ax||_2.$$

Chapter 4 considered only the special case $x \in \mathbf{R}^2$ (i.e., $n = 2$). Now we consider the general case for which $n \geq 2$.

Since $||Ax||_2^2 = x^T A^T Ax = x^T Rx$ with $R = R^T$ and $R > 0$ (if A is full rank), our problem is to maximize $x^T Rx$ subject to the equality constraint $||x||_2 = 1$ (or equivalently $x^T x = 1$). The Lagrangian is

$$L(x, \lambda) = x^T Rx - \lambda(x^T x - 1)$$

since $f(x) = x^T Rx$ and $f_0(x) = x^T x - 1$. Now

$$f(x) = x^T Rx = \sum_{i=0}^{n-1}\sum_{j=0}^{n-1} x_i x_j r_{ij}$$

$$= \sum_{i=0}^{n-1} x_i^2 r_{ii} + \sum_{\substack{i=0 \\ i \neq j}}^{n-1}\sum_{j=0}^{n-1} x_i x_j r_{ij}$$

so that (using $r_{ij} = r_{ji}$)

$$\frac{\partial f}{\partial x_k} = 2r_{kk}x_k + \sum_{\substack{j=0 \\ j \neq k}}^{n-1} x_j r_{kj} + \sum_{\substack{i=0 \\ i \neq k}}^{n-1} x_i r_{ik}$$

$$= 2r_{kk}x_k + \sum_{\substack{j=0 \\ j \neq k}}^{n-1} r_{kj}x_j + \sum_{\substack{j=0 \\ j \neq k}}^{n-1} r_{kj}x_j.$$

This reduces to

$$\frac{\partial f}{\partial x_k} = 2r_{kk}x_k + 2 \sum_{\substack{j=0 \\ j \neq k}}^{n-1} r_{kj}x_j = 2 \sum_{j=0}^{n-1} r_{kj}x_j$$

for all $k = 0, 1, \ldots, n-1$. Consequently, $\nabla_x f(x) = 2Rx$. Similarly, $\nabla_x f_0(x) = 2x$. Also, $\nabla_\lambda L(x, \lambda) = -x^T x + 1 = -f_0(x)$, and so $\nabla L(x, \lambda) = 0$ yields the equations

$$2Rx - 2\lambda x = 0,$$

$$x^T x - 1 = 0.$$

The first equation states that the maximizing solution (if it exists) must satisfy the eigenproblem

$$Rx = \lambda x$$

for which λ is an eigenvalue and x is the corresponding eigenvector. Consequently, $x^T R x = \lambda x^T x$, so

$$\lambda = \frac{x^T R x}{x^T x} = \frac{||Ax||_2^2}{||x||_2^2} = ||Ax||_2^2$$

must be chosen to be the biggest eigenvalue of R. Since $R > 0$, such an eigenvalue will exist. As before (Chapter 4), we conclude that

$$||A||_2^2 = \lambda_{n-1}$$

for which λ_{n-1} is the largest of the eigenvalues $\lambda_0, \ldots, \lambda_{n-1}$ of $R = A^T A$ ($\lambda_{n-1} \geq \lambda_{n-2} \geq \cdots \geq \lambda_0 > 0$).

APPENDIX 8.A   MATLAB CODE FOR GOLDEN SECTION SEARCH

```
%
%                          SineFit1.m
%
% This routine computes the objective function V_1(T) for user input
% T and data vector f as required by the golden section search test
% procedure golden.m.  Note that Ts (sampling period) must be consistent
% with Ts in golden.m.
%

function  V1 = SineFit1(f,T);

N = length(f); % Number of samples collected

Ts = 5*60;     % 5 minute sampling period
```

```
     % Compute the objective function V_1(T)

n = [0:N-1];
T = T*60*60;
A = [ sin(2*pi*Ts*n/T).' cos(2*pi*Ts*n/T).' ];
B = inv(A.'*A);
V1 = f.'*f - f.'*A*B*A.'*f;


%
%                             golden.m
%
% This routine tests the golden section search procedure of Chapter 8
% on the noisy sinusoid problem depicted in Fig. 8.4.
%
% This routine creates a test signal and uses SineFit1.m to compute
% the corresponding objective function V_1(T) (given by Equation (8.33)
% in Chapter 8).
%

function Topt = golden

     % Compute the test signal f

N = 500;        % Number of samples collected

Ts = 5*60;      % 5 minute sampling period

T = 24*60*60;   % 24 hr period for the sinusoid
phi = -pi/10;   % phase angle of sinusoid
a = 1.0;        % sinusoid amplitude

var = .5000;    % desired noise variance

std = sqrt(var);
eta = std*randn(1,N);
for n = 1:N
  f(n) = a*sin(((2*pi*(n-1)*Ts)/T) + phi);
  end;
f = f + eta;
f = f.';

     % Specify a starting interval and initial parameters
     % (units of hours)

xl = 16;
xu = 32;

tau = (sqrt(5)+1)/2;  % The golden ratio
tol = .05;            % Accuracy of the location of the minimum

     % Apply the golden section search procedure

I = xu - xl;    % length of the starting interval
xa = xu - I/tau;
xb = xl + I/tau;
```

```
while I > tol
  if SineFit1(f,xa) >= SineFit1(f,xb)
    I = xu - xa;
    xl = xa;
    temp = xa;
    xa = xb;
    xb = temp + I/tau;
  else
    I = xb - xl;
    xu = xb;
    temp = xb;
    xb = xa;
    xa = temp - I/tau;
  end
end

Topt = (xl + xu)/2;  % Estimate of optimum choice for T
```

REFERENCES

1. R. Fletcher, *Practical Methods of Optimization*, 2nd ed., Wiley, New York, 1987 (reprinted July 1993).

2. M. R. Schroeder, *Number Theory in Science and Communication* (*with Applications in Cryptography, Physics, Digital Information, Computing, and Self-Similarity*), 2nd (expanded) ed., Springer-Verlag, New York, 1986.

3. A. Quarteroni, R. Sacco, and F. Saleri, *Numerical Mathematics* (Texts in Applied Mathematics series, Vol. 37). Springer-Verlag, New York, 2000.

4. R. P. Brent, *Algorithms for Minimization without Derivatives*, Dover Publications, Mineola, NY, 2002.

5. S. Boyd and L. Vandenberghe, *Convex Optimization*, preprint, Dec. 2001.

PROBLEMS

8.1. Suppose $A \in \mathbf{R}^{n \times n}$, and that A is not symmetric in general. Prove that

$$\nabla(x^T A x) = (A + A^T)x,$$

where ∇ is the gradient operator [Eq. (8.3)].

8.2. This problem is about ideas from vector calculus useful in nonlinear optimization methods.

(a) If $s, x, x' \in \mathbf{R}^n$, then a line in $\mathbf{R}^n$ is defined by

$$x = x(\alpha) = x' + \alpha s, \quad \alpha \in \mathbf{R}.$$

Vector s may be interpreted as determining the direction of the line in the n-dimensional space $\mathbf{R}^n$. The notation $x(\alpha)$ implies $x(\alpha) = [x_0(\alpha) x_1(\alpha) \cdots x_{n-1}(\alpha)]^T$. Prove that the slope $df/d\alpha$ of $f(x(\alpha)) \in \mathbf{R}$ along the line at any $x(\alpha)$ is given by

$$\frac{df}{d\alpha} = s^T \nabla f,$$

where ∇ is the gradient operator [see Eq. (8.3)]. (*Hint:* Use the chain rule for derivatives.)

(b) Suppose that $u(x), v(x) \in \mathbf{R}^n$ (again $x \in \mathbf{R}^n$). Prove that

$$\nabla(u^T v) = (\nabla u^T)v + (\nabla v^T)u.$$

8.3. Use the golden ratio search method to find the global minimizer of the polynomial objective function in Fig. 8.2. Do the computations using a pocket calculator, with starting interval $[x_l^0, x_u^0] = [2, 2.5]$. Iterate 5 times.

8.4. Review Problem 7.4 (in Chapter 7).

Use a MATLAB implementation of the golden ratio search method to find detection threshold η for $P_{FA} = 0.1, 0.01,$ and 0.001. The objective function is

$$f(\eta) = \left| P_{FA} - e^{-(1/2)\eta} \sum_{k=0}^{(p/2)-1} \frac{1}{k!} \left(\frac{\eta}{2}\right)^k \right|.$$

Make reasonable choices about starting intervals.

8.5. Suppose $x = [x_0 \quad x_1]^T \in \mathbf{R}^2$, and consider

$$f(x) = x_0^4 + x_0 x_1 + (1 + x_1)^2.$$

Find general expressions for the gradient and the Hessian of $f(x)$. Is $G(0) > 0$? What does this signify? Use the Newton–Raphson method (Chapter 7) to confirm that $\hat{x} = [0.6959 \quad -1.3479]^T$ is a stationary point for $f(x)$. Select $\overline{x}_0 = [0.7000 \quad -1.3]^T$ as the starting point. Is $G(\hat{x}) > 0$? What does this signify? In this problem do all necessary computations using a pocket calculator.

8.6. If $\hat{x} \in \mathbf{R}^n$ is a local minimizer for $f(x)$, then we know that $G(\hat{x}) > 0$, that is, $G(\hat{x})$ is positive definite (pd). On the other hand, if $\hat{x}$ is a *local maximizer* of $f(x)$, then $-G(\hat{x}) > 0$. In this case we say that $G(\hat{x})$ is *negative definite* (nd). Show that

$$f(x) = (x_1 - x_0^2)^2 + x_0^5$$

has only one stationary point, and that it is neither a minimizer nor a maximizer of $f(x)$.

8.7. Take note of the criterion for a maximizer in the previous problem. For both $a = 6$ and $a = 8$, find all stationary points of the function

$$f(x) = 2x_0^3 - 3x_0^2 - ax_0 x_1 (x_0 - x_1 - 1).$$

Determine which are local minima, local maxima, or neither.

8.8. Write a MATLAB function to find the minimum of Rosenbrock's function using the Newton algorithm (basic form that does not employ a line search). Separate functions must be written to implement computation of both the gradient and the inverse of the Hessian. The function for computing the inverse of the Hessian must return an integer that indicates whether the Hessian is positive definite. The Newton algorithm must terminate if a Hessian is encountered that is not positive definite. The I/O is to be at the terminal only. The user must input the starting vector at the terminal, and the program must report the estimated minimum, or it must print an error message if the Hessian is not positive definite. Test your program out on the starting vectors

$$x^{(0)} = [-3 \quad -3]^T \quad \text{and} \quad x^{(0)} = [0 \quad 10]^T.$$

8.9. Write and test your own MATLAB routine (or routines) to verify Example 8.1.

8.10. Find the points on the ellipse

$$\frac{x^2}{a^2} + \frac{y^2}{b^2} = 1$$

that are closest to, and farthest from, the origin $(x, y) = (0, 0)$. Use the method of Lagrange multipliers.

8.11. The theory of Lagrange multipliers in Section 8.5 is a bit oversimplified. Consider the following theorem. Suppose that $\hat{x} \in \mathbf{R}^n$ gives an extremum (i.e., minimum or maximum) of $f(x) \in \mathbf{R}$ among all x satisfying $g(x) = 0$. If $f, g \in C^1[D]$ for a domain $D \subset \mathbf{R}$ containing $\hat{x}$, then either

$$g(\hat{x}) = 0 \quad \text{and} \quad \nabla g(\hat{x}) = 0, \tag{8.P.1}$$

or there is a $\lambda \in \mathbf{R}$ such that

$$g(\hat{x}) = 0 \quad \text{and} \quad \nabla f(\hat{x}) - \lambda \nabla g(\hat{x}) = 0. \tag{8.P.2}$$

From this theorem, candidate points $\hat{x}$ for extrema of $f(x)$ satisfying $g(x) = 0$ therefore are

(a) Points where f and g fail to have continuous partial derivatives.
(b) Points satisfying (8.P.1).
(c) Points satisfying (8.P.2).

In view of the theorem above and its consequences, find the minimum distance from $x = [x_0 \quad x_1 \quad x_2]^T = [0 \quad 0 \quad -1]^T$ to the surface

$$g(x) = x_0^2 + x_1^2 - x_2^5 = 0.$$

8.12. A second-order finite-impulse response (FIR) digital filter has the *frequency response* $H(e^{j\omega}) = \sum_{k=0}^{2} h_k e^{-j\omega k}$, where $h_k \in \mathbf{R}$ are the *filter parameters*. Since $H(e^{j\omega})$ is 2π−periodic we usually consider only $\omega \in [-\pi, \pi]$. The *DC response* of the filter is $H(1) = H(e^{j0}) = \sum_{k=0}^{2} h_k$. Define the energy of the filter in the band $[-\omega_p, \omega_p]$ to be

$$E = \frac{1}{2\pi} \int_{-\omega_p}^{\omega_p} |H(e^{j\omega})|^2 \, d\omega.$$

Find the filter parameters $h = [h_0 \ h_1 \ h_2]^T \in \mathbf{R}^3$ such that for $\omega_p = \pi/2$ energy E is minimized subject to the constraint that $H(1) = 1$ (i.e., the gain of the filter is unity at DC). Plot $|H(e^{j\omega})|$ for $\omega \in [-\pi, \pi]$. [*Hint:* E will have the form $E = h^T R h \in \mathbf{R}$, where $R \in \mathbf{R}^{3 \times 3}$ is a symmetric Toeplitz matrix (recall Problem 4.20). Note also that $|H(e^{j\omega})|^2 = H(e^{j\omega})H^*(e^{j\omega})$.]

8.13. This problem introduces *incremental condition estimation* (ICE) and is based on the paper C. H. Bischof, "Incremental Condition Estimation," *SIAM J. Matrix Anal. Appl.* **11**, 312–322 (April 1990). ICE can be used to estimate the condition number of a lower triangular matrix as it is generated one row at a time. Many algorithms for linear system solution produce triangular matrix factorizations one row or one column at a time. Thus, ICE may be built into such algorithms to warn the user of possible inaccuracies in the solution due to ill conditioning. Let A_n be an $n \times n$ matrix with singular values $\sigma_1(A_n) \geq \cdots \geq \sigma_n(A_n) \geq 0$. A condition number for A_n is

$$\kappa(A_n) = \frac{\sigma_1(A_n)}{\sigma_n(A_n)}.$$

Consider the order n lower triangular linear system

$$L_n x_n = d_n. \tag{8.P.3}$$

The minimum singular value, $\sigma_n(L_n)$, of L_n satisfies

$$\sigma_n(L_n) \leq \frac{||d_n||_2}{||x_n||_2}$$

($||x_n||_2^2 = \sum_i x_{n,i}^2$). Thus, an estimate (upper bound) of this singular value is

$$\hat{\sigma}_n(L_n) = \frac{||d_n||_2}{||x_n||_2}.$$

We would like to make this upper bound as small as possible. So, Bischof suggests finding x_n to satisfy (8.P.3) such that $||x_n||_2$ is maximized subject to the constraint that $||d_n||_2 = 1$. Given x_{n-1} such that $L_{n-1}x_{n-1} = d_{n-1}$

with $||d_{n-1}||_2 = 1$ [which gives us $\hat{\sigma}_{n-1}(L_{n-1}) = 1/||x_{n-1}||_2$], find s_n and c_n such that $||x_n||_2$ is maximized where

$$
L_n x_n = \begin{bmatrix} L_{n-1} & 0 \\ v_n^T & \gamma_n \end{bmatrix} x_n = \begin{bmatrix} s_n d_{n-1} \\ c_n \end{bmatrix} = d_n,
$$

$$
x_n = \begin{bmatrix} s_n x_{n-1} \\ (c_n - s_n \alpha_n)/\gamma_n \end{bmatrix}.
$$

Find α_n, c_n, s_n. Assume $\alpha_n \neq 0$. (*Comment:* The indexing of the singular values used here is different from that in Chapter 4. The present notation is more convenient in the present context.)

8.14. Assume that $A \in \mathbf{R}^{m \times n}$ with $m \geq n$, but rank(A) $< n$ is possible. As usual, $||x||_2^2 = x^T x$. Solve the following problem:

$$
\min_{x \in \mathbf{R}^n} ||Ax - b||_2^2 + \delta ||x||_2^2,
$$

where $\delta > 0$. This is often called the *Tychonov regularization problem.* It is a simple ploy to alleviate problems with ill-conditioning in least-squares applications. Since A is not necessarily of full rank, we have $A^T A \geq 0$, but not necessarily $A^T A > 0$. What is rank($A^T A + \delta I$)? Of course, I is the order n identity matrix.

9 Numerical Integration and Differentiation

9.1 INTRODUCTION

We are interested in how to compute the integral

$$I = \int_a^b f(x)\, dx \tag{9.1}$$

for which $f(x) \in \mathbf{R}$ (and, of course, $x \in \mathbf{R}$). Depending on $f(x)$, and perhaps also on $[a, b]$, the reader knows that "nice" closed-form expressions for I rarely exist. This forces us to consider numerical methods to approximate I. We have seen from Chapter 3 that one approach is to find a suitable series expansion for the integral in (9.1). For example, recall that we wished to compute the error function

$$\mathrm{erf}(x) = \frac{2}{\sqrt{\pi}} \int_0^x e^{-t^2}\, dt, \tag{9.2}$$

which has no antiderivative (i.e., "nice" formula). Recall that the error function is crucial in solving various problems in applied probability that involve the Gaussian probability density function [i.e., the function in (3.101) of Chapter 3]. The Taylor series expansion of Eq. (3.108) was suggested as a means to approximately evaluate $\mathrm{erf}(x)$, and is known to be practically effective if x is not too big. If x is large, then the asymptotic expansion of Example 3.10 was suggested. The series expansion methodology may seem to solve our problem, but there are integrals for which it is not easy to find series expansions of any kind.

A recursive approach may be attempted as an alternative. An example of this was seen in Chapter 5, where finding the norm of a Legendre polynomial required solving a recursion involving variables that were certain integrals [recall Eq. (5.96)]. As another example of this approach, consider the following case from Forsythe et al. [1]. Suppose that we wish to compute

$$E_n = \int_0^1 x^n e^{x-1}\, dx \tag{9.3}$$

An Introduction to Numerical Analysis for Electrical and Computer Engineers, by C.J. Zarowski
ISBN 0-471-46737-5 © 2004 John Wiley & Sons, Inc.

for any $n \in \mathbf{N}$ (natural numbers). Recalling integration by parts, we see that

$$\int_0^1 x^n e^{x-1} \, dx = x^n e^{x-1}|_0^1 - \int_0^1 n x^{n-1} e^{x-1} \, dx = 1 - n E_{n-1},$$

so

$$E_n = 1 - n E_{n-1} \tag{9.4}$$

for $n = 2, 3, 4, \ldots$. It is easy to confirm that

$$E_1 = \int_0^1 x e^{x-1} \, dx = \frac{1}{e}.$$

This is the initial condition for recursion (9.4). We observe that $E_n > 0$ for all n. But if, for example, MATLAB is used to compute E_{19}, we obtain computed solution $\hat{E}_{19} = -5.1930$, which is clearly wrong. Why has this happened?

Because of the need to quantize, E_1 is actually stored in the computer as

$$\hat{E}_1 = E_1 + \epsilon$$

for which ϵ is some quantization error. Assuming that the operations in (9.4) do not lead to further errors (i.e., assuming no rounding errors), we may arrive at a formula for $\hat{E}_n$:

$$\hat{E}_2 = 1 - 2\hat{E}_1 = E_2 + (-2)\epsilon,$$
$$\hat{E}_3 = 1 - 3\hat{E}_2 = E_3 + (-3)(-2)\epsilon,$$
$$\hat{E}_4 = 1 - 4\hat{E}_3 = E_4 + (-4)(-3)(-2)\epsilon,$$

and so on. In general

$$\hat{E}_n = E_n + (-1)^{n-1} 1 \cdot 2 \cdot 3 \cdots (n-1) n \epsilon, \tag{9.5}$$

so

$$\hat{E}_n - E_n = (-1)^{n-1} n! \epsilon. \tag{9.6}$$

We see that even a very tiny quantization error ϵ will grow very rapidly during the course of the computation even without any additional rounding errors at all ! Thus, (9.4) is a highly unstable numerical procedure, and so must be rejected as a method to compute (9.3). However, it is possible to arrive at a stable procedure by modifying (9.4). Now observe that

$$E_n = \int_0^1 x^n e^{x-1} \, dx \leq \int_0^1 x^n \, dx = \frac{1}{n+1}, \tag{9.7}$$

implying that $E_n \to 0$ as n increases. Instead of (9.4), consider the recursion [obtained by rearranging (9.4)]

$$E_{n-1} = \frac{1}{n}(1 - E_n). \tag{9.8}$$

If we wish to compute E_m, we may assume $E_n = 0$ for some n significantly bigger than m and apply (9.8). From (9.7) the error involved in approximating E_n by zero is not bigger than $1/(n+1)$. Thus, an algorithm for E_m is

$$E_{k-1} = \frac{1}{k}(1 - E_k)$$

for $k = n, n-1, \ldots, m+2, m+1$, where $E_n = 0$. At each stage of this algorithm the initial error is reduced by factor $1/k$ rather than being magnified as it was in the procedure of (9.4).

Other than potential numerical stability problems, which may or may not be easy to solve, it is apparent that not all integration problems may be cast into a recursive form. It is also possible that $f(x)$ is not known for all $x \in \mathbf{R}$. We might know $f(x)$ only on a finite subset of $\mathbf{R}$, or perhaps a countably infinite subset of $\mathbf{R}$. This situation might arise in the context of obtaining $f(x)$ experimentally. Such a scenario would rule out the previous suggestions.

Thus, there is much room to consider alternative methodologies. In this chapter we consider what may be collectively called *quadrature methods*. For the most part, these are based on applying some of the interpolation ideas considered in Chapter 6. But Gaussian quadrature (see Section 9.4) also employs orthogonal polynomials (material from Chapter 5).

This Chapter is dedicated mainly to the subject of numerical integration by quadratures. But the final section considers numerical approximations to derivatives (i.e., numerical differentiation). Numerical differentiation is relevant to the numerical solution of differential equations (to be considered in later chapters), and we have mentioned that it is relevant to spline interpolation (Section 6.5). In fact, it can also find a role in refined methods of numerical integration (Section 9.5).

9.2 TRAPEZOIDAL RULE

A simple approach to numerical integration is the following.

In this book we implicitly assume all functions are Riemann integrable. From elementary calculus such integrals are obtained by the limiting process

$$I = \int_a^b f(x)\,dx = \lim_{n \to \infty} \frac{b-a}{n} \sum_{k=1}^{n} f(x_k^\circ) \tag{9.9}$$

for which $x_0 = a$, $x_n = b$, and for which the value I is independent of the point $x_k^\circ \in [x_{k-1}, x_k]$. We remark that not all functions $f(x)$ satisfy this requirement and so, as was mentioned in Chapter 3, not all functions are Riemann integrable. However, we will ignore this potential problem. We may approximate I according to

$$I \approx \frac{b-a}{n} \sum_{k=1}^{n} f(x_k^{\circ}).$$

(9.10)

Such an approximation is called the *rectangular rule* (or *rectangle rule*) for numerical integration, and there are different variants depending on the choice for x_k°. Three possible choices are shown in Fig. 9.1. We mention that all variants involve assuming $f(x)$ is piecewise constant on $[x_{k-1}, x_k]$, and so amount to the *constant interpolation* of $f(x)$ (i.e., fitting a polynomial which is a constant to $f(x)$ on some interval). Define

$$h = \frac{b-a}{n}.$$

From Fig. 9.1, the *right-point rule* uses

$$x_k^{\circ} = a + kh,$$

(9.11a)

while the *left-point rule* uses

$$x_k^{\circ} = a + (k-1)h,$$

(9.11b)

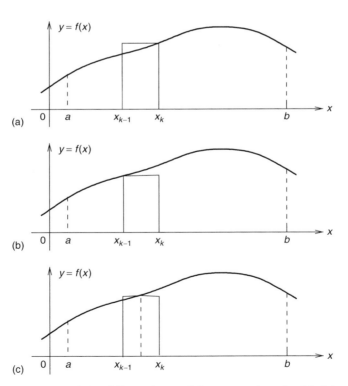

Figure 9.1   Illustration of the different forms of the rectangular rule: (a) right-point rule; (b) left-point rule; (c) midpoint rule. In all cases $x_0 = a$ and $x_n = b$.

and the *midpoint rule* uses

$$x_k^\circ = a + (k - \tfrac{1}{2})h, \tag{9.11c}$$

where in all cases $k = 1, 2, \ldots, n - 1, n$. The midpoint rule is often preferred among the three as it is usually more accurate. However, the rectangular rule is often (sometimes unfairly) regarded as too crude, and so the following (or something still "better") is chosen.

It is often better to approximate $f(x)$ with *trapezoids* as shown in Fig. 9.2. This results in the *trapezoidal rule* for numerical integration. From Fig. 9.2 we see that this rule is based on the *linear interpolation* of the function $f(x)$ on $[x_{k-1}, x_k]$. The approximation to $\int_{x_{k-1}}^{x_k} f(x)\,dx$ is given by the area of the trapezoid: this is

$$\int_{x_{k-1}}^{x_k} f(x)\,dx \approx \frac{1}{2}\left[f(x_k) + f(x_{k-1})\right](x_k - x_{k-1}) = \frac{1}{2}\left[f(x_{k-1}) + f(x_k)\right]h \tag{9.12}$$

[We have $h = x_k - x_{k-1} = (b - a)/n$.] It is intuitively plausible that this method should be more accurate than the rectangular rule, and yet not require much, if any, additional computational effort to implement it. Applying (9.12) for $k = 1$ to $k = n$, we have

$$\int_a^b f(x)\,dx \approx T(n) = \frac{h}{2}\sum_{k=1}^{n}\left[f(x_{k-1}) + f(x_k)\right], \tag{9.13}$$

and the summation expands out as

$$T(n) = \frac{h}{2}\left[f(x_0) + 2f(x_1) + 2f(x_2) + \cdots + 2f(x_{n-1}) + f(x_n)\right], \tag{9.14}$$

where $n \in \mathbf{N}$ (set of natural numbers).

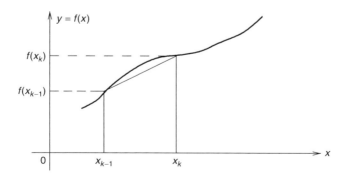

Figure 9.2   Illustration of the trapezoidal rule.

We may investigate the error behavior of the trapezoidal rule as follows. The process of analysis begins by assuming that $n = 1$; that is, we linearly interpolate $f(x)$ on $[a, b]$ using

$$p(x) = f(a)\frac{x - b}{a - b} + f(b)\frac{x - a}{b - a}, \tag{9.15}$$

which is from the Lagrange interpolation formula [recall (6.9) for $n = 1$]. It must be the case that for suitable error function $e(x)$, we have

$$f(x) = p(x) + e(x) \tag{9.16}$$

with $x \in [a, b]$. Consequently

$$\int_a^b f(x)\,dx = \int_a^b p(x)\,dx + \int_a^b e(x)\,dx$$

$$= \frac{b - a}{2}[f(a) + f(b)] + \int_a^b e(x)\,dx = T(1) + E_{T(1)}, \tag{9.17}$$

where

$$E_{T(1)} = \int_a^b e(x)\,dx, \tag{9.18}$$

which is the error involved in using the trapezoidal rule. Of course, we would like a suitable bound on this error. To obtain such a bound, we will assume that $f^{(1)}(x)$ and $f^{(2)}(x)$ both exist and are continuous on $[a, b]$. Let x be fixed at some value such that $a < x < b$, and define

$$g(t) = f(t) - p(t) - [f(x) - p(x)]\frac{(t - a)(t - b)}{(x - a)(x - b)} \tag{9.19}$$

for $t \in [a, b]$. It is not difficult to confirm that

$$g(a) = g(b) = g(x) = 0,$$

so $g(t)$ vanishes at three different places on the interval $[a, b]$. *Rolle's theorem*[1] says that there are points $\xi_1 \in (a, x), \xi_2 \in (x, b)$ such that

$$g^{(1)}(\xi_1) = g^{(1)}(\xi_2) = 0.$$

Thus, $g^{(1)}(t)$ vanishes at two different places on (a, b), so yet again by Rolle's theorem there is a $\xi \in (\xi_1, \xi_2)$ such that

$$g^{(2)}(\xi) = 0.$$

---

[1]This was proved by Bers [2], but the proof is actually rather lengthy, and so we omit it.

We note that $\xi = \xi(x)$, that is, point ξ depends on x. Therefore

$$g^{(2)}(\xi) = f^{(2)}(\xi) - p^{(2)}(\xi) - [f(x) - p(x)]\frac{2}{(x-a)(x-b)} = 0. \quad (9.20)$$

The polynomial $p(x)$ is of the first degree so $p^{(2)}(\xi) = 0$. We may use this in (9.20) and rearrange the result so that

$$f(x) = p(x) + \tfrac{1}{2}f^{(2)}(\xi(x))(x-a)(x-b) \quad (9.21)$$

for any $x \in (a, b)$. This expression also happens to be valid at $x = a$ and at $x = b$, so

$$e(x) = f(x) - p(x) = \tfrac{1}{2}f^{(2)}(\xi(x))(x-a)(x-b) \quad (9.22)$$

for $x \in [a, b]$. But we need to evaluate $E_{T(1)}$ in (9.18). The *second mean-value theorem for integrals* states that if $f(x)$ is continuous and $g(x)$ is integrable (Riemann) on $[a, b]$, and further that $g(x)$ does not change sign on $[a, b]$, then there is a point $p \in (a, b)$ such that

$$\int_a^b f(x)g(x)\,dx = f(p)\int_a^b g(x)\,dx. \quad (9.23)$$

The proof is omitted. We observe that $(x-a)(x-b)$ does not change sign on $x \in [a, b]$, so via this theorem we have

$$E_{T(1)} = \frac{1}{2}\int_a^b f^{(2)}(\xi(x))(x-a)(x-b)\,dx = -\frac{1}{12}f^{(2)}(p)(b-a)^3, \quad (9.24)$$

where $p \in (a, b)$. We emphasize that this is the error for $n = 1$ in $T(n)$. Naturally, we want an error expression for $n > 1$, too. When $n > 1$, we may refer to the integration rule as a *compound* or *composite* rule.

The error committed in numerically integrating over the kth subinterval $[x_{k-1}, x_k]$ must be [via (9.24)]

$$E_k = -\frac{1}{12}f^{(2)}(\xi_k)(x_k - x_{k-1})^3 = -\frac{h^3}{12}f^{(2)}(\xi_k) = -\frac{h^2}{12}\frac{b-a}{n}f^{(2)}(\xi_k), \quad (9.25)$$

where $\xi_k \in [x_{k-1}, x_k]$ and $k = 1, 2, \ldots, n-1, n$. Therefore, the total error committed is

$$E_{T(n)} = \int_a^b f(x)\,dx - T(n) = \sum_{k=1}^n E_k, \quad (9.26)$$

which becomes [via (9.25)]

$$E_{T(n)} = -\frac{h^2}{12}\frac{b-a}{n}\sum_{k=1}^n f^{(2)}(\xi_k). \quad (9.27)$$

The average $\frac{1}{n}\sum_{k=1}^{n} f^{(2)}(\xi_k)$ must lie between the largest and smallest values of $f^{(2)}(x)$ on $[a, b]$, so recalling that $f^{(2)}(x)$ is continuous on $[a, b]$, the *intermediate value theorem* (Theorem 7.1) yields that there is an $\xi \in (a, b)$ such that

$$f^{(2)}(\xi) = \frac{1}{n}\sum_{k=1}^{n} f^{(2)}(\xi_k).$$

Therefore

$$E_{T(n)} = -\frac{h^2}{12}(b-a)f^{(2)}(\xi), \tag{9.28}$$

where $\xi \in (a, b)$. If the maximum value of $f^{(2)}(x)$ on $[a, b]$ is known, then this may be used in (9.28) to provide an upper bound on the error. We remark that $E_{T(n)}$ is often called *truncation error*.

We see that

$$T(n) = x^T y \tag{9.29}$$

for which

$$x = h[\tfrac{1}{2} 1 \cdots 1\tfrac{1}{2}]^T \in \mathbf{R}^{n+1}, \qquad y = [f(x_0)f(x_1)\cdots f(x_{n-1})f(x_n)]^T \in \mathbf{R}^{n+1}. \tag{9.30}$$

We know that rounding errors will be committed in the computation of (9.29). The total rounding error might be denoted by E_R. We recall from Chapter 2 [Eq. (2.40)] that a bound on these errors is

$$|E_R| = |x^T y - fl[x^T y]| \leq 1.01(n+1)u|x|^T|y|, \tag{9.31}$$

where u is the unit roundoff, or else the machine epsilon. The cumulative effect of rounding errors can be expected to grow as n increases. If we suppose that

$$M = \max_{x \in [a,b]} |f^{(2)}(x)|, \tag{9.32}$$

then, from (9.28), we obtain

$$|E_{T(n)}| \leq \frac{1}{12}\frac{1}{n^2}(b-a)^3 M. \tag{9.33}$$

Thus, (9.33) is an upper bound on the truncation error for the *composite trapezoidal rule*. We see that the bound gets smaller as n increases. Thus, as we expect, truncation error is reduced as the number of trapezoids used increases. Combining (9.33) with (9.31) results in a bound on total error E:

$$|E| \leq \frac{1}{12}\frac{1}{n^2}(b-a)^3 M + 1.01(n+1)u|x|^T|y|. \tag{9.34}$$

Usually $n \gg 1$, so $n + 1 \approx n$. Thus, substituting this and (9.30) into (9.31) results in

$$|x^T y - fl[x^T y]| \leq 1.01(b - a)u \left[\frac{|f(x_0)| + |f(x_n)|}{2} + \sum_{k=1}^{n-1} |f(x_k)| \right], \quad (9.35)$$

and so (9.34) becomes

$$|E| \leq \frac{1}{12} \frac{1}{n^2} (b - a)^3 M + 1.01(b - a)u \left[\frac{|f(x_0)| + |f(x_n)|}{2} + \sum_{k=1}^{n-1} |f(x_k)| \right].$$
$$(9.36)$$

In general, the first term in the bound of (9.36) becomes smaller as n increases, while the second term becomes larger. Thus, there is a tradeoff involved in choosing the number of trapezoids to approximate a given integral, and the best choice ought to minimize the total error.

Example 9.1   We may apply the bound of (9.36) to the following problem. We wish to compute

$$I = \int_0^1 e^{-x} \, dx.$$

Thus, $[a, b] = [0, 1]$, and so $b - a = 1$. Of course, it is very easy to confirm that $I = 1 - e^{-1}$. But this is what makes it a good example to test our theory out. We also see that $f^{(2)}(x) = e^{-x}$, and so in (9.32) $M = 1$. Also

$$f(x_k) = e^{-k/n}$$

for $k = 0, 1, \ldots, n - 1, n$. It is therefore easy to see that

$$\sum_{k=1}^{n-1} f(x_k) = \frac{e^{-1/n} - e^{-1}}{1 - e^{-1/n}}.$$

We might assume that the trapezoidal rule for this problem is implemented in the C programming language using single-precision floating-point arithmetic, in which case a typical value for u would be

$$u = 1.1921 \times 10^{-7}.$$

Therefore, from (9.36) the total error is bounded according to

$$|E| \leq \frac{1}{12n^2} + 1.2040 \times 10^{-7} \left[\frac{e + 1}{2e} + \frac{e^{-1/n} - e^{-1}}{1 - e^{-1/n}} \right].$$

Figure 9.3 plots this bound versus n, and also shows the magnitude of the computed (i.e., the true or actual) total error in the trapezoidal rule approximation, which is $|T(n) - I|$. We see that the true error is always less than the bound, as we would expect. However, the bound is rather pessimistic. Also, the bound predicts

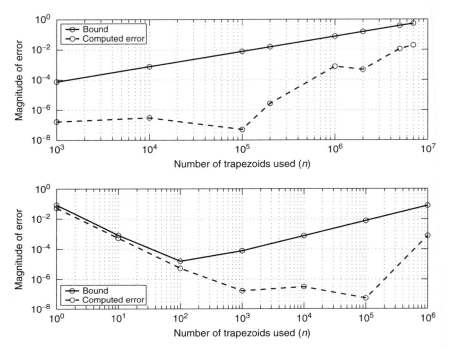

Figure 9.3   Comparison of total error (computed) to bound on total error, illustrating the tradeoff between rounding error and truncation error in numerical integration by the trapezoidal rule. The bound employed here is that of Eq. (9.36).

that the proper choice for n is much less than what the computed result predicts. Specifically, the bound suggests that we choose $n \approx 100$, while the computed result suggests that we choose $n \approx 100,000$.

What is important is that the computed result and the bound both confirm that there is a tradeoff between minimizing the truncation error and minimizing the rounding error. To minimize rounding error, we prefer a small n, but to minimize the truncation error, we prefer a large n. The best solution minimizes the total error from both sources.

In practice, attempting a detailed analysis to determine the true optimum choice for n is usually not worth the effort. What is important is to understand the fundamental tradeoffs involved in the choice of n, and from this understanding select a reasonable value for n.

9.3   SIMPSON'S RULE

The trapezoidal rule employed linear interpolation to approximate $f(x)$ between sample points x_k on the x axis. We might consider *quadratic interpolation* in the hope of improving accuracy still further. Here "accuracy" is a reference to truncation error.

Therefore, we wish to fit a quadratic curve to the points $(x_{k-1}, f(x_{k-1}))$, $(x_k, f(x_k))$ and $(x_{k+1}, f(x_{k+1}))$. We may define the quadratic to be

$$p_k(x) = a(x - x_k)^2 + b(x - x_k) + c. \tag{9.37}$$

Contrary to past practice, the subscript k now does not denote degree, but rather denotes the "centerpoint" of the interval $[x_{k-1}, x_{k+1}]$ on which we are fitting the quadratic. The situation is illustrated in Fig. 9.4 . For convenience, define $y_k = f(x_k)$. Therefore, from (9.37) we may set up three equations in the unknowns a, b, c:

$$a(x_{k-1} - x_k)^2 + b(x_{k-1} - x_k) + c = y_{k-1},$$

$$a(x_k - x_k)^2 + b(x_k - x_k) + c = y_k,$$

$$a(x_{k+1} - x_k)^2 + b(x_{k+1} - x_k) + c = y_{k+1}.$$

This is a linear system of equations, and we will assume that $h = x_k - x_{k-1} = x_{k+1} - x_k$, so therefore

$$a = \frac{y_{k+1} - 2y_k + y_{k-1}}{2h^2}, \tag{9.38a}$$

$$b = \frac{y_{k+1} - y_{k-1}}{2h}, \tag{9.38b}$$

$$c = y_k. \tag{9.38c}$$

This leads to the approximation

$$\int_{x_{k-1}}^{x_{k+1}} f(x)\, dx \approx \int_{x_{k-1}}^{x_{k+1}} p_k(x)\, dx = \frac{h}{3}[y_{k-1} + 4y_k + y_{k+1}]. \tag{9.39}$$

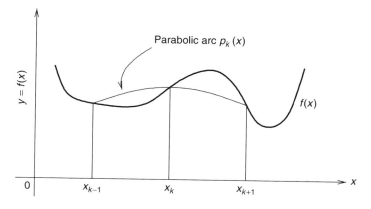

Figure 9.4   Simpson's rule for numerical integration.

Of course, some algebra has been omitted to arrive at the equality in (9.39). As in Section 9.2, we wish to integrate $f(x)$ on $[a, b]$. So, as before, $a = x_0$, and $b = x_n$. If n is an *even number*, then the number of subdivisions of $[a, b]$ is an even number, and hence we have the approximation

$$I = \int_a^b f(x)\,dx \approx \int_{x_0}^{x_2} p_1(x)\,dx + \int_{x_2}^{x_4} p_3(x)\,dx + \cdots + \int_{x_{n-2}}^{x_n} p_{n-1}(x)\,dx$$

$$= \frac{h}{3}[y_0 + 4y_1 + 2y_2 + 4y_3 + 2y_4 + \cdots + 2y_{n-2} + 4y_{n-1} + y_n]. \tag{9.40}$$

The last equality follows from applying (9.39). We define the *Simpson rule* approximation to I as

$$S(n) = \frac{h}{3}[y_0 + 4y_1 + 2y_2 + 4y_3 + 2y_4 + \cdots + 2y_{n-2} + 4y_{n-1} + y_n] \tag{9.41}$$

for which n is even and $n \geq 2$.

A truncation error analysis of Simpson's rule is more involved than that of the analysis of the trapezoidal rule seen in the previous section. Therefore, we only outline the major steps and results. We begin by using only two subintervals to approximate $I = \int_a^b f(x)\,dx$, specifically, $n = 2$. Define $c = (a + b)/2$. Denote the interpolating quadratic by $p(x)$. For a suitable error function $e(x)$, we must have

$$f(x) = p(x) + e(x). \tag{9.42}$$

Immediately we see that

$$I = \int_a^b f(x)\,dx = \int_a^b p(x)\,dx + \int_a^b e(x)\,dx$$

$$= \frac{b-a}{6}\left[f(a) + 4f\left(\frac{a+b}{2}\right) + f(b)\right]$$

$$+ \int_a^b e(x)\,dx = S(2) + E_{S(2)}. \tag{9.43}$$

So, the *truncation error* in Simpson's rule is thus

$$E_{S(2)} = \int_a^b e(x)\,dx. \tag{9.44}$$

It is clear that Simpson's rule is exact for $f(x)$ a quadratic function. Less clear is the fact that Simpson's rule is exact if $f(x)$ is a cubic polynomial. To demonstrate the truth of this claim, we need an error result from Chapter 6. We assume that $f^{(k)}(x)$ exists and is continuous for all $k = 0, 1, 2, 3, 4$ for all $x \in [a, b]$. From

Eq. (6.14) the error involved in interpolating $f(x)$ with a quadratic polynomial is given by

$$e(x) = \frac{1}{3!} f^{(3)}(\xi(x))(x-a)(x-b)(x-c) \tag{9.45}$$

for some $\xi = \xi(x) \in [a, b]$. Hence

$$E_{S(2)} = \int_a^b e(x)\,dx = \frac{1}{3!} \int_a^b f^{(3)}(\xi(x))(x-a)(x-b)(x-c)\,dx. \tag{9.46}$$

Unfortunately, polynomial $(x-a)(x-b)(x-c)$ changes sign on the interval $[a, b]$, and so we are not able to apply the second mean-value theorem for integrals as we did in Section 9.2. This is a major reason why the analysis of Simpson's rule is harder than the analysis of the trapezoidal rule. However, at this point we may still consider (9.46) for the case where $f(x)$ is a cubic polynomial. In this case we must have $f^{(3)}(x) = K$ (some constant). Consequently, from (9.46)

$$E_{S(2)} = \frac{K}{3!} \int_a^b (x-a)(x-b)(x-c)\,dx,$$

but if $z = x - c$, then, since $c = \frac{1}{2}(a+b)$, we must have

$$\begin{aligned}
E_{S(2)} &= \frac{K}{3!} \int_{-\frac{1}{2}(b-a)}^{\frac{1}{2}(b-a)} z \left[z + \frac{b-a}{2} \right] \left[z - \frac{b-a}{2} \right] dz \\
&= \frac{K}{3!} \int_{-\frac{1}{2}(b-a)}^{\frac{1}{2}(b-a)} z \left[z^2 - \left(\frac{b-a}{2} \right)^2 \right] dz.
\end{aligned} \tag{9.47}$$

The integrand is an odd function of z, and the integration limits are symmetric about the point $z = 0$. Immediately we conclude that $E_{S(2)} = 0$ in this particular case. Thus, we conclude that Simpson's rule gives the exact result when $f(x)$ is a cubic polynomial.

Hermite interpolation (considered in a general way in Section 6.4) is polynomial interpolation where not only does the interpolating polynomial match $f(x)$ at the sample points x_k but the first derivative of $f(x)$ is matched as well. It is useful to interpolate $f(x)$ with a cubic polynomial that we will denote by $r(x)$ at the points $(a, f(a))$, $(b, f(b))$, and $(c, f(c))$, and also such that $r^{(1)}(c) = f^{(1)}(c)$. A cubic polynomial is specified by four coefficients, so these constraints uniquely determine $r(x)$. In fact

$$r(x) = p(x) + \alpha(x-a)(x-b)(x-c) \tag{9.48a}$$

for which

$$\alpha = \frac{4[p^{(1)}(c) - f^{(1)}(c)]}{(b-a)^2}. \tag{9.48b}$$

Analogously to (9.19), we may define

$$g(t) = f(t) - r(t) - [f(x) - r(x)]\frac{(t-a)(t-c)^2(t-b)}{(x-a)(x-c)^2(x-b)}, \quad a \le t \le b. \tag{9.49}$$

It happens that $g^{(k)}(t)$ for $k = 0, 1, 2, 3, 4$ all exist and are continuous at all $x \in [a, b]$. Additionally, $g(a) = g(b) = g(c) = g^{(1)}(c) = g(x) = 0$. The vanishing of $g(t)$ at four distinct points on $[a, b]$, and $g^{(1)}(c) = 0$ guarantees that $g^{(4)}(\xi) = 0$ for some $\xi \in [a, b]$ by the repeated application of Rolle's theorem. Consequently, using (9.49), we obtain

$$g^{(4)}(\xi) = f^{(4)}(\xi) - r^{(4)}(\xi) - [f(x) - r(x)]\frac{4!}{(x-a)(x-c)^2(x-b)} = 0. \tag{9.50}$$

Since $r(x)$ is cubic, $r^{(4)}(\xi) = 0$, and so (9.50) can be used to say that

$$f(x) = r(x) + \frac{1}{4!}f^{(4)}(\xi(x))(x-a)(x-c)^2(x-b) \tag{9.51}$$

for $x \in (a, b)$. This is valid at the endpoints of $[a, b]$, so finally

$$e(x) = f(x) - r(x) = \frac{1}{4!}f^{(4)}(\xi(x))(x-a)(x-c)^2(x-b) \tag{9.52}$$

for $x \in [a, b]$, and $\xi(x) \in [a, b]$. Immediately, we see that

$$E_{S(2)} = \frac{1}{4!}\int_a^b f^{(4)}(\xi(x))(x-a)(x-c)^2(x-b)\,dx. \tag{9.53}$$

The polynomial in the integrand of (9.53) does not change sign on $[a, b]$. Thus, the second mean-value theorem for integrals is applicable. Hence, for some $\xi \in (a, b)$, we have

$$E_{S(2)} = \frac{f^{(4)}(\xi)}{4!}\int_a^b (x-a)(x-c)^2(x-b)\,dx, \tag{9.54}$$

which reduces to

$$E_{S(2)} = -\frac{h^5}{90}f^{(4)}(\xi) \tag{9.55}$$

again for some $\xi \in (a, b)$, where $h = (b-a)/2$.

We need an expression for $E_{S(n)}$, that is, an error expression for the *composite Simpson rule*. We will assume again that $h = (b-a)/n$, where n is an even number. Consequently

$$E_{S(n)} = \int_a^b f(x)\,dx - S(n) = \sum_{k=1}^{n/2} E_k, \tag{9.56}$$

where E_k is the error committed in the approximation for the kth subinterval $[x_{2(k-1)}, x_{2k}]$. Thus, for $\xi_k \in [x_{2(k-1)}, x_{2k}]$, with $k = 1, 2, \ldots, n/2$, we have

$$E_k = -\frac{h^5}{90} f^{(4)}(\xi_k) = -\frac{h^4}{90} \frac{b-a}{n} f^{(4)}(\xi_k). \qquad (9.57)$$

Therefore

$$E_{S(n)} = -\frac{h^4}{180} \frac{b-a}{n/2} \sum_{k=1}^{n/2} f^{(4)}(\xi_k). \qquad (9.58)$$

Applying the intermediate-value theorem to the average $\frac{1}{n/2} \sum_{k=1}^{n/2} f^{(4)}(\xi_k)$ confirms that there is a $\xi \in (a, b)$ such that

$$\frac{1}{n/2} \sum_{k=1}^{n/2} f^{(4)}(\xi_k) = f^{(4)}(\xi),$$

so therefore the truncation error expression for the composite Simpson rule becomes

$$E_{S(n)} = -\frac{h^4}{180}(b - a) f^{(4)}(\xi) \qquad (9.59)$$

for some $\xi \in (a, b)$. For convenience, we repeat the truncation error expression for the composite trapezoidal rule:

$$E_{T(n)} = -\frac{h^2}{12}(b - a) f^{(2)}(\xi). \qquad (9.60)$$

It is not really obvious which rule, trapezoidal or Simpson's, is better in general. For a particular interval $[a, b]$ and n, the two expressions depend on different derivatives of $f(x)$. It is possible that Simpson's rule may not be an improvement on the trapezoidal rule in particular cases for this reason. More specifically, a function that is not too smooth can be expected to have "big" higher derivatives. Simpson's rule has a truncation error dependent on the fourth derivative, while the trapezoidal rule has an error that depends only on the second derivative. Thus, a nonsmooth function might be better approximated by the trapezoidal rule than by Simpson's. In fact, Davis and Rabinowitz [3, p. 26] state that

> The more "refined" a rule of approximate integration is, the more certain we must be that it has been applied to a function which is sufficiently smooth. There may be little or no advantage in using a 'better' rule for a function that is not smooth.

We repeat a famous example from Ref. 3 (originally due to Salzer and Levine).

Example 9.2   The following series defines a function due to Weierstrass that happens to be continuous but is, surprisingly, *not differentiable anywhere*[2]

$$W(x) = \sum_{n=1}^{\infty} \frac{1}{2^n} \cos(7^n \pi x).$$ (9.61a)

If we assume that we may integrate this expression term by term, then

$$I(y) = \int_0^y W(x)\, dx = \frac{1}{\pi} \sum_{n=1}^{\infty} \frac{1}{7^n 2^n} \sin(7^n \pi y).$$ (9.61b)

Of course, $I = \int_a^b W(x)\, dx = I(b) - I(a)$. The series (9.61b) gives the "exact" value for $I(y)$ and so may be compared to estimates produced by the trapezoidal and Simpson rules. Assuming that $n = 100$, the following table of values is obtained [MATLAB implementation of (9.61) and the numerical integration rules]:

Interval $[a, b]$	Exact Value I	Trapezoidal $T(n)$	Error $I - T(n)$	Simpson's $S(n)$	Error $I - S(n)$
[0,.1]	0.01899291	0.01898760	0.00000531	0.01901426	−0.00002135
[.1, .2]	−0.04145650	−0.04143815	−0.00001834	−0.04146554	0.00000904
[.2, .3]	0.03084617	0.03084429	0.00000188	0.03086261	−0.00001645
[.3, .4]	0.00337701	0.00342534	−0.00004833	0.00341899	−0.00004198
[.4, .5]	−0.03298025	−0.03300674	0.00002649	−0.03303611	0.00005586

We see that the errors involved in both the trapezoidal and Simpson rules do not differ greatly from each other. So Simpson's rule has no advantage here.

It is commonplace for integration problems to involve integrals that possess oscillatory integrands. For example, the *Fourier transform* of $x(t)$ is defined to be

$$X(\omega) = \int_{-\infty}^{\infty} x(t) e^{-j\omega t}\, dt = \int_{-\infty}^{\infty} x(t) \cos(\omega t)\, dt - j \int_{-\infty}^{\infty} x(t) \sin(\omega t)\, dt.$$ (9.62)

You will likely see much of this integral in other books and associated courses (e.g., signals and systems). Also, determination of the Fourier series coefficients required computing [recall Eq. (1.45)]

$$f_n = \frac{1}{2\pi} \int_0^{2\pi} f(x) e^{-jnx}\, dx.$$ (9.63)

---

[2]Proof that $W(x)$ is continuous but not differentiable is quite difficult. There are many different Weierstrass functions possessing this property of continuity without differentiability. Another example complete with a proof appears on pp. 38–41 of Körner [4].

An integrand is said to be *rapidly oscillatory* if there are numerous (i.e., of the order of ≥ 10) local maxima and minima over the range of integration (i.e., here assumed to be the finite interval $[a, b]$). Some care is often required to compute these properly with the aid of the integration rules we have considered so far. However, we will consider only a simple idea called *integration between the zeros*. Davis and Rabinowitz have given a more detailed consideration of how to handle oscillatory integrands [3, pp. 53–68].

Relevant to the computation of (9.63) is, for example, the integral

$$I = \int_0^{2\pi} f(x) \sin(nx) \, dx. \tag{9.64}$$

It may be that $f(x)$ oscillates very little or not at all on $[0, 2\pi]$. We may therefore replace (9.64) with

$$I = \sum_{k=0}^{2n-1} \int_{k\pi/n}^{(k+1)\pi/n} f(x) \sin(nx) \, dx. \tag{9.65}$$

The endpoints of the "subintegrals"

$$I_k = \int_{k\pi/n}^{(k+1)\pi/n} f(x) \sin(nx) \, dx \tag{9.66}$$

in (9.65) are the zeros of $\sin(nx)$ on $[0, 2\pi]$. Thus, we are truly proposing integration between the zeros. At this point it is easiest to approximate (9.66) with either the trapezoidal or Simpson's rules for all k. Since the endpoints of the integrands in (9.66) are zero-valued, we can expect some savings in computation as a result because some of the terms in the rules (9.14) and (9.41) will be zero-valued.

9.4 GAUSSIAN QUADRATURE

Gaussian quadrature is a numerical integration method that uses a higher order of interpolation than do either the trapezoidal or Simpson rules. A detailed derivation of the method is rather involved as it relies on Hermite interpolation, orthogonal polynomial theory, and some aspects of the theory rely on issues relating to linear system solution. Thus, only an outline presentation is given here. However, a complete description may be found in Hildebrand [5, pp. 382–400]. Additional information is presented in Davis and Rabinowitz [3].

It helps to recall ideas from Chapter 6 here. If we know $f(x)$ for $x = x_j$, where $j = 0, 1, \ldots, n-1, n$, then the Lagrange interpolating polynomial is (with $p(x_j) = f(x_j)$)

$$p(x) = \sum_{j=0}^n f(x_j) L_j(x), \tag{9.67}$$

where

$$L_j(x) = \prod_{\substack{i=0 \\ i \neq j}}^{n} \frac{x - x_i}{x_j - x_i}. \tag{9.68}$$

Recalling (6.45b), we may similarly define

$$\pi(x) = \prod_{i=0}^{n} (x - x_i). \tag{9.69}$$

Consequently

$$\pi^{(1)}(x) = \sum_{j=0}^{n} \left\{ \prod_{\substack{k=0 \\ k \neq j}}^{n} (x - x_k) \right\} \tag{9.70}$$

so that

$$\pi^{(1)}(x_i) = \prod_{\substack{k=0 \\ k \neq i}}^{n} (x_i - x_k), \tag{9.71}$$

which allows us to rewrite (9.68) as

$$L_j(x) = \frac{\pi(x)}{\pi^{(1)}(x_j)(x - x_j)} \tag{9.72}$$

for $j = 0, 1, \ldots, n$. From (6.14)

$$e(x) = f(x) - p(x) = \frac{1}{(n+1)!} f^{(n+1)}(\xi)\pi(x) \tag{9.73}$$

for some $\xi \in [a, b]$, and $\xi = \xi(x)$.

We now summarize Hermite interpolation (recall Section 6.4 for more detail). Suppose that we have knowledge of both $f(x)$ and $f^{(1)}(x)$ at $x = x_j$ (again $j = 0, 1, \ldots, n$). We may interpolate $f(x)$ using a polynomial of degree $2n + 1$ since we must match the polynomial to both $f(x)$ and $f^{(1)}(x)$ at $x = x_j$. Thus, we need the polynomial

$$p(x) = \sum_{k=0}^{n} h_k(x) f(x_k) + \sum_{k=0}^{n} \hat{h}_k(x) f^{(1)}(x_k), \tag{9.74}$$

where $h_k(x)$ and $\hat{h}_k(x)$ are both polynomials of degree $2n + 1$ that we must determine according to the constraints of our interpolation problem.

If

$$h_i(x_j) = \delta_{i-j}, \hat{h}_i(x_j) = 0, \tag{9.75a}$$

then $p(x_j) = f(x_j)$, and if we have

$$h_i^{(1)}(x_j) = 0, \hat{h}_i^{(1)}(x_j) = \delta_{i-j}, \tag{9.75b}$$

then $p^{(1)}(x_j) = f^{(1)}(x_j)$ for all $j = 0, 1, \ldots, n$. Using (9.75), it is possible to arrive at the conclusion that

$$h_i(x) = [1 - 2L_i^{(1)}(x_i)(x - x_i)][L_i(x)]^2 \tag{9.76a}$$

and

$$\hat{h}_i(x) = (x - x_i)[L_i(x)]^2. \tag{9.76b}$$

Equation (9.74) along with (9.76) is *Hermite's interpolating formula*. [Both parts of Eq. (9.76) are derived on pp. 383–384 of Ref. 5, as well as in Theorem 6.1 of Chapter 6 (below).] It is further possible to prove that for $p(x)$ in (9.74) we have the error function

$$e(x) = f(x) - p(x) = \frac{1}{(2n+2)!} f^{(2n+2)}(\xi)[\pi(x)]^2, \tag{9.77}$$

where $\xi \in [a, b]$ and $\xi = \xi(x)$.

From (9.77) and (9.74), we obtain

$$f(x) = \sum_{k=0}^{n} h_k(x) f(x_k) + \sum_{k=0}^{n} \hat{h}_k(x) f^{(1)}(x_k) + \frac{1}{(2n+2)!} f^{(2n+2)}(\xi(x))[\pi(x)]^2. \tag{9.78}$$

Suppose that $w(x) \geq 0$ for $x \in [a, b]$. Function $w(x)$ is intended to be a weighting function such as seen in Chapter 5. Consequently, from (9.78)

$$\int_a^b w(x) f(x)\, dx = \sum_{k=0}^{n} \left[\int_a^b w(x) h_k(x)\, dx \right] f(x_k)$$

$$+ \sum_{k=0}^{n} \left[\int_a^b w(x) \hat{h}_k(x)\, dx \right] f^{(1)}(x_k)$$

$$+ \underbrace{\frac{1}{(2n+2)!} \int_a^b f^{(2n+2)}(\xi(x)) w(x)[\pi(x)]^2\, dx}_{=E}, \tag{9.79}$$

where $a < \xi(x) < b$, if $a < x_k < b$. This can be rewritten as

$$\int_a^b w(x) f(x)\, dx = \sum_{k=0}^{n} H_k f(x_k) + \sum_{k=0}^{n} \hat{H}_k f^{(1)}(x_k) + E, \tag{9.80}$$

where

$$H_k = \int_a^b w(x)h_k(x)\,dx = \int_a^b w(x)[1 - 2L_k^{(1)}(x_k)(x - x_k)][L_k(x)]^2\,dx \quad (9.81a)$$

and

$$\hat{H}_k = \int_a^b w(x)\hat{h}_k(x)\,dx = \int_a^b w(x)(x - x_k)[L_k(x)]^2\,dx. \quad (9.81b)$$

If we neglect the term E in (9.80), then the resulting approximation to $\int_a^b w(x)$ $f(x)\,dx$ is called the *Hermite quadrature formula*. Since we are assuming that $w(x) \geq 0$, the second mean-value theorem for integrals allows us to claim that

$$E = \frac{1}{(2n + 2)!} f^{(2n+2)}(\xi) \int_a^b w(x)[\pi(x)]^2\,dx \quad (9.82)$$

for some $\xi \in [a, b]$.

Now, recalling (9.72), we see that (9.81b) can be rewritten as

$$\begin{aligned}
\hat{H}_k &= \int_a^b w(x)(x - x_k)\left[\frac{\pi(x)}{\pi^{(1)}(x_k)(x - x_k)}\right]^2\,dx \\
&= \frac{1}{\pi^{(1)}(x_k)} \int_a^b w(x)\pi(x)\left[\frac{\pi(x)}{\pi^{(1)}(x_k)(x - x_k)}\right]\,dx \\
&= \frac{1}{\pi^{(1)}(x_k)} \int_a^b w(x)\pi(x)L_k(x)\,dx. \quad (9.83)
\end{aligned}$$

We recall from Chapter 5 that an inner product on $L^2[a, b]$ is (f, g are real-valued)

$$\langle f, g \rangle = \int_a^b w(x)f(x)g(x)\,dx. \quad (9.84)$$

Thus, $\hat{H}_k = 0$ for $k = 0, 1, \ldots, n$ if $\pi(x)$ is orthogonal to $L_k(x)$ over $[a, b]$ with respect to the weighting function $w(x)$. Since $\deg(L_k(x)) = n$ (all k), this will be the case if $\pi(x)$ is orthogonal to *all polynomials* of degree $\leq n$ over $[a, b]$ with respect to the weighting function $w(x)$. Note that $\deg(\pi(x)) = n + 1$ [recall (9.69)]. In fact, the polynomial $\pi(x)$ of degree $n + 1$ is orthogonal to all polynomials of degree $\leq n$ over $[a, b]$ with respect to $w(x)$, the Hermite quadrature formula reduces to the simpler form

$$\int_a^b w(x)f(x)\,dx = \sum_{k=0}^n H_k f(x_k) + E, \quad (9.85)$$

where

$$E = \frac{1}{(2n + 2)!} f^{(2n+2)}(\xi) \int_a^b w(x)[\pi(x)]^2\,dx, \quad (9.86)$$

and where $x_0, x_1, \ldots, x_n$ are the zeros of $\pi(x)$ (such that $a < x_k < b$). A formula of this type is called a *Gaussian quadrature formula*. The weights H_k are sometimes called *Christoffel numbers*. We see that this numerical integration methodology requires us to possess samples of $f(x)$ at the zeros of $\pi(x)$. Variations on this theory can be used to remove this restriction [6], but we do not consider this matter in this book. However, if $f(x)$ is known at all $x \in [a, b]$, then this is not a serious restriction.

In any case, to apply the approximation

$$\int_a^b w(x) f(x) \, dx \approx \sum_{k=0}^n H_k f(x_k), \tag{9.87}$$

it is clear that we need a method to determine the Christoffel numbers H_k. It is possible to do this using the *Christoffel–Darboux formula* [Eq. (5.11); see also Theorem 5.2]. From this it can be shown that

$$H_k = -\frac{\phi_{n+2,n+2}}{\phi_{n+1,n+1}\phi_{n+1}^{(1)}(x_k)\phi_{n+2}(x_k)}, \tag{9.88}$$

where polynomial $\phi_r(x)$ is obtained, for instance, from (5.5) and where

$$\phi_{n+1}(x) = \phi_{n+1,n+1}\pi(x). \tag{9.89}$$

Thus, we identify the zeros of $\pi(x)$ with the zeros of orthogonal polynomial $\phi_{n+1}(x)$ [recalling that $\langle \phi_i, \phi_j \rangle = \delta_{i-j}$ with respect to inner product (9.84)].

Since there are an infinite number of choices for orthogonal polynomials $\phi_k(x)$ and there is a theory for creating them (Chapter 5), it is possible to choose $\pi(x)$ in an infinite number of ways. We have implicitly assumed that $[a, b]$ is a finite length interval, but this assumption is actually entirely unnecessary. Infinite or semiinfinite intervals of integration are permitted. Thus, for example, $\pi(x)$ may be associated with the Hermite polynomials of Section 5.4, as well as with the Chebyshev or Legendre polynomials.

Let us consider as an example the case of Chebyshev polynomials of the first kind (first seen in Section 5.3). In this case

$$w(x) = \frac{1}{\sqrt{1-x^2}}$$

with $[a, b] = [-1, 1]$, and we will obtain the *Chebyshev–Gauss quadrature rule*. The Chebyshev polynomials of the first kind are $T_k(x) = \cos[k \cos^{-1} x]$, so, via (5.55) for $k > 0$, we have

$$\phi_k(x) = \sqrt{\frac{2}{\pi}} T_k(x). \tag{9.90}$$

From the recursion for Chebyshev polynomials of the first kind [Eq. (5.57)], we have $(k > 0)$

$$T_{k,k} = 2^{k-1} \qquad (9.91)$$

$\left(T_k(x) = \sum_{j=0}^{k} T_{k,j} x^j \right)$. Thus

$$\phi_{k,k} = \sqrt{\frac{2}{\pi}} 2^{k-1}. \qquad (9.92)$$

We have that $T_k(x) = 0$ for

$$x = x_i = \cos\left[\frac{2i+1}{2k} \pi \right] \qquad (9.93)$$

$(i = 0, 1, \ldots, k-1)$. Additionally

$$T_k^{(1)}(x) = k \sin[k \cos^{-1} x] \frac{1}{\sqrt{1 - x^2}}, \qquad (9.94)$$

so, if, for convenience, we define $\alpha_i = \frac{2i+1}{2k} \pi$, then $x_i = \cos \alpha_i$, and therefore

$$T_k^{(1)}(x_i) = k \frac{\sin(k\alpha_i)}{\sin \alpha_i} = k \frac{\sin\left[\frac{2i+1}{2} \pi \right]}{\sin \alpha_i} = k \frac{(-1)^i}{\sin \alpha_i}. \qquad (9.95)$$

Also

$$T_{k+1}(x_i) = \cos\left[(k+1)\left(\frac{2i+1}{2k} \pi \right) \right] = \cos\left[\frac{2i+1}{2} \pi + \frac{2i+1}{2k} \pi \right]$$

$$= \cos\left[\frac{2i+1}{2} \pi \right] \cos\left[\frac{2i+1}{2k} \pi \right] - \sin\left[\frac{2i+1}{2} \pi \right] \sin\left[\frac{2i+1}{2k} \pi \right]$$

$$= -\sin\left[\frac{2i+1}{2} \pi \right] \sin \alpha_i = (-1)^{i+1} \sin \alpha_i. \qquad (9.96)$$

Therefore, (9.88) becomes

$$H_k = -\frac{\sqrt{\frac{2}{\pi}} 2^{n+1}}{\left(\sqrt{\frac{2}{\pi}} 2^n \right) \left(\sqrt{\frac{2}{\pi}} (n+1) \frac{(-1)^k}{\sin \alpha_k} \right) \left(\sqrt{\frac{2}{\pi}} (-1)^{k+1} \sin \alpha_k \right)} = \frac{\pi}{n+1}. \qquad (9.97)$$

Thus, the weights (Christoffel numbers) are all the same in this particular case. So, (9.87) is now

$$\int_{-1}^{1} \frac{f(x)}{\sqrt{1 - x^2}} \, dx \approx \frac{\pi}{n+1} \sum_{k=0}^{n} f\left[\cos\left(\frac{2k+1}{2n+2} \pi \right) \right] = C(n). \qquad (9.98)$$

The error expression E in (9.86) can also be reduced accordingly, but we will omit this here (again, see Hildebrand [5]).

A simple example of the application of (9.98) is as follows.

Example 9.3   Suppose that $f(x) = \sqrt{1 - x^2}$, in which case

$$\int_{-1}^{1} \frac{f(x)}{\sqrt{1 - x^2}} \, dx = \int_{-1}^{1} dx = 2.$$

For this case (9.98) becomes

$$\int_{-1}^{1} dx \approx C(n) = \frac{\pi}{n + 1} \sum_{k=0}^{n} \sin\left[\frac{2k + 1}{2n + 2}\pi\right].$$

For various n, we have the following table of values:

n	$C(n)$
1	2.2214
2	2.0944
5	2.0230
10	2.0068
20	2.0019
100	2.0001

Finally, we remark that the error expression in (9.82) suggests that the method of this section is worth applying only if $f(x)$ is sufficiently smooth. This is consistent with comments made in Section 9.3 regarding how to choose between the trapezoidal and Simpson's rules. In the next example $f(x) = e^{-x}$, which is a very smooth function.

Example 9.4   Here we will consider

$$I = \int_{-1}^{1} e^{-x} \, dx = e - \frac{1}{e} = 2.350402387$$

and compare the approximation to I obtained by applying Simpson's rule and *Legendre–Gauss quadrature*. We will assume $n = 2$ in both cases.

Let us first consider application of Simpson's rule. Since $x_0 = a = -1$, $x_1 = 0$, $x_1 = b = 1$ ($h = (b - a)/n = (1 - (-1))/2 = 1$), we have via (9.41)

$$S(2) = \frac{1}{3}[e^{+1} + 4e^{0} + e^{-1}] = 2.362053757$$

for which the error is

$$E_{S(2)} = I - S(2) = -0.011651.$$

Now let us consider the Legendre–Gauss quadrature for our problem. We recall that for Legendre polynomials the weight function is $w(x) = 1$ for all $x \in [-1, 1]$ (Section 5.5). From Section 5.6 we have

$$\phi_3(x) = \frac{1}{\|P_3\|} P_3(x) = \sqrt{\frac{7}{2}} \frac{1}{2} [5x^3 - 3x],$$

$$\phi_4(x) = \frac{1}{\|P_4\|} P_4(x) = \sqrt{\frac{9}{2}} \frac{1}{8} [35x^4 - 30x^2 + 3].$$

Consequently, $\phi_{3,3} = \frac{5}{2}\sqrt{\frac{7}{2}}$, and the zeros of $\phi_3(x)$ are at $x = 0, \pm\sqrt{\frac{3}{5}}$, so now our sample points (grid points, mesh points) are

$$x_0 = -\sqrt{\frac{3}{5}}, x_1 = 0, x_2 = +\sqrt{\frac{3}{5}}.$$

Hence, since $\phi_3^{(1)}(x) = \sqrt{\frac{7}{2}} \frac{1}{2} [15x^2 - 3]$, we have

$$\phi_3^{(1)}(x_0) = 3\sqrt{\frac{7}{2}}, \phi_3^{(1)}(x_1) = -\frac{3}{2}\sqrt{\frac{7}{2}}, \phi_3^{(1)}(x_2) = 3\sqrt{\frac{7}{2}}.$$

Also, $\phi_{4,4} = \frac{35}{8}\sqrt{\frac{9}{2}}$, and

$$\phi_4(x_0) = -\frac{3}{10}\sqrt{\frac{9}{2}}, \phi_4(x_1) = \frac{3}{8}\sqrt{\frac{9}{2}}, \phi_4(x_2) = -\frac{3}{10}\sqrt{\frac{9}{2}}.$$

Therefore, from (9.88) the Christoffel numbers are

$$H_0 = \frac{5}{9}, H_1 = \frac{8}{9}, H_2 = \frac{5}{9}.$$

From (9.87) the resulting quadrature is

$$\int_{-1}^{1} e^{-x} \, dx \approx H_0 e^{-x_0} + H_1 e^{-x_1} + H_2 e^{-x_2} = L(2)$$

with

$$L(2) = 2.350336929,$$

and the corresponding error is

$$E_{L(2)} = I - L(2) = 6.5458 \times 10^{-5}.$$

Clearly, $|E_{L(2)}| \ll |E_{S(2)}|$. Thus, the Legendre-Gauss quadrature is much more accurate than Simpson's rule. Considering how small n is here, the accuracy of the Legendre–Gauss quadrature is remarkably high.

9.5   ROMBERG INTEGRATION

Romberg integration is a recursive procedure that seeks to improve on the trapezoidal and Simpson rules. But before we consider this numerical integration methodology, we will look at some more basic ideas.

Suppose that $I = \int_a^b f(x)\, dx$ and $I(n)$ is a quadrature that approximates I. For us, $I(n)$ will be either the trapezoidal rule $T(n)$ from (9.14), or else it will be Simpson's rule $S(n)$ from (9.41). It could also be the corrected trapezoidal rule $T_C(n)$, which is considered below [see either (9.104), or (9.107)].

It is possible to improve on the "basic" trapezoidal rule from Section 9.2. Begin by recalling (9.27)

$$E_{T(n)} = -\frac{1}{12}h^3 \sum_{k=1}^{n} f^{(2)}(\xi_k), \tag{9.99}$$

where $h = (b-a)/n$, $x_0 = a$, $x_n = b$, and $\xi_k \in [x_{k-1}, x_k]$. Of course, $x_k - x_{k-1} = h$ (uniform sampling grid). Assuming (as usual) that $f^{(k)}(x)$ is Riemann integrable for all $k \geq 0$, then

$$\lim_{n \to \infty} \sum_{k=1}^{n} h f^{(2)}(\xi_k) = f^{(1)}(b) - f^{(1)}(a) = \int_a^b f^{(2)}(x)\, dx. \tag{9.100}$$

Thus, we have the approximation

$$\sum_{k=1}^{n} h f^{(2)}(\xi_k) \approx f^{(1)}(b) - f^{(1)}(a). \tag{9.101}$$

Consequently,

$$E_{T(n)} = -\frac{1}{12}h^2 \sum_{k=1}^{n} h f^{(2)}(\xi_k) \approx -\frac{1}{12}h^2[f^{(1)}(b) - f^{(1)}(a)] \tag{9.102}$$

or

$$I - T(n) \approx -\frac{1}{12}h^2[f^{(1)}(b) - f^{(1)}(a)]. \tag{9.103}$$

This immediately suggests that we can improve on the trapezoidal rule by replacing $T(n)$ with the new approximation

$$T_C(n) = T(n) - \frac{1}{12}h^2[f^{(1)}(b) - f^{(1)}(a)], \qquad (9.104)$$

where $T_C(n)$ denotes the *corrected trapezoidal rule* approximation to I. Clearly, once we have $T(n)$, rather little extra effort is needed to obtain $T_C(n)$.

In fact, we do not necessarily need to know $f^{(1)}(x)$ exactly anywhere, much less at the points $x = a$ or $x = b$. In the next section we will argue that either

$$f^{(1)}(x) = \frac{1}{2h}[-3f(x) + 4f(x + h) - f(x + 2h)] + \frac{1}{3}h^2 f^{(3)}(\xi) \qquad (9.105a)$$

or that

$$f^{(1)}(x) = \frac{1}{2h}[3f(x) - 4f(x - h) + f(x - 2h)] + \frac{1}{3}h^2 f^{(3)}(\xi). \qquad (9.105b)$$

In (9.105a) $\xi \in [x, x + 2h]$, while in (9.105b) $\xi \in [x - 2h, x]$. Consequently, with $x_0 = a$ for $\xi_0 \in [a, a + 2h]$, we have [via (9.105a)]

$$f^{(1)}(x_0) = f^{(1)}(a) = \frac{1}{2h}[-3f(x_0) + 4f(x_1) - f(x_2)] + \frac{1}{3}h^2 f^{(3)}(\xi_0), \qquad (9.106a)$$

and with $x_n = b$ for $\xi_n \in [b - 2h, b]$, we have [via (9.105b)]

$$f^{(1)}(x_n) = f^{(1)}(b) = \frac{1}{2h}[3f(x_n) - 4f(x_{n-1}) + f(x_{n-2})] + \frac{1}{3}h^2 f^{(3)}(\xi_n). \qquad (9.106b)$$

Thus, (9.104) becomes (*approximate corrected trapezoidal rule*)

$$T_C(n) = T(n) - \frac{h}{24}[3f(x_n) - 4f(x_{n-1}) + f(x_{n-2}) + 3f(x_0) - 4f(x_1) + f(x_2)]$$

$$- \frac{h^4}{36}[f^{(3)}(\xi_n) - f^{(3)}(\xi_0)]. \qquad (9.107)$$

Of course, in evaluating (9.107) we would exclude the terms involving $f^{(3)}(\xi_0)$, and $f^{(3)}(\xi_n)$.

As noted in Epperson [7] for the trapezoidal and Simpson rules

$$I - I(n) \propto \frac{1}{n^p}, \qquad (9.108)$$

where $p = 2$ for $I(n) = T(n)$ and $p = 4$ for $I(n) = S(n)$. In other words, for a given rule and a suitable constant C we must have $I - I(n) \approx Cn^{-p}$. Now observe that we may define the ratio

$$r_{4n} = \frac{I(n) - I(2n)}{I(2n) - I(4n)} \approx \frac{(I - Cn^{-p}) - (I - C(2n)^{-p})}{(I - C(2n)^{-p}) - (I - C(4n)^{-p})}$$

$$= \frac{(2n)^{-p} - n^{-p}}{(4n)^{-p} - (2n)^{-p}} = \frac{2^{-p} - 1}{4^{-p} - 2^{-p}} = 2^p. \tag{9.109}$$

Immediately we conclude that

$$p \approx \log_2 r_{4n} = \frac{\log_{10} r_{4n}}{\log_{10} 2}. \tag{9.110}$$

This is useful as a check on program implementation of our quadratures. If (9.110) is not approximately satisfied when we apply the trapezoidal or Simpson rules, then (1) the integrand $f(x)$ is not smooth enough for our theories to apply, (2) there is a "bug" in the program, or (3) the error may not be decreasing quickly with n because it is already tiny to begin with, as might happen when integrating an oscillatory function using Simpson's rule.

The following examples illustrate the previous principles.

Example 9.5   In this example we consider approximating

$$I = \int_0^{\pi/2} \sin x \, dx = 1$$

using $T(n)$ [via (9.14)], $T_C(n)$ [via (9.104)], and $S(n)$ [via (9.41)]. Parameter p in (9.110) is computed for each of these cases, and is displayed in the following table (where "NaN" means "not a number"):

n	$T(n)$	p for $T(n)$	$T_C(n)$	p for $T_C(n)$	$S(n)$	p for $S(n)$
2	0.94805945	NaN	0.99946364	NaN	1.00227988	NaN
4	0.98711580	NaN	0.99996685	NaN	1.00013458	NaN
8	0.99678517	2.0141	0.99999793	4.0169	1.00000830	4.0864
16	0.99919668	2.0035	0.99999987	4.0042	1.00000052	4.0210
32	0.99979919	2.0009	0.99999999	4.0010	1.00000003	4.0052
64	0.99994980	2.0002	1.00000000	4.0003	1.00000000	4.0013
128	0.99998745	2.0001	1.00000000	4.0001	1.00000000	4.0003
256	0.99999686	2.0000	1.00000000	4.0000	1.00000000	4.0001

We see that since $f(x) = \sin x$ is a rather smooth function we obtain the values for p that we expect to see.

Example 9.6   This example is in contrast with the previous one. Here we approximate

$$I = \int_0^1 x^{1/3} \, dx = \frac{3}{4}$$

using $T(n)$ [via (9.14)], $T_C(n)$ [via (9.107)], and $S(n)$ [via (9.41)]. Again, p [via (9.110)] is computed for each of these cases, and the results are tabulated as follows:

n	$T(n)$	p for $T(n)$	$T_C(n)$	p for $T_C(n)$	$S(n)$	p for $S(n)$
2	0.64685026	NaN	0.69580035	NaN	0.69580035	NaN
4	0.70805534	NaN	0.72437494	NaN	0.72845703	NaN
8	0.73309996	1.2892	0.73980487	0.8890	0.74144817	1.3298
16	0.74322952	1.3059	0.74595297	1.3275	0.74660604	1.3327
32	0.74729720	1.3163	0.74839388	1.3327	0.74865310	1.3332
64	0.74892341	1.3227	0.74936261	1.3333	0.74946548	1.3333
128	0.74957176	1.3267	0.74974705	1.3333	0.74978788	1.3333
256	0.74982980	1.3292	0.74989962	1.3333	0.74991582	1.3333

We observe that $f(x) = x^{1/3}$, but that $f^{(1)}(x) = \frac{1}{3}x^{-2/3}$, $f^{(2)}(x) = -\frac{2}{9}x^{-5/3}$, etc. Thus, the derivatives of $f(x)$ are unbounded at $x = 0$, and so $f(x)$ is not smooth on the interval of integration. This explains why we obtain $p \approx 1.3333$ in all cases.

As a further step toward Romberg integration, consider the following. Since $I - I(2n) \approx C(2n)^{-p} = C2^{-p}n^{-p} \approx 2^{-p}(I - I(n))$, we obtain the approximate equality

$$I \approx \frac{I(2n) - 2^{-p}I(n)}{1 - 2^{-p}},$$

or

$$I \approx \frac{2^p I(2n) - I(n)}{2^p - 1} = R(2n). \tag{9.111}$$

We call $R(2n)$ *Richardson's extrapolated value* (or *Richardson's extrapolation*), which is an improvement on $I(2n)$. The *estimated error* in the extrapolation is given by

$$E_{R(2n)} = I(2n) - R(2n) = \frac{I(n) - I(2n)}{2^p - 1}. \tag{9.112}$$

Of course, p in (9.111) and (9.112) must be the proper choice for the quadrature $I(n)$. We may "confirm" that (9.111) works for $T(n)$ (as an example) by considering (9.28)

$$E_{T(n)} = -\frac{h^2}{12}(b - a)f^{(2)}(\xi) \tag{9.113}$$

(for some $\xi \in [a, b]$). Clearly

$$E_{T(2n)} = -\frac{(h/2)^2}{12}(b - a)f^{(2)}(\xi) \approx \frac{1}{4}E_{T(n)},$$

or in other words ($E_{T(n)} = I - T(n)$)

$$I - T(2n) \approx \tfrac{1}{4}[I - T(n)],$$

and hence for $n \geq 1$

$$I \approx \frac{4T(2n) - T(n)}{3} = R_T(2n), \tag{9.114}$$

which is (9.111) for case $p = 2$. Equation (9.114) is called the *Romberg integration formula* for the trapezoidal rule. Of course, a similar expression may be obtained for Simpson's rule [with $p = 4$ in (9.111)]; that is, for n even

$$I \approx \frac{16S(2n) - S(n)}{15} = R_S(2n). \tag{9.115}$$

In fact, it can also be shown that $R_T(2n) = S(2n)$:

$$S(2n) = \frac{4T(2n) - T(n)}{3}. \tag{9.116}$$

(Perhaps this is most easily seen in the special case where $n = 1$.) In other words, the Romberg procedure applied to the trapezoidal rule yields the Simpson rule.

Now, Romberg integration is really the repeated application of the Richardson extrapolation idea to the composite trapezoidal rule. A simple way to visualize the process is with the *Romberg table* (*Romberg array*):

$T(1)$				
$T(2)$	$S(2)$			
$T(4)$	$S(4)$	$R_S(4)$		
$T(8)$	$S(8)$	$R_S(8)$	$\bullet$	
$T(16)$	$S(16)$	$R_S(16)$	$\bullet$	$\bullet$
$\vdots$	$\vdots$	$\vdots$	$\vdots$	$\vdots$

In its present form the table consists of only three columns, but the recursive process may be continued to produce a complete "triangular array."

The complete Romberg integration procedure is often fully justified and developed with respect to the following theorem.

Theorem 9.1: Euler–Maclaurin Formula   Let $f \in C^{2k+2}[a, b]$ for some $k \geq 0$, and let us approximate $I = \int_a^b f(x)\,dx$ by the composite trapezoidal rule of (9.14). Letting $h_n = (b - a)/n$ for $n \geq 1$, we have

$$T(n) = I + \sum_{i=1}^{k} \frac{B_{2i}}{(2i)!} h_n^{2i} [f^{(2i-1)}(b) - f^{(2i-1)}(a)]$$

$$+ \frac{B_{2k+2}}{(2k+2)!} h_n^{2k+2} (b-a) f^{(2k+2)}(\eta), \qquad (9.117)$$

where $\eta \in [a, b]$, and for $j \geq 1$

$$B_{2j} = (-1)^{j-1} \left[\sum_{n=1}^{\infty} \frac{2}{(2\pi n)^{2j}} \right] (2j)! \qquad (9.118)$$

are the *Bernoulli numbers*.

Proof This is Property 9.3 in Quarteroni et al. [8]. A proof appears in Ralston [9]. Alternative descriptions of the Bernoulli numbers appear in Gradshteyn and Ryzhik [10]. Although not apparent from (9.118), the Bernoulli numbers are all rational numbers.

We will present the complete Romberg integration process in a more straightforward manner. Begin by considering the following theorem.

Theorem 9.2: Recursive Trapezoidal Rule   Suppose that $h = (b - a)/(2n)$; then, for $n \geq 1$ ($x_0 = a, x_n = b$)

$$T(2n) = \frac{1}{2} T(n) + h \sum_{k=1}^{n} f(x_0 + (2k-1)h). \qquad (9.119)$$

The first column in the Romberg table is given by

$$T(2^n) = \frac{1}{2} T(2^{n-1}) + \frac{x_n - x_0}{2^n} \sum_{k=1}^{2^{n-1}} f\left(x_0 + \frac{(2k-1)(x_n - x_0)}{2^n} \right) \qquad (9.120)$$

for all $n \geq 1$.

Proof Omitted, but clearly (9.120) immediately follows from (9.119).

We let $R_n^{(k)}$ denote the row n, column k entry of the Romberg table, where $k = 0, 1, \ldots, N$ and $n = k, \ldots, N$ [i.e., we construct an $(N+1) \times (N+1)$ lower triangular array, as suggested earlier]. Table entries are "blank" for $n = 0, 1, \ldots, k-1$ in column k. The first column of the table is certainly

$$R_n^{(0)} = T(2^n) \qquad (9.121)$$

for $n = 0, 1, \ldots, N$. From Theorem 9.2 we must have

$$R_0^{(0)} = \frac{b-a}{2}[f(a) + f(b)], \tag{9.122a}$$

$$R_n^{(0)} = \frac{1}{2}R_{n-1}^{(0)} + \frac{b-a}{2^n}\sum_{k=1}^{2^{n-1}} f\left(a + \frac{(2k-1)(b-a)}{2^n}\right), \tag{9.122b}$$

for $n = 1, 2, \ldots, N$. Equations (9.122) are the algorithm for constructing the first column of the Romberg table. The second column is $R_n^{(1)}$, and these numbers are given by [via (9.114)]

$$R_n^{(1)} = \frac{4R_n^{(0)} - R_{n-1}^{(0)}}{4 - 1} \tag{9.123}$$

for $n = 1, 2, \ldots, N$. Similarly, the third column is $R_n^{(2)}$, and these numbers are given by [via (9.115)]

$$R_n^{(2)} = \frac{4^2 R_n^{(1)} - R_{n-1}^{(1)}}{4^2 - 1} \tag{9.124}$$

for $n = 2, 3, \ldots, N$. The pattern suggested by (9.123) and (9.124) generalizes according to

$$R_n^{(k)} = \frac{4^k R_n^{(k-1)} - R_{n-1}^{(k-1)}}{4^k - 1} \tag{9.125}$$

for $n = k, \ldots, N$, with $k = 1, 2, \ldots, N$. Assuming that $f(x)$ is sufficiently smooth, we can estimate the error using the Richardson extrapolation method in this manner:

$$E_n^{(k)} = \frac{R_{n-1}^{(k)} - R_n^{(k)}}{4^k - 1}. \tag{9.126}$$

This can be used to stop the recursive process of table construction when $E_n^{(k)}$ is small enough. Recall that every entry $R_n^{(k)}$ in the table is an estimate of $I = \int_a^b f(x)\,dx$. In some sense $R_N^{(N)}$ is the "final estimate," and will be the best one if $f(x)$ is smooth enough. Finally, the general appearance of the Romberg table is

$R_0^{(0)}$					
$R_1^{(0)}$	$R_1^{(1)}$				
$R_2^{(0)}$	$R_2^{(1)}$	$R_2^{(2)}$			
$\vdots$	$\vdots$	$\vdots$	$\ddots$		
$R_{N-1}^{(0)}$	$R_{N-1}^{(1)}$	$R_{N-1}^{(2)}$	$\cdots$	$R_{N-1}^{(N-1)}$	
$R_N^{(0)}$	$R_N^{(1)}$	$R_N^{(2)}$	$\cdots$	$R_N^{(N-1)}$	$R_N^{(N)}$

The Romberg integration procedure is efficient. Function evaluation is confined to the construction of the first column. The remaining columns are filled in with just a fixed (and small) number of arithmetic operations per entry as determined by (9.125).

Example 9.7   Begin by considering the Romberg approximation to

$$I = \int_0^1 e^x \, dx = 1.718281828.$$

The Romberg table for this is ($N = 3$)

1.85914091			
1.75393109	1.71886115		
1.72722190	1.71831884	1.71828269	
1.72051859	1.71828415	1.71828184	1.71828183

Table entry $R_3^{(3)} = 1.71828183$ is certainly the most accurate estimate of I. Now contrast this example with the next one.

The *zeroth-order modified Bessel function of the first kind* $I_0(y) \in \mathbf{R}(y \in \mathbf{R})$ is important in applied probability. For example, it appears in the problem of computing bit error probabilities in amplitude shift keying (ASK) digital data communications [11]. There is a series expansion expression for $I_0(y)$, but there is also an integral form, which is

$$I_0(y) = \frac{1}{2\pi} \int_0^{2\pi} e^{y \cos x} \, dx. \tag{9.127}$$

For $y = 1$, we have (according to MATLAB's besseli function; $I_0(y) =$ besseli(0,y))

$$I_0(1) = 1.2660658778.$$

The Romberg table of estimates for this integral is ($N = 4$)

2.71828183				
1.54308063	1.15134690			
1.27154032	1.18102688	1.18300554		
1.26606608	1.26424133	1.26978896	1.27116647	
1.26606588	1.26606581	1.26618744	1.26613028	1.26611053

Plainly, table entry $R_4^{(4)} = 1.26611053$ is not as accurate as $R_4^{(0)} = 1.26606588$. Apparently, the integrand of (9.127) is simply not smooth enough to benefit from the Romberg approach.

9.6   NUMERICAL DIFFERENTIATION

A simple theory of numerical approximation to the derivative can be obtained via
Taylor series expansions (Chapter 3). Recall that [via (3.71)]

$$f(x + h) = \sum_{k=0}^{n} \frac{h^k}{k!} f^{(k)}(x) + \frac{h^{n+1}}{(n+1)!} f^{(n+1)}(\xi) \tag{9.128}$$

for suitable $\xi \in [x, x + h]$. As usual, $0! = 1$, and $f(x) = f^{(0)}(x)$. Since from elementary calculus

$$f^{(1)}(x) = \lim_{h \to 0} \frac{f(x + h) - f(x)}{h} \Rightarrow f^{(1)}(x) \approx \frac{f(x + h) - f(x)}{h} \tag{9.129}$$

from (9.128), we have

$$f^{(1)}(x) = \frac{f(x + h) - f(x)}{h} - \frac{1}{2!} h f^{(2)}(\xi). \tag{9.130}$$

This was obtained simply be rearranging

$$f(x + h) = f(x) + h f^{(1)}(x) + \frac{1}{2!} h^2 f^{(2)}(\xi). \tag{9.131}$$

We may write

$$f^{(1)}(x) = \underbrace{\frac{f(x + h) - f(x)}{h}}_{= \tilde{f}_f^{(1)}(x)} + \underbrace{\left[-\frac{1}{2!} h f^{(2)}(\xi) \right]}_{= e_f(x)}. \tag{9.132}$$

Approximation $\tilde{f}_f^{(1)}(x)$ is called the *forward difference approximation* to $f^{(1)}(x)$,
and the error $e_f(x)$ is seen to be approximately proportional to h. Now consider
($\xi_1 \in [x, x + h]$)

$$f(x + h) = f(x) + h f^{(1)}(x) + \frac{1}{2!} h^2 f^{(2)}(x) + \frac{1}{3!} h^3 f^{(3)}(\xi_1), \tag{9.133}$$

and clearly ($\xi_2 \in [x - h, x]$)

$$f(x - h) = f(x) - h f^{(1)}(x) + \frac{1}{2!} h^2 f^{(2)}(x) - \frac{1}{3!} h^3 f^{(3)}(\xi_2). \tag{9.134}$$

From (9.134)

$$f^{(1)}(x) = \underbrace{\frac{f(x) - f(x - h)}{h}}_{= \tilde{f}_b^{(1)}(x)} + \underbrace{\left[\frac{1}{2!} h f^{(2)}(\xi) \right]}_{= e_b(x)}. \tag{9.135}$$

Here the approximation $\tilde{f}_b^{(1)}(x)$ is called the *backward difference approximation* to $f^{(1)}(x)$, and has error $e_b(x)$ that is also approximately proportional to h. However, an improvement is possible, and this is obtained by subtracting (9.134) from (9.133)

$$f(x+h) - f(x-h) = 2hf^{(1)}(x) + \frac{1}{3!}h^3[f^{(3)}(\xi_1) + f^{(3)}(\xi_2)],$$

or on rearranging this, we have

$$f^{(1)}(x) = \underbrace{\frac{f(x+h) - f(x-h)}{2h}}_{=\tilde{f}_c^{(1)}(x)} + \underbrace{\left[-\frac{h^2}{6} \frac{f^{(3)}(\xi_1) + f^{(3)}(\xi_2)}{2} \right]}_{=e_c(x)}. \qquad (9.136)$$

Recalling the derivation of (9.28), there is a $\xi \in [x - h, x + h]$ such that

$$f^{(3)}(\xi) = \frac{1}{2}[f^{(3)}(\xi_1) + f^{(3)}(\xi_2)] \qquad (9.137)$$

($\xi_1 \in [x, x + h], \xi_2 \in [x - h, h]$). Hence, (9.136) can be rewritten as (for some $\xi \in [x - h, x + h]$)

$$f^{(1)}(x) = \underbrace{\frac{f(x+h) - f(x-h)}{2h}}_{=\tilde{f}_c^{(1)}(x)} + \underbrace{\left[-\frac{h^2}{6} f^{(3)}(\xi) \right]}_{=e_c(x)}. \qquad (9.138)$$

Clearly, the error $e_c(x)$, which is the error of the *central difference approximation* $\tilde{f}_c^{(1)}(x)$ to $f^{(1)}(x)$, is proportional to h^2. Thus, if $f(x)$ is smooth enough, the central difference approximation is more accurate than the forward or backward difference approximations.

The errors $e_f(x)$, $e_b(x)$, and $e_c(x)$ are *truncation errors* in the approximations. Of course, when implementing any of these approximations on a computer there will be rounding errors, too. Each approximation can be expressed in the form

$$\tilde{f}^{(1)}(x) = \frac{1}{h} f^T y, \qquad (9.139)$$

where for the forward difference approximation

$$f = [f(x) \quad f(x+h)]^T, y = [-1 \quad 1]^T, \qquad (9.140a)$$

for the backward difference approximation

$$f = [f(x - h) \quad f(x)]^T, y = [-1 \quad 1]^T, \qquad (9.140b)$$

and for the central difference approximation

$$f = [f(x - h) \quad f(x + h)]^T, y = \tfrac{1}{2}[-1 \quad 1]^T. \tag{9.140c}$$

Thus, the approximations are all Euclidean inner products of samples of $f(x)$ with a vector of constants y, followed by division by h. This is structurally much the same kind of computation as numerical integration. In fact, an upper bound on the size of the error due to rounding is given by the following theorem.

Theorem 9.3: Since $fl[\tilde{f}^{(1)}(x)] = fl\left[\frac{fl[f^T y]}{h}\right]$, we have $(f, y \in \mathbf{R}^m)$

$$|fl[\tilde{f}^{(1)}(x)] - \tilde{f}^{(1)}(x)| \leq \frac{u}{h}[1.01m + 1]\|f\|_2\|y\|_2. \tag{9.141}$$

Proof Our analysis here is rather similar to Example 2.4 in Chapter 2. Thus, we exploit yet again the results from Chapter 2 on rounding errors in dot product computation.

Via (2.41)

$$fl[f^T y] = f^T y(1 + \epsilon_1)$$

for which

$$|\epsilon_1| \leq 1.01mu \frac{|f|^T |y|}{|f^T y|}.$$

In addition

$$fl[\tilde{f}^{(1)}(x)] = \frac{f^T y}{h}(1 + \epsilon_1)(1 + \epsilon),$$

where $|\epsilon| \leq u$. Thus, since

$$fl[\tilde{f}^{(1)}(x)] = \frac{f^T y}{h} + \frac{f^T y}{h}(\epsilon_1 + \epsilon + \epsilon_1 \epsilon),$$

we have

$$|fl[\tilde{f}^{(1)}(x)] - \tilde{f}^{(1)}(x)| \leq \frac{|f^T y|}{h}(|\epsilon_1| + |\epsilon| + |\epsilon_1||\epsilon|) = \frac{|f^T y|}{h}$$
$$\times \left(1.01mu \frac{|f|^T |y|}{|f^T y|} + u + 1.01mu^2 \frac{|f|^T |y|}{|f^T y|}\right),$$

and since $u^2 \ll u$, we may neglect the last term, yielding

$$|fl[\tilde{f}^{(1)}(x)] - \tilde{f}^{(1)}(x)| \leq \frac{1}{h}\left(1.01mu|f|^T |y| + u|f^T y|\right).$$

But $|f|^T |y| = <|f|, |y|> \leq ||f||_2 ||y||_2$, and $|f^T y| = |\langle f, y \rangle| \leq ||f||_2 ||y||_2$ via Theorem 1.1, so finally we have

$$|fl[\tilde{f}^{(1)}(x)] - \tilde{f}^{(1)}(x)| \leq \frac{u}{h}[1.01m + 1]||f||_2 ||y||_2,$$

which is the theorem statement.

The bound in (9.141) suggests that, since we treat u, m, $||f||_2$, and $||y||_2$ are fixed, as h becomes smaller, the rounding errors will grow in size. On the other hand, as h becomes smaller, the truncation errors diminish in size. Thus, much as with numerical integration (recall Fig. 9.3), there is a tradeoff between rounding and truncation errors leading in the present case to the existence of some optimal value for the choice of h. In most practical circumstances the truncation errors will dominate, however.

We recall that interpolation theory from Chapter 6 was useful in developing theories on numerical integration in earlier sections of the present chapter. We therefore reasonably expect that interpolation ideas from Chapter 6 ought to be helpful in developing approximations to the derivative.

Recall that $p_n(x) = \sum_{k=0}^{n} p_{n,k} x^k$ interpolates $f(x)$ for $x \in [a, b]$, i.e., $p_n(x_k) = f(x_k)$ with $x_0 = a$, $x_n = b$, and $x_k \in [a, b]$ for all k, but such that $x_k \neq x_j$ for $k \neq j$. If $f(x) \in C^{n+1}[a, b]$, then, from (6.14)

$$f(x) = p_n(x) + \frac{1}{(n+1)!} f^{(n+1)}(\xi) \prod_{i=0}^{n} (x - x_i) \tag{9.142}$$

($\xi = \xi(x) \in [a, b]$). For convenience we also defined $\pi(x) = \prod_{i=0}^{n}(x - x_i)$. Thus, from (9.142)

$$f^{(1)}(x) = p_n^{(1)}(x) + \frac{1}{(n+1)!} \left[\pi^{(1)}(x) f^{(n+1)}(\xi) + \pi(x) \frac{d}{dx} f^{(n+1)}(\xi) \right]. \tag{9.143}$$

Since $\xi = \xi(x)$ is a function of x that is seldom known, it is not at all clear what $df^{(n+1)}(\xi)/dx$ is in general, and to evaluate this also assumes that $d\xi(x)/dx$ exists.[3] We may sidestep this problem by evaluating (9.143) only for $x = x_k$, and since $\pi(x_k) = 0$ for all k, Eq. (9.143) reduces to

$$f^{(1)}(x_k) = p_n^{(1)}(x_k) + \frac{1}{(n+1)!} \pi^{(1)}(x_k) f^{(n+1)}(\xi(x_k)). \tag{9.144}$$

For simplicity we will now assume that $x_k = x_0 + hk$, $h = (b - a)/n$ (i.e., uniform sampling grid). We will also suppose that $n = 2$, in which case (9.144) becomes

$$f^{(1)}(x_k) = p_2^{(1)}(x_k) + \frac{1}{6} \pi^{(1)}(x_k) f^{(3)}(\xi(x_k)). \tag{9.145}$$

---

[3]This turns out to be true, although it is not easy to prove. Fortunately, we do not need this result.

Since $\pi(x) = (x - x_0)(x - x_1)(x - x_2)$, we have $\pi^{(1)}(x) = (x - x_1)(x - x_2) + (x - x_0)(x - x_1) + (x - x_0)(x - x_2)$, and therefore

$$
\begin{aligned}
\pi^{(1)}(x_0) &= (x_0 - x_1)(x_0 - x_2) &= (-h)(-2h) &= 2h^2, \\
\pi^{(1)}(x_1) &= (x_1 - x_0)(x_1 - x_2) &= h(-h) &= -h^2, \\
\pi^{(1)}(x_2) &= (x_2 - x_0)(x_2 - x_1) &= (2h)h &= 2h^2.
\end{aligned}
\tag{9.146}
$$

From (6.9), we obtain

$$
p_2(x) = f(x_0)L_0(x) + f(x_1)L_1(x) + f(x_2)L_2(x),
\tag{9.147}
$$

where, from (6.11)

$$
L_j(x) = \prod_{\substack{i=0 \\ i \neq j}}^{2} \frac{x - x_i}{x_j - x_i},
\tag{9.148}
$$

and hence the approximation to $f^{(1)}(x_k)$ is ($k \in \{0, 1, 2\}$)

$$
p_n^{(1)}(x_k) = f(x_0)L_0^{(1)}(x_k) + f(x_1)L_1^{(1)}(x_k) + f(x_2)L_2^{(1)}(x_k).
\tag{9.149}
$$

From (9.148)

$$
\begin{aligned}
L_0(x) &= \frac{(x-x_1)(x-x_2)}{(x_0-x_1)(x_0-x_2)}, & L_0^{(1)}(x) &= \frac{(x-x_1)+(x-x_2)}{(x_0-x_1)(x_0-x_2)}, \\
L_1(x) &= \frac{(x-x_0)(x-x_2)}{(x_1-x_0)(x_1-x_2)}, & L_1^{(1)}(x) &= \frac{(x-x_0)+(x-x_2)}{(x_1-x_0)(x_1-x_2)}, \\
L_2(x) &= \frac{(x-x_0)(x-x_1)}{(x_2-x_0)(x_2-x_1)}, & L_2^{(1)}(x) &= \frac{(x-x_0)+(x-x_1)}{(x_2-x_0)(x_2-x_1)}.
\end{aligned}
\tag{9.150}
$$

Hence

$$
L_0^{(1)}(x_0) = -\frac{3}{2h}, \; L_0^{(1)}(x_1) = -\frac{1}{2h}, \; L_0^{(1)}(x_2) = +\frac{1}{2h},
$$

$$
L_1^{(1)}(x_0) = +\frac{2}{h}, \; L_1^{(1)}(x_1) = 0, \; L_1^{(1)}(x_2) = -\frac{2}{h},
\tag{9.151}
$$

$$
L_2^{(1)}(x_0) = -\frac{1}{2h}, \; L_2^{(1)}(x_1) = +\frac{1}{2h}, \; L_2^{(1)}(x_2) = +\frac{3}{2h}.
$$

We therefore have the following expressions for $f^{(1)}(x_k)$:

$$
f^{(1)}(x_0) = \frac{1}{2h}[-3f(x_0) + 4f(x_1) - f(x_2)] + \frac{1}{3}h^2 f^{(3)}(\xi(x_0)),
\tag{9.152a}
$$

$$
f^{(1)}(x_1) = \frac{1}{2h}[-f(x_0) + f(x_2)] - \frac{1}{6}h^2 f^{(3)}(\xi(x_1)),
\tag{9.152b}
$$

$$
f^{(1)}(x_2) = \frac{1}{2h}[f(x_0) - 4f(x_1) + 3f(x_2)] + \frac{1}{3}h^2 f^{(3)}(\xi(x_2)).
\tag{9.152c}
$$

We recognize that the case (9.152b) contains the central difference approximation [recall (9.138)], since we may let $x_0 = x - h$, $x_2 = x + h$ (and $x_1 = x$). If we let $x_0 = x$, $x_1 = x + h$ and $x_2 = x + 2h$, then (9.152a) yields

$$f^{(1)}(x) \approx \frac{1}{2h}[-3f(x) + 4f(x+h) - f(x+2h)], \tag{9.153}$$

and if we let $x_2 = x$, $x_1 = x - h$, and $x_0 = x - 2h$, then (9.152c) yields

$$f^{(1)}(x) \approx \frac{1}{2h}[f(x-2h) - 4f(x-h) + 3f(x)]. \tag{9.154}$$

Note that (9.153) and (9.154) were employed in obtaining $T_C(n)$ in (9.107).

REFERENCES

1. G. E. Forsythe, M. A. Malcolm, and C. B. Moler, *Computer Methods for Mathematical Computations*, Prentice-Hall, Englewood Cliffs, NJ, 1977.

2. L. Bers, *Calculus: Preliminary Edition*, Vol. 2, Holt, Rinehart, Winston, New York, 1967.

3. P. J. Davis and P. Rabinowitz, *Numerical Integration*, Blaisdell, Waltham, MA, 1967.

4. T. W. Körner, *Fourier Analysis*, Cambridge Univ. Press, New York, 1988.

5. F. B. Hildebrand, *Introduction to Numerical Analysis*, 2nd ed., McGraw-Hill, New York, 1974.

6. W. Sweldens and R. Piessens, "Quadrature Formulae and Asymptotic Error Expansions for Wavelet Approximations of Smooth Functions," *SIAM J. Numer. Anal.* **31**, 1240–1264 (Aug. 1994).

7. J. F. Epperson, *An Introduction to Numerical Methods and Analysis*, Wiley, New York, 2002.

8. A. Quarteroni, R. Sacco, and F. Saleri, *Numerical Mathematics* (Texts in Applied Mathematics series, Vol. 37), Springer-Verlag, New York, 2000.

9. A. Ralston, *A First Course in Numerical Analysis*, McGraw-Hill, New York, 1965.

10. I. S. Gradshteyn and I. M. Ryzhik, in *Table of Integrals, Series and Products*, 5th ed., A. Jeffrey, ed., Academic Press, San Diego, CA, 1994.

11. R. E. Ziemer and W. H. Tranter, *Principles of Communications: Systems, Modulation, and Noise*, Houghton Mifflin, Boston, MA, 1976.

PROBLEMS

9.1. This problem is based on an assignment problem due to I. Leonard. Consider

$$I_n = \int_0^1 \frac{x^n}{x+a}\, dx,$$

where $a \geq 1$ and $n \in \mathbf{Z}^+$. It is easy to see that for $0 < x < 1$, we have $x^{n+1} < x^n$, and $0 < I_{n+1} < I_n$ for all $n \in \mathbf{Z}^+$. For $0 < x < 1$, we have

$$\frac{x^n}{1+a} < \frac{x^n}{x+a} < \frac{x^n}{a},$$

implying that

$$\int_0^1 \frac{x^n}{1+a}\,dx < I_n < \int_0^1 \frac{x^n}{a}\,dx$$

or

$$\frac{1}{(n+1)(1+a)} < I_n < \frac{1}{(n+1)a},$$

so immediately $\lim_{n \to \infty} I_n = 0$. Also, we have the difference equation

$$I_n = \int_0^1 \frac{x^{n-1}[x+a-a]}{x+a}\,dx = \frac{1}{n} - aI_{n-1} \qquad (9.\text{P}.1)$$

for $n \in \mathbf{N}$, where $I_0 = \int_0^1 \frac{1}{x+a}\,dx = \left[\log_e(x+a)\right]_0^1 = \log_e\left(\frac{1+a}{a}\right)$.

(a) Assume that $\hat{I}_0 = I_0 + \epsilon$ is the computed value of I_0. Assume that no other errors arise in computing I_n for $n \geq 1$ using (9.P.1). Then

$$\hat{I}_n = \frac{1}{n} - a\hat{I}_{n-1}.$$

Define the error $e_n = \hat{I}_n - I_n$, and find a difference equation for e_n.

(b) Solve for e_n, and show that for large enough a we have $\lim_{n \to \infty} |e_n| = \infty$.

(c) Find a stable algorithm to compute I_n for $n \in \mathbf{Z}^+$.

9.2. (a) Find an upper bound on the magnitude of the rounding error involved in applying the trapezoidal rule to

$$\int_0^a x^2\,dx.$$

(b) Find an upper bound on the magnitude of the truncation error in applying the trapezoidal rule to the integral in (a) above.

9.3. Consider the integral

$$I(\epsilon) = \int_\epsilon^a \sqrt{x}\,dx,$$

where $a > \epsilon > 0$. Write a MATLAB routine to fill in the following table for $a = 1$, and $n = 100$:

ϵ	$|I(\epsilon) - T(n)|$	$B_{T(n)}$	$|I(\epsilon) - S(n)|$	$B_{S(n)}$
0.1000				
0.0100				
0.0010				
0.0001				

In this table $T(n)$ is from (9.14), $S(n)$ is from (9.41), and

$$|E_{T(n)}| \le B_{T(n)}, \qquad |E_{S(n)}| \le B_{S(n)},$$

where the upper bounds $B_{T(n)}$ and $B_{S(n)}$ are obtained using (9.33) and (9.59), respectively.

9.4. Consider the integral

$$I = \int_0^1 \frac{dx}{1 + x^2} = \frac{\pi}{4}.$$

(a) Use the trapezoidal rule to estimate I, assuming that $h = \frac{1}{4}$.

(b) Use Simpson's rule to estimate I, assuming that $h = \frac{1}{4}$.

(c) Use the corrected trapezoidal rule to estimate I, assuming that $h = \frac{1}{4}$.

Perform all computations using only a pocket calculator.

9.5. Consider the integral

$$I = \int_{-1/2}^{1/2} \frac{dx}{1 - x^2} = \ln 3.$$

(a) Use the trapezoidal rule to estimate I, assuming that $h = \frac{1}{4}$.

(b) Use Simpson's rule to estimate I, assuming that $h = \frac{1}{4}$.

(c) Use the corrected trapezoidal rule to estimate I, assuming that $h = \frac{1}{4}$.

Perform all computations using only a pocket calculator.

9.6. Consider the integral

$$I = \int_0^{\pi/2} \frac{\sin(3x)}{\sin x} dx = \frac{\pi}{2}.$$

(a) Use the trapezoidal rule to estimate I, assuming that $h = \pi/12$.

(b) Use Simpson's Rule to estimate I, assuming that $h = \pi/12$.

(c) Use the corrected trapezoidal rule to estimate I, assuming that $h = \pi/12$.

Perform all computations using only a pocket calculator.

9.7. The length of the curve $y = f(x)$ for $a \le x \le b$ is given by

$$L = \int_a^b \sqrt{1 + [f^{(1)}(x)]^2}\, dx.$$

Suppose $f(x) = \cos x$. Compute L for $a = -\pi/2$ and $b = \pi/2$. Use the trapezoidal rule, selecting n so that $|E_{T(n)}| \le 0.001$. Hence, using (9.33), select n such that

$$\frac{1}{12}\frac{1}{n^2}(b-a)^3 M \le 0.001.$$

Do the computations using a suitable MATLAB routine.

9.8. Recall Example 6.5. Estimate the integral

$$I = \int_{-1}^1 e^{-x^2}\, dx$$

by

(a) Integrating the natural spline interpolant
(b) Integrating the complete spline interpolant

9.9. Find the constants α and β in $x = \alpha t + \beta$, and find f in terms of g such that

$$\int_a^b g(t)\, dt = \int_{-1}^1 \frac{f(x)}{\sqrt{1-x^2}}\, dx.$$

[*Comment:* This transformation will permit you to apply the Chebyshev–Gauss quadrature rule from Eq. (9.98) to general integrands.]

9.10. Consider the midpoint rule (Section 9.2). If we approximate $I = \int_a^b f(x)\, dx$ by one rectangle, then the rule is

$$R(1) = (b-a) f\left(\frac{a+b}{2}\right),$$

so the truncation error involved in this approximation is

$$E_{R(1)} = \int_a^b f(x)\, dx - R(1).$$

Use Taylor expansion error analysis to find an approximation to $E_{R(1)}$. [*Hint:* With $\bar{x} = (a+b)/2$ and $h = b - a$, we obtain

$$f(x) = f(\bar{x}) + (x - \bar{x}) f^{(1)}(\bar{x}) + \frac{1}{2!}(x - \bar{x})^2 f^{(2)}(\bar{x}) + \cdots.$$

Consider $f(a)$, $f(b)$, and $\int_a^b f(x)\, dx$ using this series expansion.]

9.11. Use both the trapezoidal rule [Eq. (9.14)] and the Chebyshev–Gauss quadrature rule [Eq. (9.98)] to approximate $I = \frac{2}{\pi} \int_0^\pi \frac{\sin x}{x} \, dx$ for $n = 6$. Assuming that $I = 1.1790$, which rule gives an answer closer to this value for I ? Use a pocket calculator to do the computations.

9.12. Consider the integral

$$I = \int_0^\pi \cos^2 x \, dx = \frac{\pi}{2}.$$

Write a MATLAB routine to approximate I using the trapezoidal rule, and Richardson's extrapolation. The program must allow you to fill in the following table:

n	$T(n)$	$|I - T(n)|$	$R(n)$	$|I - R(n)|$
2				
4				
8				
16				
32				
64				
128				
256				
512				
1024				

Of course, the extrapolated values are obtained from (9.111), where $I(n) = T(n)$. Does extrapolation improve on the accuracy? Comment on this.

9.13. Write a MATLAB routine that allows you to make a table similar to that in Example 9.6, but for the integral

$$I = \int_0^1 x^{1/4} \, dx.$$

9.14. The *complete elliptic integral of the first kind* is

$$K(k) = \int_0^{\pi/2} \frac{d\theta}{\sqrt{1 - k^2 \sin^2 \theta}}, 0 < k < 1,$$

and the *complete elliptic integral of the second kind* is

$$E(k) = \int_0^{\pi/2} \sqrt{1 - k^2 \sin^2 \theta} \, d\theta, 0 < k < 1.$$

(a) Find a series expansion for $K(k)$. [*Hint:* Recall (3.80).]

(b) Construct a Romberg table for $N = 4$ for the integral $K(\frac{1}{2})$. Use the series expansion from (a) to find the "exact" value of $K(\frac{1}{2})$ and compare.

(c) Find a series expansion for $E(k)$. [*Hint:* Recall (3.80).]

(d) Construct a Romberg table for $N = 4$ for the integral $E(\frac{1}{2})$. Use the series expansion from (c) to find the "exact" value of $E(\frac{1}{2})$ and compare.

Use MATLAB to do all of the calculations. [*Comment:* Elliptic integrals are important in electromagnetic potential problems (e.g., finding the magnetic vector potential of a circular current loop), and are important in analog and digital filter design (e.g., elliptic filters).]

9.15. This problem is about an alternative approach to the derivation of Gauss-type quadrature rules. The problem statement is long, but the solution is short because so much information has been provided. Suppose that $w(x) \geq 0$ for $x \in [a, b]$, so $w(x)$ is some weighting function. We wish to find *weights* w_k and *sample points* x_k for $k = 0, 1, \ldots, n - 1$ such that

$$\int_a^b w(x) f(x) \, dx = \sum_{k=0}^{n-1} w_k f(x_k) \tag{9.P.2}$$

for all $f(x) = x^j$, where $j = 0, 1, \ldots, 2n - 2, 2n - 1$. This task is greatly aided by defining the *moments*

$$m_j = \int_a^b w(x) x^j \, dx,$$

where m_j is called the jth *moment of* $w(x)$. In everything that follows it is important to realize that because of (9.P.2)

$$m_j = \int_a^b x^j w(x) \, dx = \sum_{k=0}^{n-1} w_k x_k^j.$$

The method proposed here implicitly assumes that it is easy to find the moments m_j. Once the weights and sample points are found, expression (9.P.2) forms a quadrature rule according to

$$\int_a^b w(x) f(x) \, dx \approx \sum_{k=0}^{n-1} w_k f(x_k) = G(n - 1),$$

where now $f(x)$ is essentially arbitrary. Define the vectors

$$w = [w_0 w_1 \cdots w_{n-2} w_{n-1}]^T \in \mathbf{R}^n,$$
$$m = [m_0 m_1 \cdots m_{2n-2} m_{2n-1}]^T \in \mathbf{R}^{2n}.$$

Find matrix $A \in \mathbf{R}^{n \times 2n}$ such that $A^T w = m$. Matrix A turns out to be a rectangular Vandermonde matrix. If we knew the sample points x_k, then it would be possible to use $A^T w = m$ to solve for the weights in w. (This is so even though the linear system is overdetermined.) Define the polynomial

$$p_n(x) = \prod_{j=0}^{n-1}(x - x_j) = \sum_{j=0}^{n} p_{n,j} x^j$$

for which we see that $p_{n,n} = 1$. We observe that the zeros of $p_n(x)$ happen to be the sample points x_k that we are looking for. The following suggestion makes it possible to find the sample points x_k by first finding the polynomial $p_n(x)$. Using (in principle) Chapter 7 ideas, we can then find the roots of $p_n(x) = 0$, and so find the sample points. Consider the expression

$$m_{j+r} p_{n,j} = \sum_{k=0}^{n-1} w_k x_k^{j+r} p_{n,j}, \tag{9.P.3}$$

where $r = 0, 1, \ldots, n - 2, n - 1$. Consider the sum of (9.P.3) over $j = 0, 1, \ldots, n$. Use this sum to show that

$$\sum_{j=0}^{n-1} m_{j+r} p_{n,j} = -m_{n+r} \tag{9.P.4}$$

for $r = 0, 1, \ldots, n - 1$. This can be expressed in matrix form as $Mp = q$, where

$$M = [m_{i+j}]_{i,j=0,1,\ldots,n-1} \in \mathbf{R}^{n \times n}$$

which is called a *Hankel matrix*, and where

$$p = [p_{n,0} p_{n,1} \cdots p_{n,n-1}]^T \in \mathbf{R}^n$$

and

$$-q = [m_n m_{n+1} \cdots m_{2n-2} m_{2n-1}]^T.$$

Formal proof that M^{-1} always exists is possible, but is omitted from consideration. [*Comment:* The reader is warned that the approach to Gaussian quadrature suggested here is not really practical due to the ill-conditioning of the matrices involved (unless n is small).]

9.16. In the previous problem let $a = -1$, $b = 1$, and $w(x) = \sqrt{1 - x^2}$. Develop a numerically reliable algorithm to compute the moments $m_j = \int_a^b w(x) x^j \, dx$.

9.17. Derive Eq. (9.88).

9.18. Repeat Example 9.4, except that the integral is now

$$I = \int_{-1}^{1} e^{-x^2} \, dx.$$

9.19. This problem is a preview of certain aspects of probability theory. It is an example of an application for numerical integration. An experiment produces a measurable quantity denoted $x \in \mathbf{R}$. The experiment is random in that the value of x is different from one experiment to the next, but the *probability* of x lying within a particular range of values is known to be

$$P = P[a \le x \le b] = \int_{a}^{b} f_X(x) \, dx,$$

where

$$f_X(x) = \frac{1}{\sqrt{2\pi\sigma^2}} \exp\left(-\frac{x^2}{2\sigma^2}\right),$$

which is an instance of the Gaussian function mentioned in Chapter 3. This fact may be interpreted as follows. Suppose that we perform the experiment R times, where R is a "large" number. Then *on average* we expect $a \le x \le b$ a total of PR times. The function $f_X(x)$ is called a *Gaussian probability density function* (pdf) with a mean of zero, and *variance* σ^2. Recall Fig. 3.6, which shows the effect of changing σ^2. In Monte Carlo simulations of digital communications systems, or for that matter any other system where randomness is an important factor, it is necessary to write programs that generate simulated random variables such as x. The MATLAB randn function will generate zero mean Gaussian random variables with variance $\sigma^2 = 1$. For example, $x = \text{randn}(1, N)$ will load N Gaussian random variables into the row vector x. Write a MATLAB routine to generate $N = 1000$ simulated Gaussian random variables (also called Gaussian *variates*) using randn. Count the number of times x satisfies $-1 \le x \le 1$. Let this count be denoted $\hat{C}$. Your routine must also use the trapezoidal rule to estimate the probability $P = P[-1 \le x \le 1]$ using erf(x) [defined in Eq. (3.107)]. The magnitude of the error in computing P must be <0.0001. This will involve using the truncation error bound (9.33) in Chapter 9 to estimate the number of trapezoids n that you need to do this job. You are to neglect rounding error effects here. Compute $C = PN$. Your program must print $\hat{C}$ and C to a file. Of course, we expect $\hat{C} \approx C$.

9.20. Develop a MATLAB routine to fill in the following table, which uses the central difference approximation to the first derivative of a function [i.e.,

$\tilde{f}_c^{(1)}(x)]$ to estimate $f^{(1)}(x)$, where here

$$f(x) = \log_e x.$$

x	$h = 10^{-4}$	$h = 10^{-5}$	$h = 10^{-6}$	$h = 10^{-7}$	$h = 10^{-8}$	$f^{(1)}(x)$
1.0						
2.0						
3.0						
4.0						
5.0						
6.0						
7.0						
8.0						
9.0						
10.0						

Explain the results you get.

9.21. Suppose that $f(x) = e^{-x^2}$, and recall (9.132).

(a) Sketch $f^{(2)}(x)$.

(b) Let $h = 1/10$, and compute $\tilde{f}_f^{(1)}(1)$. Find an upper bound on $|e_f(1)|$. Compute $|f^{(1)}(1) - \tilde{f}_f^{(1)}(1)| = |e_f(1)|$, and compare to the bound.

(c) Let $h = \frac{1}{10}$, and compute $\tilde{f}_f^{(1)}(1/\sqrt{2})$. Find an upper bound on $|e_f(1/\sqrt{2})|$. Compute $|f^{(1)}(1/\sqrt{2}) - \tilde{f}_f^{(1)}(1/\sqrt{2})| = |e_f(1/\sqrt{2})|$, and compare to the bound.

9.22. Show that another approximation to $f^{(1)}(x)$ is given by

$$f^{(1)}(x) \approx \frac{1}{12h}[8f(x+h) - 8f(x-h) - f(x+2h) + f(x-2h)].$$

Give an expression for the error involved in using this approximation.

10 Numerical Solution of Ordinary Differential Equations

10.1 INTRODUCTION

In this chapter we consider numerical methods for the solution of *ordinary differential equations* (ODEs). We recall that in such differential equations the function that we wish to solve for is in one independent variable. By contrast *partial differential equations* (PDEs) involve solving for functions in two or more independent variables. The numerical solution of PDEs is a subject for a later chapter.

With respect to the level of importance of the subject the reader knows that all dynamic systems with physical variables that change continuously over time (or space, or both) are described in terms of differential equations, and so form the basis for a substantial portion of engineering systems analysis, and design across all branches of engineering. The reader is also well aware of the fact that it is quite easy to arrive at differential equations that completely defy attempts at an analytical solution. This remains so in spite of the existence of quite advanced methods for analytical solution (e.g., symmetry methods that use esoteric ideas from Lie group theory [1]), and so the need for this chapter is not hard to justify.

Where ODEs are concerned, differential equations arise within two broad categories of problems:

1. Initial-value problems (IVPs)
2. Boundary-value problems (BVPs)

In this chapter we shall restrict consideration to initial value problems. However, this is quite sufficient to accommodate much of electric/electronic circuit modeling, modeling the orbital dynamics of satellites around planetary bodies, and many other problems besides.

A simple example of an ODE for which no general analytical theory of solution is known is the *Duffing equation*

$$m\frac{d^2x(t)}{dt^2} + k\frac{dx(t)}{dt} + \alpha x(t) + \delta x^3(t) = F\cos(\omega t), \qquad (10.1)$$

An Introduction to Numerical Analysis for Electrical and Computer Engineers, by C.J. Zarowski
ISBN 0-471-46737-5 © 2004 John Wiley & Sons, Inc.

where $t \geq 0$. Since we are concerned with initial-value problems we would need to know, at least implicitly, $x(0)$, and $\frac{dx(0)}{dt}$ which are the initial conditions. If it were the case that $\delta = 0$ then the solution of (10.1) is straightforward because it is a particular case of a second-order linear ODE with constant coefficients. Perhaps the best method for solving (10.1) in this case would be the Laplace transform method. However, the case where $\delta \neq 0$ immediately precludes a straightforward analytical solution of this kind. It is worth noting that the Duffing equation models a forced nonlinear mechanical spring, where the restoring force of the spring is accounted for by the terms $\alpha x(t) + \delta x^3(t)$. Function $x(t)$, which we wish to solve for, is the displacement at time t of some point on the spring (e.g., the point mass m at the free end) with respect to a suitable reference frame. Term $k \frac{dx(t)}{dt}$ is the opposing friction, while $F \cos(\omega t)$ is the periodic forcing function that drives the system. An example of a recent application for (10.1) is in the modeling of micromechanical filters/resonators [2].[1]

At the outset we consider only first-order problems, specifically, how to solve (numerically)

$$\frac{dx}{dt} = f(x, t), \quad x_0 = x(0) \tag{10.2}$$

for $t \geq 0$. [From now on we shall often write x instead of $x(t)$, and dx/dt instead of $dx(t)/dt$ for brevity.] However, the example of (10.1) is a second-order problem. But it is possible to replace it with a system of equivalent first-order problems. There are many ways to do this in principle. One way is to define

$$y = \frac{dx}{dt}. \tag{10.3}$$

The functions $x(t)$ and $y(t)$ are examples of *state variables*. Since we are interpreting (10.1) as the model for a mechanical system wherein $x(t)$ is displacement, it therefore follows that we may interpret $y(t)$ as velocity. From the definition (10.3) we may use (10.1) to write

$$\frac{dy}{dt} = -\frac{\alpha}{m}x - \frac{k}{m}y - \frac{\delta}{m}x^3 + \frac{F}{m}\cos(\omega t) \tag{10.4a}$$

and

$$\frac{dx}{dt} = y. \tag{10.4b}$$

[1]The clock circuit in many present-day digital systems is built around a quartz crystal. Such crystals do not integrate onto chips. Micromechanical resonators are intended to replace the crystal since such resonators can be integrated onto chips. This is in furtherance of the goal of more compact electronic systems. This applications example is a good illustration of the rapidly growing trend to integrate nonelectrical/nonelectronic systems onto chips. The implication of this is that it is *now* very necessary for the average electrical and/or computer engineer to become very knowledgeable about most other branches of engineering, and to possess a much broader and deeper knowledge of science (physics, chemistry, biology, etc.) and mathematics.

Equations (10.4) have the forms

$$\frac{dx}{dt} = f(x, y, t) \tag{10.5a}$$

$$\frac{dy}{dt} = g(x, y, t) \tag{10.5b}$$

for the appropriate choices of f and g. The initial conditions for our example are $x(0)$ and $y(0)$ (initial position and initial velocity, respectively). These represent a coupled system of first-order ODEs. Methods applicable to the solution of (10.2) are extendable to the larger problem of solving systems of first-order ODEs, and so in this way higher-order ODEs may be solved. Thus, we shall also consider the numerical solution of initial-value problems in systems of first-order ODEs.

The next two examples illustrate how to arrive at coupled systems of first-order ODEs for electrical and electronic circuits.

Example 10.1   Consider the linear electric circuit shown in Fig. 10.1. The input to the circuit is the voltage source $v_s(t)$, while we may regard the output as the voltage drop across capacitor C, denoted $v_C(t)$. The differential equation relating the input voltage $v_s(t)$ and the output voltage $v_C(t)$ is thus

$$L_1 C \frac{d^3 v_C(t)}{dt^3} + R_1 C \frac{d^2 v_C(t)}{dt^2} + \left(\frac{L_1}{L_2} + 1 \right) \frac{d v_C(t)}{dt} + \frac{R_1}{L_2} v_C(t) = \frac{d v_s(t)}{dt}. \tag{10.6}$$

This third-order ODE may be obtained by mesh analysis of the circuit. The reader ought to attempt this derivation as an exercise. One way to replace (10.6) with a coupled system of first-order ODEs is to define the state variables $x_k(t)$ ($k \in \{0, 1, 2\}$) according to

$$x_0(t) = v_C(t), \qquad x_1(t) = \frac{d v_C(t)}{dt}, \qquad x_2(t) = \frac{d^2 v_C(t)}{dt^2}. \tag{10.7}$$

Substituting (10.7) into (10.6) yields

$$L_1 C \frac{dx_2(t)}{dt} + R_1 C x_2(t) + \left(\frac{L_1}{L_2} + 1 \right) x_1(t) + \frac{R_1}{L_2} x_0(t) = \frac{d v_s(t)}{dt}.$$

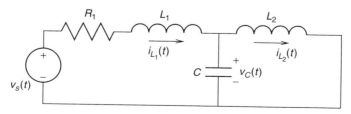

Figure 10.1   The linear electric circuit for Example 10.1.

If we recognize that

$$x_1(t) = \frac{dx_0(t)}{dt}, \qquad x_2(t) = \frac{dx_1(t)}{dt}$$

then the complete system of first order ODEs is

$$\frac{dx_0(t)}{dt} = x_1(t),$$

$$\frac{dx_1(t)}{dt} = x_2(t),$$

$$\frac{dx_2(t)}{dt} = \frac{1}{L_1 C} \frac{dv_s(t)}{dt} - \frac{R_1}{L_1} x_2(t) - \frac{1}{L_1 C}\left(\frac{L_1}{L_2} + 1\right) x_1(t) - \frac{R}{L_1 L_2 C} x_0(t).$$

(10.8)

In many ways this is not the best description for the circuit dynamics.
Instead, we may find the matrix $A \in \mathbf{R}^{3 \times 3}$ and column vector $b \in \mathbf{R}^3$ such that

$$\begin{bmatrix} \dfrac{dv_C(t)}{dt} \\ \dfrac{di_{L_1}(t)}{dt} \\ \dfrac{di_{L_2}(t)}{dt} \end{bmatrix} = A \begin{bmatrix} v_C(t) \\ i_{L_1}(t) \\ i_{L_2}(t) \end{bmatrix} + b v_s(t).$$

(10.9)

This defines a new set of state equations in terms of the new state variables $v_C(t)$, $i_{L_1}(t)$, and $i_{L_2}(t)$. The matrix A and vector b contain constants that depend only on the circuit parameters $R_1, L_1, L_2,$ and C.

Equation (10.9) is often a better representation than (10.8) because

1. There is no derivative of the forcing function $v_s(t)$ in (10.9) as there is in (10.8).
2. There is a general (linear) theory of solution to (10.9) that is in practice easy to apply, and it is based on state-space methods.
3. Inductor currents [i.e., $i_{L_1}(t), i_{L_2}(t)$] and capacitor voltages [i.e., $v_C(t)$] can be readily measured in a laboratory setting while derivatives of these are not as easily measured. Thus, it is relatively easy to compare theoretical and numerical solutions to (10.9) with laboratory experimental results.

Since

$$i_C(t) = C \frac{dv_C(t)}{dt}, \qquad v_{L_1}(t) = L_1 \frac{di_{L_1}(t)}{dt}, \qquad v_{L_2}(t) = L_2 \frac{di_{L_2}(t)}{dt}$$

on applying Kirchoff's Voltage law (KVL) and Kirchoff's Current law (KCL), we arrive at the relevant state equations as follows.

First

$$v_s(t) = R_1 i_{L_1}(t) + L_1 \frac{d i_{L_1}(t)}{dt} + L_2 \frac{d i_{L_2}(t)}{dt}. \tag{10.10}$$

We see that $v_C(t) = v_{L_2}(t)$, and so

$$v_C(t) = L_2 \frac{i_{L_2}(t)}{dt},$$

giving

$$\frac{d i_{L_2}(t)}{dt} = \frac{1}{L_2} v_C(t), \tag{10.11}$$

which is one of the required state equations. Since

$$i_{L_1}(t) = i_C(t) + i_{L_2}(t),$$

and so

$$i_{L_1}(t) = C \frac{d v_C(t)}{dt} + i_{L_2}(t),$$

we also have

$$\frac{d v_C(t)}{dt} = \frac{1}{C} i_{L_1}(t) - \frac{1}{C} i_{L_2}(t). \tag{10.12}$$

This is another of the required state equations. Substituting (10.11) into (10.10) gives the final state equation

$$\frac{d i_{L_1}(t)}{dt} = -\frac{R_1}{L_1} i_{L_1}(t) - \frac{1}{L_1} v_C(t) + \frac{1}{L_1} v_s(t). \tag{10.13}$$

The state equations may be collected together in matrix form as required:

$$\begin{bmatrix} \dfrac{d v_C(t)}{dt} \\[2mm] \dfrac{d i_{L_1}(t)}{dt} \\[2mm] \dfrac{d i_{L_2}(t)}{dt} \end{bmatrix} = \underbrace{\begin{bmatrix} 0 & \dfrac{1}{C} & -\dfrac{1}{C} \\[2mm] -\dfrac{1}{L_1} & -\dfrac{R_1}{L_1} & 0 \\[2mm] \dfrac{1}{L_2} & 0 & 0 \end{bmatrix}}_{=A} \begin{bmatrix} v_C(t) \\[2mm] i_{L_1}(t) \\[2mm] i_{L_2}(t) \end{bmatrix} + \underbrace{\begin{bmatrix} 0 \\[2mm] \dfrac{1}{L_1} \\[2mm] 0 \end{bmatrix}}_{=b} v_s(t).$$

Example 10.2 Now let us consider a more complicated third-order nonlinear electronic circuit called the *Colpitts oscillator* [9]. The electronic circuit, and its electric circuit equivalent (model) appears in Fig. 10.2. This circuit is a popular analog signal generator with a long history (it used to be built using vacuum tubes). The device Q is a three-terminal device called an *NPN-type bipolar junction transistor* (BJT). The detailed theory of operation of BJTs is beyond the scope of

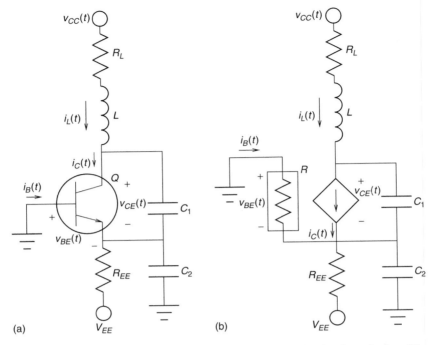

Figure 10.2   The BJT Colpitts oscillator (a) and its electric circuit equivalent (b).

this book, but may be found in basic electronics texts [10]. For present purposes it is enough to know that Q may be represented with a *nonlinear resistor R* (the resistor enclosed in the box in Fig. 10.2b), and a *current-controlled current source* (CCCS), where

$$i_C(t) = \beta_F i_B(t) \tag{10.14}$$

and

$$i_B(t) = \begin{cases} 0, & v_{BE}(t) \le V_{TH} \\ \dfrac{(v_{BE}(t) - V_{TH})}{R_{ON}}, & v_{BE}(t) > V_{TH} \end{cases} . \tag{10.15}$$

The current $i_B(t)$ is called the *base current* of Q, and flows into the base terminal of the transistor as shown. From (10.15) we observe that if the *base–emitter voltage* $v_{BE}(t)$ is below a threshold voltage V_{TH}, then there is no base current into Q (i.e., the device is *cut off*). The relationship between $v_{BE}(t) - V_{TH}$, and $i_B(t)$ obeys Ohm's law only when $v_{BE}(t)$ is above threshold, in which case Q is *active*. In either case (10.14) says the *collector current* $i_C(t)$ is directly proportional to $i_B(t)$, and the constant of proportionality β_F is called the *forward current gain* of Q. Voltage $v_{CE}(t)$ is the *collector–emitter voltage* of Q. Typically, $V_{TH} \approx 0.75 \, \text{V}$, β_F is about 100 (order of magnitude), and R_{ON} (*on resistance* of Q) is seldom more than hundreds of ohms in size.

From (10.15) $i_B(t)$ is a nonlinear function of $v_{BE}(t)$ that we may compactly write as $i_B(t) = f_R(v_{BE}(t))$. There are power supply voltages $v_{CC}(t)$, and V_{EE}. Voltage $V_{EE} < 0$ is a constant, with a typical value $V_{EE} = -5$ V. Here we treat $v_{CC}(t)$ as time-varying, but it is usually the case that (approximately) $v_{CC}(t) = V_{CC}u(t)$, where

$$u(t) = \begin{cases} 1, & t \geq 0 \\ 0, & t < 0 \end{cases}. \tag{10.16}$$

Function $u(t)$ is the *unit step function*. To say that $v_{CC}(t) = V_{CC}u(t)$ is to say that the circuit is turned on at time $t = 0$. Typically, $V_{CC} = +5$ V.

The reader may verify (again as a circuit analysis review exercise) that state equations for the Colpitts oscillator are:

$$C_1 \frac{dv_{CE}(t)}{dt} = i_L(t) - \beta_F f_R(v_{BE}(t)), \tag{10.17a}$$

$$C_2 \frac{dv_{BE}(t)}{dt} = -\frac{v_{BE}(t) + V_{EE}}{R_{EE}} - f_R(v_{BE}(t)) - i_L(t), \tag{10.17b}$$

$$L \frac{di_L(t)}{dt} = v_{CC}(t) - v_{CE}(t) + v_{BE}(t) - R_L i_L(t). \tag{10.17c}$$

Thus, the state variables are $v_{BE}(t)$, $v_{CE}(t)$, and $i_L(t)$. As previously, this circuit description is not unique, but it is convenient.

Since numerical methods only provide approximate solutions to ODEs, we are naturally concerned about the accuracy of these approximations. There are also issues about the *stability* of proposed methods, and so this matter as well will be considered in this chapter.

10.2   FIRST-ORDER ODEs

Strictly speaking, before applying a numerical method to the solution of an ODE, we must be certain that a solution exists. We are also interested in whether the solution is unique. It is worth stating that in many cases, since ODEs are often derived from problems in the physical world, existence and uniqueness are often "obvious" for physical reasons. Notwithstanding this, a mathematical statement about existence and uniqueness is worthwhile.

The following definition is needed by the succeeding theorem regarding the existence and uniqueness of solutions to first order ODE initial value problems.

Definition 10.1: The Lipschitz Condition   The function $f(x, t) \in \mathbf{R}$ satisfies a *Lipschitz condition* in x for $S \subset \mathbf{R}^2$ iff there is an $\alpha > 0$ such that

$$|f(x, t) - f(y, t)| \leq \alpha |x - y|$$

when $(x, t), (y, t) \in S$. The constant α is called a *Lipschitz constant* for $f(x, t)$.

It is apparent that if $f(x, t)$ satisfies a Lipschitz condition, then it is smooth in some sense. The following theorem is about the existence and uniqueness of the solution to

$$\frac{dx}{dt} = f(x, t) \tag{10.18}$$

for $0 \leq t \leq t_f$, with the initial condition $x_0 = x(0)$. Time $t = 0$ is the *initial time*, or *starting time*. We call constant t_f the *final time*. Essentially, we are only interested in the solution over a finite time interval. This constraint on the theory is not unreasonable since a computer can run for only a finite amount of time anyway. We also remark that interpreting the independent variable t as time is common practice, but not mandatory in general.

Theorem 10.1: Picard's Theorem   Suppose that $S = \{(x, t) \in \mathbf{R}^2 | 0 \leq t \leq t_f, -\infty < x < \infty\}$, and that $f(x, t)$ is continuous on S. If f satisfies a Lipschitz condition on set S in the variable x, then the initial-value problem (10.18) has a unique solution $x = x(t)$ for all $0 \leq t \leq t_f$.

Proof   Omitted. We simply mention that it is based on the Banach fixed-point theorem (recall Theorem 7.3).

We also mention that a proof of a somewhat different version of this theorem appears in Kreyszig [3, pp. 315–317]. It involves working with a contractive mapping on a certain closed subspace of $C(J)$, where $J = [t_0 - \beta, t_0 + \beta] \subset \mathbf{R}$ and $C(J)$ is the metric space of continuous functions on J, where the metric is that of (1.8) in Chapter 1. It was remarked in Chapter 3 [see Eq. (3.8)] that this space is complete. Thus, any closed subspace of it is complete as well (a fact that was mentioned in Chapter 7 following Corollary 7.1).

We may now consider specific numerical techniques. Define $x_n = x(t_n)$ for $n \in \mathbf{Z}^+$. Usually we assume that $t_0 = 0$, and that

$$t_{n+1} = t_n + h, \tag{10.19}$$

where $h > 0$, and we call h the *step size*. From (10.18)

$$x_n^{(1)} = \frac{dx_n}{dt} = \frac{dx(t)}{dt}\Big|_{t=t_n} = f(x_n, t_n). \tag{10.20}$$

We may expand solution $x(t)$ in a Taylor series about $t = t_n$. Therefore, since $x(t_{n+1}) = x(t_n + h) = x_{n+1}$, and with $x_n^{(k)} = x^{(k)}(t_n)$

$$x_{n+1} = x_n + h x_n^{(1)} + \frac{1}{2!} h^2 x_n^{(2)} + \frac{1}{3!} h^3 x_n^{(3)} + \cdots. \tag{10.21}$$

If we drop terms in $x_n^{(k)}$ for $k > 1$, then (10.21) and (10.20) imply

$$x_{n+1} = x_n + h x_n^{(1)} = x_n + h f(x_n, t_n).$$

Since $x_0 = x(t_0) = x(0)$ we may find (x_n) via

$$x_{n+1} = x_n + hf(x_n, t_n). \tag{10.22}$$

This is often called the *Euler method* (or *Euler's method*).[2] A more accurate description would be to call it the *explicit form* of Euler's method in order to distinguish it from the *implicit form* to be considered a little later on. The distinction matters in practice because implicit methods tend to be stable, whereas explicit methods are often prone to instability.

A few general words about stability and accuracy are now appropriate. In what follows we will assume (unless otherwise noted) that the solution to a differential equation remains bounded; that is, $|x(t)| < M < \infty$ for all $t \geq 0$. However, approximations to this solution [e.g., (x_n) from (10.22)] will not necessarily remain bounded in the limit as $n \to \infty$; that is, our numerical methods might not always be stable. Of course, in a situation like this the numerical solution will deviate greatly from the correct solution, and this is simply unacceptable. It therefore follows that we must find methods to test the stability of a proposed numerical solution. Some informal definitions relating to stability are

Stable method: The numerical solution does not grow without bound (i.e., "blow up") with any choice of parameters such as step size.

Unstable method: The numerical solution blows up with any choices of parameters (such as step size).

Conditionally stable method: For certain choices of parameters the numerical solution remains bounded.

We mention that even if the Euler method is stable, its accuracy is low because only the first two terms in the Taylor series are retained. More specifically, we say that it is a *first-order method* because only the first power of h is retained in the Taylor approximation that gave rise to it. The omission of higher-order terms causes *truncation errors*. Since h^2 (and higher power) terms are omitted we also say that the *truncation error per step* (sometimes called the *order of accuracy*) is of *order* h^2. This is often written as $O(h^2)$. (Here we follow the terminology in Kreyszig [4, pp. 793–794].) In summary, we prefer methods that are both *stable*, and *accurate*. It is important to emphasize that accuracy and stability are distinct concepts, and so must never be confused.

---

[2]Strictly speaking, in truncating the series in (10.21) we should write $\hat{x}_{n+1} = x_n + hf(x_n, t_n)$ so that Euler's method is

$$\hat{x}_{n+1} = \hat{x}_n + hf(\hat{x}_n, t_n)$$

with $\hat{x}_0 = x_0$. This is to emphasize that the method only generates approximations to $x_n = x(t_n)$. However, this kind of notation is seldom applied. It is assumed that the reader knows that the numerical method only approximates $x(t_n)$ even though the notation does not necessarily explicitly distinguish the exact value from the approximate.

We can say more about the accuracy of the Euler method:

Theorem 10.2: For $dx(t)/dt = f(x(t), t)$ let $f(x(t), t)$ be Lipschitz continuous with constant α (Definition 10.1), and assume that $x(t) \in C^2[t_0, t_f]$ ($t_f > t_0$). If $\hat{x}_n \approx x(t_n) = x_n$, where ($t_n = t_0 + nh$, and $t_n \leq t_f$)

$$\hat{x}_{n+1} = \hat{x}_n + hf(\hat{x}_n, t_n), (\hat{x}_0 \approx x(t_0))$$

then

$$|x(t_n) - \hat{x}_n| \leq e^{\alpha(t_n - t_0)}|x(t_0) - \hat{x}_0| + \frac{1}{2}hM\frac{e^{\alpha(t_n - t_0)} - 1}{\alpha}, \qquad (10.23)$$

where $M = \max_{t \in [t_0, t_f]} |x^{(2)}(t)|$.

Proof Euler's method is

$$\hat{x}_{n+1} = \hat{x}_n + hf(\hat{x}_n, t_n)$$

and from Taylor's theorem

$$x(t_{n+1}) = x(t_n) + hx^{(1)}(t_n) + \frac{1}{2}h^2x^{(2)}(\xi_n)$$

for some $\xi_n \in [t_n, t_{n+1}]$. Thus

$$x(t_{n+1}) - \hat{x}_{n+1} = x(t_n) - \hat{x}_n + h[x^{(1)}(t_n) - f(\hat{x}_n, t_n)] + \frac{1}{2}h^2x^{(2)}(\xi_n)$$

$$= x(t_n) - \hat{x}_n + h[f(x(t_n), t_n) - f(\hat{x}_n, t_n)] + \frac{1}{2}h^2x^{(2)}(\xi_n)$$

so that

$$|x(t_{n+1}) - \hat{x}_{n+1}| \leq |x(t_n) - \hat{x}_n| + \alpha h|x(t_n) - \hat{x}_n| + \frac{1}{2}h^2|x^{(2)}(\xi_n)|.$$

For convenience we will let $e_n = |x(t_n) - \hat{x}_n|$, $\lambda = 1 + \alpha h$, and $r_n = \frac{1}{2}h^2x^{(2)}(\xi_n)$, so that

$$e_{n+1} \leq \lambda e_n + r_n.$$

It is easy to see that[3]

$$
\begin{array}{rcl}
e_1 & \leq & \lambda e_0 + r_0 \\
e_2 & \leq & \lambda e_1 + r_1 = \lambda^2 e_0 + \lambda r_0 + r_1 \\
e_3 & \leq & \lambda e_2 + r_2 = \lambda^3 e_0 + \lambda^2 r_0 + \lambda r_1 + r_2 \\
& \vdots & \\
e_n & \leq & \lambda^n e_0 + \sum_{j=0}^{n-1} \lambda^j r_{n-1-j}
\end{array}
$$

[3]More formally, we may use mathematical induction.

If $M = \max_{t \in [t_0, t_f]} |x^{(2)}(t)|$ then $r_{n-1-j} \leq \frac{1}{2}h^2 M$, and hence

$$
e_n \leq \lambda^n e_0 + \frac{1}{2}h^2 M \sum_{j=0}^{n-1} \lambda^j,
$$

and since $\sum_{j=0}^{n-1} \lambda^j = \frac{\lambda^n - 1}{\lambda - 1}$, and for $x \geq -1$ we have $(1 + x)^n \leq e^{nx}$, thus

$$
\sum_{j=0}^{n-1} \lambda^j = \frac{\lambda^n - 1}{\lambda - 1} \leq \frac{e^{n\alpha h} - 1}{\alpha h} = \frac{e^{\alpha(t_n - t_0)} - 1}{\alpha h}.
$$

Consequently,

$$
e_n \leq e^{\alpha(t_n - t_0)} e_0 + \frac{1}{2}hM \frac{e^{\alpha(t_n - t_0)} - 1}{\alpha}
$$

which immediately yields the theorem statement.

We remark that $e_0 = |x(t_0) - \hat{x}_0| = 0$ only if $\hat{x}_0 = x(t_0)$ *exactly*. Where quantization errors (recall Chapter 2) are concerned, this will seldom be the case. The second term in the bound of (10.23) may be large even if h is tiny. In other words, Euler's method is not necessarily very accurate. Certainly, from (10.23) we can say that $e_n \propto h$.

As a brief digression, we also note that Theorem 10.2 needed the bound

$$
(1 + x)^n \leq e^{nx} \ (x \geq -1). \tag{10.24}
$$

We may easily establish (10.24) as follows. From the Maclaurin expansion (Chapter 3)

$$
e^x = 1 + x + \frac{1}{2}x^2 e^{\xi}
$$

(for some $\xi \in [0, x]$) so that

$$
0 \leq 1 + x \leq 1 + x + \frac{1}{2}x^2 e^{\xi} = e^x,
$$

and because $1 + x \geq 0$ (i.e., $x \geq -1$)

$$
0 \leq (1 + x)^n \leq e^{nx},
$$

thus establishing (10.24).

The stability of any method may be analyzed in the following manner. First recall the Taylor series expansion of $f(x, t)$ about the point (x_0, t_0)

$$
\begin{aligned}
f(x, t) = f(x_0, t_0) &+ (t - t_0)\frac{\partial f(x_0, t_0)}{\partial t} + (x - x_0)\frac{\partial f(x_0, t_0)}{\partial x} \\
&+ \frac{1}{2!}\left[(t - t_0)^2\frac{\partial^2 f(x_0, t_0)}{\partial t^2} + 2(t - t_0)(x - x_0)\frac{\partial^2 f(x_0, t_0)}{\partial t \partial x}\right. \\
&\left. + (x - x_0)^2\frac{\partial^2 f(x_0, t_0)}{\partial x^2}\right] + \cdots
\end{aligned}
\tag{10.25}
$$

If we retain only the linear terms of (10.25) and substitute these into (10.18), then we obtain

$$
\begin{aligned}
x^{(1)}(t) = \frac{dx}{dt} &= f(x_0, t_0) + (t - t_0)\frac{\partial f(x_0, t_0)}{\partial t} + (x - x_0)\frac{\partial f(x_0, t_0)}{\partial x} \\
&= \underbrace{\frac{\partial f(x_0, t_0)}{\partial x}}_{=\lambda}x + \underbrace{\frac{\partial f(x_0, t_0)}{\partial t}}_{=\lambda_1}t + \underbrace{\left[f(x_0, t_0) - t_0\frac{\partial f(x_0, t_0)}{\partial t} - x_0\frac{\partial f(x_0, t_0)}{\partial x}\right]}_{=\lambda_2},
\end{aligned}
\tag{10.26}
$$

so this has the general form (with λ, λ_1, and λ_2 as constants)

$$
\frac{dx}{dt} = \lambda x + \lambda_1 t + \lambda_2.
\tag{10.27}
$$

This *linearized approximation* to the original problem in (10.18) allows us to investigate the behavior of the solution in close proximity to (x_0, t_0). Equation (10.27) is often simplified still further by considering what is called the *model problem*

$$
\frac{dx}{dt} = \lambda x.
\tag{10.28}
$$

Thus, here we assume that $\lambda_1 t + \lambda_2$ in (10.27) can also be neglected. However, we do remark that (10.27) has the form

$$
\frac{dx}{dt} + P(t)x = Q(t)x^n,
\tag{10.29}
$$

where $n = 0$, $P(t) = -\lambda$, and $Q(t) = \lambda_1 t + \lambda_2$. Thus, (10.27) is an instance of *Bernoulli's differential equation* [5, p. 62] for which a general method of solution exists. But for the purpose of stability analysis it turns out to be enough (usually, but not always) to consider only (10.28). Equation (10.28) is certainly simple in that its solution is

$$
x(t) = x(0)e^{\lambda t}.
\tag{10.30}
$$

If Euler's method is applied to (10.28), then

$$x_{n+1} = x_n + h\lambda x_n = (1 + h\lambda)x_n. \tag{10.31}$$

Clearly, for $n \in \mathbf{Z}^+$

$$x_n = (1 + h\lambda)^n x_0, \tag{10.32}$$

and we may avoid $\lim_{n \to \infty} |x_n| = \infty$ if

$$|1 + h\lambda| \leq 1.$$

The model problem (10.28) with the solution (10.30) is *stable*[4] only if $\lambda < 0$. Hence Euler's method is *conditionally stable* for

$$\lambda < 0 \quad \text{and} \quad h \leq \frac{2}{|\lambda|}, \tag{10.33}$$

and is *unstable* if

$$|1 + h\lambda| > 1. \tag{10.34}$$

We see that depending on λ and h, the explicit Euler method might be unstable.

Now we consider the alternative *implicit form of Euler's method*. This method is also called the *backward Euler method*. Instead of (10.22) we use

$$x_{n+1} = x_n + hf(x_{n+1}, t_{n+1}). \tag{10.35}$$

It can be seen that a drawback of this method is the necessity to solve (10.35) for x_{n+1}. This is generally a nonlinear problem requiring the techniques of Chapter 7. However, a strength of the implicit method is enhanced stability. This may be easily seen as follows. Apply (10.35) to the model problem (10.28), yielding

$$x_{n+1} = x_n + \lambda h x_{n+1}$$

or

$$x_{n+1} = \frac{1}{1 - \lambda h} x_n. \tag{10.36}$$

Clearly

$$x_n = \left[\frac{1}{1 - \lambda h}\right]^n x_0. \tag{10.37}$$

Since we must assume as before that $\lambda < 0$, the backward Euler method is stable for all $h > 0$. In this sense we may say that the backward Euler method is *unconditionally stable*. Thus, the implicit Euler method (10.35) is certainly more stable

---

[4]For stability we usually insist that $\lambda < 0$ as opposed to allowing $\lambda = 0$. This is to accommodate a concept called *bounded-input, bounded-output* (BIBO) *stability*. However, we do not consider the details of this matter here.

than the previous explicit form (10.22). However, the implicit and explicit forms have the same accuracy as both are first-order methods.

Example 10.3   We wish to apply the implicit and explicit forms of the Euler method to

$$\frac{dx}{dt} + x = 0$$

for $x(0) = 1$. Of course, since this is a simple linear problem, we immediately know that

$$x(t) = x(0)e^{-t} = e^{-t}$$

for all $t \geq 0$. Since we have $f(x, t) = -x$, we obtain

$$\frac{\partial f(x, t)}{\partial x} = -1,$$

implying that $\lambda = -1$. Thus, the explicit Euler method (10.22) gives

$$x_{n+1} = (1 - h)x_n \tag{10.38a}$$

for which $0 < h \leq 2$ via (10.33). Similarly, (10.35) gives for the implicit method

$$x_{n+1} = x_n - hx_{n+1}$$

or

$$x_{n+1} = \frac{1}{1 + h}x_n \tag{10.38b}$$

for which $h > 0$. In both (10.38a) and (10.38b) we have $x_0 = 1$.

Some typical simulation results for (10.38) appear in Fig. 10.3. Note the instability of the explicit method for the case where $h > 2$.

It is to be noted that a small step size h is desirable to achieve good accuracy. Yet a larger h is desirable to minimize the amount of computation involved in simulating the differential equation over the desired time interval.

Example 10.4   Now consider the ODE

$$\frac{dx}{dt} + 2tx = te^{-t^2}x^3$$

[5, pp. 62–63]. The exact solution to this differential equation is

$$x^2(t) = \frac{3}{e^{-t^2} + ce^{2t^2}} \tag{10.39}$$

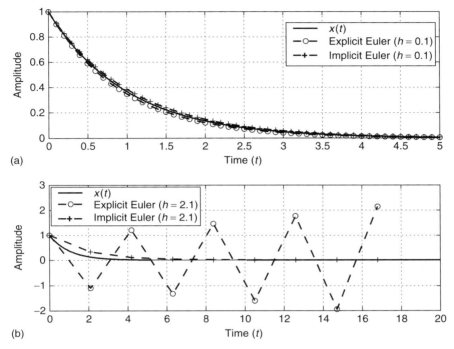

Figure 10.3 Illustration of the implicit and explicit forms of the Euler method for the differential equation in Example 10.3. In plot (a), h is small enough that the explicit method is stable. Here the implicit and explicit methods display similar accuracies. In plot (b), h is too big for the explicit method to be stable. Instability is indicated by the oscillatory behavior of the method and the growing amplitude of the oscillations with time. However, the implicit method remains stable, but because h is quite large, the accuracy is not very good.

for $t \geq 0$, where $c = \frac{3}{x_0^2} - 1$, and we assume that $c > 0$. Thus

$$f(x, t) = te^{-t^2}x^3 - 2tx$$

so

$$\frac{\partial f(x, t)}{\partial x} = 3te^{-t^2}x^2 - 2t.$$

Consequently

$$\lambda = \frac{\partial f(x_0, t_0)}{\partial x} = \frac{\partial f(x_0, 0)}{\partial x} = 0.$$

Via (10.33) we conclude that $h > 0$ is possible for both forms of the Euler method. Since stability is therefore not a problem here, we choose to simulate the differential

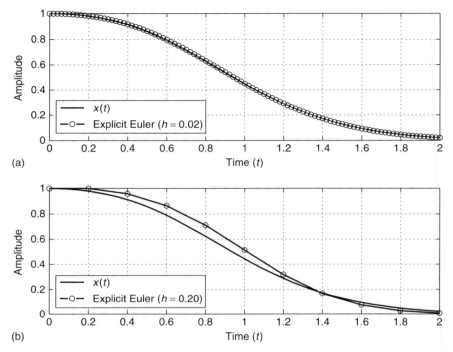

Figure 10.4   Illustration of the explicit Euler method for the differential equation in Example 10.4. Clearly, although stability is not a problem here, the accuracy of the method is better for smaller h.

equation using the explicit Euler method as this is much simpler to implement. Thus, from (10.22), we obtain

$$x_{n+1} = x_n + h[t_n e^{-t_n^2} x_n^3 - 2t_n x_n]. \tag{10.40}$$

We shall assume $x_0 = 1$ [initial condition $x(0)$]. Of course, $t_n = hn$ for $n = 0, 1, 2, \ldots$.

Figure 10.4 illustrates the exact solution from (10.39), and the simulated solution via (10.40) for $h = 0.02$ and $h = 0.20$. As expected, the result for $h = 0.02$ is more accurate.

The next example involves an ODE whose solution does not remain bounded over time. Nevertheless, our methods are applicable since we terminate the simulation after a finite time.

Example 10.5   Consider the ODE

$$\frac{dx}{dt} = t^2 - \frac{2x}{t}$$

for $t \geq t_0 > 0$ [5, pp. 60–61]. The exact solution is given by

$$x(t) = \frac{t^3}{5} + \frac{c}{5t^2}. \tag{10.41}$$

The initial condition is $x(t_0) = x_0$ with $t_0 > 0$, and so

$$x_0 = \frac{t_0^3}{5} + \frac{c}{5t_0^2},$$

implying that

$$c = 5t_0^2 x_0 - t_0^5.$$

Since $f(x, t) = t^2 - \frac{2x}{t}$, we have

$$\lambda = \frac{\partial f(x_0, t_0)}{\partial x} = -\frac{2}{t_0},$$

so via (10.33) for the explicit Euler method

$$0 < h \leq t_0.$$

However, this result is misleading here because $x(t)$ is not bounded with time. In other words, it does not really apply here. From (10.22)

$$x_{n+1} = x_n + h\left[t_n^2 - \frac{2x_n}{t_n}\right], \tag{10.42a}$$

where

$$t_n = t_0 + nh$$

for $n \in \mathbf{Z}^+$. If we consider the implicit method, then via (10.35)

$$x_{n+1} = x_n + h\left[t_{n+1}^2 - \frac{2x_{n+1}}{t_{n+1}}\right],$$

or

$$x_{n+1} = \frac{x_n + ht_{n+1}^2}{1 + \dfrac{2h}{t_{n+1}}}, \tag{10.42b}$$

where $t_{n+1} = t_0 + (n+1)h$ for $n \in \mathbf{Z}^+$.

Figure 10.5 illustrates the exact solution $x(t)$ from (10.41) along with the simulated solutions from (10.42). This is for $x_0 = 1$ and $t_0 = 0.05$ with $h = 0.025$ (a), and $h = 5$ (b). It can be seen that the implicit method is more accurate for t close to t_0. Of course, this could be very significant since startup transients are often of

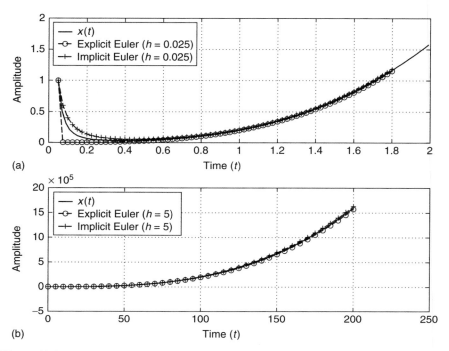

Figure 10.5   Illustration of the explicit and implicit forms of the Euler method for the ODE in Example 10.5. In plot (a), note that the implicit form tracks the true solution $x(t)$ better near $t_0 = 0.05$. In plot (b), note that both forms display a similar accuracy even though h is huge, provided $t \gg t_0$.

interest in simulations of dynamic systems. It is noteworthy that both implicit and explicit forms simulate the true solution with similar accuracy for t much bigger than t_0 even when h is large.

This example is of a "stiff system" (see Section 10.6).

Example 10.5 illustrates that stability and accuracy issues with respect to the numerical solution of ODE initial-value problems can be more subtle than our previous analysis would suggest. The reader is therefore duly cautioned about these matters.

Recalling (3.71) from Chapter 3 (or recalling Theorem 10.2), the Taylor formula for $x(t)$ about $t = t_n$ is [recall $x_n = x(t_n)$ for all n]

$$x_{n+1} = \underbrace{x_n + hf(x_n, t_n)}_{=\hat{x}_{n+1}} + \frac{1}{2}h^2 x^{(2)}(\xi) \tag{10.43}$$

for some $\xi \in [t_n, t_{n+1}]$. Thus, the *truncation error per step* in the Euler method is defined to be

$$e_{n+1} = x_{n+1} - \hat{x}_{n+1} = \tfrac{1}{2}h^2 x^{(2)}(\xi). \tag{10.44}$$

We may therefore state, as suggested earlier, that the truncation error per step is of order $O(h^2)$ because of this. The usefulness of (10.44) is somewhat limited in that it depends on the solution $x(t)$ [or rather on the second derivative $x^{(2)}(t)$], which is, of course, something we seldom know in practice.

How may we obtain more accurate methods? More specifically, this means finding methods for which the truncation error per step is of order $O(h^m)$ with $m > 2$.

One way to obtain improved accuracy is to try to improve the Euler method. More than one possibility for improvement exists. However, a popular approach is *Heun's method*. It is based on the following observation. A drawback of the Euler method in (10.22) is that $f(x_n, t_n)$ is the derivative $x^{(1)}(t)$ at the beginning of the interval $[t_n, t_{n+1}]$, and yet $x^{(1)}(t)$ varies over $[t_n, t_{n+1}]$. The implicit form of the Euler method works with $f(x_{n+1}, t_{n+1})$, namely, the derivative at $t = t_{n+1}$, and so has a similar defect. Therefore, intuitively, we may believe that we can improve the algorithm by replacing $f(x_n, t_n)$ with the average derivative

$$\tfrac{1}{2}\left[f(x_n, t_n) + f(x_n + hf(x_n, t_n), t_n + h)\right]. \tag{10.45}$$

This is *approximately* the average of $x^{(1)}(t)$ at the endpoints of interval $[t_n, t_{n+1}]$. The approximation is due to the fact that

$$f(x_{n+1}, t_{n+1}) \approx f(x_n + hf(x_n, t_n), t_n + h). \tag{10.46}$$

We see in (10.46) that we have employed (10.22) to approximate x_{n+1} according to $x_{n+1} = x_n + hf(x_n, t_n)$ (explicit Euler method). Of course, $t_{n+1} = t_n + h$ does not involve any approximation. Thus, *Heun's method* is defined by

$$x_{n+1} = x_n + \frac{h}{2}\left[f(x_n, t_n) + f(x_n + hf(x_n, t_n), t_n + h)\right]. \tag{10.47}$$

This is intended to replace (10.22) and (10.35).

However, (10.47) is an explicit method, and so we may wonder about its stability. If we apply the model problem to (10.47), we obtain

$$x_{n+1} = \left[1 + \lambda h + \tfrac{1}{2}h^2\lambda^2\right]x_n \tag{10.48}$$

for which

$$x_n = \left[1 + \lambda h + \tfrac{1}{2}h^2\lambda^2\right]^n x_0. \tag{10.49}$$

For stability we must select h such that we avoid $\lim_{n\to\infty}|x_n| = \infty$. For convenience, define

$$\sigma = 1 + \lambda h + \tfrac{1}{2}h^2\lambda^2, \tag{10.50}$$

so this requirement implies that we must have $|\sigma| \le 1$. A plot of (10.50) in terms of $h\lambda$ appears in Fig. 10.6. This makes it easy to see that we must have

$$-2 \le h\lambda \le 0. \tag{10.51}$$

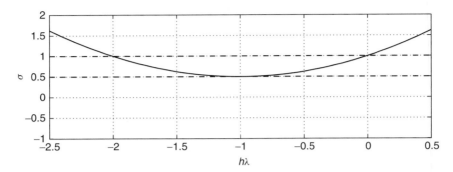

Figure 10.6 A plot of σ in terms of $h\lambda$ [see Eq. (10.50)].

Since $\lambda < 0$ is assumed [again because we assume that $x(t)$ is bounded] (10.51), implies that Heun's method is conditionally stable for the same conditions as in (10.33). Thus, the stability characteristics of the method are identical to those of the explicit Euler method, which is perhaps not such a surprize.

What about the accuracy of Heun's method? Here we may see that there is an improvement. Again via (3.71) from Chapter 3

$$x_{n+1} = x_n + hx^{(1)}(t_n) + \tfrac{1}{2}h^2 x^{(2)}(t_n) + \tfrac{1}{6}h^3 x^{(3)}(\xi) \tag{10.52}$$

for some $\xi \in [t_n, t_{n+1}]$. We may approximate $x^{(2)}(t_n)$ using a *forward difference* operation

$$x^{(2)}(t_n) \approx \frac{x^{(1)}(t_{n+1}) - x^{(1)}(t_n)}{h} \tag{10.53}$$

so that (10.52) becomes [using $x^{(1)}(t_{n+1}) = f(x_{n+1}, t_{n+1})$, and $x^{(1)}(t_n) = f(x_n, t_n)$]

$$x_{n+1} = x_n + hf(x_n, t_n) + \frac{1}{2}h^2 \left[\frac{x^{(1)}(t_{n+1}) - x^{(1)}(t_n)}{h} \right] + \frac{1}{6}h^3 x^{(3)}(\xi), \tag{10.54}$$

or upon simplifying this, we have

$$x_{n+1} = x_n + \frac{h}{2} \left[f(x_n, t_n) + f(x_{n+1}, t_{n+1}) \right] + \frac{1}{6}h^3 x^{(3)}(\xi). \tag{10.55}$$

Replacing $f(x_{n+1}, t_{n+1})$ in (10.55) with the approximation (10.46), and dropping the error term, we see that what remains is identical to (10.47), namely, Heun's method. Various approximations were made to arrive at this conclusion, but they are certainly reasonable, and so we claim that the truncation error per step for Heun's method is

$$e_{n+1} = \tfrac{1}{6}h^3 x^{(3)}(\xi), \tag{10.56}$$

where again $\xi \in [t_n, t_{n+1}]$, and so this error is of the order $O(h^3)$. In other words, although Heun's method is based on modifying the explicit Euler method, the modification has lead to a method with improved accuracy.

Example 10.6   Here we repeat Example 10.5 by applying Heun's method under the same conditions as for Fig. 10.5a. Thus, the differential equation is again

$$\frac{dx}{dt} = t^2 - \frac{2x}{t}$$

and again we choose $x_0 = 1.0$, $t_0 = 0.05$, with $h = 0.025$. The simulation result appears in Fig. 10.7.

It is very clear that Heun's method is distinctly more accurate than the Euler method, especially near $t = t_0$.

Heun's method may be viewed in a different light by considering the following. We may formally integrate (10.18) to arrive at $x(t)$ according to

$$x(t) - x(t_n) = \int_{t_n}^{t} f(x, \tau) \, d\tau. \tag{10.57}$$

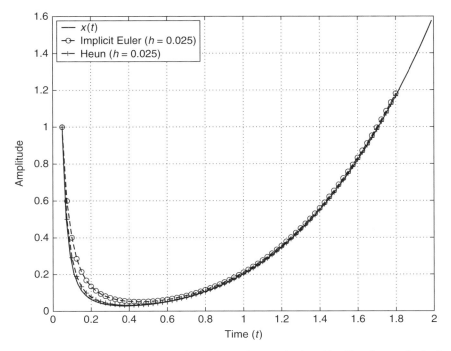

Figure 10.7 Comparison of the implicit Euler method with Heun's method for Example 10.6. We see that Heun's method is more accurate, especially for t near $t_0 = 0.05$.

So if $t = t_{n+1}$, then, from (10.57), we obtain

$$x(t_{n+1}) = x(t_n) + \int_{t_n}^{t_{n+1}} f(x, \tau)\, d\tau$$

or

$$x_{n+1} = x_n + \int_{t_n}^{t_{n+1}} f(x, \tau)\, d\tau. \qquad (10.58)$$

According to the trapezoidal rule for numerical integration (see Chapter 9, Section 9.2), we have

$$\int_{t_n}^{t_{n+1}} f(x, \tau)\, d\tau \approx \frac{h}{2}[f(x_n, t_n) + f(x_{n+1}, t_{n+1})]$$

(since $h = t_{n+1} - t_n$), and hence (10.58) becomes

$$x_{n+1} = x_n + \frac{h}{2}[f(x_n, t_n) + f(x_{n+1}, t_{n+1})], \qquad (10.59)$$

which is just the first step in the derivation of Heun's method. We may certainly call (10.59) the *trapezoidal method*. Clearly, it is an implicit method since we must solve (10.59) for x_{n+1}. Equally clearly, Eq. (10.59) appears in (10.55), and so we immediately conclude that the trapezoidal method is a second-order method with a truncation error per step of order $O(h^3)$. Thus, we may regard Heun's method as the explicit form of the trapezoidal method. Or, equivalently, the trapezoidal method can be regarded as the implicit form of Heun's method. We mention that the trapezoidal method is unconditionally stable, but will not prove this here.

The following example illustrates some more subtle issues relating to the stability of numerical solutions to ODE initial-value problems. It is an applications example from population dynamics, but the issues it raises are more broadly applicable. The example is taken from Beltrami [6].

Example 10.7 Suppose that $x(t)$ is the total size of a population (people, insects, bacteria, etc.). The members of the population exist in a habitat that can realistically support not more than N individuals. This is the *carrying capacity* for the system. The population may grow at some rate that diminishes to zero as $x(t)$ approaches N. But if the population size $x(t)$ is much smaller than the carrying capacity, the rate of growth might be considered proportional to the present population size. Consequently, a model for population growth might be

$$\frac{dx(t)}{dt} = rx(t)\left[1 - \frac{x(t)}{N}\right]. \qquad (10.60)$$

This is called the *logistic equation*. By separation of variables this equation has solution

$$x(t) = \frac{N}{1 + ce^{-rt}} \qquad (10.61)$$

for $t \geq 0$. As usual, c depends on the initial condition (initial population size) $x(0)$. The exact solution in (10.61) is clearly "well behaved." Therefore, any numerical solution to (10.60) must also be well behaved.

Suppose that we attempt to simulate (10.60) numerically using the explicit Euler method. In this case we obtain [via (10.22)]

$$x_{n+1} = (1 + hr)x_n - \frac{hr}{N}x_n^2. \tag{10.62}$$

Suppose that we transform variables according to $x_n = \alpha y_n$ in which case (10.62) can be rewritten as

$$y_{n+1} = (1 + hr)y_n \left[1 - \frac{hr\alpha}{N(1 + hr)} y_n \right]. \tag{10.63}$$

If we select

$$\alpha = \frac{N(1 + hr)}{hr},$$

then (10.63) becomes

$$y_{n+1} = \lambda y_n (1 - y_n), \tag{10.64}$$

where $\lambda = 1 + hr$. We recognize this as the *logistic map* from Chapter 7 [see Examples 7.3–7.5 and Eq. (7.83)]. From Section 7.6 in particular we recall that this map can become *chaotic* for certain choices of λ. In other words, *chaotic instability* is another possible failure mode for a numerical method that purports to solve ODEs.

The explicit form of the Euler method is actually an example of a *first-order Runge–Kutta method*. Similarly, Heun's method is an example of a *second-order Runge–Kutta method*. It is second-order essentially because the approximation involved retains the term in h^2 in the Taylor series expansion. We mention that methods of still higher order can be obtained simply by retaining more terms in the Taylor series expansion of (10.21). This is seldom done because to do so requires working with derivatives of increasing order, and this requires much computational effort. But this effort can be completely avoided by developing Runge–Kutta methods of higher order. We now outline a general approach for doing this. It is based on material from Rao [7].

All Runge–Kutta methods have a particular form that may be stated as

$$x_{n+1} = x_n + h\alpha(x_n, t_n, h), \tag{10.65}$$

where $\alpha(x_n, t_n, h)$ is called the *increment function*. The increment function is selected to represent the average slope on the interval $t \in [t_n, t_{n+1}]$. In particular, the increment function has the form

$$\alpha(x_n, t_n, h) = \sum_{j=1}^{m} c_j k_j, \tag{10.66}$$

where m is called the *order of the Runge–Kutta method*, c_j are constants, and coefficients k_j are obtained recursively according to

$$k_1 = f(x_n, t_n)$$
$$k_2 = f(x_n + a_{2,1}hk_1, t_n + p_2h)$$
$$k_3 = f(x_n + a_{3,1}hk_1 + a_{3,2}hk_2, t_n + p_3h)$$
$$\vdots$$
$$k_m = f\left(x_n + \sum_{j=1}^{m-1} a_{m,j}hk_j, t_n + p_mh\right). \tag{10.67}$$

A more compact description of the Runge–Kutta methods is

$$x_{n+1} = x_n + h\sum_{j=1}^{m} c_j k_j, \tag{10.68a}$$

where

$$k_j = f\left(x_n + h\sum_{l=1}^{j-1} a_{j,l}k_l, t_n + p_jh\right). \tag{10.68b}$$

To specify a particular method requires selecting a variety of coefficients (c_j, $a_{j,l}$, etc.). How is this to be done?

We illustrate with examples. Suppose that $m = 1$. In this case

$$x_{n+1} = x_n + hc_1k_1 = x_n + hc_1 f(x_n, t_n) \tag{10.69}$$

which gives (10.22) when $c_1 = 1$. Thus, we are justified in calling the explicit Euler method a first-order Runge–Kutta method.

Suppose that $m = 2$. In this case

$$x_{n+1} = x_n + hc_1 f(x_n, t_n) + hc_2 f(x_n + a_{2,1}hf(x_n, t_n), t_n + p_2h). \tag{10.70}$$

We observe that if we choose

$$c_2 = c_1 = \tfrac{1}{2}, a_{2,1} = 1, p_2 = 1, \tag{10.71}$$

then (10.70) reduces to

$$x_{n+1} = x_n + \tfrac{1}{2}h[f(x_n, t_n) + f(x_n + hf(x_n, t_n), t_n + h)],$$

which is Heun's method [compare this with (10.47)]. Thus, we are justified in calling Heun's method a second-order Runge–Kutta method. However, the coefficient choices in (10.71) are not unique. Other choices will lead to other second-order Runge–Kutta methods. We may arrive at a systematic approach for creating alternatives as follows.

For convenience as in (10.21), define $x_n^{(k)} = x^{(k)}(t_n)$. Since $m = 2$, we will consider the Taylor expansion

$$x_{n+1} = x_n + h x_n^{(1)} + \tfrac{1}{2} h^2 x_n^{(2)} + O(h^3) \tag{10.72}$$

[recall (10.52)] for which the term $O(h^3)$ simply denotes the higher-order terms. We recall that $x^{(1)}(t) = f(x, t)$, so $x_n^{(1)} = f(x_n, t_n)$, and via the chain rule

$$x^{(2)}(t) = \frac{\partial f}{\partial t} + \frac{\partial f}{\partial x} \frac{dx}{dt} = \frac{\partial f}{\partial t} + \frac{\partial f}{\partial x} f(x, t), \tag{10.73}$$

so (10.72) may be rewritten as

$$x_{n+1} = x_n + h f(x_n, t_n) + \frac{1}{2} h^2 \frac{\partial f(x_n, t_n)}{\partial t} + \frac{1}{2} h^2 \frac{\partial f(x_n, t_n)}{\partial x} f(x_n, t_n) + O(h^3). \tag{10.74}$$

Once again, the Runge–Kutta method for $m = 2$ is

$$x_{n+1} = x_n + h c_1 f(x_n, t_n) + h c_2 f(x_n + a_{2,1} h k_1, t_n + p_2 h). \tag{10.75}$$

Recalling (10.26), the Taylor expansion of $f(x_n + a_{2,1} h k_1, t_n + p_2 h)$ is given by

$$f(x_n + a_{2,1} h k_1, t_n + p_2 h) = f(x_n, t_n) + a_{2,1} h f(x_n, t_n) \frac{\partial f(x_n, t_n)}{\partial x}$$

$$+ p_2 h \frac{\partial f(x_n, t_n)}{\partial t} + O(h^2). \tag{10.76}$$

Now we substitute (10.76) into (10.75) to obtain

$$x_{n+1} = x_n + (c_1 + c_2) h f(x_n, t_n) + p_2 c_2 h^2 \frac{\partial f(x_n, t_n)}{\partial t}$$

$$+ a_{2,1} c_2 h^2 \frac{\partial f(x_n, t_n)}{\partial x} f(x_n, t_n) + O(h^3). \tag{10.77}$$

We may now compare like terms of (10.77) with those in (10.74) to conclude that the coefficients we seek satisfy the nonlinear system of equations

$$c_1 + c_2 = 1, \ p_2 c_2 = \tfrac{1}{2}, \ a_{2,1} c_2 = \tfrac{1}{2}. \tag{10.78}$$

To generate second-order Runge–Kutta methods, we are at liberty to choose the coefficients c_1, c_2, p_2, and $a_{2,1}$ in any way we wish so long as the choice satisfies (10.78). Clearly, Heun's method is only one choice among many possible choices. We observe from (10.78) that we have four unknowns, but possess three equations. Thus, we may select one parameter "arbitrarily" that will then determine the remaining ones. For example, we may select c_2, so then, from (10.78)

$$c_1 = 1 - c_2, \quad p_2 = \frac{1}{2c_2}, \quad \text{and} \quad a_{2,1} = \frac{1}{2c_2}. \tag{10.79}$$

Since one parameter is freely chosen, thus constraining all the rest, we say that second-order Runge–Kutta methods possess *one degree of freedom*.

It should be clear that the previous procedure may be extended to systematically generate Runge–Kutta methods of higher order (i.e., $m > 2$). However, this requires a more complete description of the Taylor expansion for a function in two variables. This is stated as follows.

Theorem 10.3: Taylor's Theorem   Suppose that $f(x, t)$, and all of its partial derivatives of order $n + 1$ or less are defined and continuous on $D = \{(x, t) | a \le t \le b, c \le x \le d\}$. Let $(x_0, t_0) \in D$; then, for all $(x, t) \in D$, there is a point $(\eta, \xi) \in D$ such that

$$f(x, t) = \sum_{r=0}^{n} \left\{ \frac{1}{r!} \sum_{k=0}^{r} \binom{r}{k} (t - t_0)^{r-k} (x - x_0)^k \frac{\partial^r f(x_0, t_0)}{\partial t^{r-k} \partial x^k} \right\}$$

$$+ \frac{1}{(n+1)!} \sum_{k=0}^{n+1} \binom{n+1}{k} (t - t_0)^{n+1-k} (x - x_0)^k \frac{\partial^{n+1} f(\eta, \xi)}{\partial t^{n+1-k} \partial x^k}$$

for which (η, ξ) is on the line segment that joins the points (x_0, t_0), and (x, t).

Proof Omitted.

The reader can now easily imagine that any attempt to apply this approach for $m > 2$ will be quite tedious. Thus, we shall not do this here. We will restrict ourselves to stating a few facts. Applying the method to $m = 3$ (i.e., the generation of third-order Runge–Kutta methods) leads to algorithm coefficients satisfying six equations with eight unknowns. There will be 2 degrees of freedom as a consequence.

Fourth-order Runge–Kutta methods (i.e., $m = 4$) also possess two degrees of freedom, and also have a truncation error per step of $O(h^5)$. One such method (attributed to Runge) in common use is

$$x_{n+1} = x_n + \frac{h}{6} [k_1 + 2k_2 + 2k_3 + k_4], \tag{10.80}$$

where

$$k_1 = f(x_n, t_n)$$

$$k_2 = f\left(x_n + \tfrac{1}{2}hk_1, t_n + \tfrac{1}{2}h\right)$$

$$k_3 = f\left(x_n + \tfrac{1}{2}hk_2, t_n + \tfrac{1}{2}h\right)$$

$$k_4 = f(x_n + hk_3, t_n + h). \tag{10.81}$$

Of course, an infinite number of other fourth-order methods are possible.

We mention that Runge–Kutta methods are explicit methods, and so in principle carry some risk of instability. However, it turns out that the higher the order of the method, the lower the risk of stability problems. In particular, users of fourth-order methods typically experience few stability problems in practice. In fact, it can be shown that on applying (10.80) and (10.81) to the model problem (10.28), we obtain

$$x_n = \left[1 + h\lambda + \frac{1}{2}h^2\lambda^2 + \frac{1}{6}h^3\lambda^3 + \frac{1}{24}h^4\lambda^4 \right]^n x_0. \tag{10.82}$$

A plot of

$$\sigma = 1 + h\lambda + \frac{1}{2}h^2\lambda^2 + \frac{1}{6}h^3\lambda^3 + \frac{1}{24}h^4\lambda^4 \tag{10.83}$$

in terms of $h\lambda$ appears in Fig. 10.8. To avoid $\lim_{n \to \infty} |x_n| = \infty$, we must have $|\sigma| \leq 1$, and so it turns out that

$$-2.785 \leq h\lambda \leq 0$$

(see Table 9.11 on p. 685 of Rao [7]). This is in agreement with Fig. 10.8. Thus

$$\lambda < 0, \qquad h \leq \frac{2.785}{|\lambda|}. \tag{10.84}$$

This represents an improvement over (10.33).

Example 10.8   Once again we repeat Example 10.5 for which

$$\frac{dx}{dt} = t^2 - \frac{2x}{t}$$

with $x_0 = 1.0$ for $t_0 = .05$, but here instead our step size is now $h = 0.05$. Additionally, our comparison is between Heun's method and the fourth-order Runge–Kutta method defined by Eqs. (10.80) and (10.81).

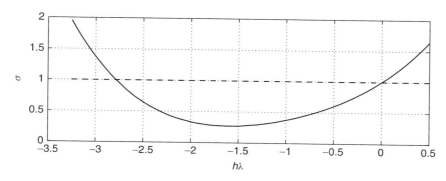

Figure 10.8   A plot of σ in terms of $h\lambda$ [see Eq. (10.83)].

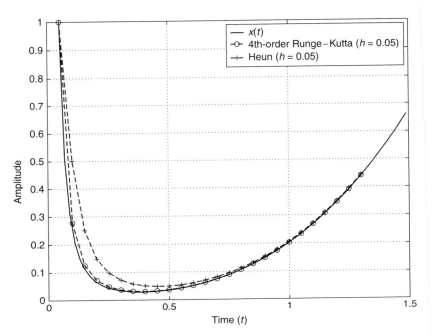

Figure 10.9 Comparison of Heun's method (a second-order Runge–Kutta method) with a fourth-order Runge–Kutta method [Eqs. (10.80) and (10.81)]. This is for the differential equation in Example 10.8.

The simulated solutions based on these methods appear in Fig. 10.9. As expected, the fourth-order method is much more accurate than Heun's method. The plot in Fig. 10.9 was generated using MATLAB, and the code for this appears in Appendix 10.A as an example. (Of course, previous plots were also produced by MATLAB codes similar to that in Appendix 10.A.)

10.3 SYSTEMS OF FIRST-ORDER ODEs

The methods of Section 10.2 may be extended to handle systems of first-order ODEs where the number of ODEs in the system is arbitrary (but finite). However, we will consider only systems of two first-order ODEs here. Specifically, we wish to solve (numerically)

$$\frac{dx}{dt} = f(x, y, t), \tag{10.85a}$$

$$\frac{dy}{dt} = g(x, y, t), \tag{10.85b}$$

where the initial condition is $x_0 = x(t_0)$ and $y_0 = y(t_0)$. This is sufficient, for example, to simulate the Duffing equation mentioned in Section 10.1.

As in Section 10.2, we will begin with Euler methods. Therefore, following (10.21) we may Taylor-expand $x(t)$ and $y(t)$ about the sampling time $t = t_n$. As before, for convenience we may define $x_n^{(k)} = x^{(k)}(t_n)$, $y_n^{(k)} = y^{(k)}(t_n)$. The relevant expansions are given by

$$x_{n+1} = x_n + hx_n^{(1)} + \tfrac{1}{2}h^2 x_n^{(2)} + \cdots, \tag{10.86a}$$

$$y_{n+1} = y_n + hy_n^{(1)} + \tfrac{1}{2}h^2 y_n^{(2)} + \cdots. \tag{10.86b}$$

The explicit Euler method follows by retaining the first two terms in each expansion in (10.86). Thus, the Euler method in this case is

$$x_{n+1} = x_n + hf(x_n, y_n, t_n), \tag{10.87a}$$

$$y_{n+1} = y_n + hg(x_n, y_n, t_n), \tag{10.87b}$$

where we have used the fact that $x_n^{(1)} = f(x_n, y_n, t_n)$ and $y_n^{(1)} = g(x_n, y_n, t_n)$. As we might expect, the implicit form of the Euler method is

$$x_{n+1} = x_n + hf(x_{n+1}, y_{n+1}, t_{n+1}), \tag{10.88a}$$

$$y_{n+1} = y_n + hg(x_{n+1}, y_{n+1}, t_{n+1}). \tag{10.88b}$$

Of course, to employ (10.88) will generally involve solving a nonlinear system of equations for x_{n+1} and y_{n+1}, necessitating the use of Chapter 7 techniques. As before, we refer to parameter h as the *step size*.

The accuracy of the explicit and implicit Euler methods for systems is the same as for individual equations; specifically, it is $O(h^2)$. However, stability analysis is more involved. Matrix methods simply cannot be avoided. This is demonstrated as follows.

The model problem for a single first-order ODE was Eq. (10.28). For a coupled system of two first-order ODEs as in (10.85), the model problem is now

$$\frac{dx}{dt} = a_{00}x + a_{01}y, \tag{10.89a}$$

$$\frac{dy}{dt} = a_{10}x + a_{11}y. \tag{10.89b}$$

Here $a_{i,j}$ are real-valued constants. We remark that this may be written in more compact matrix form

$$\frac{d\overline{x}}{dt} = A\overline{x}, \tag{10.90}$$

where $\overline{x} = \overline{x}(t) = [x(t)y(t)]^T$ and $d\overline{x}/dt = [dx(t)/dt \quad dy(t)/dt]^T$ and, of course

$$A = \begin{bmatrix} a_{00} & a_{01} \\ a_{10} & a_{11} \end{bmatrix}. \tag{10.91}$$

Powerful claims are possible using matrix methods. For example, it can be argued that if $\overline{x}(t) \in \mathbf{R}^N$ (so that $A \in \mathbf{R}^{N \times N}$), then $d\overline{x}(t)/dt = A\overline{x}(t)$ has solution

$$\overline{x}(t) = e^{At}\overline{x}(0) \tag{10.92}$$

for $t \geq 0$; that is, (10.92) for $N = 2$ is the general solution to (10.90) [and hence to (10.89)].[5]

The constants in A are related to $f(x, y, t)$ and $g(x, y, t)$ in the following manner. Recall (10.25). If we retain only the linear terms in the Taylor expansions of f and g around the point (x_0, y_0, t_0), then

$$f(x, y, t) \approx f(x_0, y_0, t_0) + (x - x_0)\frac{\partial f(x_0, y_0, t_0)}{\partial x} + (y - y_0)\frac{\partial f(x_0, y_0, t_0)}{\partial y}$$
$$+ (t - t_0)\frac{\partial f(x_0, y_0, t_0)}{\partial t}, \tag{10.93a}$$

$$g(x, y, t) \approx g(x_0, y_0, t_0) + (x - x_0)\frac{\partial g(x_0, y_0, t_0)}{\partial x} + (y - y_0)\frac{\partial g(x_0, y_0, t_0)}{\partial y}$$
$$+ (t - t_0)\frac{\partial g(x_0, y_0, t_0)}{\partial t}. \tag{10.93b}$$

As a consequence

$$A = \begin{bmatrix} \dfrac{\partial f(x_0, y_0, t_0)}{\partial x} & \dfrac{\partial f(x_0, y_0, t_0)}{\partial y} \\ \dfrac{\partial g(x_0, y_0, t_0)}{\partial x} & \dfrac{\partial g(x_0, y_0, t_0)}{\partial y} \end{bmatrix}. \tag{10.94}$$

At this point we may apply the explicit Euler method (10.87) to the model problem (10.89), which results in

$$\begin{bmatrix} x_{n+1} \\ y_{n+1} \end{bmatrix} = \begin{bmatrix} 1 + ha_{00} & ha_{01} \\ ha_{10} & 1 + ha_{11} \end{bmatrix} \begin{bmatrix} x_n \\ y_n \end{bmatrix}. \tag{10.95}$$

---

[5]Yes, as surprising as it seems, although At is a matrix $\exp(At)$ makes sense as an operation. In fact, for example, with $x(t) = [x_0(t) \cdots x_{n-1}(t)]^T$, the system of first-order ODEs

$$\frac{dx(t)}{dt} = Ax(t) + by(t)$$

$(A \in \mathbf{R}^{n \times n}, b \in \mathbf{R}^n,$ and $y(t) \in \mathbf{R})$ has the general solution

$$x(t) = e^{At}x(0^-) + \int_{0^-}^{t} e^{A(t-\tau)}by(\tau)\, d\tau.$$

The integral in this solution is an example of a *convolution integral*.

An alternative form for this is

$$\overline{x}_{n+1} = (I + hA)\overline{x}_n. \tag{10.96}$$

Here I is a 2×2 identity matrix and $\overline{x}_n = [x_n \quad y_n]^T$. With $\overline{x}_0 = [x_0 \quad y_0]^T$, we may immediately claim that

$$\overline{x}_n = (I + hA)^n \overline{x}_0, \tag{10.97}$$

where $n \in \mathbf{Z}^+$. We observe that this includes (10.32) as a special case. Naturally, we must select step size h to avoid instability; that is, we are forced to select h to prevent $\lim_{n \to \infty} ||\overline{x}_n|| = \infty$. In principle, the choice of norm is arbitrary, but 2-norms are often chosen. We recall that there is a nonsingular matrix T (matrix of eigenvectors) such that

$$T^{-1}[I + hA]T = \Lambda, \tag{10.98}$$

where Λ is the *matrix of eigenvalues*. We will assume that

$$\Lambda = \begin{bmatrix} \lambda_0 & 0 \\ 0 & \lambda_1 \end{bmatrix}. \tag{10.99}$$

In other words, we assume $I + hA$ is *diagonalizable*. This is not necessarily always the case, but is an acceptable assumption for present purposes. Since from (10.98) we have $I + hA = T \Lambda T^{-1}$, (10.96) becomes

$$\overline{x}_{n+1} = T \Lambda T^{-1} \overline{x}_n,$$

or

$$T^{-1}\overline{x}_{n+1} = \Lambda T^{-1}\overline{x}_n. \tag{10.100}$$

Let $\overline{y}_n = T^{-1}\overline{x}_n$, so therefore (10.100) becomes

$$\overline{y}_{n+1} = \Lambda \overline{y}_n. \tag{10.101}$$

In any norm $\lim_{n \to \infty} ||\overline{y}_n|| \neq \infty$, provided $|\lambda_k| \leq 1$ for all $k = 0, 1, \ldots, N - 1$ ($\Lambda \in \mathbf{R}^{N \times N}$). Consequently, $\lim_{n \to \infty} ||\overline{x}_n|| \neq \infty$ too (because $||\overline{x}_n|| = ||T\overline{y}_n|| \leq ||T|| \; ||\overline{y}_n||$ and $||T||$ is finite). We conclude that h is an acceptable step size, provided the eigenvalues of $I + hA$ do not possess a magnitude greater than unity. Note that in practice we normally insist that h result in $|\lambda_k| < 1$ for all k.

We may apply the previous stability analysis to the implicit Euler method. Specifically, apply (10.88) to model problem (10.89), giving

$$x_{n+1} = x_n + h[a_{00}x_{n+1} + a_{01}y_{n+1}]$$
$$y_{n+1} = y_n + h[a_{10}x_{n+1} + a_{11}y_{n+1}],$$

which in matrix form becomes

$$\begin{bmatrix} x_{n+1} \\ y_{n+1} \end{bmatrix} = \begin{bmatrix} x_n \\ y_n \end{bmatrix} + h \begin{bmatrix} a_{00} & a_{01} \\ a_{10} & a_{11} \end{bmatrix} \begin{bmatrix} x_{n+1} \\ y_{n+1} \end{bmatrix},$$

or more compactly as

$$\bar{x}_{n+1} = \bar{x}_n + hA\bar{x}_{n+1}. \tag{10.102}$$

Consequently, for $n \in \mathbf{Z}^+$

$$\bar{x}_n = ([I - hA]^{-1})^n \bar{x}_0. \tag{10.103}$$

For convenience we can define $B = [I - hA]^{-1}$. Superficially, (10.103) seems to have the same form as (10.97). We might be lead therefore to believe (falsely) that the implicit method can be unstable, too. However, we may assume that there exists a nonsingular matrix V such that (if $A \in \mathbf{R}^{N \times N}$)

$$V^{-1}AV = \Gamma, \tag{10.104}$$

where $\Gamma = \mathrm{diag}(\gamma_0, \gamma_1, \ldots, \gamma_{N-1})$, which is the diagonal matrix of the eigenvalues of A. (Once again, it is not necessarily the case that A is always diagonalizable, but the assumption is reasonable for our present purposes.) Immediately

$$[I - hA]^{-1} = [I - hV\Gamma V^{-1}]^{-1} = (V[V^{-1}V - h\Gamma]V^{-1})^{-1}$$

so that

$$[I - hA]^{-1} = V[I - h\Gamma]^{-1}V^{-1}. \tag{10.105}$$

Consequently $\bar{x}_{n+1} = [I - hA]^{-1}\bar{x}_n$ becomes

$$\bar{x}_{n+1} = V[I - h\Gamma]^{-1}V^{-1}\bar{x}_n. \tag{10.106}$$

Define $\bar{y}_n = V^{-1}\bar{x}_n$, and so (10.106) becomes

$$\bar{y}_{n+1} = [I - h\Gamma]^{-1}\bar{y}_n. \tag{10.107}$$

Because $I - h\Gamma$ is a diagonal matrix, a typical main diagonal element of $[I - h\Gamma]^{-1}$ is $\sigma_k = 1/(1 - h\gamma_k)$. It is a fact (which we will not prove here) that the model problem in the general case is stable provided the eigenvalues of A all possess negative-valued real parts.[6] Thus, provided $\mathrm{Re}(\gamma_k) < 0$ for all k, we are assured that $|\sigma_k| < 1$ for all k, and hence $\lim_{n \to \infty} ||\bar{y}_n|| = 0$. Thus, $\lim_{n \to \infty} ||\bar{x}_n|| = 0$, too, and so we conclude that the implicit form of the Euler method is unconditionally stable. Thus, if the model problem is stable, we may select any $h > 0$.

---

[6]The eigenvalues of A may be complex-valued, and so it is the real parts of these that truly determine system stability.

The following example is of a linear system for which a mathematically exact solution can be found. [In fact, the solution is given by (10.92).]

Example 10.9   Consider the ODE system

$$\frac{dx}{dt} = -2x + \tfrac{1}{4}y, \tag{10.108a}$$

$$\frac{dy}{dt} = -3x. \tag{10.108b}$$

The initial condition is $x_0 = x(0) = 1$, $y_0 = y(0) = -1$. From (10.94) we see that

$$A = \begin{bmatrix} -2 & \tfrac{1}{4} \\ -3 & 0 \end{bmatrix}.$$

The eigenvalues of A are $\gamma_0 = -\tfrac{1}{2}$, and $\gamma_1 = -\tfrac{3}{2}$. These eigenvalues are both negative, and so the solution to (10.108) happens to be stable. In fact, the exact solution can be shown to be

$$x(t) = -\tfrac{3}{4}e^{-t/2} + \tfrac{7}{4}e^{-3t/2}, \tag{10.109a}$$

$$y(t) = -\tfrac{9}{2}e^{-t/2} + \tfrac{7}{2}e^{-3t/2} \tag{10.109b}$$

for $t \geq 0$. Note that the eigenvalues of A appear in the exponents of the exponentials in (10.109). This is not a coincidence. The explicit Euler method has the iterations

$$x_{n+1} = x_n + h\left[-2x_n + \tfrac{1}{4}y_n\right], \tag{10.110a}$$

$$y_{n+1} = y_n - 3hx_n. \tag{10.110b}$$

Simulation results are shown in Figs. 10.10 and 10.11 for $h = 0.1$ and $h = 1.4$, respectively. This involves comparing (10.110a,b) with the exact solution (10.109a,b). Figure 10.10b shows a plot of the eigenvalues of $I + hA$ for various step sizes. We see from this that choosing $h = 1.4$ must result in an unstable simulation. This is confirmed by the result in Fig. 10.11. For comparison purposes, the eigenvalues of $I + hA$ and of $[I - hA]^{-1}$ are plotted in Fig. 10.12. This shows that, at least in this particular case, the implicit method is more stable than the explicit method.

Example 10.10   Recall the Duffing equation of Section 10.1. Also, recall the fact that this ODE can be rewritten in the form of (10.85a,b), and this was done in Eq. (10.4a,b).

Figures 10.13 and 10.14 show the result of simulating the Duffing equation using the explicit Euler method for the model parameters

$$F = 0.5, \quad \omega = 1, \quad m = 1, \quad \alpha = 1, \quad \delta = 0.1, \quad k = 0.05.$$

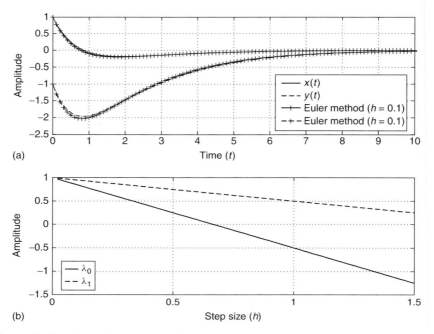

Figure 10.10 Simulation results for $h = 0.1$. Plot (b) shows the eigenvalues of $I + hA$. Both plots were obtained by applying the explicit form of the Euler method to the ODE system of Example 10.9. Clearly, the simulation is stable for $h = 0.1$.

Thus (10.1) is now

$$\frac{d^2x}{dt^2} = 0.5\cos(t) - 0.05\frac{dx}{dt} - [x + 0.1x^3].$$

We use initial condition $x(0) = y(0) = 0$. The driving function (applied force) $0.5\cos(t)$ is being opposed by the restoring force of the spring (terms in square brackets) and friction (first derivative term). Therefore, on physical grounds, we do not expect the solution $x(t)$ to grow without bound as $t \to \infty$. Thus, the simulated solution to this problem must be stable, too.

We mention that an analytical solution to the differential equation that we are simulating is not presently known.

From (10.94) for our Duffing system example we have

$$A = \begin{bmatrix} 0 & 1 \\ -\dfrac{\alpha}{m} - \dfrac{3\delta}{m}x_0^2 & -\dfrac{k}{m} \end{bmatrix}.$$

Example 10.11 According to Hydon [1, p. 61], the second-order ODE

$$\frac{d^2x}{dt^2} = \left(\frac{dx}{dt}\right)^2\frac{1}{x} + \left(x - \frac{1}{x}\right)\frac{dx}{dt} \tag{10.111}$$

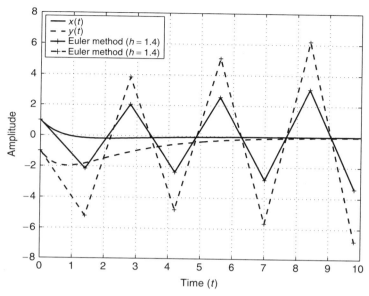

Figure 10.11 This is the result of applying the explicit form of the Euler method to the ODE system of Example 10.9. Clearly, the simulation is not stable for $h = 1.4$. This is predicted by the eigenvalue plot in Fig. 10.10b, which shows that one of the eigenvalues of $I + hA$ has a magnitude exceeding unity for this choice of h.

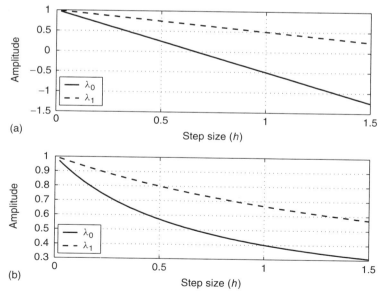

Figure 10.12 Plots of the eigenvalues of $I + hA$ (a), which determine the stability of the explicit Euler method and the eigenvalues of $[I - hA]^{-1}$ (b), which determine the stability of the implicit Euler method. This applies for the ODE system of Example 10.9.

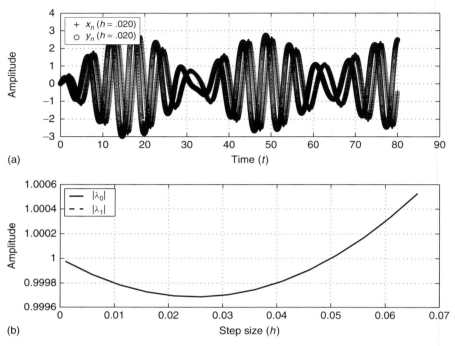

(a)

(b)

Figure 10.13 (a) Explicit Euler method simulation of the Duffing equation; (b) magnitude of the eigenvalues of $I + hA$. Both plots were obtained by applying the explicit form of the Euler method to the ODE system of Example 10.10, which is the Duffing equation expressed as a coupled system of first-order ODEs. The simulation is apparently stable for $h = 0.02$. This is in agreement with the prediction based on the eigenvalues of $I + hA$ [plot (b)], which have a magnitude of less than unity for this choice of h.

has the exact solution

$$
x(t) = \begin{cases}
c_1 - \sqrt{c_1^2 - 1} \, \tanh(\sqrt{c_1^2 - 1}\,(t + c_2)), & c_1^2 > 1 \\
\\
c_1 - (t + c_2)^{-1}, & c_1^2 = 1 \\
\\
c_1 + \sqrt{1 - c_1^2} \, \tanh(\sqrt{1 - c_1^2}\,(t + c_2)), & c_1^2 < 1
\end{cases}
\qquad (10.112)
$$

The ODE in (10.111) can be rewritten as the system of first-order ODEs

$$
\frac{dx}{dt} = y, \qquad\qquad\qquad\qquad (10.113\text{a})
$$

$$
\frac{dy}{dt} = \frac{y^2}{x} + \left(x - \frac{1}{x}\right) y. \qquad (10.113\text{b})
$$

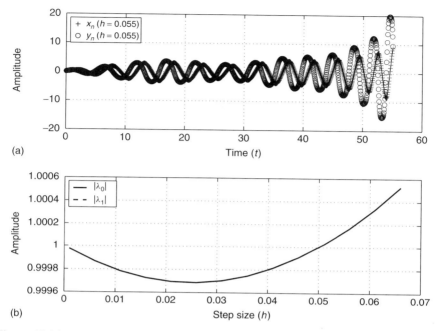

Figure 10.14   (a) Explicit Euler method simulation of the Duffing equation; (b) magnitude of the eigenvalues of $I + hA$. Both plots were obtained by applying the explicit form of the Euler method to the ODE system of Example 10.10, which is the Duffing equation expressed as a coupled system of first-order ODEs. The simulation is not stable for $h = 0.055$. This is in agreement with the prediction based on the eigenvalues of $I + hA$ [plot (b)], which have a magnitude exceeding unity for this choice of h.

The initial condition is $x_0 = x(0)$, and $y(0) = \frac{dx(t)}{dt}|_{t=0}$. From (10.94) we have

$$A = \begin{bmatrix} 0 & 1 \\ -\dfrac{y_0^2}{x_0^2} + y_0\left(1 + \dfrac{1}{x_0^2}\right) & \dfrac{2y_0}{x_0} + \left(x_0 - \dfrac{1}{x_0}\right) \end{bmatrix}. \tag{10.114}$$

Using (10.113a), we may obtain $y(t)$ from (10.112). For example, let us consider simulating the case $c_1^2 = 1$. Thus, in this case

$$y(t) = \frac{1}{(t + c_2)^2}. \tag{10.115}$$

For the choice $c_1^2 = 1$, we have

$$x_0 = c_1 - \frac{1}{c_2}, \tag{10.116a}$$

$$y_0 = \frac{1}{c_2^2}. \tag{10.116b}$$

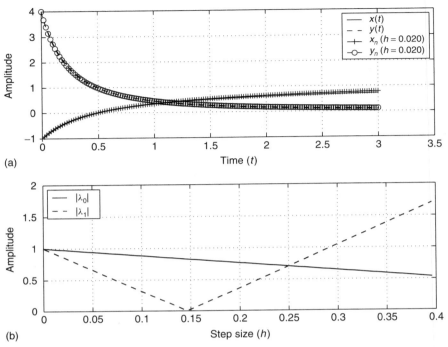

Figure 10.15   Magnitude of the eigenvalues of $I + hA$ is shown in plot (b). Both plots show the simulation results of applying the explicit Euler method to the ODE system in Example 10.11. The simulation is (as expected) stable for $h = 0.02$. Clearly, the simulated result agrees well with the exact solution.

If we select $c_1 = 1$, then

$$x_0 = 1 - \frac{1}{c_2}, \qquad y_0 = (1 - x_0)^2. \qquad (10.117)$$

Let us assume $x_0 = -1$, and so $y_0 = 4$. The result of applying the explicit Euler method to system (10.113) with these conditions is shown in Figs. 10.15 and 10.16. The magnitudes of the eigenvalues of $I + hA$ are displayed in plots (b) of both figures. We see that for Fig. 10.15, $h = 0.02$ and a stable simulation is the result, while for Fig. 10.16, we have $h = 0.35$, for which the simulation is unstable. This certainly agrees with the stability predictions based on finding the eigenvalues of matrix $I + hA$.

Examples 10.10 and 10.11 illustrate just how easy it is to arrive at differential equations that are not so simple to simulate in a stable manner with low-order explicit methods. It is possible to select an h that is "small" in some sense, yet not small enough for stability. The cubic nonlinearity in the Duffing model makes the implementation of the implicit form of Euler's method in this problem quite

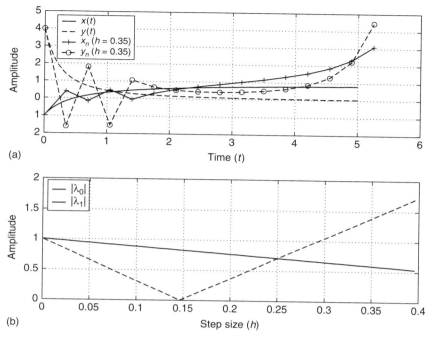

Figure 10.16   Plot (b) shows magnitude of eigenvalues of $I + hA$. Both plots show the simulation results of applying the explicit Euler method to the ODE system in Example 10.11. The simulation is (as expected) not stable for $h = 0.35$. Instability is confirmed by the fact that the simulated result deviates greatly from the exact solution when t is sufficiently large.

unattractive. So, a better approach to simulating the Duffing equation is with a higher-order explicit method.

For example, Heun's method for (10.85) may be stated as

$$x_{n+1} = x_n + \tfrac{1}{2}h[f(x_n, y_n, t_n) + f(x_n + k_1, y_n + l_1, t_n + h)], \qquad (10.118a)$$

$$y_{n+1} = y_n + \tfrac{1}{2}h[g(x_n, y_n, t_n) + g(x_n + k_1, y_n + l_1, t_n + h)], \qquad (10.118b)$$

where

$$k_1 = hf(x_n, y_n, t_n), l_1 = hg(x_n, y_n, t_n). \qquad (10.119)$$

Also, for example, Chapter 36 of Bronson [8] contains a summary of higher-order methods that may be applied to (10.85).

Example 10.12   Recall the Duffing equation simulation in Example 10.10. Figure 10.17 illustrates the simulation of the Duffing equation using both the explicit Euler method and Heun's method for a small h (i.e., $h = 0.005$ in both cases).

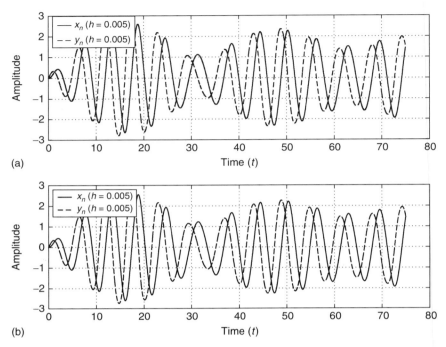

Figure 10.17 Comparison of the explicit Euler (a) and Heun (b) method simulations of the ODE in Example 10.12, which is the Duffing equation. Here the step size h is small enough that the two methods give similar results.

At this point we note that there are other ways to display the results of numerical solutions to ODEs that can lead to further insights into the behavior of the dynamic system that is modeled by those ODEs. Figure 10.18 illustrates the *phase portrait* of the Duffing system. This is obtained by plotting the points (x_n, y_n) on the Cartesian plane, yielding an approximate plot of $(x(t), y(t))$. The resulting curve is the *trajectory*, or *orbit* for the system. Periodicity of the system's response is indicated by curves that encircle a point of equilibrium, which in this case would be the center of the Cartesian plane [i.e., point $(0, 0)$]. The trajectory is tending to an approximately ellipse-shaped closed curve indicative of approximately simple harmonic motion.

The results in Fig. 10.18 are based on the parameters given in Example 10.10. However, Fig. 10.19 shows what happens when the system parameters become

$$F = 0.3, \quad \omega = 1, \quad m = 1, \quad \alpha = -1, \quad \delta = 1, \quad k = 0.22. \qquad (10.120)$$

The phase portrait displays a more complicated periodicity than what appears in Fig. 10.18. The figure is similar to Fig. 2.2.5 in Guckenheimer and Holmes [11].

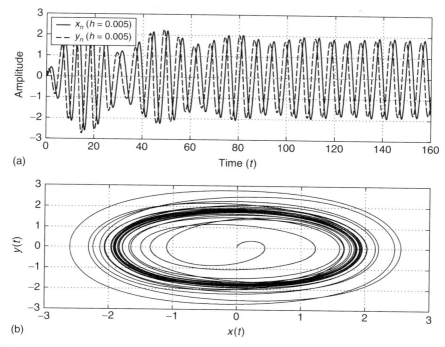

Figure 10.18 (a) The result of applying Heun's method to obtain the numerical solution of the Duffing system specified in Example 10.10; (b) the phase portrait for the system obtained by plotting the points (x_n, y_n) [from plot (a)] on the Cartesian plane, thus yielding an approximate plot of $(x(t), y(t))$.

As in the cases of the explicit and implicit Euler methods, we may obtain a theory of stability for Heun's method. As before, the approach is to apply the model problem (10.90) to (10.118) and (10.119). As an exercise, the reader should show that this yields

$$\bar{x}_{n+1} = \left[I + hA + \tfrac{1}{2}h^2 A^2 \right] \bar{x}_n, \tag{10.121}$$

where I is the 2×2 identity matrix, A is obtained by using (10.94), and, of course, $\bar{x}_n = [x_n y_n]^T$. [The similarity between (10.121) and (10.48) is no coincidence.] Criteria for the selection of step size h leading to a stable simulation can be obtained by analysis of (10.121). But the details are not considered here.

10.4   MULTISTEP METHODS FOR ODEs

The numerical ODE solvers we have considered so far were either implicit methods or explicit methods. But in all cases they were examples of so-called *single-step*

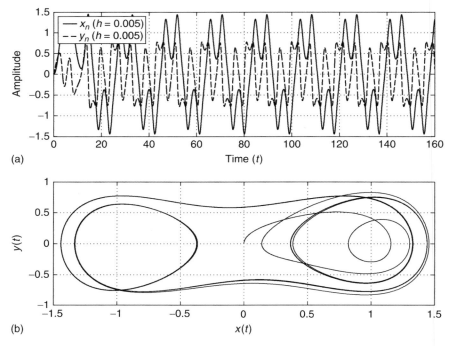

Figure 10.19   (a) The result of applying Heun's method to obtain the numerical solution of the Duffing system specified in Example 10.12 [i.e., using the parameter values in Eq. (10.120)]; (b) the phase portrait for the system obtained by plotting the points (x_n, y_n) [from plot (a)] on the Cartesian plane, thus yielding an approximate plot of $(x(t), y(t))$.

methods; that is, $\overline{x}_{n+1}$ was ultimately only a function of $\overline{x}_n$. A disadvantage of single-step methods is that to achieve good accuracy often requires the use of higher-order methods (e.g., fourth- or fifth-order Runge–Kutta). But higher-order methods need many function evaluations per step, and so are computationally expensive.

Implicit single-step methods, although inherently stable, are not more accurate than explicit methods, although they can track fast changes in the solution $\overline{x}(t)$ better than can explicit methods (recall Example 10.5). However, implicit methods may require nonlinear system solvers (i.e., Chapter 7 methods) as part of their implementation. This is a complication that is also not necessarily very efficient computationally. Furthermore, the methods in Chapter 7 possess their own stability problems. Therefore, in this section we introduce *multistep predictor–corrector methods* that overcome some of the deficiencies of the methods we have considered so far.

In this section we return to consideration of a single first-order ODE IVP

$$x^{(1)}(t) = \frac{dx(t)}{dt} = f(x(t), t), \ x_0 = x(0). \tag{10.122}$$

In reality, we have already seen a single-step predictor–corrector method in Section 10.2. Suppose that we have the following method:

$$\tilde{x}_{n+1} = x_n + hf(x_n, t_n) \quad \text{(predictor step)} \tag{10.123a}$$

$$x_{n+1} = x_n + \tfrac{1}{2}h[f(\tilde{x}_{n+1}, t_{n+1}) + f(x_n, t_n)] \quad \text{(corrector step)}. \tag{10.123b}$$

If we substitute (10.123a) into (10.123b), we again arrive at Heun's method [Eq. (10.47)], which overcame the necessity to solve for x_{n+1} in the implicit method of Eq. (10.59) (trapezoidal method). Generally, predictor–corrector methods replace implicit methods in this manner, and we will see more examples further on in this section. When a higher-order implicit method is "converted" to a predictor–corrector method, the need to solve nonlinear equations is eliminated and accuracy is preserved, but the stability characteristics of the implicit method will be lost, at least to some extent. Of course, a suitable stability theory will still allow the user to select reasonable values for the step size parameter h.

We may now consider a few simple examples of multistep methods. Perhaps the simplest multistep methods derive from the numerical differentiation ideas from Section 9.6 (of Chapter 9). Recall (9.138), for which

$$x^{(1)}(t) = \frac{1}{2h}[x(t+h) - x(t-h)] - \frac{1}{6}h^2 x^{(3)}(\xi) \tag{10.124}$$

for some $\xi \in [t-h, t+h]$. The explicit Euler method (10.22) can be replaced with the *midpoint method* derived from [using $t = t_n$ in (10.124)]

$$f(x(t_n), t_n) \approx \frac{1}{2h}[x(t_{n+1}) - x(t_{n-1})],$$

so this method is

$$x_{n+1} = x_{n-1} + 2hf(x_n, t_n). \tag{10.125}$$

This is an explicit method, but x_{n+1} depends on x_{n-1} as well as x_n. We may call it a two-step method. Similarly, via (9.153)

$$f(x(t_n), t_n) \approx \frac{1}{2h}[-3x(t_n) + 4x(t_{n+1}) - x(t_{n+2})],$$

so we have the method

$$x_{n+2} = 4x_{n+1} - 3x_n - 2hf(x_n, t_n)$$

which can be rewritten as

$$x_{n+1} = 4x_n - 3x_{n-1} - 2hf(x_{n-1}, t_{n-1}). \tag{10.126}$$

Finally, via (9.154), we obtain

$$f(x(t_n), t_n) \approx \frac{1}{2h}[x(t_{n-2}) - 4x(t_{n-1}) + 3x(t_n)],$$

yielding the method

$$x_{n+1} = \tfrac{4}{3}x_n - \tfrac{1}{3}x_{n-1} + \tfrac{2}{3}hf(x_{n+1}, t_{n+1}). \tag{10.127}$$

Method (10.126) is a two-step method that is explicit, but (10.127) is a two-step *implicit* method since we need to solve for x_{n+1}.

A problem with multistep methods is that the IVP (10.122) provides only one initial condition x_0. But for $n = 1$ in any of (10.125), (10.126), or (10.127), we need to know x_1; that is, for two-step methods we need two initial conditions, or *starting values*. A simple way out of this dilemma is to use single-step methods to provide any missing starting values. In fact, predictor–corrector methods derived from single-step concepts (e.g., Runge–Kutta methods) are often used to provide the starting values for multistep methods.

What about multistep method accuracy? Let us consider the midpoint method again. From (10.124) for some $\xi_n \in [t_{n-1}, t_{n+1}]$

$$x(t_{n+1}) = x(t_{n-1}) + 2hf(x(t_n), t_n) + \tfrac{1}{3}h^3 x^{(3)}(\xi_n), \tag{10.128}$$

so the method (10.125) has a truncation error per step that is $O(h^3)$. The midpoint method is therefore more accurate than the explicit Euler method [recall (10.43) and (10.44)]. Yet we see that both methods need only one function evaluation per step. Thus, the midpoint method is more efficient than the Euler method. We recall that Heun's method (a Runge–Kutta method) has a truncation error per step that is $O(h^3)$, too [recall (10.56)], and so Heun's method may be used to initialize (i.e., provide starting values for) the midpoint method (10.125). This specific situation holds up in general. Thus, a multistep method can often achieve comparable accuracy to single-step methods, and yet use fewer function calls, leading to reduced computational effort.

What about stability considerations? Let us continue with our midpoint method example. If we apply the model problem (10.28) to (10.125), we have the difference equation

$$x_{n+1} = x_{n-1} + 2h\lambda x_n, \tag{10.129}$$

which is a second-order difference equation. This has *characteristic equation*[7]

$$z^2 - 2h\lambda z - 1 = 0. \tag{10.130}$$

---

[7]We may rewrite (10.129) as

$$x_{n+2} - 2h\lambda x_{n+1} - x_n = 0,$$

which has the *z-transform*

$$(z^2 - 2h\lambda z - 1)X(z) = 0.$$

For convenience, let $\rho = h\lambda$, in which case the roots of (10.130) are easily seen to be

$$z_1 = \rho + \sqrt{\rho^2 + 1}, \qquad z_2 = \rho - \sqrt{\rho^2 + 1}, . \tag{10.131}$$

A general solution to (10.129) will have the form

$$x_n = c_1 z_1^n + c_2 z_2^n \tag{10.132}$$

for $n \in \mathbf{Z}^+$. Knowledge of x_0 and x_1 allows us to solve for the constants c_1 and c_2 in (10.132), if this is desired. However, more importantly, we recall that we assume $\lambda < 0$, and we seek step size $h > 0$ so that $\lim_{n\to\infty} |x_n| \neq \infty$. But in this case $\rho = h\lambda < 0$, and hence from (10.131), $|z_2| > 1$ *for all* $h > 0$. If $c_2 \neq 0$ in (10.132) (which is *practically* always the case), then we will have $\lim_{n\to\infty} |x_n| = \infty$! Thus, the midpoint method is inherently unstable under all realistic conditions ! Term $c_1 z_1^n$ in (10.132) is "harmless" since $|z_1| < 1$ for suitable h. But the term $c_2 z_2^n$, often called a *parasitic term*, will eventually "blow up" with increasing n, thus fatally corrupting the approximation to $x(t)$.

Unfortunately, parasitic terms are inherent in multistep methods. However, there are more advanced methods with stability theories designed to minimize the effects of the parasitics. We now consider a few of these improved multistep methods.

10.4.1 Adams–Bashforth Methods

Here we look at the Adams–Bashforth (AB) family of multistep ODE IVP solvers. Section 10.4.2 will look at the Adams–Moulton (AM) family. Our approach follows Epperson [12, Section 6.6]. Both families are derived using Lagrange interpolation [recall Section 6.2 from Chapter 6 (above)].

Recall (10.122) which we may integrate to obtain ($t_n = t_0 + nh$)

$$x(t_{n+1}) = x(t_n) + \int_{t_n}^{t_{n+1}} f(x(t), t) \, dt. \tag{10.133}$$

Now suppose that we had the samples $x(t_{n-k})$ for $k = 0, 1, \ldots, m$ (i.e., $m + 1$ samples of the exact solution $x(t)$). Via Lagrange interpolation theory, we may interpolate $F(t) = f(x(t), t)$ [the integrand of (10.133)] for $t \in [t_{n-m}, t_{n+1}]$[8] according to

$$p_m(t) = \sum_{k=0}^{m} L_k(t) f(x(t_{n-k}), t_{n-k}), \tag{10.134}$$

A solution to (10.129) exists only if $z^2 - 2h\lambda z - 1 = 0$. If the reader has not had a signals and systems course (or equivalent) then this reasoning must be accepted "on faith." But it may help to observe that the reasoning is similar to the theory of solution for linear ODEs in constant coefficients.

[8] The upper limit on the interval $[t_{n-m}, t_n]$ has been extended from t_n to t_{n+1} here. This is allowed under interpolation theory, and actually poses no great problem in either method development or error analysis. We are using the Lagrange interpolant to extrapolate from $t = t_n$ to t_{n+1}.

where

$$L_k(t) = \prod_{\substack{i=0 \\ i \neq k}}^{m} \frac{t - t_{n-i}}{t_{n-k} - t_{n-i}}. \tag{10.135}$$

From (6.14) for some $\xi_t \in [t_{n-m}, t_{n+1}]$

$$F(t) = p_m(t) + \frac{1}{(m+1)!} F^{(m+1)}(\xi_t) \prod_{i=0}^{m} (t - t_{n-i}). \tag{10.136}$$

However, $F(t) = f(x(t), t) = x^{(1)}(t)$ so (10.136) becomes

$$F(t) = p_m(t) + \frac{1}{(m+1)!} x^{(m+2)}(\xi_t) \prod_{i=0}^{m} (t - t_{n-i}). \tag{10.137}$$

Thus, if we now substitute (10.137) into (10.133), we obtain

$$x(t_{n+1}) = x(t_n) + \sum_{k=0}^{m} f(x(t_{n-k}), t_{n-k}) \int_{t_n}^{t_{n+1}} L_k(t)\, dt + R_m(t_{n+1}), \tag{10.138}$$

where

$$R_m(t_{n+1}) = \int_{t_n}^{t_{n+1}} \frac{1}{(m+1)!} x^{(m+2)}(\xi_t) \underbrace{\prod_{i=0}^{m} (t - t_{n-i})}_{= \pi(t)}\, dt. \tag{10.139}$$

Polynomial $\pi(t)$ does not change sign for $t \in [t_n, t_{n+1}]$ (which is the interval of integration). Thus, we can say that there is a $\xi_n \in [t_n, t_{n+1}]$ such that

$$R_m(t_{n+1}) = \frac{1}{(m+1)!} x^{(m+2)}(\xi_n) \int_{t_n}^{t_{n+1}} \pi(t)\, dt. \tag{10.140}$$

For convenience, define

$$\rho_m = \frac{1}{(m+1)!} \int_{t_n}^{t_{n+1}} \pi(t)\, dt = \frac{1}{(m+1)!} \int_{t_n}^{t_{n+1}} (t - t_n)(t - t_{n-1}) \cdots (t - t_{n-m})\, dt \tag{10.141}$$

and

$$\lambda_k = \int_{t_n}^{t_{n+1}} L_k(t)\, dt. \tag{10.142}$$

Thus, (10.138) reduces to [with $R_m(t_{n+1}) = \rho_m x^{(m+2)}(\xi_n)$]

$$x(t_{n+1}) = x(t_n) + \sum_{k=0}^{m} \lambda_k f(x(t_{n-k}), t_{n-k}) + \rho_m x^{(m+2)}(\xi_n) \tag{10.143}$$

TABLE 10.1   Adams–Bashforth Method Parameters

m	λ_0	λ_1	λ_2	λ_3	$R_m(t_{n+1})$
0	h				$\dfrac{1}{2}h^2 x^{(2)}(\xi_n)$
1	$\dfrac{3}{2}h$	$-\dfrac{1}{2}h$			$\dfrac{5}{12}h^3 x^{(3)}(\xi_n)$
2	$\dfrac{23}{12}h$	$-\dfrac{16}{12}h$	$\dfrac{5}{12}h$		$\dfrac{3}{8}h^4 x^{(4)}(\xi_n)$
3	$\dfrac{55}{24}h$	$-\dfrac{59}{24}h$	$\dfrac{37}{24}h$	$-\dfrac{9}{24}h$	$\dfrac{251}{720}h^5 x^{(5)}(\xi_n)$

for some $\xi_n \in [t_n, t_{n+1}]$. The *order $m + 1$ Adams–Bashforth method* is therefore defined to be

$$x_{n+1} = x_n + \sum_{k=0}^{m} \lambda_k f(x_{n-k}, t_{n-k}). \tag{10.144}$$

It is an explicit method involving $m + 1$ steps. Table 10.1 summarizes the method parameters for various m, and is essentially Table 6.6 from Ref. 12.

10.4.2 Adams–Moulton Methods

The Adams–Moulton methods are a modification of the Adams–Bashforth methods. The Adams–Bashforth methods interpolate using the nodes $t_n, t_{n-1}, \ldots, t_{n-m}$. On the other hand, the Adams–Moulton methods interpolate using the nodes $t_{n+1}, t_n, \ldots, t_{n-m+1}$. Note that the number of nodes is the same in both methods. Consequently, (10.138) becomes

$$x(t_{n+1}) = x(t_n) + \sum_{k=-1}^{m-1} f(x(t_{n-k}), t_{n-k}) \int_{t_n}^{t_{n+1}} L_k(t)\, dt + R_m(t_{n+1}), \tag{10.145}$$

where now

$$L_k(t) = \prod_{\substack{i=-1 \\ i \neq k}}^{m-1} \frac{t - t_{n-i}}{t_{n-k} - t_{n-i}}, \tag{10.146}$$

and $R_m(t_{n+1}) = \rho_m x^{(m+2)}(\xi_n)$ with

$$\rho_m = \frac{1}{(m+1)!} \int_{t_n}^{t_{n+1}} (t - t_{n+1})(t - t_n) \cdots (t - t_{n-m+1})\, dt. \tag{10.147}$$

Thus, the *order m + 1 Adams–Moulton method* is defined to be

$$x_{n+1} = x_n + \sum_{k=-1}^{m-1} \lambda_k f(x_{n-k}, t_{n-k}), \tag{10.148}$$

where

$$\lambda_k = \int_{t_n}^{t_{n+1}} L_k(t) \, dt \tag{10.149}$$

[same as (10.142) except $k = -1, 0, 1, \ldots, m - 1$, and $L_k(t)$ is now (10.146)]. Method (10.148) is an implicit method since it is necessary to solve for x_{n+1}. It also requires $m + 1$ steps. Table 10.2 summarizes the method parameters for various m, and is essentially Table 6.7 from Ref. 12.

10.4.3 Comments on the Adams Families

For small values of m in Tables 10.1 and 10.2, we see that the Adams families (AB family and AM family) correspond to methods seen earlier. To be specific:

1. For $m = 0$ in Table 10.1, $\lambda_0 = h$, so (10.144) yields the explicit Euler method (10.22).
2. For $m = 0$ in Table 10.2, $\lambda_{-1} = h$, so (10.148) yields the implicit Euler method (10.35).
3. For $m = 1$ in Table 10.2, $\lambda_{-1} = \lambda_0 = \frac{1}{2}h$, so (10.148) yields the trapezoidal method (10.59).

Stability analysis for members of the Adams families is performed in the usual manner. For example, when $m = 1$ in Table 10.1 (i.e., consider the second-order AB method), Eq. (10.144) becomes

$$x_{n+1} = x_n + \frac{1}{2}h[3f(x_n, t_n) - f(x_{n-1}, t_{n-1})]. \tag{10.150}$$

TABLE 10.2   Adams–Moulton Method Parameters

m	λ_{-1}	λ_0	λ_1	λ_2	$R_m(t_{n+1})$
0	h				$-\dfrac{1}{2}h^2 x^{(2)}(\xi_n)$
1	$\dfrac{1}{2}h$	$\dfrac{1}{2}h$			$-\dfrac{1}{12}h^3 x^{(3)}(\xi_n)$
2	$\dfrac{5}{12}h$	$\dfrac{8}{12}h$	$-\dfrac{1}{12}h$		$-\dfrac{1}{24}h^4 x^{(4)}(\xi_n)$
3	$\dfrac{9}{24}h$	$\dfrac{19}{24}h$	$-\dfrac{5}{24}h$	$\dfrac{1}{24}h$	$-\dfrac{19}{720}h^5 x^{(5)}(\xi_n)$

Application of the model problem (10.28) to (10.150) yields

$$x_{n+1} = \left(1 + \frac{3}{2}h\lambda\right) x_n - \frac{1}{2}h\lambda x_{n-1}$$

or

$$x_{n+2} - \left(1 + \frac{3}{2}h\lambda\right) x_{n+1} + \frac{1}{2}h\lambda x_n = 0.$$

This has characteristic equation (with $\rho = h\lambda$)

$$z^2 - \left(1 + \frac{3}{2}\rho\right) z + \frac{1}{2}\rho = 0. \tag{10.151}$$

This equation has roots

$$z_1 = \frac{1}{2}\left[\left(1 + \frac{3}{2}\rho\right) + \sqrt{1 + \rho + \frac{9}{4}\rho^2}\right],$$

$$z_2 = \frac{1}{2}\left[\left(1 + \frac{3}{2}\rho\right) - \sqrt{1 + \rho + \frac{9}{4}\rho^2}\right]. \tag{10.152}$$

We need to know what range of $h > 0$ yields $|z_1|, |z_2| < 1$. We consider only $\rho < 0$ since $\lambda < 0$. Figure 10.20 plots $|z_1|$ and $|z_2|$ versus ρ, and suggests that we may select h such that

$$-1 < h\lambda < 0. \tag{10.153}$$

Plots of *stability regions* for the other Adams families members may be seen in Figs. 6.7 and 6.8 of Epperson [12]. Note that stability regions occupy the complex plane as it is assumed in such a context that $\lambda \in \mathbf{C}$. However, we have restricted

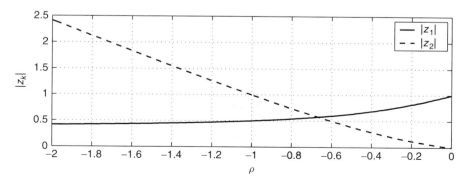

Figure 10.20   Magnitudes of the roots in (10.152) as a function of $\rho = h\lambda$.

our attention to a first-order ODE IVP here, and so it is actually enough to assume that $\lambda \in \mathbf{R}$.

Finally, observe that AB and AM methods can be combined to yield predictor–correctors. An mth-order AB method can act as a predictor for an mth-order AM method that is the corrector. A Runge–Kutta method can initialize the procedure.

10.5   VARIABLE-STEP-SIZE (ADAPTIVE) METHODS FOR ODEs

Accuracy in the numerical solution of ODEs requires either increasing the order of the method applied to the problem or decreasing the step-size parameter h. However, high-order methods (e.g., Runge–Kutta methods of order exceeding 5) are not very attractive at least in part because of the computational effort involved. To preserve accuracy while reducing computational requirements suggests that we should adaptively vary step size h.

Recall Example 10.5, where we saw that low order methods were not accurate near $t = t_0$. For a method of a given order, we would like in Example 10.5 to have a small h for t near t_0, but a larger h for t away from t_0. This would reduce the overall number of function evaluations needed to estimate $x(t)$ for $t \in [t_0, t_f]$. The idea of adaptively varying h from step to step requires monitoring the error in the solution somehow; that is, ideally, we need to infer $e_n = |x(t_n) - x_n|$ [x_n is the estimate of $x(t)$ at $t = t_n$ from some method] at step n. If e_n is small enough, h may be increased in size at the next step, but if e_n is too big, we decrease h.

Of course, we do not know $x(t_n)$, so we do not have direct access to the error e_n. However, one idea that is implemented in modern software tools (e.g., MATLAB routines ode23 and ode45) is to compute x_n for a given h using two methods, each of a different order. The method of higher order is of greater accuracy, so if x_n does not differ much between the methods, we are lead to believe that h is small enough, and so may be increased in the next step. On the other hand, if the x_n values given by the different methods significantly vary, we are then lead to believe that h is too big, and so should be reduced.

In this section we give only a basic outline of the main ideas of this process. Our emphasis is on the *Runge–Kutta–Fehlberg* (RKF) *methods*, of which MATLAB routines ode23 and ode45 are particular implementations. Routine ode23 implements second- and third-order Runge–Kutta methods, while ode45 implements fourth- and fifth-order Runge–Kutta methods. Computational efficiency is maintained by sharing intermediate results that are common to both second- and third-order methods and common to both fourth- and fifth-order methods. More specifically, Runge–Kutta methods of consecutive orders have constants such as k_j [recall (10.67)] in common with each other and so need not be computed twice. We mention that ode45 implements a method based on Dormand and Prince [14], and a more detailed account of this appears in Epperson [12]. An analysis of the RKF methods also appears in Burden and Faires [17]. The details of all of this are quite tedious, and so are not presented here. It is also worth noting that an account of MATLAB ODE solvers is given by Shampine and Reichelt [13], who present some improvements to the older MATLAB codes that make them better at solving stiff systems (next section).

A pseudocode for something like ode45 is as follows, and is based on Algorithm 6.5 in Epperson [12]:

```
Input t0, x0; { initial condition and starting time }
Input tolerance ε > 0;
Input the initial step size h > 0, and final time tf > t0;
n := 0;
while tn ≤ tf do begin
        X1 := RKF4(xn, tn, h); { 4th order RKF estimate of xn+1 }
        X2 := RKF5(xn, tn, h); { 5th order RKF estimate of xn+1 }
        E := |X1 − X2|;
        if 1/4 hε ≤ E ≤ hε then begin { h is OK }
        xn+1 := X2;
        tn+1 := tn + h;
        n := n + 1;
        else if E > hε then { h is too big }
        h := h/2; { reduce h and repeat }
        else { h is too small }
        h := 2h;
        xn+1 := X2;
        tn+1 := tn + h;
        n := n + 1;
        end;
end;
```

Of course, variations on the "theme" expressed in this pseudocode are possible. As noted in Epperson [12], a drawback of this algorithm is that it will tend to oscillate between small and large step size values. We emphasize that the method is based on considering the *local error* in going from time step t_n to t_{n+1}. However, this does not in itself guarantee that the *global error* $|x(t_n) − x_n|$ is small. It turns out that if adequate smoothness prevails [e.g., if $f(x, t)$ is Lipschitz as per Definition 10.1], then small local errors do imply small global errors (see Theorem 6.6 or 6.7 in Ref. 12).

Example 10.13   This example illustrates a typical application of MATLAB routine ode23 to the problem of simulating the Colpitts oscillator circuit of Example 10.2.

Figure 10.21 shows a typical plot of $v_{CE}(t)$, and the phase portrait for parameter values

$$V_{TH} = 0.75 \text{ V (volts)}, \quad V_{CC} = 5 \text{ (V)}, \quad V_{EE} = -5 \text{ (V)}, \quad R_{EE} = 400 \, \Omega \text{ (ohms)},$$

$$R_L = 35 \, (\Omega), \quad L = 98.5 \times 10^{-6} \text{ H (henries)}, \quad \beta_F = 200, \quad R_{ON} = 100 \, (\Omega),$$

$$C_1 = C_2 = 54 \times 10^{-9} \text{ F (farads)}. \tag{10.154}$$

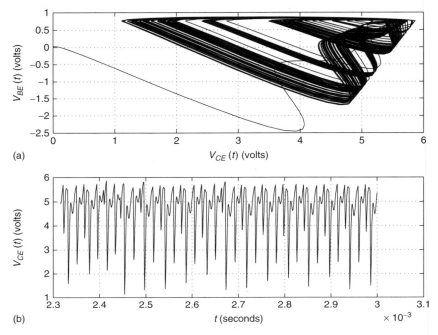

Figure 10.21   Chaotic regime (a) phase portrait of Colpitts oscillator and (b) collector–emitter voltage, showing typical results of applying MATLAB routine ode23 to the simulation of the Colpitts oscillator circuit. Equation (10.154) specifies the circuit parameters for the results shown here.

These circuit parameters were used in Kennedy [9], and the phase portrait in Fig. 10.21 is essentially that in Fig. 5 of that article [9]. The MATLAB code that generates Fig. 10.21 appears in Appendix 10.B.

For the parameters in (10.154) the circuit simulation phase portrait in Fig. 10.21 is that of a *strange attractor* [11, 16], and so is strongly indicative (although not conclusive) of chaotic dynamics in the circuit.

We note that under "normal" circumstances the Colpitts oscillator is intended to generate sinusoidal waveforms, and so the chaotic regime traditionally represents a failure mode, or abnormal operating condition for the circuit. However, Kennedy [9] suggests that the chaotic mode of operation may be useful in chaos-based data communications (e.g., chaotic–carrier communications).

The following circuit parameters lead to approximately sinusoidal circuit outputs:

$$V_{TH} = 0.75\,\text{V}, \quad V_{CC} = 5\,\text{V}, \quad V_{EE} = -5\,\text{V}, \quad R_{EE} = 100\,\Omega,$$

$$R_L = 200\,\Omega, \quad L = 100 \times 10^{-6}\,\text{H}, \quad \beta_F = 80, \quad R_{ON} = 115\,\Omega,$$

$$C_1 = 45 \times 10^{-9}\,\text{F}, \quad C_2 = 58 \times 10^{-9}\,\text{F}. \tag{10.155}$$

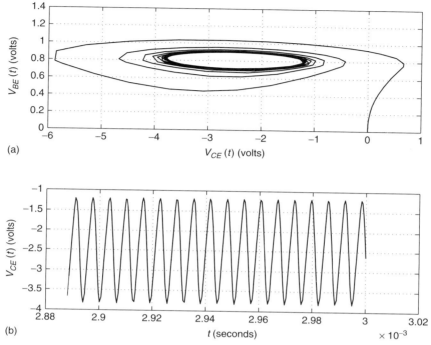

Figure 10.22 Sinusoidal operations (a) phase portrait of Colpitts operator and (b) collector–emitter voltage, showing typical results of applying MATLAB routine ode23 to the simulation of the Colpitts oscillator circuit. Equation (10.155) specifies the circuit parameters for the results shown here.

Figure 10.22 shows the phase portrait for the oscillator using these parameter values. We see that $v_{CE}(t)$ is much more sinusoidal than in Fig. 10.21. The trajectory in the phase portrait of Fig. 10.22 is tending to an elliptical closed curve indicative of simple harmonic (i.e., sinusoidal) oscillation.

10.6 STIFF SYSTEMS

Consider the general system of coupled first-order ODEs

$$
\begin{aligned}
\frac{dx_0(t)}{dt} &= f_0(x_0, x_1, \ldots, x_{m-1}, t), \\
\frac{dx_1(t)}{dt} &= f_1(x_0, x_1, \ldots, x_{m-1}, t), \\
&\vdots \\
\frac{dx_{m-1}(t)}{dt} &= f_{m-1}(x_0, x_1, \ldots, x_{m-1}, t),
\end{aligned}
\tag{10.156}
$$

which we wish to solve for $t \geq 0$ given $\overline{x}(0)$, where $\overline{x}(t) = [x_0(t) \quad x_1(t) \quad \cdots \quad x_{m-1}(t)]^T$. If we also define $\overline{f}(\overline{x}(t), t) = [f_0(\overline{x}(t), t) \quad f_1(\overline{x}(t), t) \quad \cdots \quad f_{m-1}(\overline{x}(t), t)]^T$, then we may express (10.156) in compact vector form as

$$\frac{d\overline{x}(t)}{dt} = \overline{f}(\overline{x}(t), t). \tag{10.157}$$

We have so far described a general *order m ODE IVP*.

If we now wish to consider the stability of a numerical method applied to the solution of (10.156) [or (10.157)], then we need to consider the model problem

$$\frac{d\overline{x}(t)}{dt} = A\overline{x}(t), \tag{10.158}$$

where

$$A = \begin{bmatrix} \dfrac{\partial f_0(\overline{x}(0), 0)}{\partial x_0} & \dfrac{\partial f_0(\overline{x}(0), 0)}{\partial x_1} & \cdots & \dfrac{\partial f_0(\overline{x}(0), 0)}{\partial x_{m-1}} \\[2ex] \dfrac{\partial f_1(\overline{x}(0), 0)}{\partial x_0} & \dfrac{\partial f_1(\overline{x}(0), 0)}{\partial x_1} & \cdots & \dfrac{\partial f_1(\overline{x}(0), 0)}{\partial x_{m-1}} \\[2ex] \vdots & \vdots & & \vdots \\[2ex] \dfrac{\partial f_{m-1}(\overline{x}(0), 0)}{\partial x_0} & \dfrac{\partial f_{m-1}(\overline{x}(0), 0)}{\partial x_1} & \cdots & \dfrac{\partial f_{m-1}(\overline{x}(0), 0)}{\partial x_{m-1}} \end{bmatrix} \tag{10.159}$$

[which generalizes $A \in \mathbf{R}^{2 \times 2}$ in (10.94)]. The solution to (10.158) is given by

$$\overline{x}(t) = e^{At}\overline{x}(0) \tag{10.160}$$

[recall (10.92)].[9] Ensuring the stability of the order m linear ODE system in (10.158) requires all the eigenvalues λ_k of A to have negative real parts (i.e., $\mathrm{Re}[\lambda_k] < 0$ for all $k = 0, 1, \ldots, m - 1$, where λ_k is the kth eigenvalue of A).

Of course, if $m = 1$, then with $x(t) = x_0(t)$, (10.158) reduces to

$$\frac{dx(t)}{dt} = \lambda x(t), \tag{10.161}$$

which is the model problem (10.28) again. Recall once again Example 10.5, for which we found that

$$\lambda = \frac{\partial f(x_0, t_0)}{\partial x} = -\frac{2}{t_0},$$

---

[9]Note that if we know $\overline{x}(t_0)$ (any $t_0 \in \mathbf{R}$), then we may slightly generalize our linear problem (10.158) to determining $\overline{x}(t)$ for all $t \geq t_0$, in which case

$$\overline{x}(t) = e^{A(t-t_0)}\overline{x}(t_0)$$

replaces (10.160). However, little is lost by assuming $t_0 = 0$.

as (10.41) is the solution to $dx/dt = t^2 - 2x/t$ for $t \geq t_0 > 0$. If t_0 is small, then $|\lambda|$ is large, and we saw that numerical methods, especially low-order explicit ones, had difficulty in estimating $x(t)$ accurately when t was near t_0. If we recall (10.33) (which stated that $h \leq 2/|\lambda|$) as an example, we see that large negative values for λ force us to select small step sizes h to ensure stability of the explicit Euler method. Since λ is an eigenvalue of $A = [\lambda]$ in (10.161), we might expect that this generalizes. In other words, a numerical method can be expected to have accuracy problems if A in (10.159) has eigenvalues with *large* negative real parts. In a situation like this $\bar{x}(t)$ in (10.156) has (it seems) a solution that changes so rapidly for some time intervals (e.g., fast startup transients) that accurate numerical solutions are hard to achieve. Such systems are called *stiff systems*.

So far our definition of a stiff system has not been at all rigorous. Indeed, a rigorous definition is hard to come by. Higham and Trefethen [15] argue that looking at the eigenvalues of A alone is not enough to decide on the stiffness of (10.156) in a completely reliable manner. It is possible, for example, that A may have favorable eigenvalues and yet (10.156) may still be stiff.

Stiff systems will not be discussed further here except to note that implicit methods, or higher-order predictor–corrector methods, should be used for their solution. The paper by Higham and Trefethen [15] is highly recommended reading for those readers seriously interested in the problems posed by stiff systems.

10.7 FINAL REMARKS

In the numerical solution (i.e., simulation) of ordinary differential equations (ODEs), two issues are of primary importance: accuracy and stability. The successful simulation of any system requires proper attention to both of these issues.

Computational efficiency is also an issue. Generally, we prefer to use the largest possible step size consistent with required accuracy, and as such to avoid any instability in the simulation.

APPENDIX 10.A   MATLAB CODE FOR EXAMPLE 10.8

```
%
%                              f23.m
%
% This defines function f(x,t) in the differential equation for Example 10.8
% (in Section 10.2).
%

function y = f23(x,t)

y = t*t - (2*x/t);

%
%                              Runge.m
%
% This routine simulates the Heun's, and 4th order Runge-Kutta methods as
```

```
% applied to the differential equation in Example 10.8 (Sect. 10.2), so this
% routine requires function f23.m.  It therefore generates Fig. 10.9.
%

function Runge

t0 = .05;  % initial time (starting time)
x0 = 1.0;  % initial condition (x(t0))

    % Exact solution x(t)

c = (5*t0*t0*x0) - (t0^5);
te = [t0:.02:1.5];
for k = 1:length(te)
   xe(k) = (te(k)*te(k)*te(k))/5 + c/(5*te(k)*te(k));
   end;

h = .05;

    % Heun's method simulation

xh(1) = x0;
th(1) = t0;
for n = 1:25
   fn = th(n)*th(n) - (2*xh(n)/th(n));   % f(x_n,t_n)
   th(n+1) = th(n) + h;
   xn1 = xh(n) + h*fn;
   tn1 = th(n+1);
   fn1 = tn1*tn1 - (2*xn1/tn1);          % f(x_{n+1},t_{n+1}) (approx.)
   xh(n+1) = xh(n) + (h/2)*(fn + fn1);
   end;

    % 4th order Runge-Kutta simulation

xr(1) = x0;
tr(1) = t0;
for n = 1:25
   t = tr(n);
   x = xr(n);
   k1 = f23(x,t);
   k2 = f23(x + .5*h*k1,t + .5*h);
   k3 = f23(x + .5*h*k2,t + .5*h);
   k4 = f23(x + h*k3,t + h);
   xr(n+1) = xr(n) + (h/6)*(k1 + 2*k2 + 2*k3 + k4);
   tr(n+1) = tr(n) + h;
   end;

plot(te,xe,'-',tr,xr,'--o',th,xh,'--+'), grid
legend('x(t)','4th Order Runge-Kutta (h = .05)','Heun (h = .05)',1);
xlabel(' Time (t) ')
ylabel(' Amplitude ')
```

APPENDIX 10.B   MATLAB CODE FOR EXAMPLE 10.13

```
%
%                              fR.m
%
```

```
% This is Equation (10.15) of Chapter 10 pertaining to Example 10.2.
%

function i = fR(v)

VTH = 0.75; % Threshold voltage in volts
RON = 100;  % On resistance of NPN BJT Q in Ohms

if v <= VTH
  i = 0;
else
  i = (v-VTH)/RON;
  end;
%
%                          vCC.m
%
% Supply voltage function v_CC(t) for Example 10.2 of Chapter 10.
% Here v_CC(t) = V_CC u(t) (i.e., oscillator switches on at t = 0).
%

function v = vCC(t)

VCC = 5;
if t < 0
  v = 0;
else
  v = VCC;
  end;
%
%                          Colpitts.m
%
% Computes the right-hand side of the state equations in Equation (10.17a,b,c)
% pertaining to Example 10.2 of Chapter 10.
%

function y = Colpitts(t,x)

C1 = 54e-9;
C2 = 54e-9;
REE = 400;
VEE = -5;
betaF = 200;
RL = 35;
L = 98.5e-6;

y(1) = ( x(3) - betaF*fR(x(2)) )/C1;
y(2) = ( -(x(2)+VEE)/REE - fR(x(2)) - x(3) )/C2;
y(3) = ( vCC(t) - x(1) + x(2) - RL*x(3))/L;
y = y.';
%
%                          SimulateColpitts.m
%
% This routine uses vCC.m, fR.m and Colpitts.m to simulate the Colpitts
% oscillator circuit of Example 10.2 in Chapter 10.  It produces
% Figure 10.21 in Chapter 10.
%
% The state vector x(:,:) is as follows:
%     x(:,1) = v_CE(t)
```

```
%      x(:,2) = v_BE(t)
%      x(:,3) = i_L(t)
%

function SimulateColpitts

[t,x] = ode23(@Colpitts, [ 0 0.003 ], [ 0 0 0 ]);
           % [ 0 0.003 ] ---> Simulate 0 to 3 milliseconds
           % [ 0 0 0 ]   ---> Initial state vector

clf
L = length(t);
subplot(211), plot(x(:,1),x(:,2)), grid
xlabel(' v_{CE} (t) (volts) ')
ylabel(' v_{BE} (t) (volts) ')
title(' Phase Portrait of the Colpitts Oscillator (Chaotic Regime) ')

subplot(212), plot(t(L-1999:L),x(L-1999:L,1),'-'), grid
xlabel(' t (seconds) ')
ylabel(' v_{CE} (t) (volts) ')
title(' Collector-Emitter Voltage (Chaotic Regime) ')
```

REFERENCES

1. P. E. Hydon, *Symmetry Methods for Differential Equations: A Beginner's Guide*, Cambridge Univ. Press, Cambridge, UK, 2000.
2. C. T.-C. Nguyen and R. T. Howe, "An Integrated CMOS Micromechanical Resonator High-Q Oscillator," *IEEE J. Solid-State Circuits* **34**, 450–455 (April 1999).
3. E. Kreyszig, *Introductory Functional Analysis with Applications*, Wiley, New York, 1978.
4. E. Kreyszig, *Advanced Engineering Mathematics*, 4th ed., Wiley, New York, 1979.
5. L. M. Kells, *Differential Equations: A Brief Course with Applications*, McGraw-Hill, New York, 1968.
6. E. Beltrami, *Mathematics for Dynamic Modeling*, Academic Press, Boston, MA, 1987.
7. S. S. Rao, *Applied Numerical Methods for Engineers and Scientists*, Prentice-Hall, Upper Saddle River, NJ, 2002.
8. R. Bronson, *Modern Introductory Differential Equations* (Schaum's Outline Series), McGraw-Hill, New York, 1973.
9. M. P. Kennedy, "Chaos in the Colpitts Oscillator," *IEEE Trans. Circuits Syst.* (Part I: Fundamental Theory and Applications) **41**, 771–774 (Nov. 1994).
10. A. S. Sedra and K. C. Smith, *Microelectronic Circuits*, 3rd ed., Saunders College Publ., Philadelphia, PA, 1989.
11. J. Guckenheimer and P. Holmes, *Nonlinear Oscillations, Dynamical Systems, and Bifurcations of Vector Fields*, Springer-Verlag, New York, 1983.
12. J. F. Epperson, *An Introduction to Numerical Methods and Analysis*, Wiley, New York, 2002.
13. L. F. Shampine and M. W. Reichelt, "The MATLAB ODE Suite," *SIAM J. Sci. Comput.* **18**, 1–22 (Jan. 1997).

14. J. R. Dormand and P. J. Prince, "A Family of Embedded Runge-Kutta Formulae," *J. Comput. Appl. Math.* **6**, 19–26 (1980).

15. D. J. Higham and L. N. Trefethen, "Stiffness of ODEs," *BIT* **33**, 285–303 (1993).

16. P. G. Drazin, *Nonlinear Systems*, Cambridge Univ. Press, Cambridge, UK, 1992.

17. R. L. Burden and J. D. Faires, *Numerical Analysis*, 4th ed., PWS-KENT Publ., Boston, MA, 1989.

PROBLEMS

10.1. Consider the electric circuit depicted in Fig. 10.P.1. Find matrix $A \in \mathbf{R}^{2 \times 2}$ and vector $b \in \mathbf{R}^2$ such that

$$\begin{bmatrix} \dfrac{di_{L_1}(t)}{dt} \\[2ex] \dfrac{di_{L_2}(t)}{dt} \end{bmatrix} = A \begin{bmatrix} i_{L_1}(t) \\[1ex] i_{L_2}(t) \end{bmatrix} + bv_s(t),$$

where $i_{L_k}(t)$ is the current through inductor L_k ($k \in \{1, 2, 3\}$).

(*Comment:* Although the number of energy storage elements in the circuit is 3, there are only two state variables needed to describe the circuit dynamics.)

10.2. The circuit in Fig. 10.P.2 is a simplified model for a *parametric amplifier*. The amplifier contains a reverse-biased *varactor diode* that is modeled by the parallel interconnection of linear time-invariant capacitor C_0 and linear time-varying capacitor $C(t)$. You may assume that $C(t) = 2C_1 \cos(\omega_p t)$, where C_1 is constant, and ω_p is the *pumping frequency*. Note that

$$i_C(t) = \frac{d}{dt}[C(t)v(t)].$$

The input to the amplifier is the ideal cosinusoidal current source $i_s(t) = 2I_s \cos(\omega_0 t)$, and the load is the resistor R, so the output is the current $i(t)$

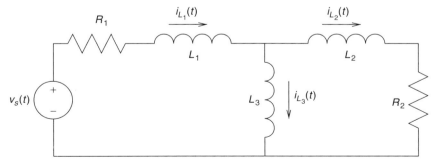

Figure 10.P.1   The linear electric circuit for Problem 10.1.

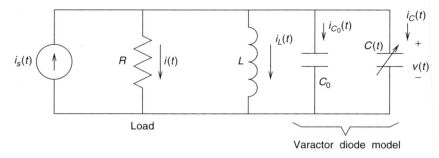

Figure 10.P.2   A model for a parametric amplifier (Problem 10.2).

into R. Write the state equations for the circuit, assuming that the state variables are $v(t)$, and $i_L(t)$. Write these equations in matrix form. (*Comment:* This problem is based on the example of a parametric amplifier as considered in C. A. Desoer and E. S. Kuh, *Basic Circuit Theory*, McGraw-Hill, New York, 1969.)

10.3. Give a detailed derivation of Eqs. (10.17).

10.4. The general linear first-order ODE is

$$\frac{dx(t)}{dt} = a(t)x(t) + b(t), \, x(t_0) = x_0.$$

Use the trapezoidal method to find an expression for x_{n+1} in terms of x_n, t_n, and t_{n+1}.

10.5. Prove that the trapezoidal method for ODEs is unconditionally stable.

10.6. Consider Theorem 10.2. Assume that $\hat{x}_0 = x(t_0)$. Use (10.23) to find an upper bound on $|x(t_n) - \hat{x}_n|/M$ for the following ODE IVPs:

(a) $\dfrac{dx}{dt} = 1 - 2x, \quad x(0) = 1.$

(b) $\dfrac{dx}{dt} = 2 \cos x, \quad x(0) = 0.$

10.7. Consider the ODE IVP

$$\frac{dx(t)}{dt} = 1 - 2x, \quad x(0) = 1.$$

(a) Approximate the solution to this problem using the explicit Euler method with $h = 0.1$ for $n = 0, 1, \ldots, 10$. Do the computations with a pocket calculator.

(b) Find the exact solution $x(t)$.

10.8. Consider the ODE IVP

$$\frac{dx(t)}{dt} = 2\cos x, \quad x(0) = 0.$$

(a) Approximate the solution to this problem using the explicit Euler method with $h = 0.1$ for $n = 0, 1, \ldots, 10$. Do the computations with a pocket calculator.

(b) Find the exact solution $x(t)$.

10.9. Write a MATLAB routine to simulate the ODE

$$\frac{dx}{dt} = \frac{x}{x+t}$$

for the initial condition $x(0) = 1$. Use both the implicit and explicit Euler methods. The program must accept as input the step size h, and the number of iterations N that are desired. Parameters h and N are the same for both methods. The program output is be written to a file in the form of a table something such as (e.g., for $h = 0.05$, and $N = 5$) the following:

```
time step       explicit x_n   implicit x_n
  .0000           value            value
  .0500           value            value
  .1000           value            value
  .1500           value            value
  .2000           value            value
  .2500           value            value
```

Test your program out on

$$h = 0.01, \quad N = 100$$

and

$$h = 0.10, \quad N = 10.$$

10.10. Consider

$$\frac{dx(t)}{dt} = \alpha(1 - 2\beta t^2)e^{-\beta t^2}. \tag{10.P.1}$$

(a) Verify that for $t \geq 0$, with $x(0) = 0$, we have the solution

$$x(t) = \alpha t e^{-\beta t^2}. \tag{10.P.2}$$

(b) For what range of step sizes h is the explicit Euler method a stable means of solving (10.P.1)?

(c) Write a MATLAB routine to simulate (10.P.1) for $x(0) = 0$ using both the explicit and implicit Euler methods. Assume that $\alpha = 10$ and $\beta = 1$. Test your program out on

$$h = 0.01, \quad N = 400 \qquad\qquad (10.P.3a)$$

and

$$h = 0.10, \quad N = 40. \qquad\qquad (10.P.3b)$$

The program must produce plots of $\{x_n | n = 0, 1, \ldots, N\}$ (for both explicit and implicit methods), and $x(t)$ (from (10.P.2)) on the same graph. This will lead to two separate plots, one for each of (10.P.3a) and (10.P.3b).

10.11. Consider

$$\frac{dx(t)}{dt} = 2tx + 1, \quad x(0) = 0.$$

(a) Find $\{x_n | n = 0, 1, \ldots, 10\}$ for $h = 0.1$ using the explicit Euler method. Do the calculations with a pocket calculator.

(b) Verify that

$$x(t) = e^{t^2} \int_0^t e^{-s^2}\, ds.$$

(c) Does the stability condition (10.33) apply here? Explain.

10.12. Prove that Runge–Kutta methods of order one have no degrees of freedom (i.e., we are forced to select c_1 in only one possible way).

10.13. A fourth-order Runge-Kutta method is

$$x_{n+1} = x_n + \tfrac{1}{6}h[k_1 + 2k_2 + 2k_3 + k_4]$$

for which

$$k_1 = f(x_n, t_n), \quad k_2 = f\left(x_n + \tfrac{1}{2}hk_1, t_n + \tfrac{1}{2}h\right),$$
$$k_3 = f\left(x_n + \tfrac{1}{2}hk_2, t_n + \tfrac{1}{2}h\right), k_4 = f(x_n + hk_3, t_n + h).$$

When applied to the model problem, we get $x_n = \sigma^n x_0$ for $n \in \mathbf{Z}^+$. Derive the expression for σ.

10.14. A third-order Runge–Kutta method is

$$x_{n+1} = x_n + \tfrac{1}{6}[k_1 + 4k_2 + k_3]$$

for which

$$k_1 = hf(x_n, t_n), \quad k_2 = hf(x_n + \tfrac{1}{2}h, t_n + \tfrac{1}{2}k_1),$$
$$k_3 = hf(x_n + h, t_n - k_1 + 2k_2).$$

(a) When applied to the model problem, we get $x_n = \sigma^n x_0$ for $n \in \mathbf{Z}^+$. Derive the expression for σ.

(b) Find the allowable range of step sizes h that ensure stability of the method.

10.15. Consider the fourth-order Runge–Kutta method in Eqs. (10.80) and (10.81). Show that if $f(x, t) = f(t)$, then the method reduces to Simpson's rule for numerical integration over the interval $[t_n, t_{n+1}]$.

10.16. Recall Eq. (10.98). Suppose that $A \in \mathbf{R}^{2 \times 2}$ has distinct eigenvalues γ_k such that $\mathrm{Re}[\gamma_k] > 0$ for at least one of the eigenvalues. Show that $I + hA$ will have at least one eigenvalue λ_k such that $|\lambda_k| > 1$.

10.17. Consider the coupled first-order ODEs

$$\frac{dx}{dt} = -y\sqrt{x^2 + y^2}, \qquad (10.P.4a)$$

$$\frac{dy}{dt} = x\sqrt{x^2 + y^2}, \qquad (10.P.4b)$$

where $(x_0, y_0) = (x(0), y(0))$ are the initial conditions.

(a) Prove that for suitable constants r_0 and θ_0, we have

$$x(t) = r_0 \cos(r_0 t + \theta_0), \quad y(t) = r_0 \sin(r_0 t + \theta_0). \qquad (10.P.5)$$

(b) Write a MATLAB routine to simulate the system represented by (10.P.4) using the explicit Euler method [which will produce $\bar{x}_n = [x_n \quad y_n]^T$ such that $x_n \approx x(t_n)$ and $y_n \approx y(t_n)$]. Assume that $h = 0.05$ and the initial condition is $\bar{x}_0 = [1 \quad 0]^T$. Plot $\bar{x}_n$ and $(x(t), y(t))$ (via (10.P.5)) on the (x, y) plane.

(c) Write a MATLAB routine to simulate the system represented by (10.P.4) using Heun's method [which will produce $\bar{x}_n = [x_n \quad y_n]^T$ such that $x_n \approx x(t_n)$ and $y_n \approx y(t_n)$]. Assume that $h = 0.05$ and the initial condition is $\bar{x}_0 = [1 \quad 0]^T$. Plot $\bar{x}_n$, and $(x(t), y(t))$ [via (10.P.5)] on the (x, y) plane.

Make reasonable choices about the number of time steps in the simulation.

10.18. In the previous problem the step size is $h = 0.05$.

(a) Determine whether the simulation using the explicit Euler method is stable for this choice of step size. (*Hint:* Recall that one must consider the eigenvalues of $I + hA$.)

(b) Determine whether the simulation using Heun's method is stable for this choice of step size. [*Hint:* Consider the implications of (10.121).]

Use the MATLAB eig function to assist you in your calculations.

10.19. A curve in $\mathbf{R}^2$ is specified parametrically according to

$$x(t) = A \cos(\omega t), \quad y(t) = \alpha A \cos(\omega t - \phi), \quad (10.\text{P}.6)$$

where $\alpha, A > 0$, and $t \in \mathbf{R}$ is the "parameter." When the points $(x(t), y(t))$ are plotted, the result is what electrical engineers often call a *Lissajous figure* (or curve), which is really just an alternative name for a *phase portrait.*

(a) Find an implicit function expression for the curve, that is, find a description of the form

$$f(x, y) = 0 \quad (10.\text{P}.7)$$

[i.e., via algebra and trigonometry eliminate t from (10.9.6) to obtain (10.9.7)].

(b) On the (x, y) plane sketch the Lissajous figures for cases $\phi = 0$, $\phi = \pm\frac{\pi}{2}$ and $\phi = \pm\frac{\pi}{4}$.

(c) An interpretation of (10.9.6) is that $x(t)$ may be the input voltage from a source in an electric circuit, while $y(t)$ may be the output voltage drop across a load element in the circuit (in the steady-state condition of course). Find a simple expression for $\sin \phi$ in terms of B such that $f(0, B) = 0$ [i.e., point(s) B on the y axis where the curve cuts the y axis], and in terms of α and A. (*Comment:* On analog oscilloscopes of olden days, it was possible to display a Lissajous figure, and so use this figure to estimate the phase angle ϕ on the lab bench.)

10.20. Verify the values for λ_k in Table 10.1 for $m = 2$.

10.21. Verify the values for λ_k in Table 10.2 for $m = 2$.

10.22. For the ODE IVP

$$\frac{dx(t)}{dt} = f(x, t), \quad x(t_0) = x_0,$$

write pseudocode for a numerical method that approximates the solution to it using an AB method for $m = 2$ as a predictor, an AM method for $m = 2$ as a corrector, and the third-order Runge–Kutta (RK) method from Problem 10.14 to perform the initialization.

10.23. From Forsythe, Malcolm, and Moler (see Ref. 5 in Chapter 2)

$$\frac{dx}{dt} = 998x + 1998y \quad (10.\text{P}.8\text{a})$$

$$\frac{dy}{dt} = -999x - 1999y \quad (10.\text{P}.8\text{b})$$

has the solution

$$x(t) = 4e^{-t} - 3e^{-1000t}, \tag{10.P.9a}$$

$$y(t) = -2e^{-t} + 3e^{-1000t}, \tag{10.P.9b}$$

where $x(0) = y(0) = 1$. Recall A as given by (10.94).

(a) By direct substitution verify that (10.P.9) is the solution to (10.P.8).

(b) If the eigenvalues of $I + hA$ are λ_0 and λ_1, plot $|\lambda_0|$ and $|\lambda_1|$ versus h (using MATLAB).

(c) If the eigenvalues of $I + hA + \frac{1}{2}h^2A^2$ [recall (10.121)] are λ_0 and λ_1, plot $|\lambda_0|$ and $|\lambda_1|$ versus h (using MATLAB).

In parts (b) and (c), what can be said about the range of values for h leading to a stable simulation of the system (10.P.8)?

10.24. Consider the coupled system of first-order ODEs

$$\frac{dx(t)}{dt} = Ax(t) + y(t), \tag{10.P.10}$$

where $A \in \mathbf{R}^{n \times n}$ and $x(t), y(t) \in \mathbf{R}^n$ for all $t \in \mathbf{R}^+$. Suppose that the eigenvalues of A are $\lambda_0, \lambda_1, \ldots, \lambda_{n-1}$ such that $\text{Re}[\lambda_k] < 0$ for all $k \in \mathbf{Z}_n$. Suppose that

$$\sigma \leq \text{Re}[\lambda_k] \leq \tau < 0$$

for all k. A *stiffness quotient* is defined to be

$$r = \frac{\sigma}{\tau}.$$

The system (10.P.10) is said to be *stiff* if $r \gg 1$ (again, we assume $\text{Re}[\lambda_k] < 0$ for all k).

For the previous problem, does A correspond to a stiff system? (*Comment:* As Higham and Trefethen [15] warned, the present stiffness definition is not entirely reliable.)

11 Numerical Methods for Eigenproblems

11.1 INTRODUCTION

In previous chapters we have seen that eigenvalues and eigenvectors are important (e.g., recall condition numbers from Chapter 4, and the stability analysis of numerical methods for ODEs in Chapter 10). In this chapter we treat the eigenproblem somewhat more formally than previously. We shall define and review the basic problem in Section 11.2, and in Section 11.3 we shall apply this understanding to the problem of computing the matrix exponential [i.e., $\exp(At)$, where $A \in \mathbf{R}^{n \times n}$ and $t \in \mathbf{R}$] since this is of central importance in many areas of electrical and computer engineering (signal processing, stability of dynamic systems, control systems, circuit simulation, etc.). In subsequent sections we will consider numerical methods to determine the eigenvalues and eigenvectors of matrices.

11.2 REVIEW OF EIGENVALUES AND EIGENVECTORS

In this section we review some basic facts relating to the determination of eigenvalues and eigenvectors of matrices. Our emphasis, with a few exceptions, is on matrices that are diagonalizable.

Definition 11.1: Eigenproblem   Let $A \in \mathbf{C}^{n \times n}$. The *eigenproblem* for A is to find solutions to the matrix equation

$$Ax = \lambda x, \tag{11.1}$$

where $\lambda \in \mathbf{C}$ and $x \in \mathbf{C}^n$ such that $x \neq 0$. A solution (λ, x) to (11.1) is called an *eigenpair*, λ is an *eigenvalue*, and x is its corresponding *eigenvector*.

Even if $A \in \mathbf{R}^{n \times n}$ (the situation we will emphasize most) it is very possible to have $\lambda \in \mathbf{C}$ and $x \in \mathbf{C}^n$. We must also emphasize that $x = 0$ is never permitted to be an eigenvector for A.

We may rewrite (11.1) as (I is the $n \times n$ identity matrix)

$$(A - \lambda I)x = 0, \tag{11.2a}$$

An Introduction to Numerical Analysis for Electrical and Computer Engineers, by C.J. Zarowski
ISBN 0-471-46737-5   © 2004 John Wiley & Sons, Inc.

or equivalently as

$$(\lambda I - A)x = 0. \tag{11.2b}$$

Equations (11.2) are *homogeneous linear systems* of n equations in n unknowns. Since $x = 0$ is never an eigenvector, an eigenvector x must be a nontrivial solution to (11.2). An $n \times n$ homogeneous linear system has a nonzero (i.e., nontrivial) solution iff the coefficient matrix is singular. Immediately, eigenvalue λ satisfies

$$\det(A - \lambda I) = 0, \tag{11.3a}$$

or equivalently

$$\det(\lambda I - A) = 0. \tag{11.3b}$$

Of course, $p(\lambda) = \det(\lambda I - A)$ is a polynomial of degree n. In principle, we may find eigenpairs by finding $p(\lambda)$ (the *characteristic polynomial* of A), then finding the zeros of $p(\lambda)$, and then substituting these into (11.2) to find x. In practice this approach really works only for small analytical examples. Properly conceived numerical methods are needed to determine eigenpairs reliably for larger matrices A.

To set the stage for what follows, consider the following examples.

Example 11.1   From Hill [1], consider the example

$$A = \begin{bmatrix} 1 & 0 & 0 \\ 2 & 1 & 0 \\ 0 & 0 & 3 \end{bmatrix}.$$

Here $p(\lambda) = \det(\lambda I - A) = (\lambda - 1)^2(\lambda - 3)$ so A has a double eigenvalue (eigenvalue of multiplicity 2) at $\lambda = 1$, and a simple eigenvalue at $\lambda = 3$. The eigenvalues may be individually denoted as $\lambda_0 = 1, \lambda_1 = 1$, and $\lambda_2 = 3$.

For $\lambda = 3$, $(3I - A)x = 0$ becomes

$$\begin{bmatrix} 2 & 0 & 0 \\ -2 & 2 & 0 \\ 0 & 0 & 0 \end{bmatrix} \begin{bmatrix} x_0 \\ x_1 \\ x_2 \end{bmatrix} = \begin{bmatrix} 0 \\ 0 \\ 0 \end{bmatrix},$$

and from the application of elementary row operations this reduces to

$$\begin{bmatrix} 1 & 0 & 0 \\ 0 & 1 & 0 \\ 0 & 0 & 0 \end{bmatrix} \begin{bmatrix} x_0 \\ x_1 \\ x_2 \end{bmatrix} = \begin{bmatrix} 0 \\ 0 \\ 0 \end{bmatrix}.$$

Immediately, $x_0 = x_1 = 0$, and x_2 is arbitrary (except, of course, it is not allowed to be zero) so that the general form of the eigenvector corresponding to $\lambda = \lambda_2$ is

$$v^{(2)} = [\; 0 \quad 0 \quad x_2 \;]^T \in \mathbf{C}^3.$$

On the other hand, now let us consider $\lambda = \lambda_0 = \lambda_1$. In this case $(I - A)x = 0$ becomes

$$\begin{bmatrix} 0 & 0 & 0 \\ -2 & 0 & 0 \\ 0 & 0 & -2 \end{bmatrix} \begin{bmatrix} x_0 \\ x_1 \\ x_2 \end{bmatrix} = \begin{bmatrix} 0 \\ 0 \\ 0 \end{bmatrix},$$

which reduces to

$$\begin{bmatrix} 1 & 0 & 0 \\ 0 & 0 & 1 \\ 0 & 0 & 0 \end{bmatrix} \begin{bmatrix} x_0 \\ x_1 \\ x_2 \end{bmatrix} = \begin{bmatrix} 0 \\ 0 \\ 0 \end{bmatrix}.$$

Immediately, $x_0 = x_2 = 0$, and x_1 is arbitrary, so an eigenvector corresponding to $\lambda = \lambda_0 = \lambda_1$ is of the general form

$$v^{(0)} = [\; 0 \quad x_1 \quad 0 \;]^T \in \mathbf{C}^3.$$

Even though $\lambda = 1$ is a double eigenvalue, we are able to find only one eigenvector for this case. In effect, one eigenvector seems to be "missing."

Example 11.2   Now consider (again from Hill [1])

$$A = \begin{bmatrix} 0 & -2 & 1 \\ 1 & 3 & -1 \\ 0 & 0 & 1 \end{bmatrix}.$$

Here $p(\lambda) = \det(\lambda I - A) = (\lambda - 1)^2(\lambda - 2)$. The eigenvalues of A are thus $\lambda_0 = 1, \lambda_1 = 1$, and $\lambda_2 = 2$.

For $\lambda = 2$, $(2I - A)x = 0$ becomes

$$\begin{bmatrix} 2 & 2 & -1 \\ -1 & -1 & 1 \\ 0 & 0 & 1 \end{bmatrix} \begin{bmatrix} x_0 \\ x_1 \\ x_2 \end{bmatrix} = \begin{bmatrix} 0 \\ 0 \\ 0 \end{bmatrix},$$

which reduces to

$$\begin{bmatrix} 1 & 1 & 0 \\ 0 & 0 & 1 \\ 0 & 0 & 0 \end{bmatrix} \begin{bmatrix} x_0 \\ x_1 \\ x_2 \end{bmatrix} = \begin{bmatrix} 0 \\ 0 \\ 0 \end{bmatrix},$$

so the general form of the eigenvector for $\lambda = \lambda_2$ is

$$v^{(2)} = x^{(2)} = [\; -x_0 \quad x_0 \quad 0 \;]^T \in \mathbf{C}^3.$$

Now, if we consider $\lambda = \lambda_0 = \lambda_1$, $(I - A)x = 0$ becomes

$$\begin{bmatrix} 1 & 2 & -1 \\ -1 & -2 & 1 \\ 0 & 0 & 0 \end{bmatrix} \begin{bmatrix} x_0 \\ x_1 \\ x_2 \end{bmatrix} = \begin{bmatrix} 0 \\ 0 \\ 0 \end{bmatrix},$$

which reduces to

$$
\begin{bmatrix} 1 & 2 & -1 \\ 0 & 0 & 0 \\ 0 & 0 & 0 \end{bmatrix}
\begin{bmatrix} x_0 \\ x_1 \\ x_2 \end{bmatrix} =
\begin{bmatrix} 0 \\ 0 \\ 0 \end{bmatrix}.
$$

Since $x_0 + 2x_1 - x_2 = 0$, we may choose any two of x_0, x_1, or x_2 as free parameters giving a general form of an eigenvector for $\lambda = \lambda_0 = \lambda_1$ as

$$
v^{(0)} = x_0 \underbrace{\begin{bmatrix} 1 \\ 0 \\ 1 \end{bmatrix}}_{=x^{(0)}} + x_1 \underbrace{\begin{bmatrix} 0 \\ 1 \\ 2 \end{bmatrix}}_{=x^{(1)}} \in \mathbf{C}^3.
$$

Thus, we have eigenpairs $(\lambda_0, x^{(0)})$, $(\lambda_1, x^{(1)})$, $(\lambda_2, x^{(2)})$.

We continue to emphasize that in all cases any free parameters are arbitrary, except that they must never be selected to give a zero-valued eigenvector. In Example 11.2 $\lambda = 1$ is an eigenvalue of multiplicity 2, and $v^{(0)}$ is a vector in a two-dimensional vector subspace of $\mathbf{C}^3$. On the other hand, in Example 11.1 $\lambda = 1$ is also of multiplicity 2, and yet $v^{(0)}$ is only a vector in a one-dimensional vector subspace of $\mathbf{C}^3$.

Definition 11.2: Defective Matrix   For any $A \in \mathbf{C}^{n \times n}$, if the multiplicity of any eigenvalue $\lambda \in \mathbf{C}$ is not equal to the dimension of the solution space (*eigenspace*) of $(\lambda I - A)x = 0$, then A is *defective*.

From this definition, A in Example 11.1 is a defective matrix, while A in Example 11.2 is not defective (i.e., is *nondefective*). In a sense soon to be made precise, defective matrices cannot be *diagonalized*. However, all matrices, diagonalizable or not, can be placed into *Jordan canonical form*, as follows.

Theorem 11.1: Jordan Decomposition   If $A \in \mathbf{C}^{n \times n}$, then there exists a nonsingular matrix $T \in \mathbf{C}^{n \times n}$ such that

$$
T^{-1} AT = \mathrm{diag}(J_0, J_1, \ldots, J_{k-1}),
$$

where

$$
J_i = \begin{bmatrix} \lambda_i & 1 & 0 & \cdots & 0 & 0 \\ 0 & \lambda_i & 1 & \cdots & 0 & 0 \\ \vdots & \vdots & \vdots & & \vdots & \vdots \\ 0 & 0 & 0 & \cdots & \lambda_i & 1 \\ 0 & 0 & 0 & \cdots & 0 & \lambda_i \end{bmatrix} \in \mathbf{C}^{m_i \times m_i},
$$

and $\sum_{i=0}^{k-1} m_i = n$.

Proof   See Halmos [2] or Horn and Johnson [3].

Of course, λ_i in the theorem is an eigenvalue of A. The submatrices J_i are called *Jordan blocks*. The number of blocks k and their dimensions m_i are unique, but their order is not unique. Note that if an eigenvalue has a multiplicity of unity (i.e., if it is simple), then the Jordan block in this case is the 1×1 matrix consisting of that eigenvalue. From the theorem statement the characteristic polynomial of $A \in \mathbf{C}^{n \times n}$ is given by

$$p(\lambda) = \det(\lambda I - A) = \prod_{i=0}^{k-1}(\lambda - \lambda_i)^{m_i}. \tag{11.4}$$

But if $A \in \mathbf{R}^{n \times n}$, and if $\lambda_i \in \mathbf{C}$ for some i, then λ_i^* must also be an eigenvalue of A, that is, a zero of $p(\lambda)$. This follows from the *fundamental theorem of algebra*, which states that complex-valued roots of polynomials with real-valued coefficients must always occur in complex–conjugate pairs.

Example 11.3   Consider (with $\theta \neq k\pi$, $k \in \mathbf{Z}$)

$$A = \begin{bmatrix} \cos\theta & -\sin\theta \\ \sin\theta & \cos\theta \end{bmatrix} \in \mathbf{R}^{2 \times 2}$$

(2×2 rotation operator from Appendix 3.A). We have the characteristic equation

$$p(\lambda) = \det(\lambda I - A) = \det\left(\begin{bmatrix} \lambda - \cos\theta & \sin\theta \\ -\sin\theta & \lambda - \cos\theta \end{bmatrix}\right)$$

$$= \lambda^2 - 2\cos\theta\lambda + 1 = 0$$

for which the roots (eigenvalues of A) are therefore $\lambda = e^{\pm j\theta}$. Define $\lambda_0 = e^{j\theta}$, $\lambda_1 = e^{-j\theta}$. Clearly $\lambda_1 = \lambda_0^*$ (i.e., the two simple eigenvalues of A are a conjugate pair).
For $\lambda = \lambda_0$, $(\lambda_0 I - A)x = 0$ is

$$\sin\theta \begin{bmatrix} j & 1 \\ -1 & j \end{bmatrix}\begin{bmatrix} x_0 \\ x_1 \end{bmatrix} = \begin{bmatrix} 0 \\ 0 \end{bmatrix},$$

which reduces to

$$\begin{bmatrix} 1 & -j \\ 0 & 0 \end{bmatrix}\begin{bmatrix} x_0 \\ x_1 \end{bmatrix} = \begin{bmatrix} 0 \\ 0 \end{bmatrix}.$$

The eigenvector for $\lambda = \lambda_0$ is therefore of the form

$$x^{(0)} = a \begin{bmatrix} 1 \\ -j \end{bmatrix}, \quad a \in \mathbf{C}.$$

Similarly, for $\lambda = \lambda_1$, the homogeneous linear system $(\lambda_1 I - A)x = 0$ is

$$\sin\theta \begin{bmatrix} -j & 1 \\ -1 & -j \end{bmatrix} \begin{bmatrix} x_0 \\ x_1 \end{bmatrix} = \begin{bmatrix} 0 \\ 0 \end{bmatrix},$$

which reduces to

$$\begin{bmatrix} 1 & j \\ 0 & 0 \end{bmatrix} \begin{bmatrix} x_0 \\ x_1 \end{bmatrix} = \begin{bmatrix} 0 \\ 0 \end{bmatrix}.$$

The eigenvector for $\lambda = \lambda_1$ is therefore of the form

$$x^{(1)} = b \begin{bmatrix} 1 \\ j \end{bmatrix}, \quad b \in \mathbf{C}.$$

Of course, free parameters a, and b are never allowed to be zero.

Computing the Jordan canonical form when $m_i > 1$ is numerically rather difficult (as noted in Golub and Van Loan [4] and Horn and Johnson [3]), and so is often avoided. However, there are important exceptions often involving state-variable (state-space) systems analysis and design (e.g., see Fairman [5]). Within the theory of Jordan forms it is possible to find supposedly "missing" eigenvectors, resulting in a theory of *generalized eigenvectors*. We will not consider this here as it is rather involved. Some of the references at the end of this chapter cover the relevant theory [3].

We will now consider a series of theorems leading to a sufficient condition for A to be nondefective. Our presentation largely follows Hill [1].

Theorem 11.2: If $A \in \mathbf{C}^{n \times n}$, then the eigenvectors corresponding to two distinct eigenvalues of A are linearly independent.

Proof We employ proof by contradiction.
Suppose that (α, x) and (β, y) are two eigenpairs for A, and $\alpha \neq \beta$. Assume $y = ax$ for some $a \neq 0$ ($a \in \mathbf{C}$). Thus

$$\beta y = Ay = aAx = a\alpha x$$

and also

$$\beta y = a\beta x.$$

Hence

$$a\beta x = a\alpha x,$$

implying that

$$a(\beta - \alpha)x = 0.$$

But $x \neq 0$ as it is an eigenvector of A, and also $a \neq 0$. Immediately, $\alpha = \beta$, contradicting our assumption that these eigenvalues are distinct. Thus, $y = ax$ is impossible; that is, we have proved that y is independent of x.

Theorem 11.2 leads us to the next theorem.

Theorem 11.3: If $A \in \mathbf{C}^{n \times n}$ has n distinct eigenvalues, then A has n linearly independent eigenvectors.

Proof Uses mathematical induction (e.g., Stewart [6]).

We have already seen the following theorem.

Theorem 11.4: If $A \in \mathbf{R}^{n \times n}$ and $A = A^T$, then all eigenvalues of A are real-valued.

Proof See Hill [1], or see the appropriate footnote in Chapter 4.

In addition to this theorem, we also have the following one.

Theorem 11.5: If $A \in \mathbf{R}^{n \times n}$, and if $A = A^T$, then eigenvectors corresponding to distinct eigenvalues of A are orthogonal.

Proof Suppose that (α, x) and (β, y) are eigenpairs of A with $\alpha \neq \beta$. We wish to show that $x^T y = y^T x = 0$ (recall Definition 1.6). Now

$$\alpha x = Ax = A^T x$$

so that

$$\alpha y^T x = y^T A^T x = (Ay)^T x = \beta y^T x,$$

implying that

$$(\alpha - \beta) y^T x = 0.$$

But $\alpha \neq \beta$ so that $y^T x = 0$, that is, $x \perp y$.

Theorem 11.5 states that eigenspaces corresponding to distinct eigenvalues of a symmetric, real-valued matrix form mutually orthogonal vector subspaces of $\mathbf{R}^n$. Any vector from one eigenspace must therefore be orthogonal to any eigenvector from another eigenspace. If we recall Definition 11.2, it is apparent that all symmetric, real-valued matrices are nondefective, *if their eigenvalues are all distinct.*[1] In fact, even if $A \in \mathbf{C}^{n \times n}$ and is not symmetric, then, as long as the eigenvalues are distinct, A will be nondefective (Theorem 11.3).

---

[1]It is possible to go even further and prove that *any* real-valued, symmetric matrix is nondefective, even if it possesses multiple eigenvalues. Thus, any real-valued, symmetric matrix is diagonalizable.

Definition 11.3: Similarity Transformation   If $A, B \in \mathbf{C}^{n \times n}$, and there is a nonsingular matrix $P \in \mathbf{C}^{n \times n}$ such that

$$B = P^{-1}AP,$$

we say that B is *similar* to A, and that P is a *similarity transformation*.

If $A \in \mathbf{C}^{n \times n}$, and A has n distinct eigenvalues forming n distinct eigenpairs $\{(\lambda_k, x^{(k)}) | k = 0, 1, \ldots, n-1\}$, then

$$Ax^{(k)} = \lambda_k x^{(k)}$$

allows us to write

$$A \underbrace{[x^{(0)} x^{(1)} \cdots x^{(n-1)}]}_{=P} = \underbrace{[x^{(0)} x^{(1)} \cdots x^{(n-1)}]}_{=P} \underbrace{\begin{bmatrix} \lambda_0 & 0 & \cdots & 0 & 0 \\ 0 & \lambda_1 & \cdots & 0 & 0 \\ \vdots & \vdots & & \vdots & \vdots \\ 0 & 0 & \cdots & 0 & \lambda_{n-1} \end{bmatrix}}_{=\Lambda},$$

(11.5)

that is, $AP = P\Lambda$, where $\Lambda = \text{diag}(\lambda_0, \lambda_1, \ldots, \lambda_{n-1}) \in \mathbf{C}^{n \times n}$ is the diagonal *matrix of eigenvalues* of A. Thus

$$P^{-1}AP = \Lambda,$$

(11.6)

and the *matrix of eigenvectors* $P \in \mathbf{C}^{n \times n}$ of A defines the similarity transformation that *diagonalizes* matrix A. More generally, we have the following theorem.

Theorem 11.6: If $A, B \in \mathbf{C}^{n \times n}$, and A and B are similar matrices, then A and B have the same eigenvalues.

Proof   Since A and B are similar, there exists a nonsingular matrix $P \in \mathbf{C}^{n \times n}$ such that $B = P^{-1}AP$. Therefore

$$\det(\lambda I - B) = \det(\lambda I - P^{-1}AP)$$

$$= \det(P^{-1}(PP^{-1}\lambda - A)P)$$

$$= \det(P^{-1}) \det(\lambda I - A) \det(P)$$

$$= \det(\lambda I - A).$$

Thus, A and B possess the same characteristic polynomial, and so possess identical eigenvalues.

In other words, similarity transformations preserve eigenvalues.[2] Note that Theorem 11.6 holds regardless of whether A and B are defective. In developing (11.6)

---

[2]This makes such transformations highly valuable in state-space control systems design, in addition to a number of other application areas.

we have seen that if $A \in \mathbf{C}^{n \times n}$ has n distinct eigenvalues, it is diagonalizable. We emphasize that this is only a sufficient condition. Example 11.2 confirms that a matrix can have eigenvalues with multiplicity greater than one, and yet still be diagonalizable.

11.3   THE MATRIX EXPONENTIAL

In Chapter 10 the problem of computing e^{At} ($A \in \mathbf{R}^{n \times n}$, and $t \in \mathbf{R}$) was associated with the stability analysis of numerical methods for systems of ODEs. It is also noteworthy that to solve

$$\frac{dx(t)}{dt} = Ax(t) + by(t) \tag{11.7}$$

$[x(t) = [x_0(t) x_1(t) \cdots x_{n-1}(t)]^T \in \mathbf{R}^n$, $A \in \mathbf{R}^{n \times n}$, and $b \in \mathbf{R}^n$ with $y(t) \in \mathbf{R}$ for all t] required us to compute e^{At} [recall Example 10.1, which involved an example of (11.7) from electric circuit analysis; see Eq. (10.9)]. Thus, we see that computing the matrix exponential is an important problem in analysis. In this section we shall gain more familiarity with the matrix exponential because of its significance.

Moler and Van Loan [7] caution that computing the matrix exponential is a numerically difficult problem. Stable, reliable, accurate, and computationally efficient algorithms are not so easy to come by. Their paper [7], as its title states, considers 19 methods, and none of them are entirely satisfactory. Indeed, this paper [7] appeared in 1978, and to this day the problem of successfully computing e^{At} for any $A \in \mathbf{C}^{n \times n}$ has not been fully resolved. We shall say something about why this is a difficult problem later.

Before considering this matter, we shall consider a general analytic (i.e., hand calculation) method for obtaining e^{At} for any $A \in \mathbf{R}^{n \times n}$, including when A is defective. In principle, this would involve working with Jordan decompositions and generalized eigenvectors, but we will avoid this by adopting the approach suggested in Leonard [8].

The matrix exponential e^{At} may be *defined* in the expected manner as

$$\Phi(t) = e^{At} = \sum_{k=0}^{\infty} \frac{1}{k!} A^k t^k, \tag{11.8}$$

so, for example, the kth derivative of the matrix exponential is

$$\Phi^{(k)}(t) = A^k e^{At} = e^{At} A^k \tag{11.9}$$

for $k \in \mathbf{Z}^+$ ($\Phi^{(0)}(t) = \Phi(t)$). To see how this works, consider the following special case $k = 1$:

$$\Phi^{(1)}(t) = \frac{d}{dt} \left\{ \sum_{k=0}^{\infty} \frac{1}{k!} A^k t^k \right\} = \frac{d}{dt} \left\{ I + \frac{1}{1!} At + \frac{1}{2!} A^2 t^2 + \cdots + \frac{1}{k!} A^k t^k + \cdots \right\}$$

$$= \frac{1}{1!}A + \frac{2}{2!}A^2 t + \cdots + \frac{k}{k!}A^k t^{k-1} + \cdots$$

$$= A\left\{ I + \frac{1}{1!}At + \cdots + \frac{1}{(k-1)!}A^{k-1}t^{k-1} + \cdots \right\}$$

$$= Ae^{At} = e^{At}A.$$

It is possible to formally verify that the series in (11.8) converges to a matrix function of $t \in \mathbf{R}$ by working with the Jordan decomposition of A. However, we will avoid this level of detail. But we will consider the situation where A is diagonalizable later on.

There is some additional background material needed to more fully appreciate [8], and we will now consider this. The main result is the Cayley–Hamilton theorem (Theorem 11.8, below).

Definition 11.4: Minors and Cofactors   Let $A \in \mathbf{C}^{n \times n}$. The *minor* m_{ij} is the determinant of the $(n-1) \times (n-1)$ submatrix of A derived from it by deleting row i and column j. The *cofactor* c_{ij} associated with m_{ij} is $c_{ij} = (-1)^{i+j} m_{ij}$ for all $i, j \in \mathbf{Z}_n$.

A formula for the inverse of A (assuming this exists) is given by Theorem 11.7.

Theorem 11.7: If $A \in \mathbf{C}^{n \times n}$ is nonsingular, then

$$A^{-1} = \frac{1}{\det(A)} \operatorname{adj}(A), \tag{11.10}$$

where $\operatorname{adj}(A)$ (*adjoint matrix* of A) is the transpose of the matrix of cofactors of A. Thus, if $C = [c_{ij}] \in \mathbf{C}^{n \times n}$ is the matrix of cofactors of A, then $\operatorname{adj}(A) = C^T$.

Proof   See Noble and Daniel [9].

Of course, the method suggested by Theorem 11.7 is useful only for the hand calculation of low-order (small n) problems. Practical matrix inversion must use ideas from Chapter 4. But Theorem 11.7 is a very useful result for theoretical purposes, such as obtaining the following theorem.

Theorem 11.8: Cayley–Hamilton Theorem   Any matrix $A \in \mathbf{C}^{n \times n}$ satisfies its own characteristic equation.

Proof   The characteristic polynomial for A is $p(\lambda) = \det(\lambda I - A)$, and can be written as

$$p(\lambda) = \lambda^n + a_1 \lambda^{n-1} + \cdots + a_{n-1}\lambda + a_n$$

for suitable constants $a_k \in \mathbf{C}$. The theorem claims that

$$A^n + a_1 A^{n-1} + \cdots + a_{n-1}A + a_n I = 0, \tag{11.11}$$

where I is the order n identity matrix. To show (11.11), we consider $\text{adj}(\mu I - A)$ whose elements are polynomials in μ of a degree that is not greater than $n - 1$, where μ is not an eigenvalue of A. Hence

$$\text{adj}(\mu I - A) = M_0 \mu^{n-1} + M_1 \mu^{n-2} + \cdots + M_{n-2}\mu + M_{n-1}$$

for suitable constant matrices $M_k \in C^{n \times n}$. Via Theorem 11.7

$$(\mu I - A)\,\text{adj}(\mu I - A) = \det(\mu I - A)I,$$

or in expanded form, this becomes

$$(\mu I - A)(M_0 \mu^{n-1} + M_1 \mu^{n-2} + \cdots + M_{n-2}\mu + M_{n-1})$$
$$= (\mu^n + a_1 \mu^{n-1} + \cdots + a_n)I.$$

If we now equate like powers of μ on both sides of this equation, we obtain

$$M_0 = I,$$
$$M_1 - AM_0 = a_1 I,$$
$$M_2 - AM_1 = a_2 I,$$
$$\vdots \tag{11.12}$$
$$M_{n-1} - AM_{n-2} = a_{n-1} I,$$
$$-AM_{n-1} = a_n I.$$

Premultiplying[3] the jth equation in (11.12) by A^{n-j} $(j = 0, 1, \ldots, n)$, and then adding all the equations that result from this, yields

$$A^n M_0 + A^{n-1}(M_1 - AM_0) + A^{n-2}(M_2 - AM_1) + \cdots + A(M_{n-1} - AM_{n-2})$$
$$- AM_{n-1} = A^n + a_1 A^{n-1} + a_2 A^{n-2} + \cdots + a_{n-1} A^{n-1} + a_n I.$$

But the left-hand side of this is seen to be zero because of cancellation of all the terms, and (11.11) immediately results.

As an exercise the reader should verify that the matrices in Examples 11.1–11.3, all satisfy their own characteristic equations.

We will now consider the approach in Leonard [8], who, however, assumes that the reader is familiar with the theory of solution of nth-order homogeneous linear ODEs in constant coefficients

$$x^{(n)}(t) + c_{n-1}x^{(n-1)}(t) + \cdots + c_1 x^{(1)}(t) + c_0 x(t) = 0, \tag{11.13}$$

---

[3]This means that we must multiply on the left.

where the initial conditions are known. In particular, the reader must know that if λ is a root of the characteristic equation

$$\lambda^n + c_{n-1}\lambda^{n-1} + \cdots + c_1\lambda + c_0 = 0, \tag{11.14}$$

then if λ has multiplicity m, its contribution to the solution of the initial-value problem (IVP) (11.13) is of the general form

$$(a_0 + a_1 t + \cdots + a_{m-1}t^{m-1})e^{\lambda t}. \tag{11.15}$$

These matters are considered by Derrick and Grossman [10] and Reid [11]. We shall be combining these facts with the results of Theorem 11.10 (below).

Leonard [8] presents two theorems that relate the solution of (11.13) to the computation of $\Phi(t) = e^{At}$.

Theorem 11.9: Leonard I   Let $A \in \mathbf{R}^{n \times n}$ be a constant matrix with characteristic polynomial

$$p(\lambda) = \det(\lambda I - A) = \lambda^n + c_{n-1}\lambda^{n-1} + \cdots + c_1\lambda + c_0.$$

$\Phi(t) = e^{At}$ is the unique solution to the nth-order matrix differential equation

$$\Phi^{(n)}(t) + c_{n-1}\Phi^{(n-1)}(t) + \cdots + c_1\Phi^{(1)}(t) + c_0\Phi(t) = 0 \tag{11.16}$$

with initial conditions

$$\Phi(0) = I, \Phi^{(1)}(0) = A, \ldots, \Phi^{(n-2)}(0) = A^{n-2}, \Phi^{(n-1)}(0) = A^{n-1}. \tag{11.17}$$

Proof   We will demonstrate uniqueness of the solution first of all.

Suppose that $\Phi_1(t)$ and $\Phi_2(t)$ are two solutions to (11.16) for the initial conditions stated in (11.17). Let $\Phi(t) = \Phi_1(t) - \Phi_2(t)$ for present purposes, in which case $\Phi(t)$ satisfies (11.16) with the initial conditions

$$\Phi(0) = \Phi^{(1)}(0) = \cdots = \Phi^{(n-2)}(0) = \Phi^{(n-1)}(0) = 0.$$

Consequently, each entry of the matrix $\Phi(t)$ satisfies a *scalar* IVP of the form

$$x^{(n)}(t) + c_{n-1}x^{(n-1)}(t) + \cdots + c_1 x^{(1)}(t) + c_0 x(t) = 0,$$
$$x(0) = x^{(1)}(0) = \cdots = x^{(n-2)}(0) = x^{(n-1)}(0) = 0,$$

where the solution is $x(t) = 0$ for all t, so that $\Phi(t) = 0$ for all $t \in \mathbf{R}^+$. Thus, $\Phi_1(t) = \Phi_2(t)$, and so the solution must be unique (if it exists).

Now we confirm that the solution is $\Phi(t) = e^{At}$ (i.e., we confirm existence in a constructive manner). Let A be a constant matrix of order n with characteristic polynomial $p(\lambda)$ as in the theorem statement. If now $\Phi(t) = e^{At}$, then we recall that

$$\Phi^{(k)}(t) = A^k e^{At}, k = 1, 2, \ldots, n \tag{11.18}$$

[see (11.9)] so that

$$\Phi^{(n)}(t) + c_{n-1}\Phi^{(n-1)}(t) + \cdots + c_1\Phi^{(1)}(t) + c_0\Phi(t)$$
$$= [A^n + c_{n-1}A^{n-1} + \cdots + c_1 A + c_0 I]e^{At}$$
$$= p(A)e^{At} = 0$$

via Theorem 11.8 (Cayley–Hamilton). From (11.18), we obtain

$$\Phi^{(0)}(0) = I, \ \Phi^{(1)}(0) = A, \ldots, \ \Phi^{(n-2)}(0) = A^{n-2}, \ \Phi^{(n-1)}(0) = A^{n-1},$$

and so $\Phi(t) = e^{At}$ is the unique solution to the IVP in the theorem statement.

Theorem 11.10: Leonard II   Let $A \in \mathbf{R}^{n \times n}$ be a constant matrix with characteristic polynomial

$$p(\lambda) = \lambda^n + c_{n-1}\lambda^{n-1} + \cdots + c_1\lambda + c_0,$$

then

$$e^{At} = x_0(t)I + x_1(t)A + x_2(t)A^2 + \cdots + x_{n-1}(t)A^{n-1},$$

where $x_k(t)$, $k \in \mathbf{Z}_n$ are the solutions to the nth-order scalar ODEs

$$x^{(n)}(t) + c_{n-1}x^{(n-1)}(t) + \cdots + c_1 x^{(1)}(t) + c_0 x(t) = 0,$$

satisfying the initial conditions

$$x_k^{(j)}(0) = \delta_{j-k}$$

for $j, k \in \mathbf{Z}_n$ $(x_k^{(0)}(t) = x_k(t))$.

Proof   Let constant matrix A have characteristic polynomial $p(\lambda)$ as in the theorem statement. Define

$$\Phi(t) = x_0(t)I + x_1(t)A + x_2(t)A^2 + \cdots + x_{n-1}(t)A^{n-1},$$

where $x_k(t)$, $k \in \mathbf{Z}_n$ are unique solutions to the nth-order scalar ODEs

$$x^{(n)}(t) + c_{n-1}x^{(n-1)}(t) + \cdots + c_1 x^{(1)}(t) + c_0 x(t) = 0,$$

satisfying the initial conditions stated in the theorem. Thus, for all $t \in \mathbf{R}^+$

$$\Phi^{(n)}(t) + c_{n-1}\Phi^{(n-1)}(t) + \cdots + c_1\Phi^{(1)}(t) + c_0\Phi(t)$$

$$= \sum_{k=0}^{n-1} \left\{ x_k^{(n)}(t) + c_{n-1}x_k^{(n-1)}(t) + \cdots + c_1x_k^{(1)}(t) + c_0x_k(t) \right\} A^k$$

$$= 0 \cdot I + 0 \cdot A + \cdots + 0 \cdot A^{n-1} = 0.$$

As well we see that

$$\begin{aligned}
\Phi(0) &= x_0(0)I + x_1(0)A + \cdots + x_{n-1}(0)A^{n-1} &= I, \\
\Phi^{(1)}(0) &= x_0^{(1)}(0)I + x_1^{(1)}(0)A + \cdots + x_{n-1}^{(1)}(0)A^{n-1} &= A, \\
&\vdots \\
\Phi^{(n-1)}(0) &= x_0^{(n-1)}(0)I + x_1^{(n-1)}(0)A + \cdots + x_{n-1}^{(n-1)}(0)A^{n-1} &= A^{n-1}.
\end{aligned}$$

Therefore

$$\Phi(t) = x_0(t)I + x_1(t)A + \cdots + x_{n-1}(t)A^{n-1}$$

satisfies the IVP

$$\Phi^{(n)}(t) + c_{n-1}\Phi^{(n-1)}(t) + \cdots + c_1\Phi^{(1)}(t) + c_0\Phi(t) = 0$$

possessing the initial conditions

$$\Phi^{(k)}(0) = A^k$$

($k \in \mathbf{Z}_n$). The solution is unique, and so we must conclude that $e^{At} = \sum_{k=0}^{n-1} x_k(t)A^k$ for all $t \in \mathbf{R}^+$, which is the central claim of the theorem.

An example of how to apply the result of Theorem 11.10 is as follows.

Example 11.4 Suppose that

$$A = \begin{bmatrix} \alpha & \gamma \\ 0 & \beta \end{bmatrix} \in \mathbf{R}^{2\times 2},$$

which clearly has the eigenvalues $\lambda = \alpha, \beta$. Begin by assuming distinct eigenvalues for A, specifically, that $\alpha \neq \beta$.

The general solution to the second-order homogeneous ODE

$$x^{(2)}(t) + c_1 x^{(1)}(t) + c_0 x(t) = 0$$

with characteristic roots α, β (eigenvalues of A) is [recall (11.15)]

$$x(t) = a_0 e^{\alpha t} + a_1 e^{\beta t}.$$

We have $x^{(1)}(t) = a_0 \alpha e^{\alpha t} + a_1 \beta e^{\beta t}$.

For the initial conditions $x(0) = 1, x^{(1)}(0) = 0$, we have the linear system of equations

$$a_0 + a_1 = 1,$$

$$\alpha a_0 + \beta a_1 = 0,$$

which solve to yield

$$a_0 = \frac{\beta}{\beta - \alpha}, \ a_1 = -\frac{\alpha}{\beta - \alpha}.$$

Thus, the solution in this case is

$$x_0(t) = \frac{1}{\beta - \alpha}[\beta e^{\alpha t} - \alpha e^{\beta t}].$$

Now, if instead the initial conditions are $x(0) = 0, \ x^{(1)}(0) = 1$, we have the linear system of equations

$$a_0 + a_1 = 0,$$

$$\alpha a_0 + \beta a_1 = 1,$$

which solve to yield

$$a_0 = -\frac{1}{\beta - \alpha}, \ a_1 = \frac{1}{\beta - \alpha}.$$

Thus, the solution in this case is

$$x_1(t) = \frac{1}{\beta - \alpha}[-e^{\alpha t} + e^{\beta t}].$$

Via Leonard II we must have

$$
\begin{aligned}
e^{At} &= x_0(t)I + x_1(t)A \\
&= \frac{1}{\beta - \alpha}\left[\begin{array}{cc} \beta e^{\alpha t} - \alpha e^{\beta t} & 0 \\ 0 & \beta e^{\alpha t} - \alpha e^{\beta t} \end{array}\right] \\
&\quad + \frac{1}{\beta - \alpha}\left[\begin{array}{cc} -\alpha e^{\alpha t} + \alpha e^{\beta t} & -\gamma e^{\alpha t} + \gamma e^{\beta t} \\ 0 & -\beta e^{\alpha t} + \beta e^{\beta t} \end{array}\right] \\
&= \frac{1}{\beta - \alpha}\left[\begin{array}{cc} (\beta - \alpha)e^{\alpha t} & -\gamma e^{\alpha t} + \gamma e^{\beta t} \\ 0 & (\beta - \alpha)e^{\beta t} \end{array}\right] \\
&= \left[\begin{array}{cc} e^{\alpha t} & \frac{\gamma}{\beta - \alpha}(e^{\beta t} - e^{\alpha t}) \\ 0 & e^{\beta t} \end{array}\right].
\end{aligned}
\tag{11.19}
$$

Now assume that $\alpha = \beta$.

The general solution to

$$x^{(2)}(t) + c_1 x^{(1)}(t) + c_0 x(t) = 0$$

with characteristic roots α, α (eigenvalues of A) is [again, recall (11.15)]

$$x(t) = (a_0 + a_1 t)e^{\alpha t}.$$

We have $x^{(1)}(t) = (a_1 + a_0\alpha + a_1\alpha t)e^{\alpha t}$.

For the initial conditions $x(0) = 1, x^{(1)}(0) = 0$ we have the linear system of equations

$$a_0 = 1,$$

$$a_1 + a_0\alpha = 0,$$

which solves to yield $a_1 = -\alpha$, so that

$$x_0(t) = (1 - \alpha t)e^{\alpha t}.$$

Now if instead the initial conditions are $x(0) = 0, x^{(1)}(0) = 1$, we have the linear system of equations

$$a_0 = 0,$$

$$a_1 + a_0\alpha = 1,$$

which solves to yield $a_1 = 1$ so that

$$x_1(t) = te^{\alpha t}.$$

If we again apply Leonard II, then we have

$$e^{At} = x_0(t)I + x_1(t)A$$

$$= \begin{bmatrix} e^{\alpha t} & \gamma t e^{\alpha t} \\ 0 & e^{\alpha t} \end{bmatrix}. \tag{11.20}$$

A good exercise for the reader is to verify that $x(t) = e^{At}x(0)$ solves $dx(t)/dt = Ax(t)$ [of course, here $x(t) = [x_0(t)x_1(t)]^T$ is a state vector] in both of the cases considered in Example 11.4. Do this by direct substitution.

Example 11.4 is considered in Moler and Van Loan [7] as it illustrates problems in computing e^{At} when the eigenvalues of A are nearly multiple. If we consider (11.19) when $\beta - \alpha$ is small, and yet is not negligible, the "divided difference"

$$\frac{e^{\beta t} - e^{\alpha t}}{\beta - \alpha}, \tag{11.21}$$

when computed directly, may result in a large relative error. In (11.19) the ratio (11.21) is multiplied by γ, so the final answer may be very inaccurate, indeed. Matrix A in Example 11.4 is of low order (i.e., $n = 2$) and is triangular. This type of problem is very difficult to detect and correct when A is larger and not triangular.

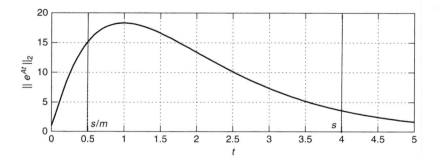

Figure 11.1   An illustration of the *hump phenomenon* in computing e^{At}.

Another difficulty noted in Moler and Van Loan [7] is sometimes called the *hump phenomenon*. It is illustrated in Fig. 11.1 for Eq. (11.19) using the parameters

$$\alpha = -1.01, \quad \beta = -1.00, \quad \gamma = 50. \tag{11.22}$$

Figure 11.1 is a plot of matrix 2-norm [spectral norm; recall Equation (4.37) with $p = 2$ in Chapter 4] $||e^{At}||_2$ versus t. (It is a version of Fig. 1 in Ref. 7.) The problem with this arises from the fact that one way or another some algorithms for the computation of e^{At} make use of the identity

$$e^{At} = (e^{At/m})^m. \tag{11.23}$$

When s/m is under the hump while s lies beyond it (e.g., in Fig. 11.1 $s = 4$ with $m = 8$), we can have

$$||e^{As}||_2 \ll ||e^{As/m}||_2^m. \tag{11.24}$$

Unfortunately, rounding errors in the mth power of $e^{As/m}$ are usually small only relative to $||e^{As/m}||_2$, rather than $||e^{As}||_2$. Thus, rounding errors may be a problem in using (11.23) to compute e^{At}.

The Taylor series expansion in (11.8) is not a good method for computing e^{At}. The reader should recall the example of catastrophic convergence in the computation of e^x ($x \in \mathbf{R}$) from Chapter 3 (Appendix 3.C). It is not difficult to imagine that the problem of catastrophic convergence in (11.8) is likely to be much worse, and much harder to contain. Indeed this is the case as shown by an example in Moler and Van Loan [7].

It was suggested earlier in this section that the series in (11.8) can be shown to converge by considering the diagonalization of A (assuming that A is nondefective). Suppose that $A \in \mathbf{C}^{n \times n}$, and that A possesses eigenvalues that are all distinct, and so we may apply (11.6). Since

$$A^k = [P \Lambda P^{-1}]^k$$
$$= \underbrace{[P \Lambda P^{-1}][P \Lambda P^{-1}] \cdots [P \Lambda P^{-1}][P \Lambda P^{-1}]}_{k \text{ factors}} = P \Lambda^k P^{-1}, \tag{11.25}$$

we have

$$
e^{At} = \sum_{k=0}^{\infty} \frac{1}{k!} A^k t^k = \sum_{k=0}^{\infty} \frac{1}{k!} P \Lambda^k P^{-1} t^k = P \left[\sum_{k=0}^{\infty} \frac{1}{k!} \Lambda^k t^k \right] P^{-1}
$$

$$
= P \operatorname{diag} \left(\sum_{k=0}^{\infty} \frac{1}{k!} \lambda_0^k t^k, \sum_{k=0}^{\infty} \frac{1}{k!} \lambda_1^k t^k, \ldots, \sum_{k=0}^{\infty} \frac{1}{k!} \lambda_{n-1}^k t^k \right) P^{-1}
$$

$$
= P \operatorname{diag}(e^{\lambda_0 t}, e^{\lambda_1 t}, \ldots, e^{\lambda_{n-1} t}) P^{-1}.
$$

If we define

$$
e^{\Lambda t} = \operatorname{diag}(e^{\lambda_0 t}, e^{\lambda_1 t}, \ldots, e^{\lambda_{n-1} t}), \tag{11.26}
$$

then clearly we can say that

$$
e^{At} = \sum_{k=0}^{\infty} \frac{1}{k!} A^k t^k = P e^{\Lambda t} P^{-1}. \tag{11.27}
$$

We know from the theory of Maclaurin series that $e^x = \sum_{k=0}^{\infty} \frac{1}{k!} x^k$ converges for all $x \in \mathbf{R}$. Thus, $e^{\Lambda t}$ converges for all $t \in \mathbf{R}$ and is well defined, and hence the series in (11.8) converges, and so e^{At} is well defined. Of course, all of this suggests that e^{At} may be numerically computed using (11.27). From Chapter 3 we infer that accurate, reliable means to compute e^x (x is a scalar) do exist. Also, reliable methods exist to find the elements of Λ, and this will be considered later. But P may be close to singular, that is, the condition number $\kappa(P)$ (recall Chapter 4) may be large, and so accurate determination of P^{-1}, which is required by (11.27), may be difficult. Additionally, the approach (11.27) lacks generality since it won't work unless A is nondefective (i.e., can be diagonalized). Matrix factorization methods to compute e^{At} (including those in the defective case) are considered in greater detail in Ref. 7, and this matter will not be mentioned further here.

In Chapter 4 a condition number $\kappa(A)$ was defined that informed us about the sensitivity of the solution to $Ax = b$ due to perturbations in A and b. It is possible to develop a similar notion for the problem of computing e^{At}. From Golub and Van Loan [4], the *matrix exponential condition number* is

$$
\nu(A, t) = \max_{||E||_2 \leq 1} \left\| \int_0^t e^{A(t-s)} E e^{As} \, ds \right\|_2 \frac{||A||_2}{||e^{At}||_2}. \tag{11.28}
$$

(The theory behind this originally appeared in Van Loan [12].) In this expression $E \in \mathbf{R}^{n \times n}$ is a perturbation matrix. The condition number (11.28) measures the sensitivity of mapping $A \to e^{At}$ for a given $t \in \mathbf{R}$. For a given t, there is a matrix E such that

$$
\frac{||e^{(A+E)t} - e^{At}||_2}{||e^{At}||_2} \approx \nu(A, t) \frac{||E||_2}{||A||_2}. \tag{11.29}
$$

We see from this that if $v(A, t)$ is large, then a small change in A (modeled by the perturbation matrix E) can cause a large change in e^{At}. In general, it is not easy to specify A leading to large values for $v(A, t)$. However, it is known that

$$v(A, t) \geq t||A||_2 \tag{11.30}$$

for $t \in \mathbf{R}^+$ with equality iff A is *normal*. (Any $A \in \mathbf{C}^{n \times n}$ is normal iff $A^H A = AA^H$.) Thus, it appears that normal matrices are generally the least troublesome with respect to computing e^{At}. From the definition of a normal matrix we see that real-valued, symmetric matrices are an important special case.

Of the less dubious means to compute e^{At}, Golub and Van Loan's algorithm 11.3.1 [4] is suggested. It is based on *Padé approximation*, which is the use of rational functions to approximate other functions. However, we will only refer the reader to Ref. 4 (or Ref. 7) for the relevant details. A version of Algorithm 11.3.1 [4] is implemented in the MATLAB expm function, and MATLAB provides other algorithm implementations for computing e^{At}.

11.4 THE POWER METHODS

In this section we consider a simple approach to determine the eigenvalues and eigenvectors of $A \in \mathbf{R}^{n \times n}$. The approach is iterative. The main result is as follows.

Theorem 11.11: Let $A \in \mathbf{R}^{n \times n}$ be such that

(a) A has n linearly independent eigenvectors $x^{(k)}$, corresponding to the eigenpairs $\{(\lambda_k, x^{(k)}) | k \in \mathbf{Z}_n\}$.

(b) The eigenvalues satisfy $\lambda_{n-1} \in \mathbf{R}$ with

$$|\lambda_{n-1}| > |\lambda_{n-2}| \geq |\lambda_{n-3}| \geq \cdots |\lambda_1| \geq |\lambda_0|$$

(λ_{n-1} is the *dominant eigenvalue*).

If $y_0 \in \mathbf{R}^n$ is a starting vector such that

$$y_0 = \sum_{j=0}^{n-1} a_j x^{(j)}$$

with $a_{n-1} \neq 0$, then for $y_{k+1} = Ay_k$ with $k \in \mathbf{Z}^+$

$$\lim_{k \to \infty} \frac{A^k y_0}{\lambda_{n-1}^k} = cx^{(n-1)} \tag{11.31}$$

for some $c \neq 0$, and (recalling that $\langle x, y \rangle = x^T y = y^T x$ for any $x, y \in \mathbf{R}^n$)

$$\lim_{k \to \infty} \frac{\langle y_0, A^k y_0 \rangle}{\langle y_0, A^{k-1} y_0 \rangle} = \lambda_{n-1}. \tag{11.32}$$

Proof   We observe that $y_k = A^k y_0$, and since $A^k x^{(j)} = \lambda_j^k x^{(j)}$, we have

$$A^k y_0 = \sum_{j=0}^{n-2} a_j \lambda_j^k x^{(j)} + a_{n-1} \lambda_{n-1}^k x^{(n-1)},$$

implying that

$$\frac{A^k y_0}{\lambda_{n-1}^k} = a_{n-1} x^{(n-1)} + \sum_{j=0}^{n-2} a_j \left(\frac{\lambda_j}{\lambda_{n-1}} \right)^k x^{(j)}. \tag{11.33}$$

Since $|\lambda_j| < |\lambda_{n-1}|$ for $j = 0, 1, \ldots, n-2$, we have $\lim_{k \to \infty} \left| \left(\frac{\lambda_j}{\lambda_{n-1}} \right)^k \right| = 0$, and hence

$$\lim_{k \to \infty} \frac{A^k y_0}{\lambda_{n-1}^k} = a_{n-1} x^{(n-1)}.$$

Now $A^k y_0 = \sum_{j=0}^{n-1} a_j \lambda_j^k x^{(j)}$, so that

$$\langle y_0, A^k y_0 \rangle = \sum_{j=0}^{n-1} a_j \lambda_j^k \langle y_0, x^{(j)} \rangle,$$

and if $\eta_j = \langle y_0, x^{(j)} \rangle$, then

$$\frac{\langle y_0, A^k y_0 \rangle}{\langle y_0, A^{k-1} y_0 \rangle} = \frac{a_{n-1} \lambda_{n-1}^k \eta_{n-1} + \sum_{j=0}^{n-2} a_j \lambda_j^k \eta_j}{a_{n-1} \lambda_{n-1}^{k-1} \eta_{n-1} + \sum_{j=0}^{n-2} a_j \lambda_j^{k-1} \eta_j}$$

$$= \lambda_{n-1} \left[\frac{a_{n-1} \eta_{n-1} + \sum_{j=0}^{n-2} a_j \eta_j \left(\frac{\lambda_j}{\lambda_{n-1}} \right)^k}{a_{n-1} \eta_{n-1} + \sum_{j=0}^{n-2} a_j \eta_j \left(\frac{\lambda_j}{\lambda_{n-1}} \right)^{k-1}} \right].$$

Again, since $\lim_{k \to \infty} \left| \left(\frac{\lambda_j}{\lambda_{n-1}} \right)^k \right| = 0$ (for $j = 0, 1, \ldots, n-2$), we have

$$\lim_{k \to \infty} \frac{\langle y_0, A^k y_0 \rangle}{\langle y_0, A^{k-1} y_0 \rangle} = \lambda_{n-1}.$$

We note that

$$1 > \left| \frac{\lambda_{n-2}}{\lambda_{n-1}} \right| \geq \left| \frac{\lambda_{n-3}}{\lambda_{n-1}} \right| \geq \cdots \geq \left| \frac{\lambda_0}{\lambda_{n-1}} \right|,$$

so the rate of convergence of $A^k y_0 / \lambda_{n-1}^k$ to $a_{n-1} x^{(n-1)}$ is, according to (11.33), dominated by the term containing $\lambda_{n-2} / \lambda_{n-1}$. This is sometimes expressed by writing

$$\left\| \frac{A^k y_0}{\lambda_{n-1}^k} - a_{n-1} x^{(n-1)} \right\| = O\left(\left(\frac{\lambda_{n-2}}{\lambda_{n-1}} \right)^k \right). \tag{11.34}$$

The choice of norm in (11.34) is arbitrary. Of course, a small value for $|\lambda_{n-2}/\lambda_{n-1}|$ implies faster convergence.

The theory of Theorem 11.11 assumes that A is nondefective. If the algorithm suggested by this theorem is applied to a defective A, it will attempt to converge. In effect, for any defective matrix A there is a nondefective matrix close to it, and so the limiting values in (11.31) and (11.32) will be for a "nearby" nondefective matrix. However, convergence can be very slow, particularly if the dependent eigenvectors of A correspond to λ_{n-1}, and λ_{n-2}. In this situation the results may not be meaningful.

If y_0 is chosen such that $a_{n-1} = 0$, rounding errors in computing $y_{k+1} = Ay_k$ will usually give a y_k with a component in the direction of the eigenvector $x^{(n-1)}$. Thus, convergence ultimately is the result. To ensure that this happens, it is best to select y_0 with noninteger components.

From (11.31), if k is big enough, then

$$y_k = A^k y_0 \approx c \lambda_{n-1}^k x^{(n-1)},$$

and so y_k is an approximation to $x^{(n-1)}$ (to within some scale factor). However, if $|\lambda_{n-1}| > 1$, we see that $\|y_k\| \to \infty$ with increasing k, while if $|\lambda_{n-1}| < 1$, we see that $\|y_k\| \to 0$ with increasing k. Either way, serious numerical problems will certainly result (overflow in the former case and rounding errors or underflow in the latter case). This difficulty may be eliminated by the proper scaling of the iterates, and leads to what is sometimes called the *scaled power algorithm*:

```
Input N; { upper limit on the number of iterations }
Input y₀; { starting vector }
y₀ := y₀/||y₀||₂; { normalize y₀ to unit norm }
k := 0;
while k < N do begin
    z_{k+1} := Ay_k;
    y_{k+1} := z_{k+1}/||z_{k+1}||₂;
    λ^{(k+1)} := y_{k+1}^T Ay_{k+1};
    k := k + 1;
    end;
```

In this algorithm y_{k+1} is the $(k+1)$th estimate of $x^{(n+1)}$, while $\lambda^{(k+1)}$ is the corresponding estimate of eigenvalue λ_{n-1}. From the pseudocode above we may easily see that

$$y_k = \frac{A^k y_0}{\|A^k y_0\|_2} \tag{11.35}$$

for all $k \geq 1$. With $y_0 = \sum_{j=0}^{n-1} a_j x^{(j)}$ ($a_j \in \mathbf{R}$ such that $||y_0||_2 = 1$)

$$A^k y_0 = \sum_{j=0}^{n-2} a_j A^k x^{(j)} + a_{n-1} A^k x^{(n-1)} = a_{n-1} \lambda_{n-1}^k x^{(n-1)} + \sum_{j=0}^{n-2} a_j \lambda_j^k x^{(j)}$$

$$= a_{n-1} \lambda_{n-1}^k \left[x^{(n-1)} + \underbrace{\sum_{j=0}^{n-2} \frac{a_j}{a_{n-1}} \left(\frac{\lambda_j}{\lambda_{n-1}} \right)^k x^{(j)}}_{=u^{(k)}} \right],$$

and as in Theorem 11.11 we see $\lim_{k \to \infty} ||u^{(k)}||_2 = 0$, and (11.35) becomes

$$y_k = \frac{a_{n-1} \lambda_{n-1}^k [x^{(n-1)} + u^{(k)}]}{||a_{n-1} \lambda_{n-1}^k [x^{(n-1)} + u^{(k)}]||_2} = \mu_k \frac{x^{(n-1)} + u^{(k)}}{||x^{(n-1)} + u^{(k)}||_2}, \qquad (11.36)$$

where μ_k is the sign of $a_{n-1} \lambda_{n-1}^k$ (i.e., $\mu_k \in \{+1, -1\}$). Clearly, as $k \to \infty$, vector y_k in (11.36) becomes a better and better approximation to eigenvector $x^{(n-1)}$. To confirm that $\lambda^{(k+1)}$ estimates λ_{n-1}, consider that from (11.36) we have (for k sufficiently large)

$$\lambda^{(k+1)} = y_{k+1}^T A y_{k+1} \approx \frac{(x^{(n-1)})^T A x^{(n-1)}}{||x^{(n-1)}||_2^2} = \lambda_{n-1}$$

[recall that $A x^{(n-1)} = \lambda_{n-1} x^{(n-1)}$].

If $|\lambda_{n-1}| = |\lambda_{n-2}| > |\lambda_j|$ for $j = 0, 1, \ldots, n-3$, we have two dominant eigenvalues. In this situation, as noted in Quarteroni et al. [13], convergence may or may not occur. If, for example, $\lambda_{n-1} = \lambda_{n-2}$, then vector sequence (y_k) converges to a vector in the subspace of $\mathbf{R}^n$ spanned by $x^{(n-1)}$, and $x^{(n-2)}$. In this case, since $A \in \mathbf{R}^{n \times n}$, we must have $\lambda_{n-1}, \lambda_{n-2} \in \mathbf{R}$, and hence for $k \geq 1$, we must have

$$A^k y_0 = \sum_{j=0}^{n-3} a_j \lambda_j^k x^{(j)} + a_{n-2} \lambda_{n-2}^k x^{(n-2)} + a_{n-1} \lambda_{n-1}^k x^{(n-1)},$$

implying that

$$\frac{A^k y_0}{\lambda_{n-1}^k} = a_{n-1} x^{(n-1)} + a_{n-2} x^{(n-2)} + \sum_{j=0}^{n-3} a_j \left(\frac{\lambda_j}{\lambda_{n-1}} \right)^k x^{(j)}$$

so that

$$\lim_{k \to \infty} \frac{A^k y_0}{\lambda_{n-1}^k} = a_{n-1} x^{(n-1)} + a_{n-2} x^{(n-2)},$$

which is a vector in a two-dimensional subspace of $\mathbf{R}^n$. On the other hand, recall Example 11.3, where $n = 2$ so that $\lambda_0 = e^{j\theta}$, $\lambda_1 = e^{-j\theta} = \lambda_0^*$. From (11.35) $||y_k||_2 = 1$, so, because A is a rotation operator, $y_k = A^k y_0 / ||A^k y_0||_2$ will always be a point on the *unit circle* $\{(x, y) | x^2 + y^2 = 1\}$. Convergence does not occur since it is generally not the same point from one iteration to the next (e.g., consider $\theta = \pi/2$ radians).

Example 11.5 Consider an example based on application of the scaled power algorithm to the matrix

$$A = \begin{bmatrix} 4 & 1 & 0 \\ 1 & 4 & 1 \\ 0 & 1 & 4 \end{bmatrix}.$$

This matrix turns out to have the eigenvalues

$$\lambda_0 = 2.58578644, \quad \lambda_1 = 4.00000000, \quad \lambda_2 = 5.41421356$$

(as may be determined in MATLAB using the eig function). If we define $y_k = [y_{k,0} \ y_{k,1} \ y_{k,2}]^T \in \mathbf{R}^3$, then, from the algorithm, we obtain the iterates:

k	$y_{k,0}$	$y_{k,1}$	$y_{k,2}$	$\lambda^{(k)}$
0	0.57735027	0.57735027	0.57735027	—
1	0.53916387	0.64699664	0.53916387	5.39534884
2	0.51916999	0.67891460	0.51916999	5.40988836
3	0.50925630	0.69376945	0.50925630	5.41322584
4	0.50444312	0.70076692	0.50444312	5.41398821
5	0.50212703	0.70408585	0.50212703	5.41416216
6	0.50101700	0.70566560	0.50101700	5.41420184
7	0.50048597	0.70641885	0.50048597	5.41421089
8	0.50023215	0.70677831	0.50023215	5.41421295
9	0.50011089	0.70694993	0.50011089	5.41421342
10	0.50005296	0.70703187	0.50005296	5.41421353
11	0.50002530	0.70707101	0.50002530	5.41421356

In only 11 iterations the power algorithm has obtained the dominant eigenvalue to an accuracy of eight decimal places.

Continue to assume that $A \in \mathbf{R}^{n \times n}$ is nondefective. Now define $A_\mu = A - \mu I$ (where I is the $n \times n$ identity matrix, as usual), and $\mu \in \mathbf{R}$ is called the *shift* (or *shift parameter*).[4] We will assume that μ always results in the existence of A_μ^{-1}. Because A is not defective, there will be a nonsingular matrix P such that $P^{-1} A P = \Lambda = \text{diag}(\lambda_0, \lambda_1, \ldots, \lambda_{n-1})$ (recall the basic facts from Section 11.2 that justify this). Consequently

$$A_\mu = P\Lambda P^{-1} - \mu I \Rightarrow P^{-1} A_\mu P = \Lambda - \mu I, \tag{11.37}$$

---

[4]The reason for introducing the shift parameter μ will be made clear a bit later.

and so A_μ^{-1} has the eigenvalues

$$\gamma_k = \frac{1}{\lambda_k - \mu}, \ k \in \mathbf{Z}_n. \tag{11.38}$$

P is a similarity transformation that diagonalizes A_μ, giving $\Lambda - \mu I$, so the eigenvalues of $(\Lambda - \mu I)^{-1}$ must be the eigenvalues of A_μ^{-1} as these are similar matrices. A modification of the previous scaled power algorithm is the following *shifted inverse power algorithm*:

Input N; { upper limit on the number of iterations }
Input y_0; { starting vector }
$y_0 := y_0/||y_0||_2$; { normalize y_0 to unit norm }
$k := 0$;
while $k < N$ do begin
    $A_\mu z_{k+1} := y_k$;
    $y_{k+1} := z_{k+1}/||z_{k+1}||_2$;
    $\gamma^{(k+1)} := y_{k+1}^T A y_{k+1}$;
    $k := k + 1$;
    end;

Assume that the eigenvalues of A satisfy

$$|\lambda_{n-1}| \geq |\lambda_{n-2}| \geq \cdots \geq |\lambda_1| > |\lambda_0|, \tag{11.39}$$

and also that $\mu = 0$, so then $|\lambda_k| = 1/|\gamma_k|$, and (11.39) yields

$$|\gamma_0| > |\gamma_1| \geq |\gamma_2| \geq \cdots \geq |\gamma_{n-2}| \geq |\gamma_{n-1}|. \tag{11.40}$$

We observe that in the shifted inverse power algorithm $A_\mu z_{k+1} = y_k$ is equivalent to $z_{k+1} = A_\mu^{-1} y_k$, and so A_μ^{-1} effectively replaces A in the statement $z_{k+1} := Ay_k$ in the scaled power algorithm. This implies that the shifted inverse power algorithm produces vector sequence (y_k) that converges to the eigenvector of $A_\mu^{-1} = A^{-1}$ corresponding to the eigenvalue γ_0 $(= 1/\lambda_0)$. Since $Ax^{(0)} = \lambda_0 x^{(0)}$ implies that $A^{-1}x^{(0)} = \frac{1}{\lambda_0}x^{(0)} = \gamma_0 x^{(0)}$, then for a sufficiently large k, the vector y_k will approximate $x^{(0)}$. The argument to verify this follows the proof of Theorem 11.11. Therefore, consider the starting vector

$$y_0 = a_0 x^{(0)} + \sum_{j=1}^{n-1} a_j x^{(j)}, a_0 \neq 0 \tag{11.41}$$

(such that $||y_0||_2 = 1$). We see that we must have

$$A^{-k} y_0 = a_0 \frac{1}{\lambda_0^k} x^{(0)} + \sum_{j=1}^{n-1} a_j \frac{1}{\lambda_j^k} x^{(j)}$$

$$= a_0 \gamma_0^k x^{(0)} + \sum_{j=1}^{n-1} a_j \gamma_j^k x^{(j)}, \tag{11.42}$$

and thus

$$\frac{A^{-k} y_0}{\gamma_0^k} = a_0 x^{(0)} + \sum_{j=1}^{n-1} a_j \left(\frac{\gamma_j}{\gamma_0}\right)^k x^{(j)}.$$

Immediately, because of (11.40), we must have

$$\lim_{k \to \infty} \frac{A^{-k} y_0}{\gamma_0^k} = a_0 x^{(0)}.$$

From (11.42), we obtain

$$A^{-k} y_0 = a_0 \gamma_0^k \left[x^{(0)} + \underbrace{\sum_{j=1}^{n-1} \frac{a_j}{a_0} \left(\frac{\gamma_j}{\gamma_0}\right)^k x^{(j)}}_{=v^{(k)}} \right]. \tag{11.43}$$

In much the same way as we arrived at (11.35), for the shifted inverse power algorithm, we must have for all $k \geq 1$

$$y_k = \frac{A_\mu^{-k} y_0}{||A_\mu^{-k} y_0||_2}. \tag{11.44}$$

From (11.43) for $\mu = 0$, this equation becomes

$$y_k = \frac{a_0 \gamma_0^k [x^{(0)} + v^{(k)}]}{||a_0 \gamma_0^k [x^{(0)} + v^{(k)}]||_2} = \nu_k \frac{x^{(0)} + v^{(k)}}{||x^{(0)} + v^{(k)}||_2}, \tag{11.45}$$

where $\nu_k \in \{+1, -1\}$. From the pseudocode for the shifted inverse power algorithm, if k is large enough, via (11.45), we obtain

$$\gamma^{(k+1)} = y_{k+1}^T A y_{k+1} \approx \frac{(x^{(0)})^T A x^{(0)}}{||x^{(0)}||_2^2} = \lambda_0. \tag{11.46}$$

Thus, $\gamma^{(k+1)}$ is an approximation to λ_0.

In summary, for a nondefective $A \in \mathbf{R}^{n \times n}$ such that (11.39) holds, the shifted inverse power algorithm will generate a sequence of increasingly better approximations to the eigenpair $(\lambda_0, x^{(0)})$, when we set $\mu = 0$.

We note that the scaled power algorithm needs $O(n^2)$ flops (recall the definition of flops from Chapter 4) at every iteration. This is due mainly to the matrix–vector product step $z_{k+1} = A y_k$. To solve the linear system $A_\mu z_{k+1} = y_k$ requires $O(n^3)$ flops in general. To save on computations in the implementation of the shifted inverse power algorithm, it is often best to LU-decompose A_μ (recall Section 4.5) only once: $A_\mu = LU$. At each iteration $LU z_{k+1} = y_k$ may be solved using forward and backward substitution, which is efficient since this needs only $O(n^2)$ flops at every iteration. Even so, because of the need to compute the LU decomposition of A_μ, the shifted inverse power algorithm still needs $O(n^3)$ flops overall. It is thus intrinsically more computation-intensive than is the scaled power algorithm.

However, it is the concept of a shift that makes the shifted inverse power algorithm attractive, at least in some circumstances. But before we consider the reason for introducing the shift parameter μ, the reader should view the following example.

Example 11.6   Let us reconsider matrix A from Example 11.5. If we apply the shifted inverse power algorithm using $\mu = 0$ to this matrix, we obtain the following iterates:

k	$y_{k,0}$	$y_{k,1}$	$y_{k,2}$	$\gamma^{(k)}$
0	0.57735027	0.57735027	0.57735027	—
1	0.63960215	0.42640143	0.63960215	5.09090909
2	0.70014004	0.14002801	0.70014004	4.39215686
3	0.69011108	−0.21792981	0.69011108	3.39841689
4	0.62357865	−0.47148630	0.62357865	2.82396484
5	0.56650154	−0.59845803	0.56650154	2.64389042
6	0.53330904	−0.65662999	0.53330904	2.59925317
7	0.51623508	−0.68337595	0.51623508	2.58886945
8	0.50782504	−0.69586455	0.50782504	2.58649025
9	0.50375306	−0.70175901	0.50375306	2.58594700
10	0.50179601	−0.70455768	0.50179601	2.58582306
11	0.50085857	−0.70589049	0.50085857	2.58579479
12	0.50041023	−0.70652615	0.50041023	2.58578834
13	0.50019597	−0.70682953	0.50019597	2.58578687
14	0.50009360	−0.70697438	0.50009360	2.58578654
15	0.50004471	−0.70704355	0.50004471	2.58578646
16	0.50002135	−0.70707658	0.50002135	2.58578644

We see that the method converges to an estimate of λ_0 (smallest eigenvalue of A) that is accurate to eight decimal places in only 16 iterations.

As an exercise, the reader should confirm that the vector y_k in the bottom row of the table above is an estimate of the eigenvector for λ_0. The reader should do the same for Example 11.5.

Recall again that A_μ^{-1} is assumed to exist (so that μ is not an eigenvalue of A). Observe that $Ax^{(j)} = \lambda_j x^{(j)}$, so that $(A - \mu I)x^{(j)} = (\lambda_j - \mu)x^{(j)}$, and therefore $A_\mu^{-1}x^{(j)} = \gamma_j x^{(j)}$. Suppose that there is an $m \in \mathbf{Z}_n$ such that

$$|\lambda_m - \mu| < |\lambda_j - \mu| \qquad (11.47)$$

for all $j \in \mathbf{Z}_n$, but that $j \neq m$; that is, λ_m is closest to μ of all the eigenvalues of A. This really says that λ_m has a multiplicity of one (i.e., is simple). Now consider the starting vector

$$y_0 = \sum_{\substack{j=0 \\ j\neq m}}^{n-1} a_j x^{(j)} + a_m x^{(m)} \tag{11.48}$$

with $a_m \neq 0$, and $\|y_0\|_2 = 1$. Clearly

$$A_\mu^{-k} y_0 = \sum_{\substack{j=0 \\ j\neq m}}^{n-1} a_j \gamma_j^k x^{(j)} + a_m \gamma_m^k x^{(m)},$$

implying that

$$\frac{A_\mu^{-k} y_0}{\gamma_m^k} = a_m x^{(m)} + \sum_{\substack{j=0 \\ j\neq m}}^{n-1} a_j \left(\frac{\gamma_j}{\gamma_m}\right)^k x^{(j)}. \tag{11.49}$$

Now via (11.47)

$$\left|\frac{\gamma_j}{\gamma_m}\right| = \left|\frac{\lambda_m - \mu}{\lambda_j - \mu}\right| < 1.$$

Therefore, via (11.49)

$$\lim_{k\to\infty} \frac{A_\mu^{-k} y_0}{\gamma_m^k} = a_m x^{(m)}.$$

This implies that the vector sequence (y_k) in the shifted inverse power algorithm converges to $x^{(m)}$. Put simply, by the proper selection of the shift parameter μ, we can extract just about any eigenpair of A that we wish to (as long as λ_m is simple). Thus, in this sense, the shifted inverse power algorithm is more general than the scaled power algorithm. The following example illustrates another important point.

Example 11.7   Once again we apply the shifted inverse power algorithm to matrix A from Example 11.5. However, now we select $\mu = 2$. The resulting sequence of iterates for this case is as follows:

k	$y_{k,0}$	$y_{k,1}$	$y_{k,2}$	$\gamma^{(k)}$
0	0.57735027	0.57735027	0.57735027	—
1	0.70710678	0.00000000	0.70710678	4.00000000
2	0.57735027	−0.57735027	0.57735027	2.66666667
3	0.51449576	−0.68599434	0.51449576	2.58823529
4	0.50251891	−0.70352647	0.50251891	2.58585859
5	0.50043309	−0.70649377	0.50043309	2.58578856
6	0.50007433	−0.70700164	0.50007433	2.58578650
7	0.50001275	−0.70708874	0.50001275	2.58578644

We see that convergence to the smallest eigenvalue of A has now occurred in only seven iterations, which is faster than the case considered in Example 11.6 (for which we used $\mu = 0$).

We see that this example illustrates the fact that a properly chosen shift parameter can greatly accelerate the convergence of iterative eigenproblem solvers. This notion of shifting to improve convergence rates is also important in practical implementations of QR iteration methods for solving eigenproblems (next section).

So far our methods extract only one eigenvalue from A at a time. One may apply a method called *deflation* to extract all the eigenvalues of A under certain conditions. Begin by noting the following elementary result.

Lemma 11.1: Suppose that $B \in \mathbf{R}^{(n-1)\times(n-1)}$, and that B^{-1} exists, and that $r \in \mathbf{R}^{n-1}$, then

$$\begin{bmatrix} 1 & r^T \\ 0 & B \end{bmatrix}^{-1} = \begin{bmatrix} 1 & -r^T B^{-1} \\ 0 & B^{-1} \end{bmatrix}. \tag{11.50}$$

Proof Exercise.

The deflation procedure is based on the following theorem.

Theorem 11.12: Deflation Suppose that $A_n \in \mathbf{R}^{n\times n}$, that eigenvalue $\lambda_i \in \mathbf{R}$ for all $i \in \mathbf{Z}_n$, and that all the eigenvalues of A_n are distinct. The dominant eigenpair of A_n is $(\lambda_{n-1}, x^{(n-1)})$, and we assume that $||x^{(n-1)}||_2 = 1$. Suppose that $Q_n \in \mathbf{R}^{n\times n}$ is an orthogonal matrix such that $Q_n x^{(n-1)} = [1 \ 0 \ \cdots \ 0 \ 0]^T = e_0$; then

$$Q_n A_n Q_n^T = \begin{bmatrix} \lambda_{n-1} & a_{n-1}^T \\ 0 & A_{n-1} \end{bmatrix}. \tag{11.51}$$

Proof Q_n exists because it can be a Householder transformation matrix (recall Section 4.6). Any eigenvector $x^{(k)}$ of A_n can always be normalized so that $||x^{(k)}||_2 = 1$.

Following (11.5), we have

$$A_n \underbrace{[x^{(n-1)} x^{(n-2)} \cdots x^{(1)} x^{(0)}]}_{=T_n}$$

$$= \underbrace{[x^{(n-1)} x^{(n-2)} \cdots x^{(1)} x^{(0)}]}_{=T_n} \underbrace{\begin{bmatrix} \lambda_{n-1} & 0 & \cdots & 0 & 0 \\ 0 & \lambda_{n-2} & \cdots & 0 & 0 \\ \vdots & \vdots & & \vdots & \vdots \\ 0 & 0 & \cdots & \lambda_1 & 0 \\ 0 & 0 & \cdots & 0 & \lambda_0 \end{bmatrix}}_{=D_n},$$

that is, $A_n T_n = T_n D_n$. Thus, $(Q_n^T = Q_n^{-1}$ via orthogonality)

$$Q_n A_n Q_n^T = Q_n T_n D_n T_n^{-1} Q_n^T = (Q_n T_n) D_n (Q_n T_n)^{-1}. \tag{11.52}$$

Now

$$Q_n T_n = [e_0 \quad Q_n x^{(n-2)} \cdots Q_n x^{(1)} Q_n x^{(0)}] = \begin{bmatrix} 1 & b_{n-1}^T \\ 0 & B_{n-1} \end{bmatrix},$$

and via Lemma 11.1, we have

$$\begin{bmatrix} 1 & b_{n-1}^T \\ 0 & B_{n-1} \end{bmatrix}^{-1} = \begin{bmatrix} 1 & -b_{n-1}^T B_{n-1}^{-1} \\ 0 & B_{n-1}^{-1} \end{bmatrix}.$$

Thus, (11.52) becomes

$$Q_n A_n Q_n^T = \begin{bmatrix} 1 & b_{n-1}^T \\ 0 & B_{n-1} \end{bmatrix} \begin{bmatrix} \lambda_{n-1} & 0 \\ 0 & D_{n-1} \end{bmatrix} \begin{bmatrix} 1 & -b_{n-1}^T B_{n-1}^{-1} \\ 0 & B_{n-1}^{-1} \end{bmatrix}$$

$$= \begin{bmatrix} \lambda_{n-1} & b_{n-1}^T (D_{n-1} - \lambda_{n-1} I_{n-1}) B_{n-1}^{-1} \\ 0 & B_{n-1} D_{n-1} B_{n-1}^{-1} \end{bmatrix},$$

which has the form given in (11.51).

From Theorem 11.6, $Q_n A_n Q_n^T$ and A_n are similar matrices, and so have the same eigenvalues. Via (11.51), A_{n-1} has the same eigenvalues as A_n, except for λ_{n-1}. Clearly, the scaled power method could be used to find the eigenpair $(\lambda_{n-1}, x^{(n-1)})$. The Householder procedure from Section 4.6 gives Q_n. From Theorem 11.12 we obtain A_{n-1}, and the deflation procedure may be repeated to find all the remaining eigenvalues of $A = A_n$.

It is important to note that the deflation procedure may be improved with respect to computational efficiency by employing instead the *Rayleigh quotient iteration method*. This replaces the power methods we have considered so far. This approach is suggested and considered in detail in Golub and Van Loan [4] and Epperson [14]; we omit the details here.

11.5 *QR* ITERATIONS

The power methods of Section 11.4 and variations thereof such as Rayleigh quotient iterations are deficient in that they are not computationally efficient methods for computing all possible eigenpairs. The power methods are really at their best when we seek only a few eigenpairs (usually corresponding to either the smallest or the largest eigenvalues). In Section 11.4 the power methods were applied only to computing real-valued eigenvalues, but it is noteworthy that power methods can be adapted to finding complex–conjugate eigenvalue pairs [19].

The *QR iterations algorithms* are, according to Watkins [15], due originally to Francis [16] and Kublanovskaya [17]. The methodology involved in *QR* iterations is based, in turn, on earlier work of H. Rutishauser performed in the 1950s. The detailed theory and rationale for the *QR* iterations are not by any means straight-forward, and even the geometric arguments in Ref. 15 (based, in turn, on the work of Parlett and Poole [18]) are not easy to follow. However, for matrices $A \in \mathbf{C}^{n \times n}$ that are *dense* (i.e., *nonsparse*; recall Section 4.7), that are not too large, and that are nondefective, the *QR* iterations are the best approach presently known for finding all possible eigenpairs of A. Indeed, the MATLAB eig function implements a modern version of the *QR* iteration methodology.[5]

Because of the highly involved nature of the *QR* iteration theory, we will only present a few of the main ideas here. Other than the references cited so far, the reader is referred to the literature [4,6,13,19] for more thorough discussions. Of course, these are not the only references available on this subject.

Eigenvalue computations such as the *QR* iterations reduce large problems into smaller problems. Golub and Van Loan [4] present two lemmas that are involved in this reduction approach. Recall from Section 4.7 that $s(A)$ denotes the set of all the eigenvalues of matrix A (and is also called the *spectrum* of A).

Lemma 11.2: If $A \in \mathbf{C}^{n \times n}$ is of the form

$$A = \begin{bmatrix} A_{00} & A_{01} \\ 0 & A_{11} \end{bmatrix},$$

where $A_{00} \in \mathbf{C}^{p \times p}$, $A_{01} \in \mathbf{C}^{p \times q}$, $A_{11} \in \mathbf{C}^{q \times q}$ ($q + p = n$), then $s(A) = s(A_{00}) \cup s(A_{11})$.

Proof Consider

$$Ax = \begin{bmatrix} A_{00} & A_{01} \\ 0 & A_{11} \end{bmatrix} \begin{bmatrix} x_1 \\ x_2 \end{bmatrix} = \lambda \begin{bmatrix} x_1 \\ x_2 \end{bmatrix}$$

($x_1 \in \mathbf{C}^p$ and $x_2 \in \mathbf{C}^q$). If $x_2 \neq 0$, then $A_{11}x_2 = \lambda x_2$, and so we conclude that $\lambda \in s(A_{11})$. On the other hand, if $x_2 = 0$, then $A_{00}x_1 = \lambda x_1$, so we must have $\lambda \in s(A_{00})$. Thus, $s(A) \in s(A_{00}) \cup s(A_{11})$. Sets $s(A)$ and $s(A_{00}) \cup s(A_{11})$ have the same *cardinality* (i.e., the same number of elements), and so $s(A) = s(A_{00}) \cup s(A_{11})$.

---

[5]If $A \in \mathbf{C}^{n \times n}$, then $[V, D] = eig(A)$ such that

$$A = VDV^{-1},$$

where $D \in \mathbf{C}^{n \times n}$ is the diagonal matrix of eigenvalues of A and $V \in \mathbf{C}^{n \times n}$ is the matrix whose columns are the corresponding eigenvectors. The eigenvectors in V are "normalized" so that each eigenvector has a 2-norm of unity.

Essentially, Lemma 11.2 states that if A is *block upper triangular*, the eigenvalues lie within the diagonal blocks.

Lemma 11.3: If $A \in \mathbf{C}^{n \times n}$, $B \in \mathbf{C}^{p \times p}$, $X \in \mathbf{C}^{n \times p}$ (with $p \le n$) satisfy

$$AX = XB, \quad \text{rank}(X) = p, \tag{11.53}$$

then there is a unitary $Q \in \mathbf{C}^{n \times n}$ (so $Q^{-1} = Q^H$) such that

$$Q^H A Q = T = \begin{bmatrix} T_{00} & T_{01} \\ 0 & T_{11} \end{bmatrix}, \tag{11.54}$$

where $T_{00} \in \mathbf{C}^{p \times p}$, $T_{01} \in \mathbf{C}^{p \times (n-p)}$, $T_{11} \in \mathbf{C}^{(n-p) \times (n-p)}$, and $s(T_{00}) = s(A) \cap s(B)$.

Proof The QR decomposition idea from Section 4.6 generalizes to any $X \in \mathbf{C}^{n \times p}$ with $p \le n$ and $\text{rank}(X) = p$; that is, complex-valued Householder matrices are available. Thus, there is a unitary matrix $Q \in \mathbf{C}^{n \times n}$ such that

$$X = Q \begin{bmatrix} R \\ 0 \end{bmatrix},$$

where $R \in \mathbf{C}^{p \times p}$. Substituting this into (11.53) yields

$$\begin{bmatrix} T_{00} & T_{01} \\ T_{10} & T_{11} \end{bmatrix} \begin{bmatrix} R \\ 0 \end{bmatrix} = \begin{bmatrix} R \\ 0 \end{bmatrix} B, \tag{11.55}$$

where

$$Q^H A Q = \begin{bmatrix} T_{00} & T_{01} \\ T_{10} & T_{11} \end{bmatrix}.$$

From (11.55) $T_{10} R = 0$, implying that $T_{10} = 0$ [yielding (11.54)], and also $T_{00} R = RB$, implying that $B = R^{-1} T_{00} R$ (R^{-1} exists because X is full-rank). T_{00} and B are similar matrices so $s(B) = s(T_{00})$. From Lemma 11.2 we have $s(A) = s(T_{00}) \cup s(T_{11})$. Thus, $s(A) = s(B) \cup s(T_{11})$. From basic properties regarding sets (distributive laws)

$$s(T_{00}) \cap s(A) = s(T_{00}) \cap [s(B) \cup s(T_{11})]$$

$$= [s(T_{00}) \cap s(B)] \cup [s(T_{00}) \cap s(T_{11})]$$

$$= s(T_{00}) \cap \emptyset,$$

implying that $s(T_{00}) = s(T_{00}) \cap s(A) = s(A) \cap s(B)$. This statement $[s(T_{00}) = s(A) \cap s(B)]$ really says that the eigenvalues of B are a subset of those of A.

Recall that a *subspace* of vector space $\mathbf{C}^n$ is a subset of $\mathbf{C}^n$ that is also a vector space. Suppose that we have the vectors $x_0, \ldots, x_{m-1} \in \mathbf{C}^n$; then we may define the *spanning set* as

$$\text{span}(x_0, \ldots, x_{m-1}) = \left\{ \sum_{j=0}^{m-1} a_j x_j | a_j \in \mathbf{C} \right\}. \tag{11.56}$$

In particular, if $S = \text{span}(x)$, where x is an eigenvector of A, then

$$y \in S \Rightarrow Ay \in S,$$

and so S is *invariant for* A, or *invariant to the action of* A. It is a subspace (eigenspace) of $\mathbf{C}^n$ that is invariant to A. Lemmas 11.2 and 11.3 can be used to establish the following important decomposition theorem (Theorem 7.4.1 in Ref. 4). We emphasize that it is for real-valued A only.

Theorem 11.13: Real Schur Decomposition   If $A \in \mathbf{R}^{n \times n}$, then there is an orthogonal matrix $Q \in \mathbf{R}^{n \times n}$ such that

$$Q^T A Q = \begin{bmatrix} R_{00} & R_{01} & \cdots & R_{0,m-1} \\ 0 & R_{11} & \cdots & R_{1,m-1} \\ \vdots & \vdots & & \vdots \\ 0 & 0 & \cdots & R_{m-1,m-1} \end{bmatrix} = \mathcal{R}, \tag{11.57}$$

where each $R_{i,i}$ is either 1×1 or 2×2. In the latter case $R_{i,i}$ will have a complex–conjugate pair of eigenvalues.

Proof   The matrix $A \in \mathbf{R}^{n \times n}$, so $\det(\lambda I - A)$ has real-valued coefficients, and so complex eigenvalues of A always occur in conjugate pairs (recall Section 11.2). Let k be the number of complex–conjugate eigenvalue pairs in $s(A)$. We will employ mathematical induction on k.

The theorem certainly holds for $k = 0$ via Lemmas 11.2 and 11.3 since real-valued matrices are only a special case. Now we assume that $k \geq 1$ (i.e., A possesses at least one conjugate pair of eigenvalues). Suppose that an eigenvalue is $\lambda = \alpha + j\beta \in s(A)$ with $\beta \neq 0$. There must be vectors $x, y \in \mathbf{R}^n$ (with $y \neq 0$) such that

$$A(x + jy) = (\alpha + j\beta)(x + jy),$$

or equivalently

$$A[xy] = [xy] \begin{bmatrix} \alpha & \beta \\ -\beta & \alpha \end{bmatrix}. \tag{11.58}$$

Since $\beta \neq 0$, vectors x and y span a two-dimensional subspace of $\mathbf{R}^n$ that is invariant to the action of A because of (11.58). From Lemma 11.3 there is an orthogonal matrix $U_1 \in \mathbf{R}^{n \times n}$ such that

$$U_1^T A U_1 = \left[\begin{array}{cc} T_{00} & T_{01} \\ 0 & T_{11} \end{array} \right],$$

where $T_{00} \in \mathbf{R}^{2 \times 2}$, $T_{01} \in \mathbf{R}^{2 \times (n-2)}$, $T_{11} \in \mathbf{R}^{(n-2) \times (n-2)}$, and $s(T_{00}) = \{\lambda, \lambda^*\}$. By induction there is another orthogonal matrix U_2 such that $U_2^T T_{11} U_2$ has the necessary structure. Equation (11.57) then follows by letting

$$Q = U_1 \left[\begin{array}{cc} I_2 & 0 \\ 0 & U_2 \end{array} \right],$$

where I_2 is the 2×2 identity matrix. Of course, this process may be repeated as often as needed.

A method that reliably gives us the blocks $R_{i,i}$ for all i in (11.57) therefore gives us all the eigenvalues of A since $R_{i,i}$ is only 1×1, or 2×2, making its eigenvalues easy to find in any case. The elements in the first subdiagonal of $Q^T A Q$ of (11.57) are not necessarily zero-valued (again because $R_{i,i}$ might be 2×2), so we say that $Q^T A Q$ is *upper quasi-triangular*.

Definition 11.5: Hessenberg Form   Matrix $A \in \mathbf{C}^{n \times n}$ is in *Hessenberg form* if $a_{i,j} = 0$ for all i, j such that $i - j > 1$.

Technically, A in this definition is *upper* Hessenberg. Matrix A in Example 4.4 is Hessenberg. All upper triangular matrices are Hessenberg. The quasi-triangular matrix $Q^T A Q$ in Theorem 11.13 is Hessenberg.

A pseudocode for the *basic QR iterations algorithm* is

```
Input N; { Upper limit on the number of iterations }
Input A ∈ R^{n×n}; { Matrix we want to eigendecompose }
H_0 := Q_0^T A Q_0; { Reduce A to Hessenberg form }
k := 1;
while k ≤ N do begin
    H_{k-1} := Q_k R_k; { QR-decomposition step }
    H_k := R_k Q_k;
    k := k + 1;
    end;
```

In this algorithm we emphasize that A is assumed to be real-valued. Generalization to the complex case is possible but omitted. The statement $H_0 = Q_0^T A Q_0$ generally involves applying orthogonal transformation $Q_0 \in \mathbf{R}^{n \times n}$ to reduce A to Hessenberg matrix H_0, although in principle this is not necessary. However, there are major advantages (discussed below) to reducing A to Hessenberg form as a first step. The basis for this initial reduction step is the following theorem, which proves that such a step is always possible.

Theorem 11.14: Hessenberg Reduction   If $A \in \mathbf{R}^{n \times n}$, there is an orthogonal matrix $Q \in \mathbf{R}^{n \times n}$ such that $Q^T A Q = H$ is Hessenberg.

Proof   From Section 4.6 in general there is an orthogonal matrix P such that $Px = \|x\|_2 e_0$ (e.g., P is a Householder matrix), where $e_0 = [1 \ 0 \ \cdots \ 0 \ 0]^T \in \mathbf{R}^n$ if $x \in \mathbf{R}^n$.

Partition A according to

$$A^{(0)} = A = \begin{bmatrix} a_{00}^{(0)} & b_0^T \\ a_0 & A_{11}^{(0)} \end{bmatrix},$$

where $a_{00}^{(0)} \in \mathbf{R}$ and $a_0, b_0 \in \mathbf{R}^{n-1}$, $A_{11}^{(0)} \in \mathbf{R}^{(n-1) \times (n-1)}$. Let P_1 be orthogonal such that $P_1 a_0 = \|a_0\|_2 e_0$ ($e_0 \in \mathbf{R}^{n-1}$). Define

$$Q_1 = \begin{bmatrix} 1 & 0 \\ 0 & P_1 \end{bmatrix},$$

and clearly $Q_1^{-1} = Q_1^T$ (i.e., Q_1 is also orthogonal). Thus

$$A^{(1)} = Q_1 A^{(0)} Q_1^T = \begin{bmatrix} a_{00}^{(0)} & b_0^T P_1^T \\ \|a_0\|_2 e_0 & P_1 A_{11}^{(0)} P_1^T \end{bmatrix}.$$

The first column of $A^{(1)}$ satisfies the Hessenberg condition since it is $[a_{00}^{(0)} \|a_0\|_2 \underbrace{0 \cdots 0}_{n-2 \text{ zeros}}]^T$. The process may be repeated again by partitioning $A^{(1)}$ according to

$$A^{(1)} = \begin{bmatrix} A_{00}^{(1)} & b_1^T \\ [0 \quad a_1] & A_{11}^{(1)} \end{bmatrix},$$

where $A_{00}^{(1)} \in \mathbf{R}^{2 \times 2}$, and $[0 \quad a_1], b_1 \in \mathbf{R}^{(n-2) \times 2}$, $A_{11}^{(1)} \in \mathbf{R}^{(n-2) \times (n-2)}$. Let P_2 be orthogonal such that $P_2 a_1 = \|a_1\|_2 e_0$ ($e_0 \in \mathbf{R}^{n-2}$). Define

$$Q_2 = \begin{bmatrix} I_2 & 0 \\ 0 & P_2 \end{bmatrix},$$

where I_2 is the 2×2 identity matrix. Thus

$$A^{(2)} = Q_2 A^{(1)} Q_2^T = Q_2 Q_1 A^{(0)} Q_1^T Q_2^T$$

$$= \begin{bmatrix} A_{00}^{(1)} & b_1^T P_2^T \\ [0 \quad \|a_1\|_2 e_0] & P_2 A_{11}^{(1)} P_2^T \end{bmatrix},$$

and the first two columns of $A^{(2)}$ satisfy the Hessenberg condition. Of course, we may continue in this fashion, finally yielding

$$A^{(n-2)} = Q_{n-2} \cdots Q_2 Q_1 A Q_1^T Q_2^T \cdots Q_{n-2}^T,$$

which is Hessenberg. We may define $Q^T = Q_{n-2} \cdots Q_2 Q_1$ and $H = A^{(n-2)}$, which is the claim made in the theorem statement.

Thus, Theorem 11.14 contains a prescription for finding $H_0 = Q_0^T A_0 Q_0$ as well as a simple proof of existence of the decomposition. Hessenberg reduction is done to facilitate reducing the amount of computation per iteration. Clearly, A and H_0 are similar matrices and so possess the same eigenvalues.

From the pseudocode for $k = 1, 2, \ldots, N$, we obtain

$$H_{k-1} = Q_k R_k,$$

$$H_k = R_k Q_k,$$

which yields

$$H_N = Q_N^T \cdots Q_2^T Q_1^T H_0 Q_1 Q_2 \cdots Q_N, \tag{11.59}$$

so therefore

$$H_N = Q_N^T \cdots Q_2^T Q_1^T Q_0^T A Q_0 Q_1 Q_2 \cdots Q_N. \tag{11.60}$$

Matrices H_N and A are similar, and so have the same eigenvalues for any N. It is important to note that if Q_k is constructed properly then H_k is Hessenberg for all k. As explained in Golub and Van Loan [4, Section 7.4.2], the use of orthogonal matrices Q_k based on the 2×2 rotation operator (matrix A from Example 11.3) is recommended. These orthogonal matrices are called *Givens matrices*, or *Givens rotations*. The result is an algorithm that needs only $O(n^2)$ flops per iteration instead of $O(n^3)$ flops. Overall computational complexity is still $O(n^3)$ flops, due to the initial Hessenberg reduction step. It is to be noted that the rounding error performance of the suggested algorithm is quite good [19].

We have already noted the desirability of the real Schur decomposition of A into $\mathcal{R}$ according to $Q^T A Q = \mathcal{R}$ in (11.57). In fact, *with proper attention to details* (many of which cannot be considered here), the QR iterations method is an excellent means to find Q and $\mathcal{R}$ as in any valid matrix norm

$$\lim_{N \to \infty} H_N = \mathcal{R} \tag{11.61}$$

and

$$\prod_{i=0}^{\infty} Q_i = Q \tag{11.62}$$

(of course, $\prod_{i=0}^{N} Q_i = Q_0 Q_1 \cdots Q_N$; the ordering of factors in the product is important since Q_i is a matrix for all i). The formal proof of this is rather difficult, and so it is omitted.

Suppose that we have

$$A = \begin{bmatrix} 0 & 0 & 0 & \cdots & 0 & -a_n \\ 1 & 0 & 0 & \cdots & 0 & -a_{n-1} \\ 0 & 1 & 0 & \cdots & 0 & -a_{n-2} \\ \vdots & \vdots & \vdots & & \vdots & \vdots \\ 0 & 0 & 0 & \cdots & 0 & -a_2 \\ 0 & 0 & 0 & \cdots & 1 & -a_1 \end{bmatrix} \in \mathbf{R}^{n \times n}. \tag{11.63}$$

It can be shown that

$$p(\lambda) = \det(\lambda I - A) = \lambda^n + a_1 \lambda^{n-1} + a_2 \lambda^{n-2} + \cdots + a_{n-1}\lambda + a_n. \tag{11.64}$$

Matrix A is called a *companion matrix*. We see that it is easy to obtain (11.64) from (11.63), or vice versa. We also see that A is Hessenberg. Because of (11.61), we may conceivably input A of (11.63) into the basic QR iterations algorithm (omitting the initial Hessenberg reduction step), and so determine the roots of $p(\lambda) = 0$ as these are the eigenvalues of A. Since $p(\lambda)$ is essentially arbitrary, except that it should not yield a defective A, we have an algorithm to solve the polynomial zero-finding problem that was mentioned in Chapter 7. Unfortunately, it has been noted [4,19] that this is not necessarily a stable method for finding polynomial zeros.

Example 11.8 Suppose that

$$p(\lambda) = (\lambda^2 - 2\lambda + 2)(\lambda^2 - \sqrt{2}\lambda + 1)$$
$$= \lambda^4 - (2 + \sqrt{2})\lambda^3 + (2\sqrt{2} + 3)\lambda^2 - 2(1 + \sqrt{2})\lambda + 2,$$

which has zeros for

$$\lambda \in \left\{ 1 \pm j, \frac{1}{\sqrt{2}} \pm \frac{1}{\sqrt{2}} j \right\}.$$

After 50 iterations of the basic QR iterations algorithm, we obtain

$$H_{50} = \begin{bmatrix} 0.5000 & -6.0355 & 0.9239 & 5.3848 \\ 0.2071 & 1.5000 & -0.3827 & -2.2304 \\ 0.0000 & 0.0000 & 0.7071 & -0.7071 \\ 0.0000 & 0.0000 & 0.7071 & 0.7071 \end{bmatrix} = \begin{bmatrix} R_{0,0} & R_{0,1} \\ 0 & R_{1,1} \end{bmatrix},$$

where $R_{i,j} \in \mathbf{R}^{2 \times 2}$. The reader may wish to confirm that $R_{0,0}$ has the eigenvalues $1 \pm j$, and that $R_{1,1}$ has the eigenvalues $\frac{1}{\sqrt{2}}(1 \pm j)$.

On the other hand, for

$$p(\lambda) = (\lambda + 1)(\lambda^2 - 2\lambda + 2)(\lambda^2 - \sqrt{2}\lambda + 1),$$

which is a slight modification of the previous example, the basic QR iterations algorithm will fail to converge.

The following point is also important. In (11.60), define $Q = Q_0 Q_1 \cdots Q_N$, so that $H_N = Q^T A Q$. Now suppose that $A = A^T$. Clearly, $H_N^T = Q^T A^T Q = Q^T A Q = H_N$. This implies that H_N will be *tridiagonal* (defined in Section 6.5) for a real and symmetric A. Because of (11.61), we must now have

$$\lim_{N \to \infty} H_N = \mathrm{diag}(R_{0,0}, R_{1,1}, \ldots, R_{m-1,m-1}) = D,$$

where each $R_{i,i}$ is 1×1. Thus, D is the diagonal matrix of eigenvalues of A. Also, we must have $\prod_{i=0}^{\infty} Q_i$ as the corresponding matrix of eigenvectors of A.

Example 11.9   Suppose that A is the matrix from Example 11.5:

$$A = \begin{bmatrix} 4 & 1 & 0 \\ 1 & 4 & 1 \\ 0 & 1 & 4 \end{bmatrix}.$$

We see that A is in Hessenberg form already. After 36 iterations of the basic QR iteration algorithm, we obtain

$$H_{36} = \begin{bmatrix} 5.4142 & 0.0000 & 0.0000 \\ 0.0000 & 4.0000 & 0.0000 \\ 0.0000 & 0.0000 & 2.5858 \end{bmatrix},$$

so the matrix is diagonal and reveals all eigenvalues of A. Additionally, we have

$$\prod_{i=0}^{36} Q_i = \begin{bmatrix} 0.5000 & -0.7071 & 0.5000 \\ 0.7071 & 0.0000 & -0.7071 \\ 0.5000 & 0.7071 & 0.5000 \end{bmatrix},$$

which is a good approximation to the eigenvectors of A.

We noted in Section 11.4 that shifting can be used to accelerate convergence in power methods. Similarly, shifting can be employed in QR iterations to achieve the same result. Indeed, all modern implementations of QR iterations incorporate some form of shifting for this reason. The previous basic QR iteration algorithm may be modified to incorporate the shift parameter $\mu \in \mathbf{R}$. The overall structure of the result is described by the following pseudocode:

Input N; { Upper limit on the number of iterations }
Input $A \in \mathbf{R}^{n \times n}$; { Matrix we want to eigendecompose }
$H_0 := Q_0^T A Q_0$; { Reduce A to Hessenberg form }
$k := 1$;
while $k \leq N$ do begin
    Determine the shift parameter $\mu \in \mathbf{R}$;
    $H_{k-1} - \mu I := Q_k R_k$; { QR-decomposition step }
    $H_k := R_k Q_k + \mu I$;
    $k := k + 1$;
    end;

The reader may readily confirm that we still have

$$H_N = Q_N^T \cdots Q_1^T Q_0^T A Q_0 Q_1 \cdots Q_N$$

just as we had in (11.60) for the basic *QR* iteration algorithm. Thus, we again find that H_k is similar to A for all k. Perhaps the simplest means to generate μ is the *single-shift QR iterations algorithm*:

Input N; { Upper limit on the number of iterations }
Input $A \in \mathbf{R}^{n \times n}$; { Matrix we want to eigendecompose }
$H_0 := Q_0^T A Q_0$; { Reduce A to Hessenberg form }
$k := 1$;
while $k \leq N$ do begin
    $\mu_k := H_{k-1}(n-1, n-1)$; { μ_k is the lower right corner element of H_{k-1} }
    $H_{k-1} - \mu_k I := Q_k R_k$; { QR-decomposition step }
    $H_k := R_k Q_k + \mu_k I$;
    $k := k + 1$;
    end;

We note that μ is not fixed in general from one iteration to the next. Basically, μ varies from iteration to iteration in order to account for new information about $s(A)$ as the subdiagonal entries of H_k converge to zero. We will avoid the technicalities involved in a full justification of this approach except to mention that it is flawed, and that more sophisticated shifting methods are needed for an acceptable algorithm (e.g., the double shift [4,19]). However, the following example shows that shifting in this way does speed convergence.

Example 11.10 If we apply the single-shift *QR* iterations algorithm to A in Example 11.9, we obtain the following matrix in only one iteration:

$$H_1 = \begin{bmatrix} 4.0000 & -1.4142 & 0.0000 \\ -1.4142 & 4.0000 & 0.0000 \\ 0.0000 & 0.0000 & 4.0000 \end{bmatrix}.$$

This matrix certainly does not have the structure of H_{36} in Example 11.9, but the eigenvalues of the submatrix

$$\begin{bmatrix} 4.0000 & -1.4142 \\ -1.4142 & 4.0000 \end{bmatrix}$$

are 5.4142, 2.5858.

Finally, we mention that our pseudocodes assume a user-specified number of iterations N. This is not convenient, and is inefficient in practice. Criteria to automatically terminate the QR iterations without user intervention are available, but a discussion of this matter is beyond our scope.

REFERENCES

1. D. R. Hill, *Experiments in Computational Matrix Algebra* (C. B. Moler, consulting ed.), Random House, New York, 1988.

2. P. Halmos, *Finite Dimensional Vector Spaces*, Van Nostrand, New York, 1958.

3. R. A. Horn and C. R. Johnson, *Matrix Analysis*, Cambridge Univ. Press, Cambridge, UK, 1985.

4. G. H. Golub and C. F. Van Loan, *Matrix Computations*, 2nd ed., Johns Hopkins Univ. Press, Baltimore, MD, 1989.

5. F. W. Fairman, "On Using Singular Value Decomposition to Obtain Irreducible Jordan Realizations," in *Linear Circuits, Systems and Signal Processing: Theory and Application*, C. I. Byrnes, C. F. Martin, and R. E. Saeks, eds., North-Holland, Amsterdam, 1988, pp. 35–40.

6. G. Stewart, *Introduction to Matrix Computations*, Academic Press, New York, 1973.

7. C. Moler and C. Van Loan, "Nineteen Dubious Ways to Compute the Exponential of a Matrix," *SIAM Rev.* **20**, 801–836 (Oct. 1978).

8. I. E. Leonard, "The Matrix Exponential," *SIAM Rev.* **38**, 507–512 (Sept. 1996).

9. B. Noble and J. W. Daniel, *Applied Linear Algebra*, Prentice-Hall, Englewood Cliffs, NJ, 1977.

10. W. R. Derrick and S. I. Grossman, *Elementary Differential Equations with Applications*, 2nd ed., Addison-Wesley, Reading, MA, 1981.

11. W. T. Reid, *Ordinary Differential Equations*, Wiley, New York, 1971.

12. C. Van Loan, "The Sensitivity of the Matrix Exponential," *SIAM J. Numer. Anal.* **14**, 971–981 (Dec. 1977).

13. A. Quarteroni, R. Sacco, and F. Saleri, *Numerical Mathematics* (Texts in Applied Mathematics series, Vol. 37). Springer-Verlag, New York, 2000.

14. J. F. Epperson, *An Introduction to Numerical Methods and Analysis*, Wiley, New York, 2002.

15. D. S. Watkins, "Understanding the QR Algorithm," *SIAM Rev.* **24**, 427–440 (Oct. 1982).

16. J. G. F. Francis, "The QR Transformation: A Unitary Analogue to the LR Transformations, Parts I and II," *Comput. J.* **4**, 265–272, 332–345 (1961).

17. V. N. Kublanovskaya, "On Some Algorithms for the Solution of the Complete Eigenvalue Problem," *USSR Comput. Math. Phys.* **3**, 637–657, (1961).

18. B. N. Parlett and W. G. Poole, Jr., "A Geometric Theory for the QR, LU, and Power Iterations," *SIAM J. Numer. Anal.* **10**, 389–412 (1973).

19. J. H. Wilkinson, *The Algebraic Eigenvalue Problem*, Clarendon Press, Oxford, UK, 1965.

PROBLEMS

11.1. Aided with at most a pocket calculator, find all the eigenvalues and eigenvectors of the following matrices:

(a) $A = \begin{bmatrix} 4 & 1 \\ 1 & 4 \end{bmatrix}$.

(b) $B = \begin{bmatrix} 0 & 0 & -2 \\ 1 & 0 & 1 \\ 0 & 1 & 2 \end{bmatrix}$.

(c) $C = \begin{bmatrix} 0 & 1 & \frac{1}{4} \\ 0 & 0 & -\frac{1}{4} \\ 1 & 0 & 1 \end{bmatrix}$.

(d) $D = \begin{bmatrix} 0 & 1-j \\ -j & 2 \end{bmatrix}$.

11.2. A conic section in $\mathbf{R}^2$ is described in general by

$$\alpha x_0^2 + 2\beta x_0 x_1 + \gamma x_1^2 + \delta x_0 + \epsilon x_1 + \rho = 0. \tag{11.P.1}$$

(a) Show that (11.P.1) can be rewritten as a quadratic form:

$$x^T A x + g^T x + \rho = 0, \tag{11.P.2}$$

where $x = [x_0 \quad x_1]^T$, and $A = A^T$.

(b) For a conic section in *standard form* A is diagonal. Suppose that A is diagonal. State the conditions on the diagonal elements that result in (11.P.2) describing an ellipse, a parabola, and a hyperbola.

(c) Suppose that (11.P.2) is not in standard form (i.e., A is not a diagonal matrix). Explain how similarity transformations might be used to place (11.P.2) in standard form.

11.3. Consider the *companion matrix*

$$C = \begin{bmatrix} 0 & 0 & 0 & \cdots & 0 & -c_n \\ 1 & 0 & 0 & \cdots & 0 & -c_{n-1} \\ 0 & 1 & 0 & \cdots & 0 & -c_{n-2} \\ \vdots & \vdots & \vdots & & \vdots & \vdots \\ 0 & 0 & 0 & \cdots & 0 & -c_2 \\ 0 & 0 & 0 & \cdots & 1 & -c_1 \end{bmatrix} \in \mathbf{C}^{n \times n}.$$

Prove that

$$p_n(\lambda) = \det(\lambda I - C) = \lambda^n + c_1\lambda^{n-1} + c_2\lambda^{n-2} + \cdots + c_{n-1}\lambda + c_n.$$

11.4. Suppose that the eigenvalues of $A \in \mathbf{C}^{n \times n}$ are $\lambda_0, \lambda_1, \ldots, \lambda_{n-1}$. Find the eigenvalues of $A + \alpha I$, where I is the order n identity matrix and $\alpha \in \mathbf{C}$ is a constant.

11.5. Suppose that $A \in \mathbf{R}^{n \times n}$ is orthogonal (i.e., $A^{-1} = A^T$); then, if λ is an eigenvalue of A, show that we must have $|\lambda| = 1$.

11.6. Find all the eigenvalues of

$$A = \begin{bmatrix} \cos\theta & 0 & -\sin\theta & 0 \\ 0 & \cos\phi & 0 & -\sin\phi \\ \sin\theta & 0 & \cos\theta & 0 \\ 0 & \sin\phi & 0 & \cos\phi \end{bmatrix}.$$

(*Hint:* The problem is simplified by using permutation matrices.)

11.7. Consider the following definition: $A, B \in \mathbf{C}^{n \times n}$ are *simultaneously diagonalizable* if there is a similarity matrix $S \in \mathbf{C}^{n \times n}$ such that $S^{-1}AS$, and $S^{-1}BS$ are both diagonal matrices. Show that if $A, B \in \mathbf{C}^{n \times n}$ are simultaneously diagonalizable, then they commute (i.e., $AB = BA$).

11.8. Prove the following theorem. Let $A, B \in \mathbf{C}^{n \times n}$ be diagonalizable. Therefore, A and B commute iff they are simultaneously diagonalizable.

11.9. Matrix $A \in \mathbf{C}^{n \times n}$ is a *square root* of $B \in \mathbf{C}^{n \times n}$ if $A^2 = B$. Show that every diagonalizable matrix in $\mathbf{C}^{n \times n}$ has a square root.

11.10. Prove the following (Bauer–Fike) theorem (which says something about how perturbations of matrices affect their eigenvalues). If γ is an eigenvalue of $A + E \in \mathbf{R}^{n \times n}$, and $T^{-1}AT = D = \text{diag}(\lambda_0, \lambda_1, \ldots, \lambda_{n-1})$, then

$$\min_{\lambda \in s(A)} |\lambda - \gamma| \leq \kappa_2(T)\|E\|_2.$$

Recall that $s(A)$ denotes the set of eigenvalues of A (i.e., the spectrum of A). [*Hint:* If $\gamma \in s(A)$, the result is certainly true, so we need consider only the situation where $\gamma \notin s(A)$. Confirm that if $T^{-1}(A + E - \gamma I)T$ is singular, then so is $I + (D - \gamma I)^{-1}(T^{-1}ET)$. Note that if for some $B \in \mathbf{R}^{n \times n}$ the matrix $I + B$ is singular, then $(I + B)x = 0$ for some $x \in \mathbf{R}^n$ that is nonzero, so $\|x\|_2 = \|Bx\|_2$, and so $\|B\|_2 \geq 1$. Consider upper and lower bounds on the norm $\|(D - \gamma I)^{-1}(T^{-1}ET)\|_2$.]

11.11. The Daubechies 4-tap scaling function $\phi(t)$ satisfies the two-scale difference equation

$$\phi(t) = p_0\phi(2t) + p_1\phi(2t - 1) + p_2\phi(2t - 2) + p_3\phi(2t - 3), \quad (11.\text{P}.3)$$

where $\text{supp}\phi(t) = [0, 3] \subset \mathbf{R}$ [i.e., $\phi(t)$ is nonzero only on the interval $[0, 3]$], and where

$$p_0 = \tfrac{1}{4}(1 + \sqrt{3}), \qquad\qquad p_1 = \tfrac{1}{4}(3 + \sqrt{3}),$$

$$p_2 = \tfrac{1}{4}(3 - \sqrt{3}), \qquad\qquad p_3 = \tfrac{1}{4}(1 - \sqrt{3}).$$

Note that the solution to (11.P.3) is continuous (an important fact).

(a) Find the matrix M such that

$$M\phi = \phi, \tag{11.P.4}$$

where $\phi = [\phi(1)\phi(2)]^T \in \mathbf{R}^2$.

(b) Find the eigenvalues of M. Find the solution ϕ to (11.P.4).

(c) Take ϕ from (b) and multiply it by constant α such that $\alpha \sum_{k=1}^{2} \phi(k) = 1$ (i.e., replace ϕ by the normalized form $\alpha\phi$).

(d) Using the normalized vector from (c) (i.e., $\alpha\phi$), find $\phi(k/2)$ for all $k \in \mathbf{Z}$.

[*Comment:* Computation of the Daubechies 4-tap scaling function is the first major step in computing the Daubechies 4-tap wavelet. The process suggested in (d) may be continued to compute $\phi(k/2^J)$ for any $k \in \mathbf{Z}$, and for any positive integer J. The algorithm suggested by this is often called the *interpolatory graphical display algorithm* (IGDA).]

11.12. Prove the following theorem. Suppose $A \in \mathbf{R}^{n \times n}$ and $A = A^T$. Then $A > 0$ iff $A = P^T P$ for some nonsingular matrix $P \in \mathbf{R}^{n \times n}$.

11.13. Let $A \in \mathbf{R}^{n \times n}$ be symmetric with eigenvalues

$$\lambda_0 \leq \lambda_1 \leq \cdots \leq \lambda_{n-2} \leq \lambda_{n-1}.$$

Show that for all $x \in \mathbf{R}^n$

$$\lambda_0 x^T x \leq x^T A x \leq \lambda_{n-1} x^T x.$$

[*Hint:* Use the fact that there is an orthogonal matrix P such that $P^T A P = \Lambda$ (diagonal matrix of eigenvalues of A). Partition P in terms of its row vectors.]

11.14. Section 11.3 presented a method of computing e^{At} "by hand." Use this method to

(a) Derive (10.109) (in Example 10.9).

(b) Derive (10.P.9) in Problem 10.23.

(c) Find a closed-form expression for e^{At}, where

$$A = \begin{bmatrix} \lambda & 1 & 0 \\ 0 & \lambda & 1 \\ 0 & 0 & \lambda \end{bmatrix}.$$

11.15. This exercise confirms that eigenvalues and singular values are definitely not the same thing. Consider the matrix

$$A = \begin{bmatrix} 0 & 2 \\ 1 & 1 \end{bmatrix}.$$

Use the MATLAB eig and svd functions to find the eigenvalues and singular values of A.

11.16. Show that $e^{(A+B)t} = e^{At}e^{Bt}$ for all $t \in \mathbf{R}$ if $AB = BA$. Does $e^{(A+B)t} = e^{At}e^{Bt}$ always hold for all $t \in \mathbf{R}$ when $AB \neq BA$? Justify your answer.

11.17. This problem is an introduction to *Floquet theory*. Consider a linear system with state vector $x(t) \in \mathbf{R}^n$ for all $t \in \mathbf{R}$ such that

$$\frac{dx(t)}{dt} = A(t)x(t) \tag{11.P.5}$$

for some $A(t) \in \mathbf{R}^{n \times n}$ (all $t \in \mathbf{R}$), and such that $A(t + T) = A(t)$ for some $T > 0$ [so that $A(t)$ is periodic with period T]. Let $\Phi(t)$ be the *fundamental matrix* of the system such that

$$\frac{d\Phi(t)}{dt} = A(t)\Phi(t), \ \Phi(0) = I. \tag{11.P.6}$$

(a) Let $\Psi(t) = \Phi(t + T)$, and show that

$$\frac{d\Psi(t)}{dt} = A(t)\Psi(t). \tag{11.P.7}$$

(b) Show that $\Phi(t + T) = \Phi(t)C$, where $C = \Phi(T)$. [*Hint:* Equations (11.P.6) and (11.P.7) differ only in their initial conditions [i.e., $\Psi(0) = $ what ?].]

(c) Assume that C^{-1} exists for some $C \in \mathbf{R}^{n \times n}$, and that there exists some $R \in \mathbf{R}^{n \times n}$ such that $C = e^{TR}$. Define $P(t) = \Phi(t)e^{-tR}$, and show that $P(t + T) = P(t)$. [Thus, $\Phi(t) = P(t)e^{tR}$, which is the general form of the solution to (11.P.5).]

(*Comment:* Further details of the theory of solution of (11.P.5) based on working with (11.P.6) may be found in E. A. Coddington and N. Levinson,

Theory of Ordinary Differential Equations, McGraw-Hill, New York, 1955. The main thing for the student to notice is that the theory involves matrix exponentials.)

11.18. Consider the matrix

$$A = \begin{bmatrix} 4 & 2 & 0 & 0 \\ 1 & 4 & 1 & 0 \\ 0 & 1 & 4 & 1 \\ 0 & 0 & 2 & 4 \end{bmatrix}.$$

(a) Create a MATLAB routine that implements the scaled power algorithm, and use your routine to find the largest eigenvalue of A.

(b) Create a MATLAB routine that implements the shifted inverse power algorithm, and use your routine to find the smallest eigenvalue of A.

11.19. For A in the previous problem, find $\kappa_2(A)$ using the MATLAB routines that you created to solve the problem.

11.20. If $A \in \mathbf{R}^{n \times n}$, and $A = A^T$, then, for some $x \in \mathbf{R}^n$ such that $x \neq 0$, we define the *Rayleigh quotient* of A and x to be the ratio $x^T A x / x^T x$ ($= \langle x, Ax \rangle / \langle x, x \rangle$). The *Rayleigh quotient iterative algorithm* is described as

```
k := 0;
while k < N do begin
    μk := zkᵀAzk/zkᵀzk;
    (A − μkl)yk := zk;
    zk+1 := yk/||yk||2;
    k := k + 1;
    end;
```

The user inputs z_0, which is the initial guess about the eigenvector. Note that the shift μ_k is changed (i.e., updated) at every iteration. This has the effect of accelerating convergence (i.e., of reducing the number of iterations needed to achieve an accurate solution). However, $A_{\mu_k} = A - \mu_k I$ needs to be factored anew with every iteration as a result. Prove the following theorem. Let $A \in \mathbf{R}^{n \times n}$ be symmetric and (λ, x) be an eigenpair for A. If $y \approx x$, $\mu = y^T A y / y^T y$ with $||x||_2 = ||y||_2 = 1$, then

$$|\lambda - \mu| \leq ||A - \lambda I||_2 ||x - y||_2^2.$$

[*Comment:* The norm $||A - \lambda I||_2$ is an A-dependent constant, while $||x - y||_2^2 = ||e||_2^2$ is the square of the size of the error between eigenvector x and the estimate y of it. So the size of the error between λ and the estimate μ (i.e., $|\lambda - \mu|$) are proportional to $||e||_2^2$ at the worst. This explains the fast convergence of the method (i.e., only a relatively small N is usually needed). Note that the proof uses (4.31).]

11.21. Write a MATLAB routine that implements the basic QR iteration algorithm. You may use the MATLAB function qr to perform QR factorizations. Test your program out on the following matrices:

(a)

$$A = \begin{bmatrix} 4 & 2 & 0 & 0 \\ 1 & 4 & 1 & 0 \\ 0 & 1 & 4 & 1 \\ 0 & 0 & 2 & 4 \end{bmatrix}.$$

(b)

$$B = \begin{bmatrix} 0 & 0 & 0 & -\frac{1}{2} \\ 1 & 0 & 0 & 1+\frac{1}{\sqrt{2}} \\ 0 & 1 & 0 & -\frac{3}{2}-\sqrt{2} \\ 0 & 0 & 1 & 1+\sqrt{2} \end{bmatrix}.$$

Use other built-in MATLAB functions (e.g., roots or eig) to verify your answers. Iterate enough to obtain entries for H_N that are accurate to four decimal places.

11.22. Repeat the previous problem using your own MATLAB implementation of the single-shift QR iteration algorithm. Compare the number of iterations needed to obtain four decimal places of accuracy with the result from the previous problem.

11.23. Suppose that $X, Y \in \mathbf{R}^{n \times n}$, and we define the matrices

$$A = X + jY, B = \begin{bmatrix} X & -Y \\ Y & X \end{bmatrix}.$$

Show that if $\lambda \in s(A)$ is real-valued, then $\lambda \in s(B)$. Find a relationship between the corresponding eigenvectors.

12 Numerical Solution of Partial Differential Equations

12.1 INTRODUCTION

The subject of *partial differential equations* (PDEs) with respect to the matter of their numerical solution is impossibly large to properly cover within a single chapter (or, for that matter, even within a single textbook). Furthermore, the development of numerical methods for PDEs is a highly active area of research, and so it continues to be a challenge to decide what is truly "fundamental" material to cover at an introductory level. In this chapter we shall place emphasis on wave propagation problems modeled by *hyperbolic PDEs* (defined in Section 12.2). We will consider especially the *finite-difference time-domain* (FDTD) *method* [8], as this appears to be gaining importance in such application areas as modeling of the scattering of electromagnetic waves from particles and objects and modeling of optoelectronic systems. We will only illustrate the method with respect to planar electromagnetic wave propagation problems at normal incidence. However, prior to this we shall give an overview of PDEs, including how they are classified into elliptic, parabolic, and hyperbolic types.

12.2 A BRIEF OVERVIEW OF PARTIAL DIFFERENTIAL EQUATIONS

In this section we define some notation and terminology that is used throughout the chapter. We explain how second-order PDEs are classified. We also summarize some problems that will not be covered within this book, simply citing references where the interested reader can find out more.

We will consider only two-dimensional functions $u(x, t) \in \mathbf{R}$ (or $u(x, y) \in \mathbf{R}$), where the independent variable x is interpreted as a space variable, and independent variable t is interpreted as time. The *order* of a PDE is the order of the highest derivative. For our purposes, we will never consider PDEs of an order greater than 2. Common shorthand notation for partial derivatives includes

$$u_x = \frac{\partial u}{\partial x}, \quad u_t = \frac{\partial u}{\partial t}, \quad u_{xt} = \frac{\partial^2 u}{\partial t \, \partial x}, \quad u_{xx} = \frac{\partial^2 u}{\partial x^2}.$$

An Introduction to Numerical Analysis for Electrical and Computer Engineers, by C.J. Zarowski
ISBN 0-471-46737-5 © 2004 John Wiley & Sons, Inc.

If the PDE has solution $u(x, y)$, where x and y are spatial variables, then often we are interested only in approximating the solution on a bounded region (i.e., a bounded subset of $\mathbf{R}^2$). However, we will consider mainly a PDE with solution $u(x, t)$, where, as already noted, t is time, and so we consider $u(x, t)$ only for $x \in [a, b] = [x_0, x_f]$ and $t \in [t_0, t_f]$. Commonly, $t_0 = 0$, $x_0 = 0$, and $x_f = L$, with $t_f = T$. We wish to approximate the PDE solution $u(x, t)$ at grid points (mesh points) much as we did in the problem of numerically solving ODEs as considered in Chapter 10. Thus, we wish to approximate $u(x_k, t_n)$, where

$$x_k = x_0 + hk, \quad t_n = t_0 + \tau n, \tag{12.1}$$

such that

$$h = \frac{1}{M}(x_f - x_0), \quad \tau = \frac{1}{N}(t_f - t_0), \tag{12.2}$$

and so $k = 0, 1, \ldots, M$, and $n = 0, 1, \ldots, N$. This implies that we assume sampling on a uniform two-dimensional grid defined on the xt plane [(x, t) plane]. Commonly, the numerical approximation to $u(x_k, t_n)$ is denoted by $u_{k,n}$. The index k is then a *space index*, and n is a *time index*.

There is a classification scheme for *second-order linear PDEs*. According to Kreyszig [1], Myint-U and Debnath [2], and Courant and Hilbert [9], a PDE of the form

$$Au_{xx} + 2Bu_{xy} + Cu_{yy} = F(x, y, u, u_x, u_y) \tag{12.3}$$

is *elliptic* if $AC - B^2 > 0$, *parabolic* if $AC - B^2 = 0$, and is *hyperbolic* if $AC - B^2 < 0$.[1] It is possible for A, B, and C to be functions of x and y, in which case (12.3) may be of a type (elliptic, parabolic, hyperbolic) that varies with x and y. For example, (12.3) might be hyperbolic in one region of $\mathbf{R}^2$ but parabolic in another. Of course, the terminology as to type remains the same when space variable y is replaced by time variable t.

An example of an elliptic PDE is the *Poisson equation* from electrostatics [10]

$$V_{xx} + V_{yy} = -\frac{1}{\epsilon}\rho(x, y), \tag{12.4}$$

where the solution $V(x, y)$ is the *electrical potential* (e.g., in units of volts) at the spatial location (x, y) in $\mathbf{R}^2$, constant ϵ is the *permittivity* of the medium (e.g., in units of farads per meter), and $\rho(x, y)$ is the *charge density* (e.g., in units of coulombs per square meter) at the spatial point (x, y). We have assumed that the permittivity is a constant, but it can vary spatially as well. Certainly, (12.4)

---

[1] This classification scheme is related to the classification of *conic sections* on the Cartesian plane. The general equation for such a conic on $\mathbf{R}^2$ is

$$Ax^2 + Bxy + Cy^2 + Dx + Ey + F = 0.$$

The conic is *hyperbolic* if $B^2 - 4AC > 0$, *parabolic* if $B^2 - 4AC = 0$, and is *elliptic* if $B^2 - 4AC < 0$.

has the form of (12.3), where for $u(x, y) = V(x, y)$ we have $B = 0$, $A = C = 1$, and $F(x, y, u, u_x, u_y) = -\rho(x, y)/\epsilon$. Therefore, $AC - B^2 = 1 > 0$, confirming that (12.4) is elliptic.

To develop an approximate method of solving (12.4), one may employ finite differences. For example, from Taylor series theory

$$\frac{\partial^2 V(x_k, y_n)}{\partial x^2} = \frac{V(x_{k+1}, y_n) - 2V(x_k, y_n) + V(x_{k-1}, y_n)}{h^2} - \frac{h^2}{12} \frac{\partial^4 V(\xi_k, y_n)}{\partial x^4}$$

(12.5a)

for some $\xi_k \in [x_{k-1}, x_{k+1}]$, and

$$\frac{\partial^2 V(x_k, y_n)}{\partial y^2} = \frac{V(x_k, y_{n+1}) - 2V(x_k, y_n) + V(x_k, y_{n-1})}{\tau^2} - \frac{\tau^2}{12} \frac{\partial^4 V(x_k, \eta_n)}{\partial y^4}$$

(12.5b)

for some $\eta_n \in [y_{n-1}, y_{n+1}]$, where $x_k = x_0 + hk$, $y_n = y_0 + \tau n$ [recall (12.1), and (12.2)]. The finite-difference approximation to (12.4) is thus

$$\frac{V_{k+1,n} - 2V_{k,n} + V_{k-1,n}}{h^2} + \frac{V_{k,n+1} - 2V_{k,n} + V_{k,n-1}}{\tau^2} = -\frac{\rho(x_k, y_n)}{\epsilon}.$$

(12.6)

Here $k = 0, 1, \ldots, M$, and $n = 0, 1, \ldots, N$. Depending on $\rho(x, y)$ and boundary conditions, it is possible to rewrite (12.6) as a linear system of equations in the unknown (approximate) potentials $V_{k,n}$. In practice, N and M may be large, and so the linear system of equations will be of high order consisting of $O(NM)$ unknowns to solve for. Because of the structure of (12.6), the linear system is a sparse one, too. It has therefore been pointed out [3,7,11] that iterative solution methods are preferred, such as the Jacobi or Gauss-Seidel methods (recall Section 4.7). This avoids the problems inherent in storing and manipulating large dense matrices. Epperson [11] notes that in recent years *conjugate gradient methods* have begun to displace Gauss–Seidel/Jacobi approaches to solving large and sparse linear systems such as are generated from (12.6). In part this is due to the difficulties inherent in obtaining the optimal value for the relaxation parameter ω (recall the definition from Section 4.7).

An example of a parabolic PDE is sometimes called the *heat equation*, or *diffusion equation* [2–4, 11] since it models one-dimensional diffusion processes such as the flow of heat through a metal bar. The general form of the basic parabolic PDE is

$$u_t = \alpha^2 u_{xx}$$

(12.7)

for $x \in [0, L]$, $t \in \mathbf{R}^+$. Here $u(x, t)$ could be the temperature of some material at (x, t). It could also be the concentration of some chemical substance that is diffusing out from some source. (Of course, other physical interpretations are possible.) Typical *boundary conditions* are

$$u(0, t) = 0, \quad u(L, t) = 0 \text{ for } t > 0,$$

(12.8a)

and the *initial condition* is

$$u(x, 0) = f(x). \tag{12.8b}$$

The initial condition might be an initial temperature distribution, or chemical concentration. If $u(x, t)$ is interpreted as temperature, then the boundary conditions (12.8a) state that the ends of the one-dimensional medium are held at a constant temperature of 0 (e.g., degrees Celsius).

Equation (12.7) has the form of (12.3) with $B = C = 0$, and hence $AC - B^2 = 0$, which is the criterion for a parabolic PDE. Finite-difference schemes analogous to the case for elliptic PDEs may be developed. Classically, perhaps the most popular choice is the *Crank–Nicolson method* summarized by Burden and Faires [3], but given a more detailed treatment by Epperson [11].

Another popular numerical solution technique for PDEs is the *finite-element* (FEL) *method*. It applies to a broad class of PDEs, and there are many commercially available software packages that implement this approach for various applications such as structural vibration analysis, or electromagnetics. However, we will not consider the FEL method as it deserves its own textbook. The interested reader can see Strang and Fix [5] or Brenner and Scott [6] for details. A brief introduction appears in Burden and Faires [3].

As stated earlier, the emphasis in this book will be on wave propagation problems as modeled by hyperbolic PDEs. We now turn our attention to this class of PDEs.

12.3 APPLICATIONS OF HYPERBOLIC PDEs

In this section we summarize two problems that illustrate how hyperbolic PDEs arise in practice. In later sections we will see that although both involve the modeling of waves propagating in physical systems, the numerical methods for their solution are different in the two cases, and yet they have in common the application of finite-difference schemes.

12.3.1 The Vibrating String

Consider an elastic string with its ends fixed at the points $x = 0$ and $x = L$ (so that the string is of length L unstretched). If the string is plucked at position $x = x_P$ ($x_P \in (0, L)$) at time $t = 0$ such as shown in Fig. 12.1, then it will vibrate for $t > 0$. The PDE describing $u(x, t)$, which is the displacement of the string at position x and time t, is given by

$$u_{tt} = c^2 u_{xx}. \tag{12.9}$$

The system of Fig. 12.1 is also characterized by the *boundary conditions*

$$u(0, t) = 0, \quad u(L, t) = 0 \text{ for all } t \in \mathbf{R}^+, \tag{12.10}$$

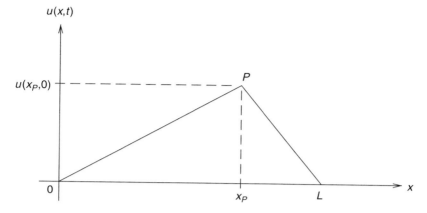

Figure 12.1   An elastic string plucked at time $t = 0$ at point P, which is located at $x = x_P$.

which specify that the string's ends are fixed, and we have the *initial conditions*

$$u(x, 0) = f(x), \quad \left.\frac{\partial u(x, t)}{\partial t}\right|_{t=0} = g(x), \tag{12.11}$$

which describes the initial displacement, and velocity of the string, respectively. As explained, for example, in Kreyszig [1] or in Elmore and Heald [12], the PDE (12.9) is derived from elementary Newtonian mechanics based on the following assumptions:

1. The mass of the string per unit of length is a constant.
2. The string is perfectly elastic, offers no resistance to bending, and there is no friction.
3. The tension in stretching the string before fixing its ends is large enough to neglect the action of gravity.
4. The motion of the string is purely a vibration in the vertical plane (i.e., the y direction), and the deflection and slope are small in absolute value.

We will omit the details of the derivation of (12.9), as this would carry us too far off course. However, note that constant c^2 in (12.9) is

$$c^2 = \frac{\mathcal{T}}{\rho}, \tag{12.12}$$

where $\mathcal{T}$ is the tension in the string (e.g., units of newtons) and ρ is the density of the string (e.g., units of kilograms per meter). A dimensional analysis of (12.12) quickly reveals that c has the units of speed (e.g, meters per second). It specifies the speed at which waves propagate on the string.

It is easy to confirm that (12.9) is a hyperbolic PDE since, on comparing (12.9) with (12.3), we have $A = c^2$, $B = 0$, and $C = -1$. Thus, $AC - B^2 = -c^2 < 0$, which meets the definition of a hyperbolic PDE.

At this point we summarize a standard method for obtaining series-based analytical solutions to PDEs. This is the *method of separation of variables* (also called the *product method*). We shall also find that Fourier series expansions (recall Chapters 1 and 3) have an important role to play in the solution method. The solutions we obtain yield test cases that we can use to gauge the accuracy of numerical methods that we consider later on.

Assume that the solution[2] to (12.9) can be rewritten in the form

$$u(x, t) = X(x)T(t). \tag{12.13}$$

Clearly

$$u_{xx} = X_{xx}T, \quad u_{tt} = XT_{tt}, \tag{12.14}$$

which may be substituted into (12.9), yielding

$$XT_{tt} = c^2 T X_{xx}, \tag{12.15}$$

or equivalently

$$\frac{T_{tt}}{c^2 T} = \frac{X_{xx}}{X}. \tag{12.16}$$

The expression on the left-hand side is a function of t only, while that on the right-hand side is a function only of x. Thus, both sides must equal some constant, say, κ:

$$\frac{T_{tt}}{c^2 T} = \frac{X_{xx}}{X} = \kappa. \tag{12.17}$$

From (12.17) we obtain two second-order linear ODEs in constant coefficients

$$X_{xx} - \kappa X = 0 \tag{12.18}$$

and

$$T_{tt} - c^2 \kappa T = 0. \tag{12.19}$$

We will now ascertain the general form of the solutions to (12.18) and (12.19), based on the conditions (12.10) and (12.11). From (12.10) substituted into (12.13), we obtain

$$u(0, t) = X(0)T(t) = 0, \quad u(L, t) = X(L)T(t) = 0.$$

---

[2]Theories about the existence and uniqueness of solutions to PDEs are often highly involved, and so we completely ignore this matter here. The reader is advised to consult books dedicated to PDEs and their solution for such information.

If $T(t) = 0$ (all t), then $u(x, t) = 0$ for all x and t. This is the trivial solution, and we reject it. Thus, we must have

$$X(0) = X(L) = 0. \tag{12.20}$$

For $\kappa = 0$, Eq. (12.18) is $X_{xx} = 0$, which has the general solution $X(x) = ax + b$, but from (12.20) we conclude that $a = b = 0$, and so $X(x) = 0$ for all x. This is the trivial solution and so is rejected. If $\kappa = \mu^2 > 0$, we have ODE $X_{xx} - \mu^2 X = 0$, which has a characteristic equation possessing roots at $\pm\mu$. Consequently, $X(x) = ae^{\mu x} + be^{-\mu x}$. If we apply (12.20) to this, we conclude that $a = b = 0$, once again giving the trivial solution $X(x) = 0$ for all x. Now finally suppose that $\kappa = -\beta^2 < 0$, in which case (12.18) becomes

$$X_{xx} + \beta^2 X = 0, \tag{12.21}$$

which has the characteristic equation $s^2 + \beta^2 = 0$. Thus, the general solution to (12.21) is of the form

$$X(x) = a\cos(\beta x) + b\sin(\beta x). \tag{12.22}$$

Applying (12.20) yields

$$X(0) = a = 0, \quad X(L) = b\sin(\beta L) = 0. \tag{12.23}$$

Clearly, to avoid encountering the trivial solution, we must assume that $b \neq 0$. Thus, we must have

$$\sin(\beta L) = 0,$$

implying that we have

$$\beta = \frac{n\pi}{L}, \quad n \in \mathbf{Z}. \tag{12.24}$$

However, we avoid $\beta = 0$ (for $n = 0$) to prevent $X(x) = 0$ for all x; and we consider only $n \in \{1, 2, 3, \ldots\} = \mathbf{N}$ because $\sin(-x) = -\sin x$, and the minus sign can be absorbed into the constant b. Thus, in general

$$X(x) = X_n(x) = b_n \sin\left(\frac{n\pi}{L}x\right) \tag{12.25}$$

for $n \in \mathbf{N}$, and where $x \in [0, L]$. So now we have found that

$$\kappa = -\beta^2 = -\left(\frac{n\pi}{L}\right)^2,$$

in which case (12.19) takes on the form

$$T_{tt} + \lambda_n^2 T = 0, \quad \text{where} \quad \lambda_n = \frac{n\pi c}{L}. \tag{12.26}$$

This has a general solution of the form

$$T_n(t) = A_n \cos(\lambda_n t) + B_n \sin(\lambda_n t) \tag{12.27}$$

again for $n \in \mathbf{N}$. Consequently, $u_n(x, t) = X_n(x)T_n(t)$ is a solution to (12.9) for all $n \in \mathbf{N}$, and

$$u_n(x, t) = [A_n \cos(\lambda_n t) + B_n \sin(\lambda_n t)] \sin\left(\frac{n\pi}{L}x\right). \tag{12.28}$$

The functions (12.28) are *eigenfunctions* with *eigenvalues* $\lambda_n = n\pi c/L$ for the PDE in (12.9). The set $\{\lambda_n | n \in \mathbf{N}\}$ is the *spectrum*. Each $u_n(x, t)$ represents harmonic motion of the string with frequency $\lambda_n/(2\pi)$ cycles per unit of time, and is also called the nth *normal mode* for the string. The first mode for $n = 1$ is called the *fundamental mode*, and the others (for $n > 1$) are called *overtones*, or *harmonics*. It is clear that $u_n(x, t)$ in (12.28) satisfies PDE (12.9), and the boundary conditions (12.10). However, $u_n(x, t)$ by itself will not satisfy (12.9), (12.10) *and* (12.11) all simultaneously. In general, the *complete solution* is [using superposition as the PDE (12.9) is linear]

$$u(x, t) = \sum_{n=1}^{\infty} u_n(x, t) = \sum_{n=1}^{\infty} [A_n \cos(\lambda_n t) + B_n \sin(\lambda_n t)] \sin\left(\frac{n\pi}{L}x\right), \tag{12.29}$$

where the initial conditions (12.11) are employed to find the series coefficients A_n and B_n for all n. We will now consider how this is done in general.

From (12.11) we obtain

$$u(x, 0) = \sum_{n=1}^{\infty} A_n \sin\left(\frac{n\pi}{L}x\right) = f(x) \tag{12.30}$$

and

$$\frac{\partial u(x, t)}{\partial t}|_{t=0} = \left[\sum_{n=1}^{\infty} [-A_n\lambda_n \sin(\lambda_n t) + B_n\lambda_n \cos(\lambda_n t)] \sin\left(\frac{n\pi}{L}x\right)\right]_{t=0} = g(x)$$

or

$$\sum_{n=1}^{\infty} B_n\lambda_n \sin\left(\frac{n\pi}{L}x\right) = g(x). \tag{12.31}$$

The orthogonality properties of sinusoids can be used to determine A_n and B_n for all n. Note that (12.30) and (12.31) are particular instances of Fourier series expansions. In particular, observe that for $k, n \in \mathbf{N}$

$$\int_0^L \sin\left(\frac{n\pi}{L}x\right) \sin\left(\frac{k\pi}{L}x\right) dx = \begin{cases} 0, & n \neq k \\ L/2, & n = k \end{cases}. \tag{12.32}$$

Plainly, set $\{\sin(n\pi x/L)|n \in \mathbf{N}\}$ is an orthogonal set. Thus, from (12.30)

$$\frac{2}{L}\int_0^L \sin\left(\frac{k\pi}{L}x\right)\left\{\sum_{n=1}^{\infty} A_n \sin\left(\frac{n\pi}{L}x\right)\right\} dx = \frac{2}{L}\int_0^L f(x)\sin\left(\frac{k\pi}{L}x\right) dx,$$

and via (12.32) this reduces to

$$A_k = \frac{2}{L}\int_0^L f(x)\sin\left(\frac{k\pi}{L}x\right) dx, \tag{12.33}$$

and similarly from (12.31), and (12.32)

$$B_k = \frac{2}{\lambda_k L}\int_0^L g(x)\sin\left(\frac{k\pi}{L}x\right) dx. \tag{12.34}$$

Example 12.1 Suppose that $g(x) = 0$ for all x. Thus, the initial velocity of the string is zero. Let the initial position (deflection) of the plucked string be triangular such that

$$f(x) = \begin{cases} \dfrac{2H}{L}x, & 0 < x \le \dfrac{L}{2} \\[2mm] \dfrac{2H}{L}(L-x), & \dfrac{L}{2} < x < L \end{cases}. \tag{12.35}$$

In Fig. 12.1 this corresponds to $x_P = L/2$ and $u(x_P, 0) = H$. Since $g(x) = 0$ for all x via (12.34), we must have $B_k = 0$ for all k. From (12.33) we have

$$A_k = \frac{4H}{L^2}\left[\int_0^{L/2} x \sin\left(\frac{k\pi}{L}x\right) dx + \int_{L/2}^L (L-x)\sin\left(\frac{k\pi}{L}x\right) dx\right].$$

Since

$$\int \sin(ax)\,dx = -\frac{1}{a}\cos(ax) + C$$

and

$$\int x\sin(ax)\,dx = -\frac{1}{a}x\cos(ax) + \frac{1}{a^2}\sin(ax) + C$$

(C is a constant of integration), on simplification we have

$$A_k = \frac{8H}{k^2\pi^2}\sin\left(\frac{k\pi}{2}\right). \tag{12.36}$$

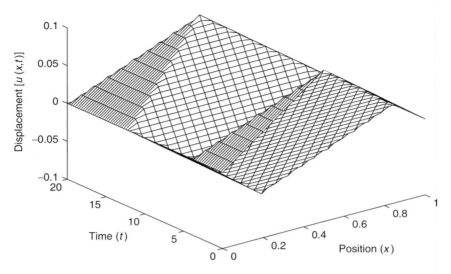

Figure 12.2   Fourier series solution to the vibrating string problem. A mesh plot of $u(x, t)$ as given by Eq. (12.37) for the parameters $L = 1$, $c/L = \frac{1}{8}$, and $H = \frac{1}{10}$. The plot employed the first 100 terms of the series expansion.

Thus, substituting (12.36) into (12.29) yields the general solution for our example, namely

$$u(x, t) = \frac{8H}{\pi^2} \sum_{k=1}^{\infty} \frac{1}{k^2} \sin\left(\frac{k\pi}{2}\right) \cos\left(\frac{k\pi c}{L}t\right) \sin\left(\frac{k\pi}{L}x\right),$$

but $\sin(k\pi/2) = 0$ for even k, and so this expression reduces to

$$u(x, t) = \frac{8H}{\pi^2} \sum_{n=1}^{\infty} \frac{(-1)^{n-1}}{(2n-1)^2} \cos\left(\frac{(2n-1)\pi c}{L}t\right) \sin\left(\frac{(2n-1)\pi}{L}x\right). \quad (12.37)$$

Figure 12.2 shows a typical plot of the function $u(x, t)$ as given by (12.37) (for the parameters stated in the figure caption).

The reader is encouraged to think about Fig. 12.2, and to ask if the picture is a reasonable one on the basis of his/her intuitive understanding of how a plucked string (say, that on a stringed musical instrument) behaves. Of course, this question should be considered with respect to the modeling assumptions that lead to (12.9), and that were listed earlier.

12.3.2   Plane Electromagnetic Waves

An electromagnetic wave (e.g., radio wave or light) in three-dimensional space $\mathbf{R}^3$ within some material is described by the *vector magnetic field intensity* $\overline{H}(x, y, z, t)$

[e.g., in units of amperes per meter, (A/m)] and the *vector electric field intensity* $\overline{E}(x, y, z, t)$ [e.g., in units of volts per meter (V/m)] such that

$$\overline{H}(x, y, z, t) = H_x(x, y, z, t)\hat{x} + H_y(x, y, z, t)\hat{y} + H_z(x, y, z, t)\hat{z},$$

$$\overline{E}(x, y, z, t) = E_x(x, y, z, t)\hat{x} + E_y(x, y, z, t)\hat{y} + E_z(x, y, z, t)\hat{z},$$

where $\hat{x}$, $\hat{y}$, and $\hat{z}$ are the unit vectors in the x, y, and z directions of $\mathbf{R}^3$, respectively. The dynamic equations that $\overline{H}$ and $\overline{E}$ both satisfy are *Maxwell's equations*:

$$\nabla \times \overline{E} = -\frac{\partial \overline{B}}{\partial t} \quad \text{(Faraday's law)}, \tag{12.38}$$

$$\nabla \times \overline{H} = \frac{\partial \overline{D}}{\partial t} \quad \text{(Ampere's law)}. \tag{12.39}$$

Here the material in which the wave propagates contains no charges or current sources. The *magnetic flux density* $\overline{B}(x, y, z, t)$, and the *electric flux density* $\overline{D}(x, y, z, t)$ are assumed to satisfy

$$\overline{D} = \epsilon \overline{E}, \quad \overline{B} = \mu \overline{H}. \tag{12.40}$$

These relations assume that the material is linear, *isotropic* [i.e., the same in all directions, and *homogeneous* i.e., the parameters ϵ and μ do not vary with (x, y, z)]. Constant ϵ is the material's *permittivity* [units of farads per meter (F/m)], and constant μ is the material's *permeability* [units of henries per meter (H/m)]. The permittivity and permeability of *free space* (i.e., a vacuum) are often denoted by ϵ_0 and μ_0, respectively, and

$$\epsilon = \epsilon_r \epsilon_0, \quad \mu = \mu_r \mu_0, \tag{12.41}$$

where ϵ_r is the *relative permittivity* and μ_r is the *relative permeability* of the material. Note that

$$\epsilon_0 = 8.854185 \times 10^{-12} \text{ F/m}, \quad \mu_0 = 400\pi \times 10^{-9} \text{ H/m}. \tag{12.42}$$

If the material is air, then $\epsilon_r \approx 1$ and $\mu_r \approx 1$ to very good approximation, and so air is not practically distinguished (usually) from free space. For commonly occurring dielectric materials (i.e., insulators), we have $\mu_r \approx 1$, also to excellent approximation. On the other hand, for magnetic materials (e.g., iron, cobalt, nickel, various alloys and mixtures), μ_r will be very different from unity, and in fact the relationship $\overline{B} = \mu \overline{H}$ must often be replaced by sometimes quite complicated nonlinear relationships, often involving the phenomenon of *hysteresis*. But we will completely avoid this situation here.

In general, for the vector field $\overline{A} = A_x \hat{x} + A_y \hat{y} + A_z \hat{z}$, the *curl* is the determinant

$$\nabla \times \overline{A} = \begin{vmatrix} \hat{x} & \hat{y} & \hat{z} \\ \dfrac{\partial}{\partial x} & \dfrac{\partial}{\partial y} & \dfrac{\partial}{\partial z} \\ A_x & A_y & A_z \end{vmatrix}, \tag{12.43}$$

so, expanding this expression with $\overline{A} = \overline{E}$ and $\overline{A} = \overline{H}$ gives (respectively)

$$\nabla \times \overline{E} = \hat{x}\left(\frac{\partial E_z}{\partial y} - \frac{\partial E_y}{\partial z}\right) + \hat{y}\left(\frac{\partial E_x}{\partial z} - \frac{\partial E_z}{\partial x}\right) + \hat{z}\left(\frac{\partial E_y}{\partial x} - \frac{\partial E_x}{\partial y}\right), \tag{12.44}$$

and

$$\nabla \times \overline{H} = \hat{x}\left(\frac{\partial H_z}{\partial y} - \frac{\partial H_y}{\partial z}\right) + \hat{y}\left(\frac{\partial H_x}{\partial z} - \frac{\partial H_z}{\partial x}\right) + \hat{z}\left(\frac{\partial H_y}{\partial x} - \frac{\partial H_x}{\partial y}\right). \tag{12.45}$$

We will consider only *transverse electromagnetic* (TEM) *waves* (i.e., *plane waves*). If such a wave propagates in the x direction, then we may assume that $E_x = E_z = 0$, and so $H_x = H_y = 0$. Note that the electric and magnetic field components E_y and H_z are orthogonal to each other. They lie within the (y, z) plane, which itself is orthogonal to the direction of travel of the plane wave. From (12.44) and (12.45), Maxwell's equations (12.38) and (12.39) reduce to

$$\frac{\partial E_y}{\partial x} = -\mu \frac{\partial H_z}{\partial t}, \quad \frac{\partial H_z}{\partial x} = -\epsilon \frac{\partial E_y}{\partial t}, \tag{12.46}$$

where we have used (12.40). Combining the two equations in (12.46), we obtain either

$$\frac{\partial^2 H_z}{\partial x^2} = \mu\epsilon \frac{\partial^2 H_z}{\partial t^2}, \tag{12.47}$$

which is the *wave equation for the magnetic field*, or

$$\frac{\partial^2 E_y}{\partial x^2} = \mu\epsilon \frac{\partial^2 E_y}{\partial t^2}, \tag{12.48}$$

which is the *wave equation for the electric field*. If we define

$$v = \frac{1}{\sqrt{\mu\epsilon}}, \tag{12.49}$$

then the general solution to (12.48) (for example) can be expressed in the form

$$E_y(x, t) = E_{y_r}(x - vt) + E_{y_l}(x + vt), \tag{12.50}$$

where the first term is a wave propagating in the $+x$ direction (i.e., to the right) with speed v and the second term is a wave propagating in the $-x$ direction (i.e., to the left) with speed v.[3] Equation (12.50) is the classical *D'Alembert solution* to the scalar wave equation (12.48). Clearly, similar reasoning applies to (12.47). Of course, using (12.49) in (12.48), we can write

$$\frac{\partial^2 E_y}{\partial t^2} = v^2 \frac{\partial^2 E_y}{\partial x^2} \tag{12.51}$$

which has the same form as (12.9). In short, the mathematics describing the vibrations of mechanical systems is much the same as that describing electromagnetic systems, only the physical interpretations differ. Of course, (12.51) clearly implies that (12.47) and (12.48) are hyperbolic PDEs.

Example 12.2   It is easy to confirm that (12.37) can be rewritten in the form of (12.50). Via the identity

$$\cos A \, \sin B = \tfrac{1}{2}[\sin(A + B) - \sin(A - B)] \tag{12.52}$$

we see that

$$\cos\left[\frac{(2n-1)\pi c}{L}t\right]\sin\left[\frac{(2n-1)\pi}{L}x\right] = \frac{1}{2}\sin\left[\frac{(2n-1)\pi}{L}(x+ct)\right]$$
$$+ \frac{1}{2}\sin\left[\frac{(2n-1)\pi}{L}(x-ct)\right].$$

Thus, (12.37) may immediately be rewritten as

$$u(x,t) = \underbrace{\frac{4H}{\pi^2}\sum_{n=1}^{\infty}\frac{(-1)^{n-1}}{(2n-1)^2}\sin\left[\frac{(2n-1)\pi}{L}(x-ct)\right]}_{=u_1(x-ct)}$$
$$+ \underbrace{\frac{4H}{\pi^2}\sum_{n=1}^{\infty}\frac{(-1)^{n-1}}{(2n-1)^2}\sin\left[\frac{(2n-1)\pi}{L}(x+ct)\right]}_{=u_1(x+ct)}.$$

We note that when $\epsilon_r = \mu_r = 1$, we have $v = c$, where

$$c = \frac{1}{\sqrt{\mu_0 \epsilon_0}}. \tag{12.53}$$

---

[3]Readers are invited to draw a simple sketch and convince themselves that this interpretation is correct. This interpretation is vital in understanding the propagation of electromagnetic waves in layered materials, such as thin optical films.

This is the *speed of light* in a vacuum. Since for real materials $\mu_r > 1$ and $\epsilon_r > 1$, we have $v < c$, so an electromagnetic wave cannot travel at a speed exceeding that of light in a vacuum.

Now we will assume sinusoidal solutions to the wave equations such as would originate from sinusoidal sources.[4] Specifically, let us assume that

$$E_y(x, t) = E_0 \sin(\omega t - \beta x), \qquad (12.54)$$

where $\beta = \omega\sqrt{\mu\epsilon} = 2\pi/\lambda$, and λ is the *wavelength* (e.g., in units of meters). The frequency of the source is ω, a fixed constant, and so the wavelength will vary depending on the medium. If the free-space wavelength is denoted λ_0, then

$$\frac{2\pi}{\lambda_0} = \frac{\omega}{c}, \qquad (12.55)$$

where c is from (12.53). If the free-space wave then propagates into a denser material, then

$$\frac{2\pi}{\lambda} = \frac{\omega}{v} \qquad (12.56)$$

for v given by (12.49). From (12.55) and (12.56), we obtain

$$\lambda = \frac{v}{c}\lambda_0. \qquad (12.57)$$

Since $v \leq c$, we always have $\lambda \leq \lambda_0$; that is, the wavelength will shorten. This observation is useful in checking numerical methods that model the propagation of sinusoidal waves across interfaces between different materials (e.g., layered structures such as thin optical films).

From (12.54) we have

$$\frac{\partial E_y}{\partial x} = -\beta E_0 \cos(\omega t - \beta x)$$

so that from the first equation in (12.46) we have

$$H_z = \int \frac{\beta E_0}{\mu} \cos(\omega t - \beta x)\, dt = \frac{\beta E_0}{\mu\omega} \sin(\omega t - \beta x). \qquad (12.58)$$

The *characteristic impedance* of the medium in which the sinusoidal electromagnetic wave travels is defined to be

$$Z = \frac{E_y}{H_z} = \frac{\mu\omega}{\beta} = \sqrt{\frac{\mu}{\epsilon}} = \sqrt{\frac{\mu_r}{\epsilon_r}}Z_0, \qquad (12.59)$$

---

[4]For example, the Colpitts oscillator from Chapter 10 could operate as a sinusoidal signal generator to drive an antenna, thus producing a sinusoidal electromagnetic wave (radio wave) in space.

where $Z_0 = \sqrt{\mu_0/\epsilon_0}$ is the *characteristic impedance of free space*. The units of Z are in ohms (Ω). We see that Z is analogous to the concept of impedance that arises in electric circuit analysis.

The analogy between our present problem and phasor analysis in basic electric circuit theory can be exploited. For suitable $E(x) \in \mathbf{R}$

$$E_y = E_y(x, t) = E(x)e^{j\omega t} \tag{12.60}$$

so that

$$\frac{\partial^2 E_y}{\partial x^2} = \frac{d^2 E(x)}{dx^2}e^{j\omega t}, \qquad \frac{\partial^2 E_y}{\partial t^2} = -\omega^2 E(x)e^{j\omega t}. \tag{12.61}$$

Substituting (12.61) into (12.48) yields

$$\frac{d^2 E(x)}{dx^2}e^{j\omega t} = -\mu\epsilon\omega^2 E(x)e^{j\omega t},$$

which reduces to the second-order linear ODE

$$\frac{d^2 E(x)}{dx^2} + \mu\epsilon\omega^2 E(x) = 0. \tag{12.62}$$

For convenience, we define the *propagation constant*

$$\gamma = j\omega\sqrt{\mu\epsilon} = j\beta, \tag{12.63}$$

so $-\gamma^2 = \mu\epsilon\omega^2$, and (12.62) is now

$$\frac{d^2 E}{dx^2} - \gamma^2 E = 0 \tag{12.64}$$

($E = E(x)$). This ODE has a general solution of the form

$$E(x) = E_0 e^{-\gamma x} + E_1 e^{\gamma x}, \tag{12.65}$$

where E_0 and E_1 are constants. Recalling (12.60), it follows that

$$E_y(x, t) = E_0 e^{j\omega t - \gamma x} + E_1 e^{j\omega t + \gamma x}$$
$$= E_0 e^{-j(\beta x - \omega t)} + E_1 e^{j(\beta x + \omega t)}. \tag{12.66}$$

Of course, the first term is a wave propagating to the right, and the second term is a wave propagating to the left.

So far we have assumed wave propagation in lossless materials since this is the easiest case to consider at the outset. We shall now consider the effects of lossy materials on propagation. This will be important in that it is a more realistic assumption in practice, and it is important in designing *perfectly matched layers* (PMLs) in the *finite-difference time-domain* (FDTD) *method*, as will be seen later.

We may define *electrical conductivity* σ [units of amperes per volt-meter [A/(V·m)] or mhos/meter], and *magnetic conductivity* σ^* [units of volts per ampere-meter [V/(A·m)] or ohms/meter]. In this case (12.38) and (12.39) take on the more general forms.

$$\nabla \times \overline{E} = -\mu \frac{\partial \overline{H}}{\partial t} - \sigma^* \overline{H}, \tag{12.67}$$

$$\nabla \times \overline{H} = \epsilon \frac{\partial \overline{E}}{\partial t} + \sigma \overline{E}. \tag{12.68}$$

Note that σ^* is not the complex conjugate of σ. In fact, $\sigma, \sigma^* \in \mathbf{R}$ with $\sigma > 0$ for a lossy material (i.e., an imperfect insulator, or a conductor), and $\sigma^* \geq 0$. As before we will assume $\overline{E}$ possesses only a y component, and $\overline{H}$ possesses only a z component. Since we again have propagation only in the x direction, via (12.44), (12.45) in (12.67), and (12.68), we have

$$\frac{\partial E_y}{\partial x} = -\mu \frac{\partial H_z}{\partial t} - \sigma^* H_z, \tag{12.69}$$

$$-\frac{\partial H_z}{\partial x} = \epsilon \frac{\partial E_y}{\partial t} + \sigma E_y. \tag{12.70}$$

If $E_y = E_y(x, t)$ possesses a term that propagates only to the right, then (using phasors again)

$$\overline{E} = E_y(x, t)\hat{y} = E_0 e^{j\omega t} e^{-\gamma x} \hat{y} \tag{12.71}$$

for some suitable propagation constant γ. For suitable characteristic impedance Z, we must have

$$\overline{H} = H_z(x, t)\hat{z} = \frac{1}{Z} E_0 e^{j\omega t} e^{-\gamma x} \hat{z}. \tag{12.72}$$

We may use (12.69) and (12.70) to determine γ and Z. Substituting (12.71) and (12.72) into (12.69) and (12.70) and solving for γ and Z yields

$$Z^2 = \frac{j\omega\mu + \sigma^*}{j\omega\epsilon + \sigma}, \quad \gamma^2 = (j\omega\epsilon + \sigma)(j\omega\mu + \sigma^*). \tag{12.73}$$

How to handle the complex square roots needed to obtain Z and γ will be dealt with below. Observe that the equations in (12.73) reduce to the previous cases (12.59), and (12.63) when $\sigma = \sigma^* = 0$. It is noteworthy that when we have the condition

$$\frac{\sigma^*}{\mu} = \frac{\sigma}{\epsilon}, \tag{12.74}$$

then we have

$$Z^2 = \frac{j\omega\mu + \frac{\mu}{\epsilon}\sigma}{j\omega\epsilon + \sigma} = \frac{\mu}{\epsilon} \frac{j\omega + \frac{\sigma}{\epsilon}}{j\omega + \frac{\sigma}{\epsilon}} = \frac{\mu}{\epsilon}. \tag{12.75}$$

Condition (12.74) is what makes the creation of a PML possible, as will be considered later in this section, and will be demonstrated in Section 12.5.

Now we must investigate what happens to waves when they encounter a sudden change in the material properties, specifically, an *interface* between layers. This situation is depicted in Fig. 12.3. Assume that medium 1 has physical parameters $\epsilon_1, \mu_1, \sigma_1$, and σ_1^*, while medium 2 has physical parameters $\epsilon_2, \mu_2, \sigma_2$, and σ_2^*. The corresponding characteristic impedance and propagation constant for medium 1 is thus [via (12.73)]

$$Z_1^2 = \frac{j\omega\mu_1 + \sigma_1^*}{j\omega\epsilon_1 + \sigma_1}, \quad \gamma_1^2 = (j\omega\epsilon_1 + \sigma_1)(j\omega\mu_1 + \sigma_1^*), \tag{12.76}$$

while for medium 2 we have

$$Z_2^2 = \frac{j\omega\mu_2 + \sigma_2^*}{j\omega\epsilon_2 + \sigma_2}, \quad \gamma_2^2 = (j\omega\epsilon_2 + \sigma_2)(j\omega\mu_2 + \sigma_2^*). \tag{12.77}$$

In Fig. 12.3 for some constants E and H we have for the incident field

$$\overline{E}_i = E_i\hat{y} = Ee^{j\omega t}e^{-\gamma_1 x}\hat{y}, \quad \overline{H}_i = H_i\hat{z} = He^{j\omega t}e^{-\gamma_1 x}\hat{z}$$

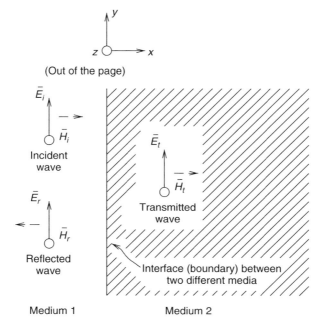

Figure 12.3 A plane wave normally incident on an interface (boundary) between two different media. The magnetic field components are directed orthogonal to the page. The interface is at $x = 0$ and is the yz-plane.

so that

$$\frac{\partial H_i}{\partial t} = j\omega H_i, \quad \frac{\partial E_i}{\partial x} = -\gamma_1 E_i,$$

and via (12.69)

$$-\gamma_1 E_i = -j\omega\mu_1 H_i - \sigma_1^* H_i,$$

implying that

$$\frac{E_i}{H_i} = \frac{j\omega\mu_1 + \sigma_1^*}{\gamma_1} = Z_1. \tag{12.78}$$

Similarly, for the transmitted field, we must have

$$\frac{E_t}{H_t} = \frac{j\omega\mu_2 + \sigma_2^*}{\gamma_2} = Z_2. \tag{12.79}$$

But, for the reflected field, again for suitable constants E', and H' we have

$$\overline{E}_r = E_r \hat{y} = E' e^{j\omega t} e^{\gamma_1 x} \hat{y}, \quad \overline{H}_r = H_r \hat{z} = H' e^{j\omega t} e^{\gamma_1 x} \hat{z}$$

so that

$$\frac{\partial H_r}{\partial t} = j\omega H_r, \quad \frac{\partial E_r}{\partial x} = \gamma_1 E_r,$$

and via (12.69)

$$\gamma_1 E_r = -j\omega\mu_1 H_r - \sigma_1^* H_r,$$

implying

$$\frac{E_r}{H_r} = -\frac{j\omega\mu_1 + \sigma_1^*}{\gamma_1} = -Z_1. \tag{12.80}$$

The electric and magnetic field components are tangential to the interface, and so must be continuous across it. This implies that at $x = 0$, and for all t (in Fig. 12.3)

$$H_i + H_r = H_t, \quad E_i + E_r = E_t. \tag{12.81}$$

If we substitute (12.78)–(12.80) into (12.81), then after a bit of algebra, we have

$$\frac{E_t}{E_i} = \frac{2Z_2}{Z_2 + Z_1} = \tau \tag{12.82}$$

and

$$\frac{E_r}{E_i} = \frac{Z_2 - Z_1}{Z_2 + Z_1} = \rho. \tag{12.83}$$

It is easy to confirm that

$$\tau = 1 + \rho. \tag{12.84}$$

We call τ the *transmission coefficient* from medium 1 into medium 2, and ρ is the *reflection coefficient* from medium 1 into medium 2. The coefficients τ and ρ are often called *Fresnel coefficients*, especially in the field of optics.

If $\sigma^* = 0$ then from (12.73) the propagation constant for a sinusoidal wave in a conductor is obtained from

$$\gamma^2 = -\mu\epsilon\omega^2 + j\omega\mu\sigma, \tag{12.85}$$

where $\mu\epsilon\omega^2 > 0$ and $\omega\mu\sigma \geq 0$. We may express γ^2 in polar form: $\gamma^2 = r_1 e^{j\theta_1}$. But $\gamma = r e^{j\theta} = [r_1 e^{j\theta_1}]^{1/2}$. In general, if $z = re^{j\theta}$, then

$$z^{1/2} = r^{1/2} e^{j\theta/2}, \qquad \text{or} \qquad z^{1/2} = r^{1/2} e^{j(\theta/2+\pi)}. \tag{12.86}$$

From (12.85)

$$r_1 = |\gamma^2| = \omega\mu\sqrt{\sigma^2 + \epsilon^2\omega^2}, \quad \theta_1 = \frac{\pi}{2} + \tan^{-1}\left(\frac{\epsilon}{\sigma}\omega\right). \tag{12.87}$$

Consequently

$$\gamma = \pm[\omega\mu\sqrt{\sigma^2 + \epsilon^2\omega^2}]^{1/2} e^{j[\frac{\pi}{4}+\frac{1}{2}\tan^{-1}(\epsilon\omega/\sigma)]}. \tag{12.88}$$

A special case is the perfect insulator (perfect dielectric) for which $\sigma = 0$. Since $\tan^{-1}(\infty) = \pi/2$, Eq. (12.88) reduces to $\gamma = \pm j\sqrt{\mu\epsilon}\omega = \pm j\beta$. More generally ($\sigma$ not necessarily zero) we have $\gamma = \pm(\alpha + j\beta)$, where

$$\alpha = [\omega\mu\sqrt{\sigma^2 + \epsilon^2\omega^2}]^{1/2} \cos\left[\frac{\pi}{4} + \frac{1}{2}\tan^{-1}\left(\frac{\epsilon}{\sigma}\omega\right)\right], \tag{12.89a}$$

$$\beta = [\omega\mu\sqrt{\sigma^2 + \epsilon^2\omega^2}]^{1/2} \sin\left[\frac{\pi}{4} + \frac{1}{2}\tan^{-1}\left(\frac{\epsilon}{\sigma}\omega\right)\right]. \tag{12.89b}$$

Example 12.3   Assume that $\mu = \mu_0$, $\epsilon = \epsilon_0$, and that $\sigma = 0.0001$ mhos/meter. For $\omega = 2\pi f$ [f is the sinusoid's frequency in Hertz (H$_z$)] from (12.89), we obtain the following table:

f (Hz)	α	β
1×10^6	.015236	0.025911
1×10^7	.018762	0.210423
1×10^8	.018836	2.095929
1×10^9	.018837	20.958455

Keeping the parameters the same, except that now $\sigma = 0.01$ mhos/meter, we have the following table:

f (Hz)	α	β
1×10^6	.198140	0.199245
1×10^7	.611091	0.646032
1×10^8	1.523602	2.591125
1×10^9	1.876150	21.042254

In general, from (12.89a) we have $\alpha \geq 0$. If in Fig. 12.3 medium 1 is free space while medium 2 is a conductor with $0 < \sigma < \infty$, then $\overline{E}_t$ has the form [using (12.82)] for $x \geq 0$

$$\overline{E}_t = \tau E e^{j\omega t} e^{-\gamma_2 x} \hat{y} = \tau E e^{j\omega t} e^{-\alpha_2 x} e^{-j\beta_2 x} \hat{y}. \tag{12.90}$$

Of course, we also have $\overline{E}_i = E e^{j\omega t} e^{-j\beta_1 x} \hat{y}$ for $x \leq 0$ since $\gamma_1 = j\beta_1$ (i.e., $\alpha_1 = 0$ in free space), and $\overline{E}_r = \rho E e^{j\omega t} e^{j\beta_1 x} \hat{y}$ for $x \leq 0$ [using (12.83)]. Since $\alpha_2 > 0$, the factor $e^{-\alpha_2 x}$ will go to zero as $x \to \infty$. The amplitude of the wave must decay as it progresses from free space (medium 1) into the conductive medium (medium 2). The rate of decay certainly depends on the size of α_2.

Now suppose that medium 1 is again free space, but that medium 2 has both $\sigma_2 > 0$ and $\sigma_2^* > 0$ such that condition (12.74) holds with $\mu_2 = \mu_0$, and $\epsilon_2 = \epsilon_0$, specifically

$$\frac{\sigma_2^*}{\mu_0} = \frac{\sigma_2}{\epsilon_0} \tag{12.91}$$

which implies [via (12.75)] that $Z_2 = \sqrt{\mu_0/\epsilon_0}$. Since medium 1 is free space $Z_1 = \sqrt{\mu_0/\epsilon_0}$, too. The reflection coefficient from medium 1 into medium 2 is [via (12.83)]

$$\rho = \frac{Z_2 - Z_1}{Z_2 + Z_1} = \frac{Z_0 - Z_0}{2Z_0} = 0.$$

When wave $\overline{E}_i$ in medium 1 encounters the interface (at $x = 0$ in Fig. 12.3), there will be no reflected component, that is, we will have $\overline{E}_r = 0$. From (12.73) we obtain

$$\gamma_2^2 = (\sigma_2 \sigma_2^* - \omega^2 \mu_0 \epsilon_0) + j\omega(\sigma_2 \mu_0 + \sigma_2^* \epsilon_0), \tag{12.92}$$

and we select the medium 2 parameters so that for $\gamma_2 = \alpha_2 + j\beta_2$ we obtain $\alpha_2 > 0$, and α_2 is large enough so that the wave is rapidly attenuated in that $e^{-\alpha_2 x}$ is small for relatively small x. In this case we may define medium 2 to be a *perfectly matched layer* (PML). It is perfectly matched in the sense that its characteristic impedance is the same as that of medium 1, thus eliminating reflections at the interface. Because it is lossy, it absorbs radiation incident on it. The layer dissipates energy without reflection. It thus simulates the walls of an *anechoic chamber*. In other words, an anechoic chamber has walls that approximately realize condition (12.74). The necessity to simulate the walls of an anechoic chamber will become clearer in Section 12.5 when we look at the FDTD method.

Finally, we remark on the similarities between the vibrating string problem and the problem considered here. The analytical solution method employed in Section 12.3.1 was separation of variables, and we have employed the same approach here since all of our electromagnetic field solutions are of the form $u(x, t) = X(x)T(t)$. The main difference is that in the vibrating string problem we have boundary conditions defined by the ends of the string being tied down somewhere, while in the electromagnetic wave propagation problem as we have considered it here there are no boundaries, or rather, the boundaries are at $x = \pm\infty$.

12.4   THE FINITE-DIFFERENCE (FD) METHOD

We now consider a classical approach to the numerical solution of (12.9) that we call the *finite-difference* (FD) *method*. Note that the method to follow is by no means the only approach. Indeed, the FDTD method to be considered in Section 12.5 is an alternative, and there are still others.

Following (12.5)

$$\frac{\partial^2 u(x_k, t_n)}{\partial t^2} = \frac{u(x_k, t_{n+1}) - 2u(x_k, t_n) + u(x_k, t_{n-1})}{\tau^2} - \frac{\tau^2}{12}\frac{\partial^4 u(x_k, \eta_n)}{\partial t^4} \quad (12.93)$$

for some $\eta_n \in [t_{n-1}, t_{n+1}]$, and

$$\frac{\partial^2 u(x_k, t_n)}{\partial x^2} = \frac{u(x_{k+1}, t_n) - 2u(x_k, t_n) + u(x_{k-1}, t_n)}{h^2} - \frac{h^2}{12}\frac{\partial^4 u(\xi_k, t_n)}{\partial x^4} \quad (12.94)$$

for some $\xi_k \in [x_{k-1}, x_{k+1}]$, where

$$x_k = kh, \quad t_n = n\tau \quad (12.95)$$

for $k = 0, 1, \ldots, M$, and $n \in \mathbf{Z}^+$. On substitution of (12.93) and (12.94) into (12.9), we have

$$\frac{u(x_k, t_{n+1}) - 2u(x_k, t_n) + u(x_k, t_{n-1})}{\tau^2} - c^2\frac{u(x_{k+1}, t_n) - 2u(x_k, t_n) + u(x_{k-1}, t_n)}{h^2}$$
$$= \underbrace{\frac{1}{12}\left[\tau^2\frac{\partial^4 u(x_k, \eta_n)}{\partial t^4} - c^2h^2\frac{\partial^4 u(\xi_k, t_n)}{\partial x^4}\right]}_{=e_{k,n}}, \quad (12.96)$$

where $e_{k,n}$ is the *local truncation error*. Since $u_{k,n} \approx u(x_k, t_n)$ from (12.96), we obtain the difference equation

$$u_{k,n+1} - 2u_{k,n} + u_{k,n-1} - e^2 u_{k+1,n} + 2e^2 u_{k,n} - e^2 u_{k-1,n} = 0, \quad (12.97)$$

where

$$e = \frac{\tau}{h}c \quad (12.98)$$

is sometimes called the *Courant parameter*. It has a crucial role to play in determining the stability of the FD method (and of the FDTD method, too). If we solve (12.97) for $u_{k,n+1}$ we obtain

$$u_{k,n+1} = 2(1 - e^2)u_{k,n} + e^2(u_{k+1,n} + u_{k-1,n}) - u_{k,n-1}, \qquad (12.99)$$

where $k = 1, 2, \ldots, M - 1$, and $n = 1, 2, 3, \ldots$. Equation (12.99) is the main recursion in the FD algorithm. However, we need to account for the initial and boundary conditions in order to initialize this recursion. Before we consider this matter note that, in the language of Section 10.2, the FD algorithm has a *truncation error of* $O(\tau^2 + h^2)$ *per step* [via $e_{k,n}$ in (12.96)].

Immediately on applying (12.10), since $L = Mh$, we have

$$u_{0,n} = u_{M,n} = 0 \qquad \text{for} \qquad n \in \mathbf{Z}^+. \qquad (12.100)$$

Since from (12.11) $u(x, 0) = f(x)$, we also have

$$u_{k,0} = f(x_k) \qquad (12.101)$$

for $k = 0, 1, \ldots, M$. From (12.101) we have $u_{k,0}$, but we also need $u_{k,1}$ [consider (12.99) for $n = 1$]. To obtain a suitable expression first observe that from (12.9)

$$\frac{\partial^2 u(x, t)}{\partial t^2} = c^2 \frac{\partial^2 u(x, t)}{\partial x^2} \Rightarrow \frac{\partial^2 u(x, 0)}{\partial t^2} = c^2 \frac{\partial^2 u(x, 0)}{\partial x^2} = c^2 f^{(2)}(x). \quad (12.102)$$

Now, on applying the Taylor series expansion, we see that for some $\mu \in [0, t_1] = [0, \tau]$ (and μ may depend on x)

$$u(x, t_1) = u(x, 0) + \tau \frac{\partial u(x, 0)}{\partial t} + \frac{1}{2}\tau^2 \frac{\partial^2 u(x, 0)}{\partial t^2} + \frac{1}{6}\tau^3 \frac{\partial^3 u(x, \mu)}{\partial t^3},$$

and on applying (12.11) and (12.102), this becomes

$$u(x, t_1) = u(x, 0) + \tau g(x) + \frac{1}{2}\tau^2 c^2 f^{(2)}(x) + \frac{1}{6}\tau^3 \frac{\partial^3 u(x, \mu)}{\partial t^3}. \qquad (12.103)$$

In particular, for $x = x_k$, this yields

$$u(x_k, t_1) = u(x_k, 0) + \tau g(x_k) + \frac{1}{2}\tau^2 c^2 f^{(2)}(x_k) + \frac{1}{6}\tau^3 \frac{\partial^3 u(x_k, \mu_k)}{\partial t^3}. \qquad (12.104)$$

If $f(x) \in C^4[0, L]$, then, for some $\zeta_k \in [x_{k-1}, x_{k+1}]$, we have

$$f^{(2)}(x_k) = \frac{f(x_{k+1}) - 2f(x_k) + f(x_{k-1})}{h^2} - \frac{h^2}{12}f^{(4)}(\zeta_k). \qquad (12.105)$$

Since $u(x_k, 0) = f(x_k)$ [recall (12.11) again], and if we substitute (12.105) into (12.104), we obtain

$$u(x_k, t_1) = f(x_k) + \tau g(x_k) + \frac{1}{2}\frac{c^2\tau^2}{h^2}[f(x_{k+1}) - 2f(x_k) + f(x_{k-1})]$$

$$+ O(\tau^3 + \tau^2 h^2). \tag{12.106}$$

Via (12.98) this yields the required approximation

$$u_{k,1} = f(x_k) + \tau g(x_k) + \tfrac{1}{2}e^2[f(x_{k+1}) - 2f(x_k) + f(x_{k-1})]$$

or

$$u_{k,1} = (1 - e^2)f(x_k) + \tfrac{1}{2}e^2[f(x_{k+1}) + f(x_{k-1})] + \tau g(x_k). \tag{12.107}$$

Taking account of the boundary conditions (12.100), Eq. (12.99) can be expressed in matrix form as

$$
\begin{bmatrix}
u_{1,n+1} \\
u_{2,n+1} \\
\vdots \\
u_{M-2,n+1} \\
u_{M-1,n+1}
\end{bmatrix}
=
\begin{bmatrix}
2(1-e^2) & e^2 & 0 & \cdots & 0 & 0 \\
e^2 & 2(1-e^2) & e^2 & \cdots & 0 & 0 \\
\vdots & \vdots & \vdots & & \vdots & \vdots \\
0 & 0 & 0 & \cdots & 2(1-e^2) & e^2 \\
0 & 0 & 0 & \cdots & e^2 & 2(1-e^2)
\end{bmatrix}
$$

$$
\times
\begin{bmatrix}
u_{1,n} \\
u_{2,n} \\
\vdots \\
u_{M-2,n} \\
u_{M-1,n}
\end{bmatrix}
-
\begin{bmatrix}
u_{1,n-1} \\
u_{2,n-1} \\
\vdots \\
u_{M-2,n-1} \\
u_{M-1,n-1}
\end{bmatrix}. \tag{12.108}
$$

This matrix recursion is run for $n = 1, 2, 3, \ldots$, and the initial conditions are provided by (12.101) and (12.107).

We recall from Chapter 10 that numerical methods for the solution of ODE IVPs can be unstable. The same problem can arise in the numerical solution of PDEs. In particular, as the FD method is effectively an explicit method, it can certainly become unstable if h and τ are inappropriately selected.

As noted by others [2, 13], we will have

$$\lim_{h,\tau \to 0} u_{k,n} = u(hk, \tau n)$$

provided that $0 < e \leq 1$. This is the famous *Courant–Friedrichs–Lewy* (CFL) *condition* for the stability of the FD method, and is originally due to Courant et al. [23]. The special case where $e = 1$ is interesting and easy to analyze. In this case (12.99) reduces to

$$u_{k,n+1} = u_{k+1,n} - u_{k,n-1} + u_{k-1,n}. \tag{12.109}$$

We recall that $u(x, t)$ has the form

$$u(x, t) = v(x - ct) + w(x + ct)$$

[see (12.50)]. Hence $u(x_k, t_n) = v(x_k - ct_n) + w(x_k + ct_n)$. Observe that, since $h = c\tau$ (as $e = 1$), we have

$$
\begin{aligned}
u(x_{k+1}, t_n) - u(x_k, t_{n-1}) + u(x_{k-1}, t_n) &= v(x_k + h - ct_n) + w(x_k + h + ct_n) \\
&\quad - v(x_k - ct_n + c\tau) - w(x_k + ct_n - c\tau) \\
&\quad + v(x_k - h - ct_n) + w(x_k - h + ct_n) \\
&= v(x_k - h - ct_n) + w(x_k + h + ct_n) \\
&= v(x_k - ct_n - c\tau) + w(x_k + ct_n + c\tau) \\
&= v(x_k - ct_{n+1}) + w(x_k + ct_{n+1}) \\
&= u(x_k, t_{n+1}),
\end{aligned}
$$

or

$$u(x_k, t_{n+1}) = u(x_{k+1}, t_n) - u(x_k, t_{n-1}) + u(x_{k-1}, t_n). \tag{12.110}$$

Equation (12.110) has a form that is identical to that of (12.109). In other words, the algorithm (12.109) gives the exact solution to (12.9), but only at $x = hk$ and $t = \tau n$ with $h = c\tau$ (which is a rather restrictive situation).

A more general approach to error analysis that confirms the CFL condition is sometimes called *von Neumann stability analysis* [2]. We outline the approach as follows. We begin by defining the *global truncation error*

$$\epsilon_{k,n} = u(x_k, t_n) - u_{k,n}. \tag{12.111}$$

Via (12.96)

$$
\begin{aligned}
u(x_k, t_{n+1}) - 2u(x_k, t_n) + u(x_k, t_{n-1}) - e^2[u(x_{k+1}, t_n) - 2u(x_k, t_n) + u(x_{k-1}, t_n)] \\
= \tau^2 e_{k,n}. \tag{12.112}
\end{aligned}
$$

If we subtract (12.97) from (12.112) and simplify the result using (12.111), we obtain

$$\epsilon_{k,n+1} = 2(1 - e^2)\epsilon_{k,n} + e^2[\epsilon_{k+1,n} + \epsilon_{k-1,n}] - \epsilon_{k,n-1} + \tau^2 e_{k,n} \tag{12.113}$$

for $k = 0, 1, \ldots, M$ ($L = Mh$), and $n = 1, 2, 3, \ldots$. Equation (12.113) is a two-dimensional difference equation for the global error sequence $(\epsilon_{k,n})$. The term $\tau^2 e_{k,n}$ is a *forcing term*, and if $u(x, t)$ is smooth enough, the forcing term will be bounded for all k and n. Basically, we can show that the CFL condition $0 < e \leq 1$ prevents $\lim_{n \to \infty} |\epsilon_{k,n}| = \infty$ for all $k = 0, 1, \ldots, M$. Analogously to our stability analysis approach for ODE IVPs from Chapter 10, we may consider the homogeneous problem

$$\epsilon_{k,n+1} = 2(1 - e^2)\epsilon_{k,n} + e^2[\epsilon_{k+1,n} + \epsilon_{k-1,n}] - \epsilon_{k,n-1}, \tag{12.114}$$

which is just (12.113) with the forcing term made identically zero for all k and n. In Section 12.3.1 we learned that separation of variables was a useful means to solve (12.9). We therefore believe that a discrete version of this approach is helpful at solving (12.114). To this end we postulate a typical solution of (12.114) of the form

$$\epsilon_{k,n} = \exp[j\alpha kh + \beta n\tau] \tag{12.115}$$

for suitable constants $\alpha \in \mathbf{R}$ and $\beta \in \mathbf{C}$. We note that (12.115) has similarities to (12.28) and is really a term in a discrete form of Fourier series expansion. We also see that

$$|\epsilon_{k,n}| = |\exp(\beta n\tau)| = |s^n|.$$

Thus, if $|s| \leq 1$, we will not have unbounded growth of the error sequence $(\epsilon_{k,n})$ as n increases. If we now substitute (12.115) into (12.114), we obtain (after simplification) the *characteristic equation*

$$s^2 - \underbrace{[2(1 - e^2) + 2\cos(\alpha h)e^2]}_{=2b}s + 1 = 0. \tag{12.116}$$

Using the identity $2\sin^2 x = 1 - \cos(2x)$, we obtain

$$b = 1 - 2e^2 \sin^2\left(\frac{\alpha h}{2}\right). \tag{12.117}$$

It is easy to confirm that $|b| \leq 1$ for all e such that $0 \leq e \leq 1$ because $0 \leq \sin^2(\alpha h/2) \leq 1$ for all all $\alpha h \in \mathbf{R}$. We note that $s^2 - 2bs + 1 = 0$ for $s = s_1, s_2$, where

$$s_1 = b + \sqrt{b^2 - 1}, \quad s_2 = b - \sqrt{b^2 - 1}. \tag{12.118}$$

If $|b| > 1$, then $|s_k| > 1$ for some $k \in \{1, 2\}$, which can happen if we permit $e > 1$. Naturally we reject this choice as it yields unbounded growth in the size of $\epsilon_{k,n}$ as $n \to \infty$. If $|b| \leq 1$, then clearly $|s_k| = 1$ for all k. (To see this, consider the product $s_1 s_2 = s_1 s_1^* = |s_1|^2$.) This prevents unbounded growth of $\epsilon_{k,n}$. Thus, we have validated the CFL condition for the selection of Courant parameter e (i.e., we must always choose e to satisfy $0 < e \leq 1$).

Example 12.4   Figure 12.4 illustrates the application of the recursion (12.108) to the vibrating string problem of Example 12.1. The simulation parameters are stated in the figure caption. The reader should compare the approximate solution

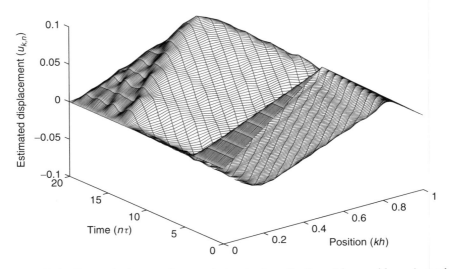

Figure 12.4 FD method approximate solution to the vibrating string problem. A mesh plot of $u_{k,n}$ as given by Eq. (12.108) for the parameters $L = 1, c/L = \frac{1}{8}$, and $H = \frac{1}{10}$. Additionally, $h = 0.05$ and $\tau = 0.1$, which meets the CFL criterion for stability of the simulation.

of Fig. 12.4 to the exact solution of Fig. 12.2. The apparent loss of accuracy as the number of time steps increases (i.e., with increasing $n\tau$) is due to the phenomenon of *numerical dispersion* [22], a topic considered in the next section in the context of the FDTD method. Of course, simulation accuracy improves as $h, \tau \to 0$ for fixed $T = N\tau$, and $L = Mh$.

12.5 THE FINITE-DIFFERENCE TIME-DOMAIN (FDTD) METHOD

The FDTD method is often attributed to Yee [14]. It is a finite-difference scheme just as the FD method of Section 12.4 is a finite-difference scheme. However, it is of such a nature as to be particularly useful in solving hyperbolic PDEs where the boundary conditions are at infinity (i.e., wave propagation problems of the kind considered in Section 12.3.2).

The FDTD method considers approximations to $H_z(x, t)$ and $E_y(x, t)$ given by applying the central difference and forward difference approximations to the first derivatives in the PDEs (12.69) and (12.70). We will use the following notation for the sampling of continuous functions such as $f(x, t)$:

$$f_{k,n} \approx f(k\Delta x, n\Delta t), \quad f_{k+\frac{1}{2}, n+\frac{1}{2}} \approx f((k + \tfrac{1}{2})\Delta x, (n + \tfrac{1}{2})\Delta t) \quad (12.119)$$

(so Δx replaces h, and Δt replaces τ here, where h and τ were the grid spacings used in previous sections). For convenience, let $E = E_y$ and $H = H_z$ (i.e., we drop the subscripts on the field components). We approximate the derivatives in (12.69) and (12.70) specifically according to

$$\frac{\partial H}{\partial t} \approx \frac{1}{\Delta t}[H_{k+\frac{1}{2},n+\frac{1}{2}} - H_{k+\frac{1}{2},n-\frac{1}{2}}], \qquad (12.120a)$$

$$\frac{\partial E}{\partial t} \approx \frac{1}{\Delta t}[E_{k,n+1} - E_{k,n}], \qquad (12.120b)$$

$$\frac{\partial H}{\partial x} \approx \frac{1}{\Delta x}[H_{k+\frac{1}{2},n+\frac{1}{2}} - H_{k-\frac{1}{2},n+\frac{1}{2}}], \qquad (12.120c)$$

$$\frac{\partial E}{\partial x} \approx \frac{1}{\Delta t}[E_{k+1,n} - E_{k,n}]. \qquad (12.120d)$$

Define $\epsilon_k = \epsilon(k\Delta x)$, $\mu_k = \mu((k + \frac{1}{2})\Delta x)$, $\sigma_k = \sigma(k\Delta x)$, and $\sigma_k^* = \sigma^*((k + \frac{1}{2})$ $\Delta x)$, which assumes the general situation where the material parameters vary with $x \in [0, L]$ (*computational region*). Substituting these discretized material parameters, and (12.120) into (12.69) and (12.70), we obtain the following algorithm:

$$H_{k+\frac{1}{2},n+\frac{1}{2}} = \left[1 - \frac{\sigma_k^*}{\mu_k}\Delta t\right]H_{k+\frac{1}{2},n-\frac{1}{2}} - \frac{1}{\mu_k}\frac{\Delta t}{\Delta x}[E_{k+1,n} - E_{k,n}], \qquad (12.121a)$$

$$E_{k,n+1} = \left[1 - \frac{\sigma_k}{\epsilon_k}\Delta t\right]E_{k,n} - \frac{1}{\epsilon_k}\frac{\Delta t}{\Delta x}[H_{k+\frac{1}{2},n+\frac{1}{2}} - H_{k-\frac{1}{2},n+\frac{1}{2}}]. \qquad (12.121b)$$

This is sometimes called the *leapfrog algorithm*. The dependencies between the estimated field components in (12.121) are illustrated in Fig. 12.5. If we assume that

$$H(-\tfrac{1}{2}\Delta x, t) = H((M + \tfrac{1}{2})\Delta x, t) = 0 \qquad (12.122)$$

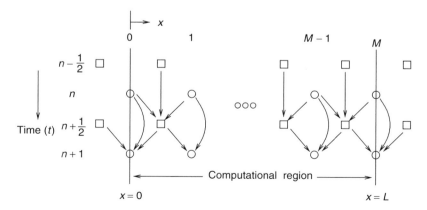

Figure 12.5   An illustration of the dependencies between the approximate field components given by (12.121a,b); the lines with arrows denote the "flow" of these dependencies [○ electric field component (E_y); □ magnetic field component (H_z)].

for all $t \in \mathbf{R}^+$, then a more detailed pseudocode description of the *FDTD algorithm* is

$H_{k+\frac{1}{2},-\frac{1}{2}} := 0$ for $k = 0, 1, \ldots, M-1$;
$E_{k,0} := 0$ for $k = 0, 1, \ldots, M$;
for $n := 0$ to $N-1$ do begin
  $E_{m,n} := E_0 \sin(\omega n \Delta t)$; $\{0 < m < M\}$
  for $k := 0$ to $M-1$ do begin

$$H_{k+\frac{1}{2},n+\frac{1}{2}} := \left[1 - \frac{\sigma_k^*}{\mu_k}\Delta t\right] H_{k+\frac{1}{2},n-\frac{1}{2}} - \frac{1}{\mu_k}\frac{\Delta t}{\Delta x}[E_{k+1,n} - E_{k,n}];$$

  end;
  for $k := 0$ to M do begin

$$E_{k,n+1} := \left[1 - \frac{\sigma_k}{\epsilon_k}\Delta t\right] E_{k,n} - \frac{1}{\epsilon_k}\frac{\Delta t}{\Delta x}[H_{k+\frac{1}{2},n+\frac{1}{2}} - H_{k-\frac{1}{2},n+\frac{1}{2}}];$$

  end;
end;

The statement $E_{m,n} := E_0 \sin(\omega n \Delta t)$ simulates an antenna that broadcasts a sinusoidal electromagnetic wave from the location $x = m\Delta x \in (0, L)$. Of course, the antenna must be located in free space.

The FDTD algorithm is an explicit difference scheme, and so it may have stability problems. However, it can be shown that the algorithm is stable provided we have

$$e = c\frac{\Delta t}{\Delta x} \leq 1 \left(\text{or } \Delta t \leq \frac{\Delta x}{c}\right), \tag{12.123}$$

Thus, the CFL condition of Section 12.4 applies to the FDTD algorithm as well. A justification of this claim appears in Taflove [8]. A MATLAB implementation of the FDTD algorithm may be found in Appendix 12.A (see routine FDTD.m). In this implementation we have introduced the parameters s_x and s_t $(0 < s_x, s_t < 1)$ such that

$$\Delta x = s_x \lambda_0, \quad \Delta t = s_t \frac{\Delta x}{c}. \tag{12.124}$$

Clearly, $c\Delta t/\Delta x = s_t$, and so the CFL condition is met. Also, spatial sampling is determined by $\Delta x = s_x \lambda_0$, which is some fraction of a free-space wavelength λ_0 [recall (12.55)]. Note that the algorithm simulates the field for all $t \in [0, T]$, where $T = N\Delta t$. If the wave is propagating only through free space, then the wave will travel a distance

$$D = cT = Ns_t s_x \lambda_0, \tag{12.125}$$

that is, the distance traveled is $Ns_t s_x$ free-space wavelengths. Since $L = Ms_x \lambda_0$ (i.e., the computational region spans Ms_x free-space wavelengths), this allows us to make a reasonable choice for N.

A problem with the FDTD algorithm is that even if the computational region is only free-space, a wave launched from location $x = m\Delta x \in (0, L)$ will eventually strike the boundaries at $x = 0$ and/or $x = L$, and so will be reflected back toward the source. These reflections will cause very large errors in the estimates of H, and

E. But we know from Section 12.3.2 that we may design absorbing layers called *perfectly matched layers* (PMLs) that suppress these reflections.

Suppose that the PML has physical parameters μ, ϵ, σ, and σ^*, then, from (12.73), the PML will have a propagation constant given by

$$\gamma^2 = (\sigma\sigma^* - \omega^2\mu\epsilon) + j\omega(\sigma\mu + \sigma^*\epsilon). \tag{12.126}$$

If we enforce the condition (12.74), namely

$$\frac{\sigma^*}{\mu} = \frac{\sigma}{\epsilon}, \tag{12.127}$$

then (12.126) becomes

$$\gamma^2 = \frac{\mu}{\epsilon}(\sigma^2 - \omega^2\epsilon^2) + 2j\omega\sigma\mu. \tag{12.128}$$

If we enforce $\sigma^2 - \omega^2\epsilon^2 \le 0$, then $\gamma = \alpha + j\beta$, where

$$\alpha = \sqrt{\frac{\mu}{\epsilon}}[\sigma^2 + \omega^2\epsilon^2]^{1/2} \cos\left[\frac{\pi}{4} + \frac{1}{2}\mathrm{Tan}^{-1}\left(\frac{\omega^2\epsilon^2 - \sigma^2}{2\omega\sigma\epsilon}\right)\right]. \tag{12.129}$$

Equation (12.129) is obtained by the same arguments that yielded (12.89). As $\alpha > 0$, then a wave on entering the PML will be attenuated by a factor $e^{-\alpha x}$, where x is the depth of penetration of the wave into the PML. A particularly simple choice for α is to let $\sigma^2 = \omega^2\epsilon^2$, in which case

$$\alpha = \omega\sqrt{\mu\epsilon}. \tag{12.130}$$

Since $\omega = \frac{2\pi}{\lambda_0}c$ and $c = \frac{1}{\sqrt{\mu_0\epsilon_0}}$, with $\mu = \mu_r\mu_0$ and $\epsilon = \epsilon_r\epsilon_0$, we can rewrite (12.130) as

$$\alpha = \sqrt{\mu_r\epsilon_r}\frac{2\pi}{\lambda_0}. \tag{12.131}$$

If we are matching the PML to free space, then $\mu_r = \epsilon_r = 1$, and so $\alpha = \frac{2\pi}{\lambda_0}$, in which case

$$e^{-\alpha x} = e^{-\frac{2\pi}{\lambda_0}x}. \tag{12.132}$$

If the PML is of thickness $x = 2\lambda_0$, then, from (12.132) we have $e^{-\alpha x} = e^{-4\pi} \approx 3.5 \times 10^{-6}$. A PML that is two free-space wavelengths thick will therefore absorb very nearly all of the radiation incident on it at the wavelength λ_0. Since we have chosen $\sigma^2 = \omega^2\epsilon^2$, it is easy to confirm that

$$\sigma = \frac{2\pi}{\lambda_0}\frac{\epsilon_r}{Z_0}, \qquad \sigma^* = \frac{2\pi}{\lambda_0}\mu_r Z_0 \tag{12.133}$$

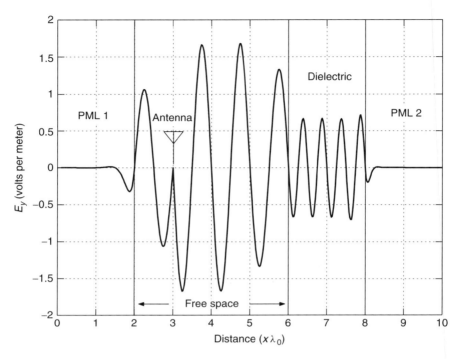

Figure 12.6 Typical output from FDTD.m (see Appendix 12.A) for the system described in Example 12.5. The antenna broadcasts a sinusoid of wavelength $\lambda_0 = 500$ nm (nanometers) from the location $x = 3\lambda_0$. The transmitted field strength is $E_0 = 1$ V/m.

[via (12.127)], where we recall that $Z_0 = \sqrt{\mu_0/\epsilon_0}$. Routine FDTD.m in Appendix 12.A implements PMLs according to this approach.

Example 12.5   Figure 12.6 illustrates a typical output from FDTD.m (Appendix 12.A). The system shown in the figure occupies a computational region of length $10\lambda_0$ (i.e., $x \in [0, 10\lambda_0]$). Somewhat arbitrarily we have $\lambda_0 = 500$ nm (nanometers). The antenna (which, given the wavelength, could be a laser) is located at index $m = 150$ (i.e., is at $x = m\Delta x = m s_x \lambda_0 = 3\lambda_0$, since $s_x = .02$). The free-space region is for $x \in (2\lambda_0, 6\lambda_0)$. The lossless dielectric occupies $x \in [6\lambda_0, 8\lambda_0]$, and has a relative permittivity of $\epsilon_r = 4$. The entire computational region is nonmagnetic, and so we have $\mu = \mu_0$ everywhere. Clearly, PML 1 is matched to free space, while PML 2 is matched to the dielectric.

Since $\epsilon_r = 4$, according to (12.82), the transmission coefficient from free space into the dielectric is

$$\tau = \frac{2\sqrt{\dfrac{\mu_0}{\epsilon_r \epsilon_0}}}{\sqrt{\dfrac{\mu_0}{\epsilon_r \epsilon_0}} + \sqrt{\dfrac{\mu_0}{\epsilon_0}}} = \frac{2}{1 + \sqrt{\epsilon_r}} = \frac{2}{3}.$$

Since $E_0 = 1$ V/m, the amplitude of the electric field within the dielectric must be $\tau E_0 = \frac{2}{3}$ V/m. From Fig. 12.6 the reader can see that the electric field within the dielectric does indeed have an amplitude of about $\frac{2}{3}$ V/m to a good approximation. From (12.57) the wavelength within the dielectric material is

$$\lambda = \frac{1}{\sqrt{\epsilon_r}}\lambda_0 = \frac{1}{2}\lambda_0.$$

Again from Fig. 12.6 we see that the wavelength of the transmitted field is indeed close to $\frac{1}{2}\lambda_0$ within the dielectric.

We observe that the PMLs in Example 12.5 do not perfectly suppress reflections at their boundaries. For example, the wave crest closest to the interface between PML 2 and the dielectric, and that lies within the dielectric, is somewhat higher than it should be. It is the discretization of a continuous space that has lead to these *residual reflections*.

The theory of PMLs presented here does not easily extend from electromagnetic wave propagation problems in one spatial dimension into propagation problems in two or three spatial dimensions. It appears that the first truly successful extension of PML theory to higher spatial dimensions is due to Bérenger [15,16]. Wu and Fang [17] claim to have improved the theory still further by improving the suppression of the residual reflections noted above.

The problem of *numerical dispersion* was mentioned in Example 12.4 in the application of the FD method to the simulation of a vibrating string. We conclude this chapter with an account of the problem based mainly on the work of Trefethen [22]. We will assume lossless propagation, so $\sigma = \sigma^* = 0$ in (12.121). We will also assume that the computational region is free space, so $\mu = \mu_0$, and $\epsilon = \epsilon_0$ everywhere. If we now substitute $E(x, t) = E_0 \sin(\omega t - \beta x)$ and $H(x, t) = H_0 \sin(\omega t - \beta x)$ into either of (12.121a) or (12.121b), apply the appropriate trigonometric identities, and then cancel out common factors, we obtain the identity

$$\sin\left(\frac{\omega \Delta t}{2}\right) = e \sin\left(\frac{\beta \Delta x}{2}\right). \qquad (12.134)$$

We may use (12.134) and (12.123) to obtain

$$v_p = \frac{\omega}{\beta} = \frac{c}{e}\frac{2}{\beta \Delta x}\sin^{-1}\left[e \sin\left(\frac{\beta \Delta x}{2}\right)\right] \qquad (12.135)$$

which is the *phase speed* of the wave of wavelength λ_0 (recall $\beta = 2\pi/\lambda_0$) in the FDTD method. For the *continuous wave* $E(x, t)$ [or, for that matter, $H(x, t)$], recall from Section 12.3.2 that $\omega/\beta = c$, so without spatial or temporal discretization effects, a sinusoid will propagate through free space at the speed c regardless of its wavelength (or, equivalently, its frequency). However, (12.135) suggests that the speed of an FDTD-simulated sinusoidal wave will vary with the wavelength. As explained in Ref. 22 (or see Ref. 12), the *group speed*

$$v_g = \frac{d\omega}{d\beta} = \frac{d}{d\beta}(\beta v_p) = c \frac{\cos\left(\frac{\beta \Delta x}{2}\right)}{\sqrt{1 - e^2 \sin^2\left(\frac{\beta \Delta x}{2}\right)}} \tag{12.136}$$

is more relevant to assessing how propagation speed varies with the wavelength. Again, in free space the continuous wave propagates at the group speed $\frac{d\omega}{d\beta} = \frac{d}{d\beta}(\beta c) = c$. A plot of v_g/c versus $\lambda_0/\Delta x$ [with v_g/c given by (12.136)] appears in Fig. 2 of Represa et al. [19]. It shows that short-wavelength sinusoids travel at slower speeds than do long-wavelength sinusoids when simulated using the FDTD method.

We have seen that nonsinusoidal waveforms (e.g., the triangle wave of Example 12.1) are a superposition of sinusoids of varying frequency. Thus, if we use the FDTD method, the FD method, or indeed *any* numerical method to simulate wave propagation, we will see the effects of *numerical dispersion*. In other words, the various frequency components in the wave will travel at different speeds, and so the original shape of the wave will become lost as the simulation progresses in time (i.e., as $n\Delta t$ increases). Figure 12.7 illustrates this for the case of two

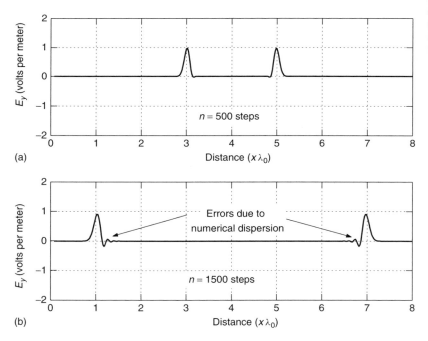

(a)

(b)

Figure 12.7 Numerical dispersion in the FDTD method as illustrated by the propagation of two Gaussian pulses. The medium is free space.

Gaussian pulses traveling in opposite directions. The two pulses originally appeared at $x = 4\lambda_0$, and the medium is free space. For $N = 1500$ time steps, there is a very noticeable error due to the "breakup" of the pulses as their constituent frequency components separate out as a result of the numerical dispersion.

In closing, note that more examples of numerical dispersion may be found in Luebbers et al. [18]. Shin and Nevels [21] explain how to work with Gaussian test pulses to reduce numerical dispersion. We mention that Represa et al. [19] use absorbing boundary conditions based on the theory in Mur [20], which is a different method from the PML approach we have used in this book.

APPENDIX 12.A   MATLAB CODE FOR EXAMPLE 12.5

```
%
%                       permittivity.m
%
% This routine specifies the permittivity profile of the
% computational region [0,L], and is needed by FDTD.m
%

function epsilon = permittivity(k,sx,lambda0,M)

epsilon0 = 8.854185*1e-12; % free-space permittivity
er1 = 4;                   % relative permittivity of the dielectric
Dx = sx*lambda0;           % this is Delta x
x = k*Dx;                  % position at which we determine epsilon
L = M*Dx;                  % location of right end of computational
                           % region

if ((x >= 0) & (x < (L-4*lambda0)))
  epsilon = epsilon0;
else
  epsilon = er1*epsilon0;
  end;
%
%                       permeability.m
%
% This routine specifies the permeability profile of the
% computational region [0,L], and is needed by FDTD.m
%

function  mu = permeability(k,sx,lambda0,M)

mu0 = 400*pi*1e-9;        % free-space permeability
Dx = sx*lambda0;          % this is Delta x
x = k*Dx;                 % position at which we determine mu
L = M*Dx;                 % location of right end of computational
                          % region

mu = mu0;
%
%                       econductivity.m
%
% This routine specifies the electrical conductivity profile
```

```
% of the computational region [0,L], and is needed by FDTD.m
%

function sigma = econductivity(k,sx,lambda0,M)

epsilon0 = 8.854185*1e-12; % free-space permittivity
mu0 = 400*pi*1e-9;         % free-space permeability
er1 = 4;                   % dielectric relative permittivity
epsilon1 = er1*epsilon0;   % dielectric permittivity
Z0 = sqrt(mu0/epsilon0);   % free-space impedance
Dx = sx*lambda0;           % this is Delta x
x = k*Dx;                  % position at which we determine sigma
L = M*Dx;                  % location of right end of computational
                           % region

star1 = (2*pi/lambda0)*(1/Z0); % conductivity of PML 1 (at x = 0 end)
star2 = er1*star1;             % conductivity of PML 2 (at x = L end)
if ((x > 2*lambda0) & (x < L-2*lambda0))
  sigma = 0;
elseif (x <= 2*lambda0)
  sigma = star1;
elseif (x >= (L - 2*lambda0))
  sigma = star2;
  end;
%
%                   mconductivity.m
%
% This routine specifies the magnetic conductivity profile of the
% computational region [0,L], and is needed by FDTD.m
%

function sigmastar = mconductivity(k,sx,lambda0,M)

epsilon0 = 8.854185*1e-12; % free-space permittivity
mu0 = 400*pi*1e-9;         % free-space permeability
Z0 = sqrt(mu0/epsilon0);   % free-space impedance
Dx = sx*lambda0;           % this is Delta x
x = (k+.5)*Dx;             % position at which we determine sigmastar
L = M*Dx;                  % location of right end of computational
                           % region

star = (2*pi/lambda0)*Z0;
if ((x > 2*lambda0) & (x < L-2*lambda0))
  sigmastar = 0;
else
  sigmastar = star;
  end;
%
%                         FDTD.m
%
% This routine produces the plot in Fig. 12.6  which is associated with
% Example 12.5. Thus, it illustrates the FDTD method.
%
% The routine returns the total electric field component Ey to the
% caller.
%
```

```
function Ey = FDTD

mu0 = 400*pi*1e-9;          % free-space permeability
epsilon0 = 8.854185*1e-12;  % free-space permittivity
c = 1/sqrt(mu0*epsilon0);   % speed of light in free-space

lambda0 = 500;              % free-space wavelength of the source
                            % in nanometers
lambda0 = lambda0*1e-9;     % free-space wavelength of the source
                            % in meters
beta0 = (2*pi)/lambda0;     % free-space beta (wavenumber)

sx = .02;                   % fraction of a free-space wavelength
                            % used to determine Delta x
st = .10;                   % scale factor used to determine time-step
                            % size Delta t
Dx = sx*lambda0;            % Delta x
Dt = (st/c)*Dx;             % application of the CFL condition
                            % to determine time-step Delta t

E0 = 1;                     % amplitude of the electric field (V/m)
                            % generated by the source
m = 150;                    % source (antenna) location index
omega = beta0*c;            % source frequency (radians/second)

M = 500;                    % number of spatial grid points is (M+1)
N = 4000;                   % the number of time steps in the simulation

E = zeros(1,M+1);           % initial electric field
H = zeros(1,M+2);           % initial magnetic field

    % Specify the material properties in the computational region
    % (which is x in [0,M*Dx])

for k = 0:M
  epsilon(k+1) = permittivity(k,sx,lambda0,M);
  ce(k+1) = 1 - Dt*econductivity(k,sx,lambda0,M)/epsilon(k+1);
  h(k+1) = 1/epsilon(k+1);
  end;
h = h*Dt/Dx;
for k = 0:(M-1)
  mu(k+1) = permeability(k,sx,lambda0,M);
  ch(k+1) = 1 - Dt*mconductivity(k,sx,lambda0,M)/mu(k+1);
  e(k+1) = 1/mu(k+1);
  end;
e = e*Dt/Dx;

    % Run the simulation for N time steps

for n = 1:N
  E(m+1) = E0*sin(omega*(n-1)*Dt);  % Antenna is at index m
  H(2:M+1) = ch.*H(2:M+1) - e.*(E(2:M+1) - E(1:M));
  E(1:M+1) = ce.*E(1:M+1) - h.*(H(2:M+2) - H(1:M+1));
  E(m+1) = E0*sin(omega*n*Dt);
  Ey(n,:) = E;  % Save the total electric field at time step n
  end;
```

REFERENCES

1. E. Kreyszig, *Advanced Engineering Mathematics*, 4th ed., Wiley, New York, 1979.

2. T. Myint-U and L. Debnath, *Partial Differential Equations for Scientists and Engineers*, 3rd ed., North-Holland, New York, 1987.

3. R. L. Burden and J. D. Faires, *Numerical Analysis*, 4th ed., PWS-KENT Publ., Boston, MA, 1989.

4. A. Quarteroni, R. Sacco and F. Saleri, *Numerical Mathematics* (Texts in Applied Mathematics series, Vol. 37), Springer-Verlag, New York, 2000.

5. G. Strang and G. Fix, *An Analysis of the Finite Element Method*, Prentice-Hall, Englewood Cliffs, NJ, 1973.

6. S. Brenner and R. Scott, *The Mathematical Theory of Finite Element Methods*, Springer-Verlag, New York, 1994.

7. C. T. Kelley, *Iterative Methods for Linear and Nonlinear Equations*, SIAM, Philadelphia, PA, 1995.

8. A. Taflove, *Computational Electrodynamics: The Finite-Difference Time-Domain Method*, Artech House, Norwood, MA, 1995.

9. R. Courant and D. Hilbert, *Methods of Mathematical Physics*, Vol. II. *Partial Differential Equations*, Wiley, New York, 1962.

10. J. D. Kraus and K. R. Carver, *Electromagnetics*, 2nd ed., McGraw-Hill, New York, 1973.

11. J. F. Epperson, *An Introduction to Numerical Methods and Analysis*, Wiley, New York, 2002.

12. W. C. Elmore and M. A. Heald, *Physics of Waves*, Dover Publ., New York, 1969.

13. E. Isaacson and H. B. Keller, *Analysis of Numerical Methods*, Wiley, New York, 1966.

14. K. S. Yee, "Numerical Solution of Initial Boundary Value Problems Involving Maxwell's Equations in Isotropic Media," *IEEE Trans. Antennas Propag.* **AP-14**, 302–307 (May 1966).

15. J.-P. Bérenger, "A Perfectly Matched Layer for the Absorption of Electromagnetic Waves," *J. Comput. Phys.* **114**, 185–200 (1994).

16. J.-P. Bérenger, "Perfectly Matched Layer for the FDTD Solution of Wave-Structure Interaction Problems," *IEEE Trans. Antennas Propag.* **44**, 110–117 (Jan. 1996).

17. Z. Wu and J. Fang, "High-Performance PML Algorithms," *IEEE Microwave Guided Wave Lett.* **6**, 335–337 (Sept. 1996).

18. R. J. Luebbers, K. S. Kunz and K. A. Chamberlin, "An Interactive Demonstration of Electromagnetic Wave Propagation Using Time-Domain Finite Differences," *IEEE Trans. Educ.* **33**, 60–68 (Feb. 1990).

19. J. Represa, C. Pereira, M. Panizo and F. Tadeo, "A Simple Demonstration of Numerical Dispersion under FDTD," *IEEE Trans. Educ.* **40**, 98–102 (Feb. 1997).

20. G. Mur, "Absorbing Boundary Conditions for the Finite-Difference Approximation of the Time-Domain Electromagnetic-Field Equations," *IEEE Trans. Electromagn. Compat.* **EMC-23**, 377–382 (Nov. 1981).

21. C.-S. Shin and R. Nevels, "Optimizing the Gaussian Excitation Function in the Finite Difference Time Domain Method," *IEEE Trans. Educ.* **45**, 15–18 (Feb. 2002).

22. L. N. Trefethen, "Group Velocity in Finite Difference Schemes," *SIAM Rev.* **24**, 113–136 (April 1982).

23. R. Courant, K. Friedrichs and H. Lewy, "Über die Partiellen Differenzengleichungen der Mathematischen Physik," *Math. Ann.* **100**, 32–74 (1928).

PROBLEMS

12.1. Classify the following PDEs into elliptic, parabolic, and hyperbolic types (or a combination of types).

(a) $3u_{xx} + 5u_{xy} + u_{yy} = x + y$

(b) $u_{xx} - u_{xy} + 2u_{yy} = u_x + u$

(c) $yu_{xx} + u_{yy} = 0$

(d) $y^2 u_{xx} - 2xy u_{xy} + x^2 u_{yy} = 0$

(e) $4x^2 u_{xx} + u_{yy} = u$

12.2. Derive (12.5) (both equations).

12.3. In (12.6) let $h = \tau$, and let $N = M = 4$, and for convenience let $f_{k,n} = \rho(x_k, y_n)/\epsilon$. Define the vectors

$$V = [V_{1,1} \; V_{1,2} \; V_{1,3} \; V_{2,1} \; V_{2,2} \; V_{2,3} \; V_{3,1} \; V_{3,2} \; V_{3,3}],$$

$$f = [f_{1,1} \; f_{1,2} \; f_{1,3} \; f_{2,1} \; f_{2,2} \; f_{2,3} \; f_{3,1} \; f_{3,2} \; f_{3,3}].$$

Find the matrix A such that $AV = h^2 f$. Assume that

$$V_{0,n} = V_{k,0} = V_{k,4} = V_{4,n} = 0$$

for all k, and n.

12.4. In (12.6) let $h = \tau = \frac{1}{4}$, and $N = M = 4$, and assume that $\rho(x_k, y_n) = 0$ for all k, and n. Let $x_0 = y_0 = 0$. Suppose that

$$V(0, y) = V(x, 0) = 0, \; V(x, 1) = x, \; V(1, y) = y.$$

Find the linear system of equations for $V_{k,n}$ with $1 \le k, n \le 3$, and put it in matrix form.

12.5. Recall the previous problem.

(a) Write a MATLAB routine to implement the Gauss–Seidel method (recall Section 4.7). Use your routine to solve the linear system of equations in the previous problem.

(b) Find the exact solution to the PDE

$$V_{xx} + V_{yy} = 0$$

for the boundary conditions stated in the previous problem.

(c) Use the solution in (b) to find $V(k/4, n/4)$ at $1 \leq k, n \leq 3$, and compare to the results obtained from part (a). They should be the same. Explain why. [*Hint:* Consider the error terms in (12.5).]

12.6. The previous two problems suggest that the linear systems that arise in the numerical solution of elliptic PDEs are sparse, and so it is worth considering their solution using the iterative methods from Section 4.7. Recall that the iterative methods of Section 4.7 have the general form

$$x^{(k+1)} = Bx^{(k)} + f$$

[from (4.155)]. Show that

$$\frac{||x^{(k+1)} - x^{(k)}||_\infty}{||x^{(k)} - x^{(k-1)}||_\infty} \leq ||B||_\infty.$$

[*Comment:* Recalling (4.36c), it can be shown that $\rho(A) \leq ||A||_p$ [which is really another way of expressing (4.158)]. For example, this result can be used to estimate the spectral radius of B_J [Eq. (4.171)].]

12.7. Consider Eq. (12.9), with the initial and boundary conditions in (12.10) and (12.11), respectively.

(a) Consider the change of variables

$$\xi = x + ct, \quad \eta = x - ct$$

and $\phi(\xi, \eta)$ replaces $u(x, t)$ according to

$$\phi(\xi, \eta) = u\left(\frac{1}{2}(\xi + \eta), \frac{1}{2c}(\xi - \eta)\right). \tag{12.P.1}$$

Verify the derivative operator equivalences

$$\frac{\partial}{\partial x} \equiv \frac{\partial}{\partial \xi} + \frac{\partial}{\partial \eta}, \frac{1}{c}\frac{\partial}{\partial t} \equiv \frac{\partial}{\partial \xi} - \frac{\partial}{\partial \eta}. \tag{12.P.2}$$

[*Hint:* $\frac{\partial \phi}{\partial \xi} = \frac{\partial u}{\partial x}\frac{\partial x}{\partial \xi} + \frac{\partial u}{\partial t}\frac{\partial t}{\partial \xi}, \frac{\partial \phi}{\partial \eta} = \frac{\partial u}{\partial x}\frac{\partial x}{\partial \eta} + \frac{\partial u}{\partial t}\frac{\partial t}{\partial \eta}.$]

(b) Show that (12.9) is replaceable with

$$\frac{\partial^2 \phi}{\partial \xi \partial \eta} = 0. \tag{12.P.3}$$

(c) Show that the solution to (12.P.3) is of the form

$$\phi(\xi, \eta) = P(\xi) + Q(\eta),$$

and hence

$$u(x, t) = P(x + ct) + Q(x - ct),$$

where P and Q are arbitrary twice continuously differentiable functions.

(d) Show that

$$P(x) + Q(x) = f(x),$$

$$P^{(1)}(x) - Q^{(1)}(x) = \frac{1}{c}g(x).$$

(e) Use the facts from (d) to show that

$$u(x, t) = \frac{1}{2}[f(x + ct) + f(x - ct)] + \frac{1}{2c}\int_{x-ct}^{x+ct} g(s)\,ds.$$

12.8. In (12.11) suppose that

$$f(x) = \begin{cases} -\frac{H}{d}\left(x - \frac{L}{2} - d\right), & \frac{L}{2} \leq x \leq \frac{L}{2} + d \\ \frac{H}{d}\left(x - \frac{L}{2} + d\right), & \frac{L}{2} - d \leq x \leq \frac{L}{2} \\ 0, & \text{elsewhere} \end{cases},$$

where $0 < d \leq L/2$. Assume that $g(x) = 0$ for all x.

(a) Sketch $f(x)$.

(b) Write a MATLAB routine to implement the FD algorithm (12.108) for computing $u_{k,n}$. Write the routine in such a way that it is easy to change the parameters c, H, h, τ, N, M, and d. The routine must produce a mesh plot similar to Fig. 12.4 and a plot similar to Fig. 12.7 on the same page (i.e., make use of the subplot). The latter plot is to be of $u_{k,N}$ versus k. Try out your routine using the parameters $c = \frac{1}{8}$, $h = 0.05$, $\tau = 0.025$, $H = 0.1$, $d = L/10$, $M = 200$, and $N = 1100$ (recalling that $L = Mh$). Do you find numerical dispersion effects?

12.9. Repeat Example 12.1 for $f(x)$ and $g(x)$ in the previous problem.

12.10. Example 12.1 is about the "plucked string." Repeat Example 12.1 assuming that $f(x) = 0$ and

$$g(x) = \begin{cases} \frac{2V}{L}x, & 0 \leq x \leq \frac{L}{2} \\ \frac{2V}{L}(L - x), & \frac{L}{2} \leq x \leq L \end{cases}$$

This describes a "struck string."

12.11. The MATLAB routines in Appendix 12.A implement the FDTD method, and generate information for the plot in Fig. 12.6. However, the reflected field component for $2\lambda_0 \leq x \leq 6\lambda_0$ is not computed or displayed.

Modify the code(s) in Appendix 12.A to compute the reflected field component in the free-space region and to plot it. Verify that at the interface between the free-space region and the dielectric that $|\rho| = \frac{1}{3}$ (magnitude of the reflection coefficient). Of course, you will need to read the amplitude of the reflected component from your plot to do this.

12.12. Derive Eq. (12.134).

12.13. Modify the MATLAB code(s) in Appendix 12.A to generate a plot similar to Fig. 12.7. [*Hint:* $E_{k,0} = E_0 \exp\left[-\left(\frac{k-m}{i_w}\right)^2 \right]$ for $k = 0, 1, \ldots, M$ is the initial electric field. Set the initial magnetic field to zero for all k.]

12.14. Plot v_g/c versus $\lambda_0/\Delta x$ for v_g given by (12.136). Choose $e = 0.1, 0.5$, and 0.8, and plot all curves on the same graph. Use these curves to explain why the errors due to numerical dispersion (see Fig. 12.7) are worse on the side of the pulse opposite to its direction of travel.

13 An Introduction to MATLAB

13.1 INTRODUCTION

MATLAB is short for "matrix laboratory," and is an extremely powerful software tool[1] for the development and testing of algorithms over a wide range of fields including, but not limited to, control systems, signal processing, optimization, image processing, wavelet methods, probability and statistics, and symbolic computing. These various applications are generally divided up into toolboxes that typically must be licensed separately from the core package.

Many books have already been written that cover MATLAB in varying degrees of detail. Some, such as Nakamura [3], Quarteroni et al. [4], and Recktenwald [5], emphasize MATLAB with respect to numerical analysis and methods, but are otherwise fairly general. Other books emphasize MATLAB with respect to particular areas such as matrix analysis and methods (e.g., see Golub and Van Loan [1] or Hill [2]). Some books implicitly assume that the reader already knows MATLAB [1,4]. Others assume little or no previous knowledge on the part of the reader [2,3,5].

This chapter is certainly not a comprehensive treatment of the MATLAB tool, and is nothing more than a quick introduction to it. Thus, the reader will have to obtain other books on the subject, or consult the appropriate manuals for further information. MATLAB's online help facility is quite useful, too.

13.2 STARTUP

Once properly installed, MATLAB is often invoked (e.g., on a UNIX workstation with a cmdtool window open) by typing matlab, and hitting RETURN. A window under which MATLAB runs will appear. The MATLAB prompt also appears:

```
>>
```

MATLAB commands may then be entered and executed interactively.

If you wish to work with commands in M-files (discussed further below), then having two cmdtool windows open to the same working directory is usually desirable. One window would be used to run MATLAB, and the other would be used

---

[1]MATLAB is written in C, but is effectively a language on its own.

An Introduction to Numerical Analysis for Electrical and Computer Engineers, by C.J. Zarowski
ISBN 0-471-46737-5 © 2004 John Wiley & Sons, Inc.

to edit the M-files as needed (to either develop the algorithm in the file or to debug it).

13.3   SOME BASIC OPERATORS, OPERATIONS, AND FUNCTIONS

The MATLAB command

```
>> diary filename
```

will create the file *filename*, and every command you type in and run, and every result of this, will be stored in filename. This is useful for making a permanent record of a MATLAB session which can help in documentation and sometimes in debugging. When writing MATLAB M-files, always make sure to document your programs. The examples in Section 13.6 illustrate this.

MATLAB tends to work in the more or less intuitive way where matrix and/or vector operations are concerned. Of course, it is in the nature of this software tool to assume the user is already familiar with matrix analysis and methods before attempting to use it.

When a MATLAB command creates a vector as the output from some operation, it may be in the form of a column or a row vector, depending on the command. A typical MATLAB row vector is

```
>>x = [ 1 1 1];
>>
```

The semicolon at the end of a line prevents the printout of the result of the command at that line (this is useful in preventing display clutter). If you wish to turn it into a column vector, then type:

```
>>x = x.'
x =

    1
    1
    1
```

Making this conversion is sometimes necessary as the inputs to some routines need the vectors in either row or column format, and some routines do not care. Routines that do not care whether the input vector is row or column make the conversion to a consistent form internally. For example, the following command sequence (which can be stored in a file called an *M-file*) converts vector x into a row vector if it is not one already:

```
>> [N,M] = size(x);
>> if N ~= 1
>>    x = x.';
>>    end;
>>
```

In this routine N is the number of rows and M is the number of columns in x. (The size command also accepts matrices.) The related command length(x) will return the length of vector x. This can be very useful in FOR loops (below).

The addition of vectors works in the obvious way:

```
>> x = [ 1 1 1];
>> y = [ 2 -1 2 ];
>> x + y

ans =

      3     0    3
>>
```

In this routine, the answer might be saved in vector z by typing $>> z = x + y;$. Clearly, to add vectors without error means that they must be of the same size. MATLAB will generate an error message if matrix and vector objects are not dimensioned properly when operations are performed on them. The mismatching of array sizes is a very common error in MATLAB programming.

Matrices can be entered as follows:

```
>> A = [ 1 1 ; 1 2 ]

A =

      1     1
      1     2
>>
```

Again, addition or subtraction would occur in the expected manner. We can invert a matrix as follows:

```
>> inv(A)

ans =

      2    -1
     -1     1
>>
```

Operation det(A) will give the determinant of A, and $[L, U] = $ lu(A) will give the LU factorization of A (if it exists). Of course, there are many other routines for common matrix operations, and decompositions (QR decomposition, singular value decomposition, eigendecompositions, etc.). Compute $y = Ax + b$:

```
>> x = [ 1 -1 ].';
>> b = [ 2 3 ].';
>> y = A*x + b
```

```
y =

    2
    2

>>
```

The colon operator can extract parts of matrices and vectors. For example, to place the elements in rows j to k of column n of matrix B into vector x, use $>> x = B(j : k, n);$. To extract the element from row k and column n of matrix A, use $>> x = A(k, n);$. To raise something (including matrices) to a specific power, use $>> C = A^\wedge p$, for which p is the desired power. (This computes $C = A^p$.)

(*Note:* MATLAB indexes vectors and matrices beginning with 1.)

Unless the user overrides the defaults, variables i and j denote the square root of -1:

```
>> sqrt(-1)

ans =
        0 + 1.0000i

>>
```

Here, i and j are built-in constants. So, to enter a complex number, say, $z = 3 - 2j$, type

```
>> z = 3 - 2*i;
```

Observe

```
>> x = [ 1   1+i ];
>> x'

ans =

    1.0000
    1.0000 - 1.0000i

>>
```

So the transposition operator without the period gives the complex–conjugate transpose (Hermitian transpose). Note that besides i and j, another useful built-in constant is pi $(= \pi)$.

Floating-point numbers are entered as, for instance, $1.5e - 3$ (which is 1.5×10^{-3}). MATLAB agrees with IEEE floating-point conventions, and so 0/0 will result in NaN ("not a number") to more clearly indicate an undefined operation. An operation like 1/0 will result in Inf as an output.

We may summarize a few important operators, functions, and other terms:

Relational Operators	
$<$	less than
$<=$	less than or equal to
$>$	greater than
$>=$	greater than or equal to
$==$	equal to
$\sim=$	not equal to

Logical Operators	
&	AND
|	OR
$\sim$	NOT

Trigonometric Functions	
sin	sine
cos	cosine
tan	tangent
asin	arcsine
acos	arccosine
atan	arctangent
atan2	four quadrant arctangent
sinh	hyperbolic sine
cosh	hyperbolic cosine
tanh	hyperbolic tangent
asinh	hyperbolic arcsine
acosh	hyperbolic arccosine
atanh	hyperbolic arctangent

Elementary Mathematical Functions	
abs	absolute value
angle	phase angle (argument of a complex number)
sqrt	square root

Elementary Mathematical Functions	
real	real part of a complex number
imag	imaginary part of a complex number
conj	complex conjugate
rem	remainder or modulus
exp	exponential to base e
log	natural logarithm
log10	base-10 logarithm
round	round to nearest integer
fix	round toward zero
floor	round toward $-\infty$
ceil	round toward $+\infty$

In setting up the time axis for plotting things (discussed below), a useful command is illustrated by

```
>> y = [0:.2:1]

y =

      0    0.2000   0.4000   0.6000   0.8000   1.0000

>>
```

Thus, $[x:y:z]$ creates a row vector whose first element is x and whose last element is z (depending on step size y), where the elements in between are of the form $x + ky$ (where k is a positive integer).

It can also be useful to create vectors of zeros:

```
>> zeros(size([1:4]))

ans =

      0    0    0    0

>>
```

Or, alternatively, a simpler way is

```
>> zeros(1,4)

ans =

      0    0    0    0

>>
```

Using "zeros(n,m)" will result in an $n \times m$ matrix of zeros. Similarly, a vector (or matrix) containing only ones would be obtained using the MATLAB function called "ones."

13.4   WORKING WITH POLYNOMIALS

We have seen on many occasions that polynomials are vital to numerical analysis and methods. MATLAB has nice tools for dealing with these objects.

Suppose that we have polynomials $P_1(s) = s + 2$ and $P_2(s) = 3s + 4$ and wish to multiply them. In this case type the following command sequence:

```
>> P1 = [ 1 2 ];
>> P2 = [ 3 4 ];
>> conv(P1,P2)

ans =

      3  10  8

>>
```

This is the correct answer since $P_1(s)P_2(s) = 3s^2 + 10s + 8$. From this we see polynomials are represented as vectors of the polynomial coefficients, where the highest-degree coefficient is the first element in the vector. This rule is followed pretty consistently. (*Note:* "conv" is the MATLAB convolution function, so if you don't already know this, convolution is mathematically essentially the same as polynomial multiplication.)

Suppose $P(s) = s^2 + s - 2$, and we want the roots. In this case you may type

```
>> P = [ 1  1  -2 ];
>> roots(P)

ans =

      -2
       1

>>
```

MATLAB (version 5 and later) has "mroots," which is a root finder that does a better job of computing multiple roots.

The MATLAB function "polyval" is used to evaluate polynomials. For example, suppose $P(s) = s^2 + 3s + 5$, and we wanted to compute $P(-3)$. The command sequence is

```
>> P = [ 1  3  5 ];
>> polyval(P,-3)
```

```
ans =

      5

>>
```

13.5   LOOPS

We may illustrate the simplest loop construct in MATLAB as follows:

```
>> t = [0:.1:1];
>> for k = 1:length(t)
>>    x(k) = 5*sin( (pi/3) * t(k) ) + 2;
>>    end;
>>
```

This command sequence computes $x(t) = 5\sin(\frac{\pi}{3}t) + 2$ for $t = 0.1k$, where $k = 0, 1, \ldots, 10$. The result is saved in the (row) vector x. However, an alternative approach is to *vectorize* the calculation according to

```
>> t = [0:.1:1];
>> x = 5*sin( pi*t/3 ) + 2*ones(1,length(t));
>>
```

This yields the same result. Vectorizing calculations leads to faster code (in terms of runtime).

A potentially useful method to add (append) elements to a vector is

```
>> x = [];
>> for k = 1:2:6
>>    x = [ x k ];
>>    end;
>> x

x =

      1   3   5

>>
```

where $x = [\ ]$ defines x to be initially empty, while the FOR loop appends 1, 3, and 5 to the vector one element at a time.

The format of numerical outputs can be controlled using MATLAB fprintf (which has many similarities to the ANSI C fprintf function). For example

```
>> for k = 0:9
      fprintf('%12.8f\n',sqrt(k));
      end;
 0.00000000
```

```
1.00000000
1.41421356
1.73205081
2.00000000
2.23606798
2.44948974
2.64575131
2.82842712
3.00000000
>>
```

The use of a file identifier can force the result to be printed to a specific file instead of to the terminal (which is the result in this example). MATLAB also has save and load commands that can save variables and arrays to memory, and read them back, respectively.

Certainly, FOR loops may be nested in the expected manner. Of course, MATLAB also supports a "while" statement. For information on conditional statements (i.e., "if" statements), use ">> help if."

13.6 PLOTTING AND M-FILES

Let's illustrate plotting and the use of M-files with an example. Note that M-files are also called *script files* (use *script* as the keyword when using help for more information on this feature).

As an exercise the reader may wish to create a file called "stepH.m" (open and edit it in the manner you are accustomed to). In this file place the following lines:

```
%
%                    stepH.m
%
% This routine computes the unit-step response of the
% LTI system with system function H(s) given by
%
%                         K
%         H(s)  =   -------------
%                    s^2 + 3s + K
%
% for user input parameter K.  The result is plotted.
%

function stepH(K)

b = [ 0 0 K ];
a = [ 1 3 K ];

clf             % Clear any existing plots from the screen
step(b,a);      % Compute the step response and plot it
grid            % plot the grid
```

This M-file becomes a MATLAB command, and for $K = 0.1$ may be executed using

```
>> stepH(.1);
>>
```

This will result in another window opening where the plot of the step response will appear. To save this file for printing, use

```
>> print -dps filename.ps
```

which will save the plot as a postscript file called "filename.ps." Other printing formats are available. As usual, the details are available from online help. Figure 13.1 is the plot produced by stepH.m for the specified value of K.

Another example of an M-file that computes the frequency response (both magnitude and phase) of the linear time-invariant (LTI) system with Laplace transfer function is

$$H(s) = \frac{1}{s^2 + \frac{1}{2}s + 1}.$$

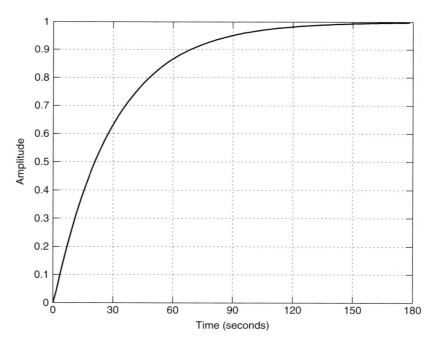

Figure 13.1   Step response: typical output from stepH.m.

If you are not familiar with Laplace transforms, recall phasor analysis from basic electric circuit theory. You may, for instance, interpret $H(j\omega)$ as the ratio of two phasor voltages (frequency ω), such as

$$H(j\omega) = \frac{V_2(j\omega)}{V_1(j\omega)}.$$

The numerator phasor $V_2(j\omega)$ is the output of the system, and the denominator phasor $V_1(j\omega)$ is the input to the system (a sinusoidal voltage source).

```
%
%                        freqresp.m
%
% This routine plots the frequency response of the Laplace
% transfer function
%
%                           1
%            H(s) =  ----------------
%                    s^2 + .5s  +  1
%
% The magnitude response (in dB) and the phase response (in
% degrees) are returned in the vectors mag, and pha,
% respectively.  The places on the frequency axis where the
% response is computed are returned in vector f.
%

function [mag,pha,f] = freqresp

b = [ 0 0 1];
a = [ 1 .5 1 ];

w = logspace(-2,1,50); % Compute the frequency response for 10^(-2) to 10^1
                       % radians per second at 50 points in this range
h = freqs(b,a,w);      % h is the frequency response

mag = abs(h);          % magnitude response
pha = angle(h);        % phase response

f = w/(2*pi);          % setup frequency axis in Hz
pha = pha*180/pi;      % phase now in degrees
mag = 20*log10(mag);   % magnitude response now in dB

clf

subplot(211), semilogx(f,mag,'-'), grid
xlabel(' Frequency (Hz) ')
ylabel(' Amplitude (dB) ')
title(' Magnitude Response ')

subplot(212), semilogx(f,pha,'-'), grid
xlabel(' Frequency (Hz) ')
ylabel(' Phase Angle (Degrees) ')
title(' Phase Response ')
```

Executing the command

```
>> [mag,phase,f] = freqresp;
>>
```

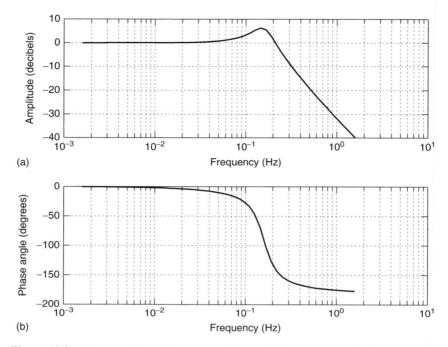

Figure 13.2 The output from freqresp.m: (a) magnitude response; (b) phase response.

will result in the plot of Fig. 13.2, and will also give the vectors mag, phase, and *f*. Vector mag contains the magnitude response (in decibels), and vector phase contains the phase response (in degrees) at the sample values in the vector *f*, which defines the frequency axis for the plots (in hertz). In other words, "freqresp.m" is a Bode plotting routine. Note that the MATLAB command "bode" does Bode plots as well.

Additional labeling may be applied to plots using the MATLAB command text (or via a mouse using "gtext"; see online help). As well, the legend statement is useful in producing labels for a plot with different curves on the same graph. For example

```
function ShowLegend

ul = 2.5*pi;
ll = -pi/4;
N = 200;

dt = (ul - ll)/N;
for k = 0:N-1
  t(k+1) = ll + dt*k;
  x(k+1) = exp(-t(k+1));
  y(k+1) = sin(t(k+1));
  end;
```

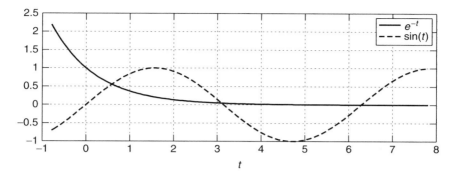

Figure 13.3   Illustration of the MATLAB legend statement.

```
subplot(211), plot(t,x,'-',t,y,'--'), grid
xlabel(' t ')
legend(' e^{-t} ', ' sin(t) ')
```

When this code is run, it gives Fig. 13.3. Note that the label syntax is similar to that in LaTeX [6].

Other sample MATLAB codes have appeared as appendixes in earlier chapters, and the reader may wish to view these as additional examples.

REFERENCES

1. G. H. Golub and C. F. Van Loan, *Matrix Computations*, 2nd ed. Johns Hopkins Univ. Press, Baltimore, MD, 1989.

2. D. R. Hill, *Experiments in Computational Matrix Algebra* (C. B. Moler, consulting ed.), Random House, New York, 1988.

3. S. Nakamura, *Numerical Analysis and Graphic Visualization with MATLAB*, 2nd ed. Prentice-Hall, Upper Saddle River, NJ, 2002.

4. A. Quarteroni, R. Sacco, and F. Saleri, *Numerical Mathematics* (Texts in Applied Mathematics series, Vol. 37). Springer-Verlag, New York, 2000.

5. G. Recktenwald, *Numerical Methods with MATLAB: Implementation and Application*, Prentice-Hall, Upper Saddle River, NJ, 2000.

6. M. Goossens, F. Mittelbach, and A. Samarin, *The LaTeX Companion*, Addison-Wesley, Reading, MA, 1994.

INDEX

An Introduction to Numerical Analysis for Electrical and Computer Engineers, by C.J. Zarowski
ISBN 0-471-46737-5   © 2004 John Wiley & Sons, Inc.